THE BANACH–TARSKI PARADOX

Second Edition

The Banach–Tarski Paradox is a most striking mathematical construction: it asserts that a solid ball can be taken apart into finitely many pieces that can be rearranged using rigid motions to form a ball twice as large. This volume explores the consequences of the paradox for measure theory and its connections with group theory, geometry, set theory, and logic.

This new edition of a classic book unifies contemporary research on the paradox. It has been updated with many new proofs and results and discussions of the many problems that remain unsolved. Among the new results presented are several unusual paradoxes in the hyperbolic plane, one of which involves the shapes of Escher's famous "Angel and Devils" woodcut. A new chapter is devoted to a complete proof of the remarkable result that the circle can be squared using set theory, a problem that had been open for over sixty years.

Grzegorz Tomkowicz is a self-educated Polish mathematician who has made several important contributions to the theory of paradoxical decompositions and invariant measures.

Stan Wagon is a Professor of Mathematics at Macalester College. He is a winner of the Wolfram Research Innovator Award, as well as numerous writing awards including the Ford, Evans, and the Allendoerfer Awards. His previous work includes *A Course in Computational Number Theory* (2000), *The SIAM 100-Digit Challenge* (2004) and *Mathematica® in Action* (3rd Ed. 2010).

Encyclopedia of Mathematics and its Applications

This series is devoted to significant topics or themes that have wide application in mathematics or mathematical science and for which a detailed development of the abstract theory is less important than a thorough and concrete exploration of the implications and applications.

Books in the **Encyclopedia of Mathematics and its Applications** cover their subjects comprehensively. Less important results may be summarized as exercises at the ends of chapters. For technicalities, readers can be referred to the bibliography, which is expected to be comprehensive. As a result, volumes are encyclopedic references or manageable guides to major subjects.

Encyclopedia of Mathematics and its Applications

114 J. Beck *Combinatorial Games*
115 L. Barreira and Y. Pesin *Nonuniform Hyperbolicity*
116 D. Z. Arov and H. Dym *J-Contractive Matrix Valued Functions and Related Topics*
117 R. Glowinski, J.-L. Lions, and J. He *Exact and Approximate Controllability for Distributed Parameter Systems*
118 A. A. Borovkov and K. A. Borovkov *Asymptotic Analysis of Random Walks*
119 M. Deza and M. Dutour Sikirić *Geometry of Chemical Graphs*
120 T. Nishiura *Absolute Measurable Spaces*
121 M. Prest *Purity, Spectra and Localisation*
122 S. Khrushchev *Orthogonal Polynomials and Continued Fractions*
123 H. Nagamochi and T. Ibaraki *Algorithmic Aspects of Graph Connectivity*
124 F. W. King *Hilbert Transforms I*
125 F. W. King *Hilbert Transforms II*
126 O. Calin and D.-C. Chang *Sub-Riemannian Geometry*
127 M. Grabisch *et al. Aggregation Functions*
128 L. W. Beineke and R. J. Wilson (eds.) with J. L. Gross and T. W. Tucker *Topics in Topological Graph Theory*
129 J. Berstel, D. Perrin, and C. Reutenauer *Codes and Automata*
130 T. G. Faticoni *Modules over Endomorphism Rings*
131 H. Morimoto *Stochastic Control and Mathematical Modeling*
132 G. Schmidt *Relational Mathematics*
133 P. Kornerup and D. W. Matula *Finite Precision Number Systems and Arithmetic*
134 Y. Crama and P. L. Hammer (eds.) *Boolean Models and Methods in Mathematics, Computer Science, and Engineering*
135 V. Berthé and M. Rigo (eds.) *Combinatorics, Automata and Number Theory*
136 A. Kristály, V. D. Rădulescu, and C. Varga *Variational Principles in Mathematical Physics, Geometry, and Economics*
137 J. Berstel and C. Reutenauer *Noncommutative Rational Series with Applications*
138 B. Courcelle and J. Engelfriet *Graph Structure and Monadic Second-Order Logic*
139 M. Fiedler *Matrices and Graphs in Geometry*
140 N. Vakil *Real Analysis through Modern Infinitesimals*
141 R. B. Paris *Hadamard Expansions and Hyperasymptotic Evaluation*
142 Y. Crama and P. L. Hammer *Boolean Functions*
143 A. Arapostathis, V. S. Borkar, and M. K. Ghosh *Ergodic Control of Diffusion Processes*
144 N. Caspard, B. Leclerc, and B. Monjardet *Finite Ordered Sets*
145 D. Z. Arov and H. Dym *Bitangential Direct and Inverse Problems for Systems of Integral and Differential Equations*
146 G. Dassios *Ellipsoidal Harmonics*
147 L. W. Beineke and R. J. Wilson (eds.) with O. R. Oellermann *Topics in Structural Graph Theory*
148 L. Berlyand, A. G. Kolpakov, and A. Novikov *Introduction to the Network Approximation Method for Materials Modeling*
149 M. Baake and U. Grimm *Aperiodic Order I: A Mathematical Invitation*
150 J. Borwein *et al. Lattice Sums Then and Now*
151 R. Schneider *Convex Bodies: The Brunn–Minkowski Theory (Second Edition)*
152 G. Da Prato and J. Zabczyk *Stochastic Equations in Infinite Dimensions (Second Edition)*
153 D. Hofmann, G. J. Seal, and W. Tholen (eds.) *Monoidal Topology*
154 M. Cabrera García and Á. Rodríguez Palacios *Non-Associative Normed Algebras I: The Vidav–Palmer and Gelfand–Naimark Theorems*
155 C. F. Dunkl and Y. Xu *Orthogonal Polynomials of Several Variables (Second Edition)*
156 L. W. Beineke and R. J. Wilson (eds.) with B. Toft *Topics in Chromatic Graph Theory*
157 T. Mora *Solving Polynomial Equation Systems III: Algebraic Solving*
158 T. Mora *Solving Polynomial Equation Systems IV: Buchberger Theory and Beyond*
159 V. Berthé and M. Rigo (eds.) *Combinatorics, Words and Symbolic Dynamics*
160 B. Rubin *Introduction to Radon Transforms: With Elements of Fractional Calculus and Harmonic Analysis*
161 M. Ghergu and S. D. Taliaferro *Isolated Singularities in Partial Differential Inequalities*
162 G. M. Bisci, V. Radulescu, and R. Servadei *Variational Methods for Nonlocal Fractional Problems*
163 G. Tomkowicz and S. Wagon *The Banach–Tarski Paradox (Second Edition)*

ENCYCLOPEDIA OF MATHEMATICS AND ITS APPLICATIONS

The Banach–Tarski Paradox

Second Edition

GRZEGORZ TOMKOWICZ

Centrum Edukacji G2, Bytom, Poland

STAN WAGON

Macalester College, St. Paul, Minnesota

CAMBRIDGE
UNIVERSITY PRESS

University Printing House, Cambridge CB2 8BS, United Kingdom

One Liberty Plaza, 20th Floor, New York, NY 10006, USA

477 Williamstown Road, Port Melbourne, VIC 3207, Australia

314-321, 3rd Floor, Plot 3, Splendor Forum, Jasola District Centre, New Delhi - 110025, India

79 Anson Road, #06-04/06, Singapore 079906

Cambridge University Press is part of the University of Cambridge.

It furthers the University's mission by disseminating knowledge in the pursuit of education, learning and research at the highest international levels of excellence.

www.cambridge.org
Information on this title: www.cambridge.org/9781107617315

First published 2016
First paperback edition 2019

A catalogue record for this publication is available from the British Library

Library of Congress Cataloging in Publication data
Names: Tomkowicz, Grzegorz. | Wagon, S.
Title: The Banach-Tarski paradox.
Description: Second edition / Grzegorz Tomkowicz, Centrum Edukacji G2, Bytom, Poland, Stan Wagon, Macalester College, St. Paul, Minnesota. | New York, NY : Cambridge University Press, [2016] | Series: Encyclopedia of mathematics and its applications ; 163 | Previous edition: The Banach-Tarski paradox / Stan Wagon (Cambridge : Cambridge University Press, 1985). | Includes bibliographical references and index.
Identifiers: LCCN 2015046410 | ISBN 9781107042599 (hardback : alk. paper)
Subjects: LCSH: Banach-Tarski paradox. | Measure theory. | Decomposition (Mathematics)
Classification: LCC QA248 .W22 2016 | DDC 511.3/22–dc23
LC record available at http://lccn.loc.gov/2015046410

ISBN 978-1-107-04259-9 Hardback
ISBN 978-1-107-61731-5 Paperback

To Jan Mycielski, whose enthusiasm and knowledge strongly influenced both of us

DELIANS: *How can we be rid of the plague?*
DELPHIC ORACLE: *Construct a cubic altar having double the size of the existing one.*
BANACH AND TARSKI: *Can we use the Axiom of Choice?*

Contents

Foreword

This book is motivated by the following theorem of Hausdorff, Banach, and Tarski: Given any two bounded sets A and B in three-dimensional space $\mathbb{R}^3$, each having nonempty interior, one can partition A into finitely many disjoint parts and rearrange them by rigid motions to form B. This, I believe, is the most surprising result of theoretical mathematics. It shows the imaginary character of the unrestricted idea of a set in $\mathbb{R}^3$. It precludes the existence of finitely additive, congruence-invariant measures over all bounded subsets of $\mathbb{R}^3$, and it shows the necessity of more restricted constructions, such as Lebesgue measure.

In the 1950s, the years of my mathematical education in Poland, this result was often discussed. J. F. Adams, T. J. Dekker, J. von Neumann, R. M. Robinson, and W. Sierpiński wrote about it; my PhD thesis was motivated by it. (All this is referenced in this book.) Thus it is a great pleasure to introduce you to this book, where this striking theorem and many related results in geometry and measure theory, and the underlying tools of group theory, are presented with care and enthusiasm. The reader will also find some applications of the most recent advances of group theory to measure theory: the work of Gromov, Margulis, Rosenblatt, Sullivan, Tits, and others.

But to me the interest of mathematics lies no more in its theorems and theories than in the challenge of its surprising problems. And, on the pages of this book, you will find many old and new open problems. So let me conclude this foreword by turning your attention to one of them, from my teacher E. Marczewski (before 1939, he published under the name Szpilrajn): Does there exist a finite sequence $A_1, \ldots, A_n$ of pairwise disjoint open subsets of the unit cube and isometries $\sigma_1, \ldots, \sigma_n$ of $\mathbb{R}^3$ such that the unit cube is a proper subset of the topological closure of the union $\sigma_1(A_1) \cup \ldots \cup \sigma_n(A_n)$? This remarkable problem is discussed in Chapters 3, 9, 11, and 13 of this book.

I wish you the most pleasant reading and many fruitful thoughts.

Addendum to the Foreword

Several spectacular results have been proved since the 1985 first edition of this book. Two of them are particularly striking.

A. The answer to Marczewski's problem mentioned earlier is yes. R. Dougherty and M. Foreman have shown (Thm. 11.16) that any two bounded nonempty open sets A and B in the Euclidean space $\mathbb{R}^n$ ($n \geq 3$) are equivalent in the following sense: A has finitely many disjoint regular-open subsets whose union is everywhere dense in A and which can be moved by isometries into disjoint subsets of B whose union is everywhere dense in B. Similar results hold for spheres $\mathbb{S}^n$ and the hyperbolic spaces $\mathbb{H}^n$ ($n \geq 2$).
B. The answer to Tarski's "squaring the circle" problem is also yes. M. Laczkovich proved that if a circle and a square in $\mathbb{R}^2$ have the same area, then they are equivalent by finite decomposition, and the isometries of the corresponding pieces are translations (i.e., simple vector addition). And the same is true for many other pairs of sets in $\mathbb{R}^n$. A proof of this is in Chapter 9.

Some outstanding problems are still open:

1. **The Banach–Ulam problem 2 from the Scottish Book.** Does every compact metric space admit a finitely additive, congruence-invariant probability measure on its Borel sets? (Question 3.13)
2. **Exotic Borel measures.** Is Lebesgue measure the only finitely additive, isometry-invariant measure on the Borel sets of $\mathbb{R}^n$ that normalizes the unit cube? (It is not the only translation-invariant one. In the minimal model for set theory that contains all real and all ordinal numbers (usually called $L(\mathbb{R})$), and assuming a certain large cardinal exists, all sets are Lebesgue measurable and have the Property of Baire, and in this model the answer is yes.) (Question 13.15)

3. **Borel circle-squaring.** Can the pieces of the circle-squaring decomposition of Laczkovich be taken to be Borel sets? (§9.3)

All these results and problems are presented in a penetrating and lucid way in this new edition.

Jan Mycielski
Boulder, Colorado
August 2015

Preface

Although many properties of infinite sets and their subsets were considered to be paradoxical when they were discovered, the development of paradoxical decompositions really began with the formalization of measure theory at the beginning of the twentieth century. The classic example (Vitali, 1905) of a non–Lebesgue measurable set was the first instance of the use of a paradoxical decomposition to show the nonexistence of a certain type of measure. Ten years later, Hausdorff constructed a much more surprising paradox on the surface of the sphere (again, to show the nonexistence of a measure), and this inspired some important work in the 1920s. Namely, there was Banach's construction of invariant measures on the line and in the plane (which required the discovery of the main ideas of the Hahn–Banach Theorem) and the famous Banach–Tarski Paradox on duplicating, or enlarging, spheres and balls. This latter result, which at first seems patently impossible, is often stated as follows: It is possible to cut up a pea into finitely many pieces that can be rearranged to form a ball the size of the sun!

Their construction has turned out to be much more than a curiosity. Ideas arising from the Banach–Tarski Paradox have become the foundation of a theory of finitely additive measures, a theory that involves much interplay between analysis (measure theory and linear functionals), algebra (combinatorial group theory), geometry (isometry groups), and topology (locally compact topological groups). Moreover, the Banach–Tarski Paradox itself has been useful in important work on the uniqueness of Lebesgue measure: It shows that certain measures necessarily vanish on the sets of Lebesgue measure zero.

The purpose of this volume is twofold. The first aim is to present proofs that are as simple as possible of the two main classical results—the Banach–Tarski Paradox in $\mathbb{R}^3$ (and $\mathbb{R}^n$, $n > 3$) and Banach's theorem that no such paradox exists in $\mathbb{R}^1$ or $\mathbb{R}^2$. The first three chapters are devoted to the paradox and are accessible to anyone familiar with the rudiments of linear algebra, group theory, and countable sets. (Background related to the Euclidean isometry groups is included in Appendix A.) Chapter 12, which contains Banach's theorem in $\mathbb{R}^1$ and $\mathbb{R}^2$, can be read independently of Chapters 4–11 but requires a little more background in measure theory (Lebesgue measure) and general topology (Tychonoff Compactness Theorem). Although isolated proofs use some special techniques, such as transfinite induction or analytic functions, most of the material is accessible to a first- or second-year graduate student.

The book's other purpose is to serve as a background source for those interested in current research that has a connection to paradoxical decompositions. The period since 1980 has been especially active, and several classic problems in this area have been solved, some by using the deepest techniques of modern mathematics. This volume contains a unified and modern treatment of the fundamental results about amenable groups, finitely additive measures, and free groups of isometries and so should prove useful to someone in any field who is interested in these modern results and their historical context.

The group theory connections arise from the difference between the isometry groups of $\mathbb{R}^2$ and $\mathbb{R}^3$, a difference that explains the presence of the Banach–Tarski Paradox in $\mathbb{R}^3$ and its absence in the plane. This distinction led to the study of the class of groups that are not paradoxical, that is, groups that cannot be duplicated by left translation of finitely many pairwise disjoint subsets. This class, denoted by AG for amenable groups, contains all solvable and finite groups but excludes free non-Abelian groups. A famous problem is whether AG equals NF, the class of groups not having a free non-Abelian subgroup. This was solved in 1980 (Ol'shanskii [Ols80]), using ideas connected with growth conditions in groups (Cohen [Coh82]) and the solution of Burnside's Problem (Adian [Adi79]). However, the classes AG and NF do coincide when restricted to linear groups (a deep result of Tits [Tit72]) or to connected, locally compact topological groups (Balcerzyk and Mycielski [BM57]). Growth conditions in groups, first studied in depth by Milnor and Wolf [Mil68b, Mil68c, Wol68], also elucidate a weaker sort of paradox, the Sierpiński–Mazurkiewicz Paradox, which exists in $\mathbb{R}^2$ but not in $\mathbb{R}^1$. The class AG has also led to the study of (topological) amenability in topological groups (where only Borel sets are considered). Amenability and the related notion of an invariant mean have proven to be useful tools in the study of topological groups (Greenleaf [Gre69]). Chapter 12 contains an introduction to the theory of amenable groups, and Chapter 14 discusses the relevance of growth conditions for the theory of amenability and paradoxical decompositions.

In analysis, important work solving the Ruziewicz Problem has its roots in Banach's results about $\mathbb{R}^1$ and $\mathbb{R}^2$. Banach showed that Lebesgue measure is not the only finitely additive, isometry-invariant measure on the bounded, measurable subsets of the plane (or line) that normalizes the unit square (or interval). The analogous problem for $\mathbb{R}^3$ and beyond was unsolved for over fifty years. But using Kazhdan's Property T and techniques of functional analysis, Margulis [Mar80, Mar82], Sullivan [Sul81], Rosenblatt [Ros81], and Drinfeld [Dri85] settled this question in the expected way: No "exotic" measures exist, except in the cases Banach considered. The construction of exotic measures in $\mathbb{R}^1$ and $\mathbb{R}^2$ and a discussion of the higher-dimensional situation are presented in Chapter 13.

Actions of free groups are central to the whole theory, and a general problem is to determine which naturally occurring groups have certain sorts of free subgroups. Chapters 4, 6, 7, and 8 present the classical results on the isometry groups of spheres, Euclidean spaces. Some problems were solved relatively recently. Until the work of Deligne and Sullivan [DS83], it was not known that $SO_6(\mathbb{R})$ (or $SO_{4n+2}(\mathbb{R})$) contained a free non-Abelian subgroup, no element of which (except the identity) has $+1$ as an eigenvalue. And a similar problem about locally commutative free subgroups was solved in all $SO_n(\mathbb{R})$, except $SO_5(\mathbb{R})$, in 1956 (Dekker [Dek56b]), with the remaining case solved by A. Borel [Bor83] in a paper generalizing the work of Deligne and Sullivan. Chapter 7 presents a technique for improving this type of result to get uncountable free subgroups and discusses the geometrical consequences of these larger free groups of isometries.

There has been a remarkable amount of progress on famous problems since the first edition of this book appeared in 1985. The most striking discoveries are the solutions to two very famous open questions. Laczkovich solved the famous Tarski circle-squaring problem by showing that a disk and square in the plane having the same area are equidecomposable; and Dougherty and Foreman proved that a Banach–Tarski-type paradox exists with pieces having the Property of Baire. Other noteworthy results are T. Wilson's solution to the de Groot problem—he showed that the pieces in the classic paradox could be chosen so that the moves to the new positions preserve disjointness at every instant—and the work of Sherman and Just on bounded paradoxical sets in the plane. This new edition contains all the essential details of the work of Laczkovich, Wilson, and Sherman. Another addition is a presentation of Følner's Condition and amenability through the use of pseudogroups in Chapter 12.

Also new are Chapter 4, which collects diverse results about the hyperbolic plane, and Chapter 8, which focuses on the Euclidean plane and the group of area-preserving linear transformations. A main theme in many of these is that counterintuitive paradoxes can be constructed without requiring the Axiom of Choice.

The book is divided into two parts. The first deals with the construction of paradoxical decompositions (which imply that certain sorts of finitely additive measures do not exist), and the second deals with the construction of measures (which show why certain paradoxical decompositions do not exist). Chapter 11 ties the two parts together by presenting a theorem of Tarski that asserts that the existence of a paradoxical decomposition is equivalent to the nonexistence of an invariant, finitely additive measure. The final chapter, Chapter 15, discusses some technical and philosophical points relevant to the foundational discussion engendered by the use of the Axiom of Choice in the Banach–Tarski Paradox and related results.

Here is an observation of Bertrand Russell from 1918 [Rus10]: "The point of philosophy is to start with something so simple as not to seem worth stating, and to end with something so paradoxical that no one will believe it." Although paradoxes are by no means the point of the study of mathematics, there is no question that counterintuitive results can clarify our understanding and provide motivation for more detailed study. The Banach–Tarski Paradox certainly plays such a role, and we both have found its study to be intensely rewarding. We hope you enjoy learning about it.

The authors would like to express their gratitude to Jan Mycielski and Joseph Rosenblatt for their support and advice during the preparation of this volume. Moreover, a book such as this, touching on many mathematical disciplines, would not have been possible without the willingness of many people to share their expertise. The help of the following mathematicians is gratefully acknowledged: Bill Barker, Curtis Bennett, Armand Borel, Rotislav Grigorchuk, Branko Grünbaum, William Hanf, Joan Hutchinson, Victor Klee, Miklos Laczkovich, Rich Laver, Robert Macrae, Dave Morris, Arlan Ramsay, Robert Riley, Robert Solovay, Dennis Sullivan, Alan Taylor, and Trevor Wilson.

Grzegorz Tomkowicz
Centrum Edukacji G2
Moniuszki 9
41-902 Bytom, Poland
gtomko@vp.pl

Stan Wagon
Macalester College
1600 Grand Avenue
St. Paul, Minnesota, USA
wagon@macalester.edu

PART ONE

Paradoxical Decompositions, or the Nonexistence of Finitely Additive Measures

1

Introduction

It has been known since antiquity that the notion of infinity leads very quickly to seemingly paradoxical constructions, many of which seem to change the size of objects by operations that appear to preserve size. In a famous example, Galileo observed that the set of positive integers can be put into a one-one correspondence with the set of square integers, even though the set of nonsquares, and hence the set of all integers, seems more numerous than the squares. He deduced from this that "the attributes 'equal,' 'greater' and 'less' are not applicable to infinite . . . quantities," anticipating developments in the twentieth century, when paradoxes of this sort were used to prove the nonexistence of certain measures.

An important feature of Galileo's observation is its resemblance to a duplicating machine; his construction shows how, starting with the positive integers, one can produce two sets, each of which has the same size as the set of positive integers. The idea of duplication inherent in this example will be the main object of study in this book. The reason that this concept is so fascinating is that, soon after paradoxes such as Galileo's were being clarified by Cantor's theory of cardinality, it was discovered that even more bizarre duplications could be produced using rigid motions, which *are* distance-preserving (and hence also area-preserving) transformations. We refer to the Banach–Tarski Paradox on duplicating spheres or balls, which is often stated in the following fanciful form: a pea may be taken apart into finitely many pieces that may be rearranged using rotations and translations to form a ball the size of the sun. The fact that the Axiom of Choice is used in the construction makes it quite distant from physical reality, though there are interesting examples that do not need the Axiom of Choice (see Thm. 1.7, §§4.2, 4.3, 11.2).

Two distinct themes arise when considering the refinements and ramifications of the Banach–Tarski Paradox. First is the use of ingenious geometric and algebraic methods to construct paradoxes in situations where they seem impossible and thereby getting proofs of the nonexistence of certain measures. Second, and this comprises Part II of this book, is the construction of measures and their use in showing that some paradoxical decompositions are not possible.

We begin with a formal definition of the idea of duplicating a set using certain transformations. The general theory is much simplified if the transformations used are all bijections of a single set, and the easiest way to do this is to work in the context of group actions. Recall that a group G is said to act on a set X if to each $g \in G$ there corresponds a function (necessarily a bijection) from X to X, also denoted by g, such that for any $g, h \in G$ and $x \in X$, $g(h(x)) = (gh)(x)$ and $e(x) = x$, where e denotes the identity of G.

Definition 1.1. *Let G be a group acting on a set X and suppose $E \subseteq X$ is a nonempty subset of X. Then E is G-*paradoxical *(or paradoxical with respect to G) if, for some positive integers m, n, there are pairwise disjoint subsets $A_1, \ldots, A_n, B_1, \ldots, B_m$ of E and $g_1, \ldots, g_n, h_1, \ldots, h_m \in G$ such that $E = \bigcup g_i(A_i)$ and $E = \bigcup h_j(B_j)$.*

Loosely speaking, the set E has two disjoint subsets $(\bigcup A_i, \bigcup B_j)$ each of which can be taken apart and rearranged via G to cover all of E. If E is G-paradoxical, then the sets witnessing that may be chosen so that $\{g_i(A_i)\}$, $\{h_j(B_j)\}$, and $\{A_i\} \cup \{B_j\}$ are each partitions of E. For the first two, one need only replace A_i, B_j by smaller sets to ensure pairwise disjointness of $\{g_i(A_i)\}$ and $\{h_j(B_j)\}$, but the proof that, in addition, $\{A_i\} \cup \{B_j\}$ may be taken to be all of E is more intricate and will be given in Corollary 3.7.

1.1 Examples of Paradoxical Actions

1.1.1 The Banach–Tarski Paradox

Any ball in $\mathbb{R}^3$ is paradoxical with respect to G_3, the group of isometries of $\mathbb{R}^3$.

This result, a paradigm of the whole theory, will be proved in Chapter 3. More generally, we shall consider the possibility of paradoxes when X is a metric space and G is a subgroup of the group of isometries of X (an *isometry* is a bijection from X to X that preserves distance). In the case that G is the group of all isometries of X, we shall suppress G, using simply, E is *paradoxical*. We shall be concerned mostly with the case that X is one of the Euclidean spaces $\mathbb{R}^n$.

1.1.2 Free Non-Abelian Groups

Any group acts naturally on itself by left translation. The question of which groups are paradoxical with respect to this action turns out to be quite fascinating and is discussed in Chapter 12. In this context, the central example is the free group on two generators. Recall that the free group F with generating set M is the group of all finite words using letters from $\{\sigma, \sigma^{-1} : \sigma \in M\}$, where two words are equivalent if one can be transformed to the other by the removal or addition of finite pairs of adjacent letters of the form $\sigma\sigma^{-1}$ or $\sigma^{-1}\sigma$. A word with no such adjacent pairs is called a reduced word, and to avoid the use of equivalence classes, F may be taken to consist of all reduced words, with the group operation being

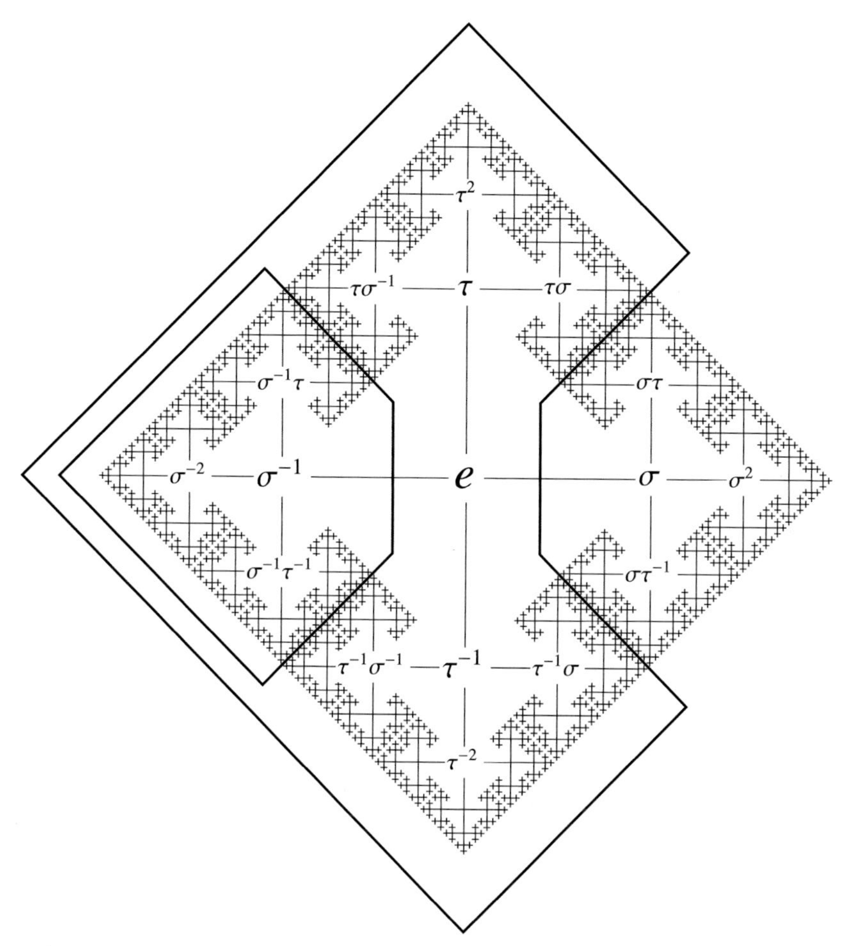

Figure 1.1. The free group of rank 2. The small enclosed region represents $W(\sigma^{-1})$, and left translation of this by σ gives the words in the larger enclosed region. Thus $W(\sigma) \cup \sigma W(\sigma^{-1}) = F$.

concatenation; the concatenation of two words is equivalent to a unique reduced word. (From now on, all words will be assumed to be reduced.) The identity of F, which is denoted by e, is the empty word. A subset S of a group is called *free* if no nonidentity reduced word using elements of S gives the identity. Any two free generating sets for a free group have the same size, which is called the rank of the free group. Free groups of the same rank are isomorphic; any group that is isomorphic to a free group will also be called a free group. See [MKS66] for further details about free groups and their properties.

Theorem 1.2. *A free group F of rank 2 is F-paradoxical, where F acts on itself by left multiplication.*

Proof. Suppose σ, τ are free generators of F. If ρ is one of $\sigma^{\pm 1}$, $\tau^{\pm 1}$, let $W(\rho)$ be the set of elements of F whose representation as a word in $\sigma, \sigma^{-1}, \tau, \tau^{-1}$ begins, on the left, with ρ. Then $F = \{e\} \cup W(\sigma) \cup W(\sigma^{-1}) \cup W(\tau) \cup W(\tau^{-1})$, and these subsets are pairwise disjoint. Furthermore, $W(\sigma) \cup \sigma W(\sigma^{-1}) = F$ (see Fig. 1.1) and $W(\tau) \cup \tau W(\tau^{-1}) = F$. For if $h \in F \setminus W(\sigma)$, then $\sigma^{-1}h \in W(\sigma^{-1})$ and $h = \sigma(\sigma^{-1}h) \in \sigma W(\sigma^{-1})$. Note that this proof uses only four pieces. □

The preceding proof can be improved so that the four sets in the paradoxical decomposition cover all of F rather than just $F \setminus \{e\}$. The reader might enjoy trying to find such a neat four-piece paradoxical decomposition of a rank 2 free group (or see Fig. 3.2). When we say that a group is paradoxical, we shall be referring to the action of left translation; this should cause no confusion with the usage mentioned in Example 1.1.1.

1.1.3 Free Semigroups

We shall on occasion be interested in the action of a semigroup S (a set with an associative binary operation and an identity) on a set X. Because of the lack of inverses in a semigroup, the function on X induced by some $\sigma \in S$ may not be a bijection; thus it is inappropriate to apply Definition 1.1 to such actions. Nonetheless, there are similarities between free semigroups and free groups, as the following proposition shows. A free semigroup with free generating set T is simply the set of all words using elements of T as letters, with concatenation being the semigroup operation. The rank of a free semigroup is the number of elements in T. A free subsemigroup of a group is a subset of the group that contains e and is closed under the group operation such that the semigroup is isomorphic to a free semigroup.

Proposition 1.3. *A free semigroup S, with free generators σ and τ, contains two disjoint sets A and B such that $\sigma S = A$ and $\tau S = B$. Any group having a free subsemigroup of rank 2 contains a paradoxical set.*

Proof. Let A be the set of words whose leftmost term is σ and B the same using τ. Then $\sigma S = A$ and $\tau S = B$ (see Fig. 1.2). If S is embedded in a group, then S itself is a paradoxical subset of the group because $\sigma^{-1}(A) = S = \tau^{-1}(B)$. □

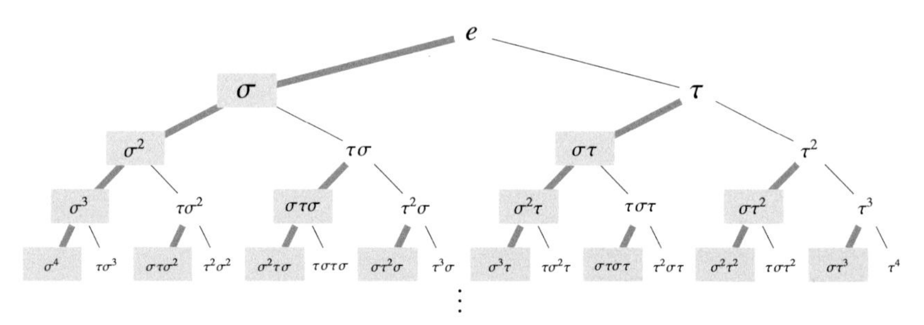

Figure 1.2. A paradox in a group having a free subsemigroup S of rank 2. If A is the set of words with σ on the left (gray background) and B are those with τ on the left (not gray and not e), then $\sigma^{-1}(A) = S = \tau^{-1}(B)$. The thicker edges indicate left multiplication by σ.

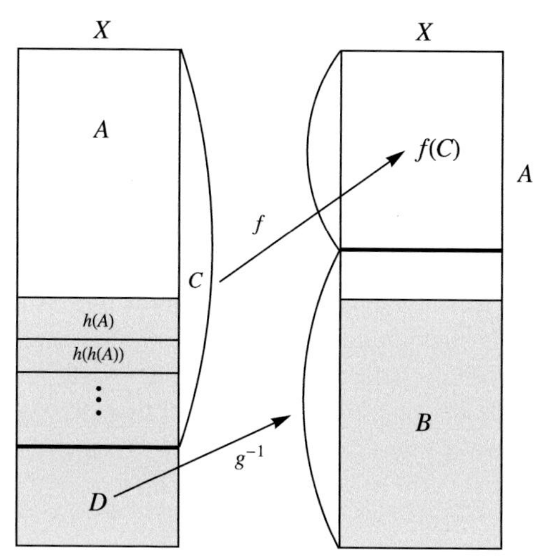

Figure 1.3. The method of constructing the permutation f_1 of X from two bijections f, g from X to subsets of X.

1.1.4 Arbitrary Bijections

The following result, showing that any infinite set is paradoxical using arbitrary bijections, is the modern version of Galileo's observation about the integers. The implications with (c) as hypothesis use the Axiom of Choice (AC). Recall that, in the presence of AC, the cardinality of an infinite set X, $|X|$, is the unique cardinal $\aleph_\alpha$ for which there is a bijection with X. In the absence of AC, $|X|$ is used only in the context of the equivalence relation: $|X| = |Y|$ iff there is a bijection from X to Y.

Theorem 1.4. *The following are equivalent:*

(a) $|X| = 2|X|$.
(b) X *is paradoxical with respect to the group of all permutations of* X, *that is, all bijections from* X *to* X.
(c) X *is infinite or empty.*

Proof. We will show (a) $\Rightarrow$ (b) $\Rightarrow$ (c) $\Rightarrow$ (a).

(b) $\Rightarrow$ (c) is clear because finite sets do not admit paradoxes.

(a) $\Rightarrow$ (b). This proof uses the classic back-and-forth idea of the Schröder–Bernstein Theorem (see Thm. 3.6). We start with a partition of X into A and B and bijections $f: X \to A$ and $g: X \to B$. We need bijections f_1 and f_2 from X to X so that f_1 agrees with f on A and f_2 agrees with f on B. To get f_1, let $h = g \circ f: X \to B$ and let $C = A \cup h(A) \cup h(h(A)) \cup \ldots$, which is a disjoint union because h is one-to-one (see Fig. 1.3). Let $D = X \setminus C$. Then f maps C onto $f(C) \subseteq A$, and $g(f(C)) = C \setminus A$. Therefore g is a bijection from $X \setminus f(C)$ to D. So let f_1 be f on C and g^{-1} on D.

The construction of f_2 is similar. Let $h = f \circ f : X \to A$ and $C = B \cup h(B) \cup h(h(B)) \cup \ldots$. Then f maps C to $X \setminus C$ bijectively and f^{-1} is a bijection of $X \setminus C$ with C. So let f_2 be f on C and f^{-1} on $X \setminus C$.

A similar argument gets g_i so that g_1 agrees with g on A and g_2 agrees with g on B.

These functions now give us $X = f_1^{-1}(f(A)) \cup f_2^{-1}(f(B))$. Similarly, X can be realized as a union of permutations restricted to $g(A)$ and $g(B)$. Because $f(A)$, $f(B)$, $g(A)$, $g(B)$ are pairwise disjoint, and this shows that X is paradoxical.

(c) $\Rightarrow$ (a). This is a consequence of the Axiom of Choice. First, it is proved for cardinals by transfinite induction, and then AC (in the form: every set may be mapped bijectively onto a cardinal) is invoked (see [KM68, Chap. 8]). Alternatively, one can give a more direct proof using Zorn's Lemma (see [End77, p. 163]). □

1.2 Geometrical Paradoxes

The first example of a geometrical paradox, that is, one using isometries, arose in connection with the existence of a non–Lebesgue measurable set. The well-known construction of such a set fits into our context if Definition 1.1 is modified to allow countably many pieces. Thus E is *countably G-paradoxical* means that

$$E = \bigcup_{i=1}^{\infty} g_i A_i = \bigcup_{i=1}^{\infty} h_i B_i,$$

where $\{A_1, A_2, \ldots, B_1, B_2, \ldots\}$ is a countable collection of pairwise disjoint subsets of E and $g_i, h_i \in G$. Recall that $\mathbb{S}^1$ denotes the unit circle and $SO_2(\mathbb{R})$ denotes the group of rotations of the circle.

Theorem 1.5 (AC).* *$\mathbb{S}^1$ is countably $SO_2(\mathbb{R})$-paradoxical. If G denotes the group of translations modulo 1 acting on* $[0, 1)$*, then* $[0, 1)$ *is countably G-paradoxical.*

Proof. Let M be a choice set for the equivalence classes of the relation on $\mathbb{S}^1$ given by calling two points equivalent if one is obtainable from the other by a rotation about the origin through a (positive or negative) rational multiple of 2π radians. Because the rationals are countable, these rotations may be enumerated as $\{\rho_i : i = 1, 2, \ldots\}$; let $M_i = \rho_i(M)$. Then $\{M_i\}$ partitions $\mathbb{S}^1$ and, because any two of the M_i are congruent by rotation, the even-indexed of these sets may be (individually) rotated to yield all the M_i, that is, to cover the whole circle. The same is true of $\{M_i : i \text{ odd}\}$. This construction is easily transferred to $[0, 1)$ using the bijection taking $(\cos\theta, \sin\theta)$ to $\theta/2\pi$, which induces an isomorphism of $SO_2(\mathbb{R})$ with G. □

Corollary 1.6 (AC). *(a) There is no countably additive, rotation-invariant measure of total measure 1, defined for all subsets of* $\mathbb{S}^1$.

* In the sequel, theorems whose proof uses the Axiom of Choice will be followed by (AC).

(b) There is a subset of $[0, 1]$ *that is not Lebesgue measurable.*

(c) There is no countably additive, translation-invariant measure defined on all subsets of* $\mathbb{R}^n$ *and normalizing* $[0, 1]^n$.

Proof. (a) Suppose μ is such a measure and let A and B be disjoint subsets of the circle that witness the paradox of Theorem 1.5; then the properties of μ give $1 \geq \mu(A \cup B) = \mu(A) + \mu(B) = 2$, a contradiction.

(b) This follows from (c); in fact, $\{\alpha \in [0, 1) : (\cos\alpha, \sin\alpha) \in M\}$ is not Lebesgue measurable.

(c) For $\mathbb{R}^1$, such a measure cannot exist because its restriction to subsets of $[0, 1]$ would be invariant under translations modulo 1, contradicting Theorem 1.5. Such a measure in $\mathbb{R}^n$ would induce one on the subsets of $\mathbb{R}$, by the correspondence $A \leftrightarrow A \times [0, 1]^{n-1}$. □

The connection between the Axiom of Choice and the existence of nonmeasurable sets is complex, involving the theory of large cardinals and forcing—two branches of contemporary set theory. We consider these connections in more detail in Chapter 15. For now, we note only that (without assuming Choice) the following two assertions are *not* equivalent:

- All sets of reals are Lebesgue measurable.
- There is a countably additive, translation-invariant extension of Lebesgue measure to all sets of reals.

It is known that the second assertion does not imply the first.

It comes as a bit of a surprise that even with the restriction to finitely many pieces, paradoxes can be constructed using isometries. The following construction, the first of its kind, does not require any form of the Axiom of Choice, which adds some weight to the comment of Eves [Eve63] that the result is "contrary to the dictates of common sense." Recall that when no group is explicitly mentioned, it is understood that the isometry group is being used.

Theorem 1.7 (Sierpiński–Mazurkiewicz Paradox). *There is a paradoxical subset of the plane* $\mathbb{R}^2$.

The reason this paradox exists is that the planar isometry group G_2 has a free non-Abelian subsemigroup that acts in a particularly nice way (Thm. 1.8). The single most important idea in constructing a paradoxical decomposition is the transfer of an algebraic paradox from a group or semigroup (as in Prop. 1.3) to a set on which it acts. This technique was first used, independently, by Hausdorff and by Sierpiński and Mazurkiewicz. The next theorem shows that a free subsemigroup exists for plane isometries.

* Measures are allowed to have values in $[0, \infty]$.

Theorem 1.8. *There are two isometries, σ, τ, of $\mathbb{R}^2$ that generate a free subsemigroup of G_2. Moreover, σ and τ can be chosen so that for any two words w_1 and w_2 in σ, τ having leftmost terms σ, τ, respectively, $w_1(0,0) \neq w_2(0,0)$.*

Proof. Choose θ so that $\beta = e^{i\theta}$ is transcendental; $\theta = 1$ works, but it is simpler just to use the fact that the unit circle is uncountable whereas the set of algebraic numbers is countable. Then let σ be rotation by θ and let τ be translation by $(1,0)$. In $\mathbb{C}$, σ is multiplication by β and τ is addition of 1. We need only prove that σ and τ satisfy the second assertion, because freeness follows from that. For if $w_1 = w_2$, where w_1 and w_2 are distinct semigroup words and one of them is (the identity or) an initial segment of the other, then left cancellation yields $v = e$ for a nontrivial word v. If v has σ on the left, then $v\tau(0) = \tau(0)$, and if v has τ on the left, then $v\sigma(0) = \sigma(0)$, contradicting the second assertion in either case. And if neither is an initial segment of the other, then left cancellation yields w_1 and w_2, which are equal in G_2 but have different leftmost terms.

So, suppose $w_1 = \tau^{j_1}\rho^{j_2}\cdots\tau^{j_m}$ and $w_2 = \rho^{k_1}\tau^{k_2}\cdots\tau^{k_\ell}$, where $m, \ell \geqslant 1$ and each exponent is a positive integer; because $\rho(0) = 0$, it is all right to assume that w_1 and w_2 both end in a power of τ, unless w_2 is simply ρ^{k_1}. Then

$$w_1(0) = j_1 + j_3u^{j_2} + j_5u^{j_2+j_4} + \cdots + j_mu^{j_2+j_4+\cdots+j_{m-1}}$$

and

$$w_2(0) = k_2u^{k_1} + k_4u^{k_1+k_3} + \cdots + k_\ell u^{k_1+k_3+\cdots+k_{\ell-1}} \; (= 0 \text{ if } w_2 = \rho^{k_1}).$$

If $w_1(0) = w_2(0)$, these two expressions may be subtracted to yield a nonconstant polynomial with integer coefficients that vanishes for the value $e^{i\theta}$, and this contradicts the choice of θ. □

Using the isometries of Theorem 1.8 (and working in $\mathbb{C}$), we can prove Theorem 1.7 by directly constructing a paradoxical set in the plane. Let E be the orbit of 0 under the free subsemigroup of Theorem 1.8. Then let $A = \sigma(E)$ and $B = \tau(E)$. Figure 1.4 shows the orbit of 0 in $\mathbb{C}$, where σ is replaced by multiplication by β and τ by addition of 1. The framed numbers form A, and the others are B; we have $E = A/\beta = B - 1$.

Another way of saying this is that E is the set of complex numbers of the form $a_0 + a_1\beta + \cdots + a_n\beta^n$ where n and the coefficients are nonnegative integers. Then A is the set of such numbers for which $a_0 = 0$, and B consists of the others.

We can state this construction in a more abstract form as follows.

Proposition 1.9. *Suppose a group G acting on X contains σ, τ such that for some $x \in X$, any two words in σ, τ beginning with σ, τ, respectively, disagree when applied to x. Then there is a nonempty G-paradoxical subset of X.*

Proof. Let S be the subsemigroup of G generated by τ and ρ, and let E be the S-orbit of x. Then $E \supseteq \tau(E), \rho(E)$, and the hypothesis on x implies that

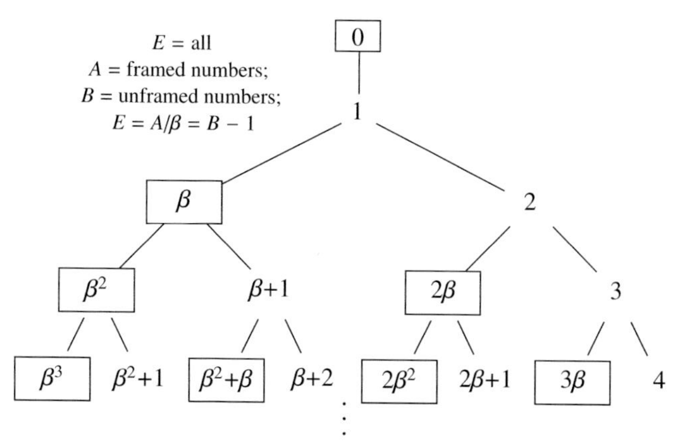

Figure 1.4. The Sierpiński–Mazurkiewicz paradox in the plane, where $\beta = e^i$. The set of numbers at the vertices arises from the right half of Figure 1.1, with each word replaced by what it does to 0.

$\tau w_1(x) \neq \rho w_2(x)$ for any words $w_1, w_2 \in S$. Hence $\tau(E) \cap \rho(E) = \varnothing$. Because $\tau^{-1}(\tau E) = E = \rho^{-1}(\rho E)$, this shows that E is G-paradoxical. □

The fact that the plane admits a paradoxical set E is by no means contradictory. After all, E is countable and so has measure 0; the fact that E is paradoxical implies only that $2 \cdot 0 = 0$. Still, this construction raises many questions about the sorts of planar sets that are paradoxical and about the possibilities in other dimensions. To give some flavor of what is to come, we list some of these related results:

- No nonempty subset of $\mathbb{R}^1$ is paradoxical (Thm. 14.25).
- There are uncountable paradoxical subsets of $\mathbb{R}^2$ (Thm. 7.14; see also Thm. 14.15).
- Any bounded subset of $\mathbb{R}^3$ (or $\mathbb{R}^n$, $n \geqslant 3$) with nonempty interior is paradoxical (Thm. 3.12). This is a generalization of the Banach–Tarski Paradox.
- No subset of $\mathbb{R}^2$ with nonempty interior is paradoxical (Cor. 14.9).
- There are bounded subsets of $\mathbb{R}^2$ that are paradoxical, but none can be paradoxical using two pieces (Thms. 14.16, 14.18).
- The subgroups G of G_2 such that a nonempty G-paradoxical subset of the plane exists are precisely the subgroups having a free non-Abelian subsemigroup (Thm. 14.30).

The ideas of the Sierpiński–Mazurkiewicz construction form the foundation of much of the early history of geometrical paradoxes, though more in the context of groups rather than semigroups. The analogue of Proposition 1.9 is that a paradoxical decomposition of a group is easily lifted to a set on which the group acts *without nontrivial fixed points* (by which is meant that no nonidentity element of the group fixes a point of the set). The conclusion of the following proposition is

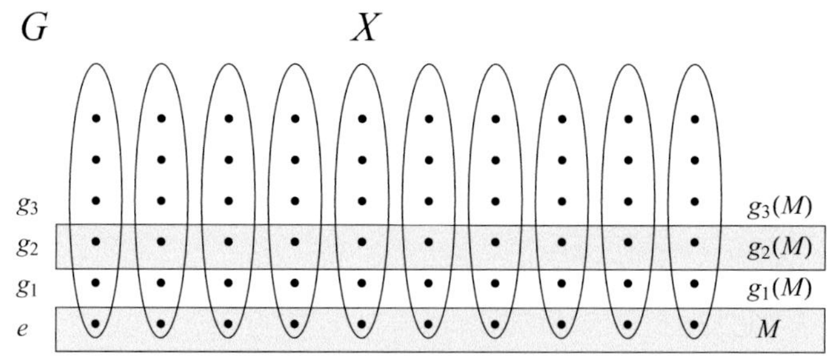

Figure 1.5. A choice set M for the group orbits allows structures in the group to be lifted to the set on which the group acts.

stronger than that of Proposition 1.9 because it states explicitly which set is paradoxical, namely, the whole set upon which the group acts. Unlike Proposition 1.9, the Axiom of Choice is required.

Proposition 1.10 (AC). *If G is paradoxical and acts on X without nontrivial fixed points, then X is G-paradoxical. Hence X is F-paradoxical whenever F, a free group of rank 2, acts on X with no nontrivial fixed points.*

Proof. Suppose $A_i, B_j \subseteq G$ and g_i, h_j witness that G is paradoxical. By the Axiom of Choice, there is a set M containing exactly one element from each G-orbit in X. Then $\{g(M) : g \in G\}$ is a partition of X (see Fig. 1.5, which shows each $g(M)$ as a horizontal array); pairwise disjointness of the family is an easy consequence of the lack of fixed points in G's action. Now this partition serves as a way to transform any subset S of G to a subset of X by $S^* = \{g(M) : g \in S\}$. When we transform the sets of the paradox in G, they yield a paradox in X. For example, $G = \bigcup g_i A_i$ becomes $X = \bigcup g_i(A_i^*)$, and disjointness is preserved; similarly for B_j. So the sets A_i^*, B_j^* form a paradoxical decomposition of X. The assertion about F follows from Theorem 1.2. Note that the number of sets used for X is the same as the number of A_i, B_j originally given for G. □

If the action of G on X is transitive (as is the case if, as in Prop. 1.9, X is replaced by a single G-orbit in X), then the Axiom of Choice is not needed to define M. As an exercise, the reader can show that the converse of Proposition 1.10 is valid for all actions: If X is G-paradoxical, then G is G-paradoxical (transfer the paradoxical decomposition of a single orbit to G). This result has an important measure-theoretic interpretation (see Thm. 10.3).

The main example of a paradoxical group is a free group of rank 2. Theorem 1.2 and constructions such as the Banach–Tarski Paradox are based on the realization of such a group as a group of isometries of $\mathbb{R}^n$. But actions of isometries on $\mathbb{R}^n$ have, in general, many fixed points, and so the main applications of Proposition 1.10 involve figuring out some way to deal with them. Nonetheless, the idea of lifting a paradox from a group to a set upon which it acts is, by itself, sufficient to obtain interesting pseudo-paradoxical results that have important implications in measure theory (see Thm. 2.6).

Notes

Galileo's observations on the problems of infinity can be found in [Gal14]. For more information on nineteenth-century thoughts on these sorts of paradoxes, see [Bol50]. The definition of G-paradoxical appears in [Sie54], but has its roots in earlier work of Hausdorff, Banach and Tarski, and of Sierpiński and Mazurkiewicz. Banach and Tarski [BT24] published their paradoxical construction in $\mathbb{R}^3$ in 1924, and many papers discussing or simplifying their work have appeared since. The initial discovery that free groups (actually, free products) bear paradoxes that can be given a geometric interpretation is due to Hausdorff (see notes to Chap. 2; also §4.2).

The question of whether the assertion of Theorem 1.4(a) (or (b)) implies the Axiom of Choice was posed by Tarski [Tar24] and solved by Sageev [Sag75]. The classical construction of a non-Lebesgue measurable subset of the circle in Theorem 1.5 was given by Vitali [Vit05] in 1905. The slightly more general Corollary 1.6 was pointed out by Hausdorff [Hau14b].

Sierpiński raised the possibility of a paradoxical subset of the plane (Thm. 1.7), and the existence of such a set was initially proved by Mazurkiewicz, whose proof was, in turn, simplified by Sierpiński (see [MS14]). The results in this chapter related to semigroups (Thm. 1.8 and Props. 1.3 and 1.9) are derived from their work. The Sierpiński–Mazurkiewicz Paradox is discussed in [Eve63, p. 277], [HDK64, p. 26], and [Mes66, p. 155]. These three books contain discussions of the Banach–Tarski Paradox as well. Generalizations of the choiceless Sierpiński–Mazurkiewicz Paradox are discussed in Chapter 7.

2

The Hausdorff Paradox

It was shown in Chapter 1 (Prop. 1.10) how free non-Abelian groups can be used to generate paradoxes. The first opportunity for such a free group to appear among the Euclidean isometry groups is in 3-space. This is because G_1 and G_2 are solvable (App. A; see also Thm. 12.4) and hence cannot contain any free non-Abelian subgroup. In this chapter we explicitly construct a free non-Abelian subgroup of G_3 and describe a paradoxical decomposition that this causes. A subset S of a group G will be called *independent* if S is a free set of generators of H, the subgroup of G generated by S; H is then a free group of rank $|S|$.

There are many ways of getting a pair of independent rotations of $\mathbb{S}^2$, the unit sphere in 3-space. We present a simple pair found by K. Satô. Let

$$\sigma = \frac{1}{7}\begin{bmatrix} 6 & 2 & 3 \\ 2 & 3 & -6 \\ -3 & 6 & 2 \end{bmatrix} \quad \text{and} \quad \tau = \frac{1}{7}\begin{bmatrix} 2 & -6 & 3 \\ 6 & 3 & 2 \\ -3 & 2 & 6 \end{bmatrix},$$

which we call the *Satô rotations*. They will reappear in Theorem 2.8. We use F_n to denote the free group of rank n.

Theorem 2.1. *The two Satô rotations are independent. Hence, if $n \geq 3$, $SO_n(\mathbb{Q})$ has a free subroup of rank 2.*

Proof. We wish to show that no nontrivial reduced word in $\sigma^{\pm 1}$, $\tau^{\pm 1}$ equals the identity. Assume w is such a word and w equals the identity. Conjugating by a sufficiently high power of σ and, if necessary, inverting, we can assume that w has σ as its rightmost term.

Define four matrices by

$$M_\sigma = \begin{bmatrix} 6 & 2 & 3 \\ 2 & 3 & -6 \\ -3 & 6 & 2 \end{bmatrix}, \quad M_\tau = \begin{bmatrix} 2 & -6 & 3 \\ 6 & 3 & 2 \\ -3 & 2 & 6 \end{bmatrix},$$

$$M_\sigma^- = \begin{bmatrix} 6 & 2 & -3 \\ 2 & 3 & 6 \\ 3 & -6 & 2 \end{bmatrix}, \quad M_\tau^- = \begin{bmatrix} 2 & 6 & -3 \\ -6 & 3 & 2 \\ 3 & 2 & 6 \end{bmatrix}.$$

It is simpler to work with these matrices instead of the true matrices corresponding to the rotations; the only difference is an integer power of 7 that would be added at the end. We will analyze $w(1, 0, 0)$, the first column of the matrix $M_\sigma^{k_n} M_\tau^{k_{n-1}} \cdots M_\tau^{k_2} M_\sigma^{k_1} M_\sigma$, where $k_1 \geq 0$, $k_n \in \mathbb{Z}$, and the other ks are nonzero. A negative power refers to a power of $M^-_{\sigma \text{ or } \tau}$. We will work modulo 7 throughout, which suffices because an integer that is not divisible by 7 is nonzero. Define four sets of vectors:

$$V_\sigma = \{(3, 1, 2), (5, 4, 1), (6, 2, 4)\},$$

$$V_\tau = \{(3, 2, 6), (5, 1, 3), (6, 4, 5)\},$$

$$V_\tau^- = \{(3, 5, 1), (5, 6, 4), (6, 3, 2)\},$$

$$V_\sigma^- = \{(1, 5, 4), (2, 3, 1), (4, 6, 2)\}.$$

To start, we have $M_\sigma(1, 0, 0) = (6, 2, -3) \equiv (6, 2, 4) \pmod 7$, and this vector is V_σ. The following four properties, which show that the V-sets act as invariants, are easily verified by simple matrix computations:

1. For any $v \in V_\sigma \cup V_\tau \cup V_\tau^-$, $\sigma v \in V_\sigma$.
2. For any $v \in V_\sigma^- \cup V_\tau \cup V_\tau^-$, $\sigma^- v \in V_\sigma^-$.
3. For any $v \in V_\tau \cup V_\sigma \cup V_\sigma^-$, $\tau v \in V_\tau$.
4. For any $v \in V_\tau^- \cup V_\sigma \cup V_\sigma^-$, $\tau^- v \in V_\tau^-$.

Now work left through the word. By (1), because the first step gives us $(6, 2, 4)$, $M_\sigma^{k_1}$ leaves the vector in V_σ. By (3) or (4) (depending on the sign of k_2), the second power $M_\tau^{k_2}$ places the vector in $V_\tau \cup V_\tau^-$. And (1) and (2) (depending on the sign of k_3) mean that the third power leaves the vector in $V_\sigma \cup V_\sigma^-$. This alternation continues as we move left through the word. Therefore the vector ends up in one of the V-sets and so is not $(1, 0, 0)$. □

The preceding proof is just one of many constructions of a free non-Abelian group of rotations in $\mathbb{R}^3$. Hausdorff gave the first such construction in 1914; he showed that if ϕ and ρ are rotations through 180° and 120°, respectively, about axes containing the origin, and if $\cos 2\theta$ is transcendental where θ is the angle between the axes, then ϕ and ρ are free generators of $\mathbb{Z}_2 * \mathbb{Z}_3$. Because $\mathbb{Z}_2 * \mathbb{Z}_3$ has a free subgroup of rank 2 ($\rho\phi\rho$, $\phi\rho\phi\rho\phi$ freely generate such a subgroup), this leads to an F_2 within the rotation group. In 1978, Osofsky simplified Hausdorff's approach by showing that ϕ and ρ generate $\mathbb{Z}_2 * \mathbb{Z}_3$ even if $\theta = 45°$.

There are other results that yield specific independent pairs of rotations. We omit the proof of the following theorem (see [Gro56, Swi58]), because from the point of view of paradoxical decompositions of $\mathbb{S}^2$, the additional information it provides is not necessary. (However, a special case of part (b) is implicit in the proof of Thm. 6.7.)

Theorem 2.2. *Suppose ϕ and ρ are rotations of $\mathbb{S}^2$ with the same angle, θ, of rotation. Then ϕ and ρ are independent if either* (*a*) *the axes are perpendicular* *and* $\cos\theta$ *is a rational different from* $0, \pm\frac{1}{2}, \pm 1$ *or* (*b*) *the axes are distinct and the cosine of the angle formed by the axes is transcendental.*

This theorem yields the nicely simple pair of independent rotations given by

$$\frac{1}{5}\begin{bmatrix} 3 & -4 & 0 \\ 4 & 3 & 0 \\ 0 & 0 & 1 \end{bmatrix} \text{ and } \frac{1}{5}\begin{bmatrix} 1 & 0 & 0 \\ 0 & 3 & -4 \\ 0 & 4 & 3 \end{bmatrix}.$$

One can prove directly, using a mod-5 argument in a similar way to the proof of Theorem 2.1, that these two matrices are independent (see [Tao04]). Theorem 2.2 leads to the following problem: Do there exist two independent rotations of the rational sphere $\mathbb{S}^2 \cap \mathbb{Q}^3$ such that the group they generate acts on the rational sphere is without fixed points? The existence of such a group would lead to a paradoxical decomposition without requiring the Axiom of Choice. In this chapter we will see that the answer for this question is positive.

While Theorems 2.1 and 2.2 give specific examples of free groups, it turns out that, from a topological point of view, almost any pair of rotations is independent. More precisely, if $SO_3(\mathbb{R}) \times SO_3(\mathbb{R})$ is given the product topology, then $\{(\phi, \rho) \in SO_3(\mathbb{R}) \times SO_3(\mathbb{R}) : \phi \text{ and } \rho \text{ are independent}\}$ is comeager (and hence dense); this is a consequence of the discussion following Corollary 7.6.

It was pointed out that the properties of being solvable and of containing a free non-Abelian group are mutually exclusive. Because of nonsolvable finite groups, the two conditions are not exhaustive, but one can ask whether every group either has a solvable subgroup of finite index or has a free non-Abelian subgroup. Though not true for all groups, it is valid for a wide class of groups. This is a deep result proved by Tits in 1972 (see Thm. 12.6 for a fuller discussion). In particular, Tits showed the property to be valid for all subgroups of $GL_n(\mathbb{R})$, the nonsingular linear transformations of $\mathbb{R}^n$, from which it follows for all groups of Euclidean isometries. Thus the fact that as soon as $SO_n(\mathbb{R})$ loses its solvability, it gains a free non-Abelian subgroup, is a special case of a far-reaching result in the theory of matrix groups.

Later on we shall see that the existence of larger free groups has implications for paradoxical decompositions. Thus, despite the power of Tits's Theorem, we shall investigate (in Chaps. 6, 7, and 8) other sorts of free groups of isometries. Of particular interest will be the fact that there is an independent set of rotations in $SO_3(\mathbb{R})$ with the same cardinality as the continuum; in fact, the free product of any sequence of continuum many cyclic groups is representable in $SO_3(\mathbb{R})$. Also, there are continuum many independent isometries of $\mathbb{R}^3$ such that the group they generate acts on $\mathbb{R}^3$ without nontrivial fixed points (see Thm. 7.4).

Each element of the free group of rotations (call it F) constructed in Theorem 2.1 fixes all points on some line in $\mathbb{R}^3$, and so Proposition 1.10 cannot yet be applied. A naive approach to this difficulty turns out to be fruitful. Each nonidentity rotation in F has two fixed points on $\mathbb{S}^2$, the unit sphere, namely, the intersection of the rotation's axis with the sphere. Let D be the collection of all such points; because F is countable, so is D. Now, if $P \in \mathbb{S}^2 \setminus D$ and $g \in F$, then $g(P)$ lies in $\mathbb{S}^2 \setminus D$ as well: If h fixed $g(P)$, then P would be a fixed point of $g^{-1}hg$. Hence F acts on $\mathbb{S}^2 \setminus D$ without nontrivial fixed points, and Proposition 1.10 may be applied to this action to obtain the following result.

Theorem 2.3 (Hausdorff Paradox) (AC). *There is a countable subset D of $\mathbb{S}^2$ such that $\mathbb{S}^2 \setminus D$ is $SO_3(\mathbb{R})$-paradoxical.*

A countable subset of the sphere can be dense, and so the paradoxical nature of Theorem 2.3 is not immediately evident. Still, countable sets are very small in size compared to the whole (uncountable) sphere. We shall see in the next chapter how the smallness of D allows it to be eliminated completely, yielding the Banach–Tarski Paradox: $\mathbb{S}^2$ is $SO_3(\mathbb{R})$-paradoxical. But even without eliminating D, Hausdorff's Paradox has an important measure-theoretic consequence.

The fact (1.6) that there is no countably additive, isometry-invariant measure on $\mathcal{P}(\mathbb{R}^n)$ normalizing the unit cube naturally leads to the question of whether measures exist satisfying a weaker set of conditions. The usual approach is to allow the measure to assign values to a smaller collection of sets, that is, use Lebesgue measure and live with the fact that some sets are not Lebesgue measurable. Finitely additive measures had been studied prior to Lebesgue, and it is natural to ask whether there might be a finitely additive, isometry-invariant measure defined for all subsets of $\mathbb{R}^n$. It was this question that motivated Hausdorff to carry out his groundbreaking construction on $\mathbb{S}^2$, because he was able to use it to provide an answer in all dimensions except one and two. First we give a definition and proposition that state precisely the fundamental connection between paradoxical decompositions and the nonexistence of finitely additive measures. Suppose a group G acts on a set X, and $E \subseteq X$.

Definition 2.4. *E is called G-*negligible *if $\mu(E) = 0$ whenever μ is a finitely additive, G-invariant measure on $\mathcal{P}(X)$ with $\mu(E) < \infty$.*

Proposition 2.5. *If E is G-paradoxical, then E is G-negligible.*

Proof. Suppose μ is a finitely additive, G-invariant measure on $\mathcal{P}(X)$ and $\mu(E) < \infty$. Let the fact that E is G-paradoxical be witnessed by A_i, g_i, B_j, h_j. Then $\mu(E) \geqslant \sum \mu(A_i) + \sum \mu(B_j) = \sum \mu(g_i A_i) + \sum \mu(h_j B_j) \geqslant \mu(\bigcup g_i A_i) + \mu(\bigcup h_j B_j) = \mu(E) + \mu(E) = 2\mu(E)$. Because $\mu(E) < \infty$, this means $\mu(E) = 0$. □

One of the more noteworthy results of the theory of finitely additive measures is Tarski's theorem that the converse of Proposition 2.5 is valid: If E is not G-paradoxical, then a finitely additive, G-invariant measure on $\mathcal{P}(X)$ normalizing E must exist. This will be proved in §11.1.

The next theorem deduces the $SO_3(\mathbb{R})$-negligibility of $\mathbb{S}^2$ from the Hausdorff Paradox by proving that countable sets are negligible with respect to finite measures on $\mathcal{P}(\mathbb{S}^2)$.

Theorem 2.6 (AC). *The sphere $\mathbb{S}^2$ is $SO_3(\mathbb{R})$-negligible. Hence there is no finitely additive, rotation-invariant measure on $\mathcal{P}(\mathbb{S}^2)$ having total measure one. Moreover, for any $n \geqslant 3$, there is no finitely additive, isometry-invariant measure on $\mathcal{P}(\mathbb{R}^n)$ normalizing the unit cube.*

Proof. Suppose μ is a finitely additive, $SO_3(\mathbb{R})$-invariant measure on $\mathcal{P}(\mathbb{S}^2)$ with $\mu(\mathbb{S}^2) < \infty$. If D is the countable set in the Hausdorff Paradox then, by Proposition 2.5, $\mu(\mathbb{S}^2 \setminus D) = 0$. Hence it suffices to show that $\mu(D) = 0$. Let ℓ be a line through the origin that is disjoint from D. For each point $P \in D$, let $A(P)$ be the set of angles θ such that the rotation of P around ℓ through θ^j radians, for any positive integer j, takes P to another point in D. The countability of D and the set of possible j implies that $A(P)$ is countable and hence that $A = \bigcup\{A(P) : P \in D\}$ is countable. If ρ is chosen to be a rotation around ℓ through one of the uncountably many angles not in A, then ρ has the property that, for any j, $\rho^j(D)$ is disjoint from D. It follows that $D \cup \rho(D) \cup \rho^2(D) \cup \ldots$ is a pairwise disjoint union. Now suppose that $\mu(D) > 0$. Then we can choose an integer k so that $k\mu(D) > 1$. This means $\mu(D) + \mu(\rho(D)) + \cdots + \mu(\rho^k(D)) > 1 = \mu(\mathbb{S}^2)$, a contradiction. So $\mu(D) = 0$.

To prove the assertion about $\mathbb{R}^n$, it suffices to consider $n = 3$, because a measure in a higher dimension induces one in $\mathbb{R}^3$ as described in the proof of Corollary 1.6. Now, if μ is an isometry-invariant measure on $\mathbb{R}^3$ normalizing the unit cube, then μ must vanish on singletons. This is because any two singletons are congruent and so receive the same measure; hence, if a singleton's measure were positive, then the measure of the unit cube would be infinite. Moreover, translation invariance implies that any cube has finite, nonzero measure, and it follows that $0 < \mu(B) < \infty$, where B denotes the unit ball. Define a measure ν on $\mathcal{P}(\mathbb{S}^2)$ by the adjunction of radii, that is, $\nu(A) = \mu\{\alpha P : P \in A, 0 < \alpha \leqslant 1\}$. Because $\mu(\{0\}) = 0$, $\nu(\mathbb{S}^2) = \mu(B)$. Moreover, ν is finitely additive and $SO_3(\mathbb{R})$-invariant because μ is. This contradicts the $SO_3(\mathbb{R})$-negligibility of $\mathbb{S}^2$. □

The Hausdorff Paradox in Theorem 2.3 uses the Axiom of Choice, and that cannot be avoided when working in $\mathbb{S}^2$. But the two rotations underlying the paradox—the Satô rotations of Theorem 2.1—have the pleasing property that they take rational points on the sphere to rational points on the sphere; that is, they act on the rational sphere, $\mathbb{S}^2 \cap \mathbb{Q}^3$. And, even nicer, the group generated by the two rotations acts on the rational sphere with no fixed points. This nice property is not essential for the classic Banach–Tarski Paradox to be presented in Chapter 3, but we give the proof here because it immediately yields a paradoxical decomposition of the rational sphere that does not require the Axiom of Choice.

First we point out that the rational sphere does indeed look just like the real sphere.

Proposition 2.7. *The rational sphere $\mathbb{S}^2 \cap \mathbb{Q}^3$ is dense in the unit sphere $\mathbb{S}^2$.*

Proof. Let $F : \mathbb{R}^2 \to \mathbb{S}^2$ be stereographic projection: $F(P)$ is the point on the unit sphere that is on the line connecting P to the north pole, $(0, 0, 1)$. Easy algebra shows that $F(p, q) = (2p, 2q, p^2 + q^2 - 1)/(p^2 + q^2 + 1))$ and so F takes rational points in the plane to rational points on the upper hemisphere. Because F is continuous, the rational points that arise are dense in the hemisphere, and hence the rational points in the complete sphere are dense in the sphere. □

Theorem 2.8. *The Satô rotations of Theorem 2.1 act on the rational sphere with no nontrivial fixed points.*

Proof. An eigenvector computation shows that the axis of σ (resp., τ) is $(2, 1, 0)$ (resp., $(0, 1, 2)$). Suppose the result is false and w is a nontrivial word in $\{\sigma^{\pm 1}, \tau^{\pm 1}\}$ (the *atoms*) of minimal length that has a rational fixed point. Because the axes just given strike the unit sphere in the nonrational point $\frac{(2,1,0)}{\sqrt{5}}$ or $\frac{(0,1,2)}{\sqrt{5}}$, w cannot be a pure power of an atom. Because the fixed-point property is preserved under conjugation, w must be of the form $\tau^{\pm 1} \ldots \sigma^{\pm 1}$ or $\sigma^{\pm 1} \ldots \tau^{\pm 1}$. Then inversion allows us to assume that $w = \tau^{\pm 1} \ldots \sigma^{\pm 1}$.

The key tool is the quaternion representation of rotations as $(c, \vec{s})$. Here $c = \cos(\theta/2)$, $\vec{s}$ determines the rotation axis, and $c^2 + |\vec{s}|^2 = 1$. Working out the axes (by eigenvectors) and the angles (by looking at the image of $(1, 0, 0)$) yields that the representation of $\sigma^{\pm 1}$ is $\frac{1}{\sqrt{14}}(3, \pm(2, 1, 0))$ and $\tau^{\pm}$ has the representation $\frac{1}{\sqrt{14}}(3, \pm(0, 1, 2))$.

Any quaternion $(c, \vec{s})$ determines a unique (up to sign) quaternion of norm 1. So we can think of any quaternion as representing a unique rotation. This allows us to use, say, $(3, \pm(2, 1, 0))$ to reperesent σ, and this transformation to integers simplifies the algebra. So when looking at the quaternion representation of a word w of length k, we will always multiply by $\sqrt{14}^k$, thus making all components of quaternion integers. The key point is that quaternion multiplication $(c_1, \vec{s}_1) \cdot (c, \vec{s})$ gives the representation of the composition of the two underlying rotations. That multiplication formula is $(c_1 c - \vec{s}_1 \cdot \vec{s}, c\vec{s}_1 + c_1\vec{s} + \vec{s}_1 \times \vec{s})$, where the dot and cross are the usual dot and cross products, respectively (see App. A).

Let q denote the integer-valued quaternion for the rotation corresponding to a word: $q(w) = (c_w, (X_w, Y_w, Z_w))$. The rotation axis intersects $\mathbb{S}^2$ at $\pm(X_w, Y_w, Z_w)/\sqrt{X_w^2 + Y_w^2 + Z_w^2}$, so it suffices to prove that $X_W^2 + Y_W^2 + Z_W^2$ is not a perfect square. This will be done by showing that $X_W^2 + Y_W^2 + Z_W^2 \equiv 3$ or 5 or 6 (mod 7); these values are not squares modulo 7, so the unreduced sum of squares will be proved to be not a perfect square.

So now we will reduce all integers modulo 7, letting $\overline{q}(w)$ be the mod-7 reduction of $q(w)$. But we need to make one more critical reduction. If we multiply a reduced representation $\overline{q}(w)$ by some integer $1, 2, 3, 4, 5$ or 6, reducing mod 7, the result is the same as far as the quadratic character of $X^2 + Y^2 + Z^2$ goes. This is because, for any such scalar m, $X^2 + Y^2 + Z^2$ is a mod-7 square iff

$m^2(X^2 + Y^2 + Z^2)$ is. So when working with $\overline{q}(w)$, we may always assume, by appropriate multiplication by a modular inverse, that $c = 1$; and we will use $\sim$ to denote the underlying equivalence relation. This reduction allows us to show that any power of an atom is the same as the atom itself, as far as the quadratic character of $X^2 + Y^2 + Z^2 \pmod 7$. □

Claim. For k a positive integer, $\overline{q}(\sigma^k) \sim (1, (3, 5, 0))$, $\overline{q}(\sigma^{-k}) \sim (1, (4, 2, 0))$, $\overline{q}(\tau^k) \sim (1, (0, 5, 3))$, and $\overline{q}(\sigma^{-k}) \sim (1, (0, 2, 4))$.

Proof of claim. The base case is $\overline{q}(\sigma) \sim (3, (2, 1, 0)) \sim (1, (3, 5, 0))$. It suffices to show that $\overline{q}(\sigma) \cdot \overline{q}(\sigma) = (1, (3, 5, 0))$, which is easily verified by one quaternion multiplication and reduction modulo 7. This equation means that inductively moving up through the powers leads to no change. The other three cases are identical.

The claim means that any power of an atom in a word may be replaced by the corresponding atom without affecting the mod-7 quadratic character we care about. So we may assume $w = \tau^{\pm 1}\sigma^{\pm 1} \ldots \tau^{\pm 1}\sigma^{\pm 1}$.

Define $V = \{((1, (1, 1, 5)), (1, (5, 1, 1)), (1, (4, 3, 4)), (1, (6, 5, 6))\}$. This is an invariant set for words of the form $\tau^{\pm 1}\sigma^{\pm 1}$. By this we mean $\overline{q}(\tau^{\pm 1}\sigma^{\pm 1}V) \subseteq V$. This concludes the proof, because inductively moving left through w, we have that $\overline{q}(w) \in V$. The sums of squares of the four vectors from V are all 6 (mod 7). Because we are using $\sim$-equivalence classes, the actual sum of squares that arises from a given word w will be one of 3, 5, or 6, because the quadratic character does not change.

To prove invariance, observe first that it holds for the four basic words of the form $\tau^{\pm 1}\sigma^{\pm 1}$. For example, consider $\tau\sigma^{-1}$:

$$\begin{aligned}\overline{q}(\tau) \cdot \overline{q}(\sigma^{-1}) &= (1, (0, 5, 3)) \cdot (1, (4, 2, 0)) = (10, (-4, -4, 8)) \\ &\equiv (3, (3, 3, 1)) \sim (1, (1, 1, 5)).\end{aligned}$$

The other three are similar. So it remains only to show that application of any of the four words of length two leaves V invariant. This requires 16 quaternion multiplications for the four possibilities in V and four words of length two. A simple computation shows that the set is invariant as claimed. □

Corollary 2.9. *The rational sphere is paradoxical.*

Proof. By Theorem 2.1 and Proposition 1.10. □

As observed at the beginning of this chapter, there are no free non-Abelian subgroups in the lower-dimensional Euclidean isometry groups. This is why the ideas of the Hausdorff Paradox cannot be used to decide the existence of isometry-invariant, finitely additive measures defined on $\mathcal{P}(\mathbb{R}^1)$ or $\mathcal{P}(\mathbb{R}^2)$. In Chapter 12 we shall show that invariant measures always exist with respect to groups satisfying certain abstract conditions. Because these conditions include solvability, it will follow that isometry-invariant, finitely additive measures defined on $\mathcal{P}(\mathbb{R})$ and $\mathcal{P}(\mathbb{R}^2)$ exist.

Notes

The matrices that Satô used in Theorem 2.1 are from [Sat95]. The proof presented here was found with the help of *Mathematica*. There are many ways to get an independent pair of rotations; but the ones of Satô also yield the stronger rational result of Theorem 2.8.

Hausdorff's original embedding of $\mathbb{Z}_2 * \mathbb{Z}_3$ in a rotation group appears in [Hau14b, Hau14a, p. 469]; for a modern treatment, see [Str79]. Osofsky's simpler construction appears as a problem in the *American Mathematical Monthly* [Oso76, Oso78] (see also [Har83]). The fact that $\mathbb{Z}_2 * \mathbb{Z}_3$, and hence $SO_3(\mathbb{R})$, contains a free subgroup of rank 2 is first stated in [Neu29, p. 80n22]. In fact, $G * H$ always has a free subgroup of rank 2 unless one of G or H is trivial, or each of G, H has order 2 (see [MKS66, p. 195, Ex. 19]). In an addendum to [Oso78], Lyndon shows, using quaternions, that the three rotations through 120° around the three orthogonal axes in $\mathbb{R}^3$ freely generate $\mathbb{Z}_3 * \mathbb{Z}_3 * \mathbb{Z}_3$.

It is curious that Hausdorff's embedding of a free product in $SO_3(\mathbb{R})$ appeared in the same year as the Sierpiński–Mazurkiewicz Paradox (Thm. 1.7) that involves the embedding of a free semigroup in G_2. Perhaps Hausdorff was motivated by Felix Klein's representation of $\mathbb{Z}_2 * \mathbb{Z}_3$ as linear fractional transformations and hence as isometries of the hyperbolic plane [KF90]; see the discussion of the hyperbolic plane in Chapter 4.

Many papers on the representation of free groups using rotations or isometries [BM61, Bot57, Gro56, GD54, Dek58b, Dek59b, MS58] appeared in the Netherlands and Poland in the 1950s, largely inspired by work of Sierpiński and de Groot, and many results were obtained independently in both countries. Thus part (b) of Theorem 2.2, originally conjectured by de Groot [Gro56], was proved by Dekker [Dek59b], while a similar result was proved by Balcerzyk and Mycielski [BM61]. Part (a) of Theorem 2.2 is stated in [Swi58]. The existence of a free group of rotations of rank $2^{\aleph_0}$ was essentially proved by Sierpiński [Sie45b] in 1945 and rediscovered by de Groot and Dekker [Gro56, GD54] in 1954. De Groot [Gro56] conjectured that any free product of continuum many subgroups of $SO_3(\mathbb{R})$, each of size strictly less than $2^{\aleph_0}$, is also a subgroup of $SO_3(\mathbb{R})$. This was proved by Balcerzyk and Mycielski [BM61], although Dekker [Dek59b] independently proved the special case where all the groups are cyclic. The existence of free groups of isometries acting on $\mathbb{R}^3$ without nontrivial fixed points was proved independently by Mycielski and Świerczkowski [MS58] and Dekker [Dek59b]. Dekker also studied the isometry groups of higher-dimensional Euclidean and non-Euclidean spaces (see Chap. 6).

The result of Tits on matrix groups, which answered a conjecture of Bass and Serre, appears in [Tit72].

The Hausdorff Paradox (Thm 2.3) is, of course, due to Hausdorff [Hau14b, Hau14a, p. 469], who used it to prove Theorem 2.6. Hausdorff's formulation of Theorem 2.3 was slightly different though. He showed that, modulo a countable set of points, a "half" of a sphere could be congruent to a "third" of a sphere, rather than that a near-sphere could be duplicated. From the point of view of disproving the existence of measures, this is all that is required.

Theorem 2.8 and the corollary that follows are due to Satô [Sat95].

3

The Banach–Tarski Paradox: Duplicating Spheres and Balls

The idea of cutting a figure into pieces and rearranging them to form another figure goes back at least to Greek geometry, where this method was used to derive area formulas for regions such as parallelograms. When forming such rearrangements, one totally ignores the boundaries of the pieces. The consideration of a notion of dissection in which every single point is taken into account, that is, a set-theoretic generalization of the classical geometric definition, leads to an interesting, and very general, equivalence relation. By studying the abstract properties of this new relation, Banach and Tarski were able to improve on Hausdorff's Paradox (Thm. 2.3) by eliminating the need to exclude a countable subset of the sphere. Because geometric rearrangements will be useful too, we start with the classical definition in the plane.

Definition 3.1. *Two polygons in the plane are* congruent by dissection *if one of them can be decomposed into finitely many polygonal pieces that can be rearranged using isometries (and ignoring boundaries) to form the other polygon.*

It is clear that polygons that are congruent by dissection have the same area. The converse was proved in the early nineteenth century, and a simple proof can be given by efficiently making use of the fact that congruence by dissection is an equivalence relation (transitivity is easily proved by superposition, using the fact that the intersection of two polygons is a polygon (see Boltianskii [Bol78] or Eves [Eve63, p. 233]).

Theorem 3.2 (Bolyai–Gerwien Theorem). *Two polygons are congruent by dissection if and only if they have the same area.*

Proof. To prove the reverse direction, it suffices, because of transitivity, to show that any polygon is congruent by dissection into a single square. We do this first for a triangle. Figure 3.1(a) shows that any triangle is congruent by dissection to a rectangle. Figure 3.1(b) shows how a rectangle whose length is at most 4 times its width can be transformed to a square: The triangles to be moved are clearly similar to their images, and the fact that the area of the square equals that of the

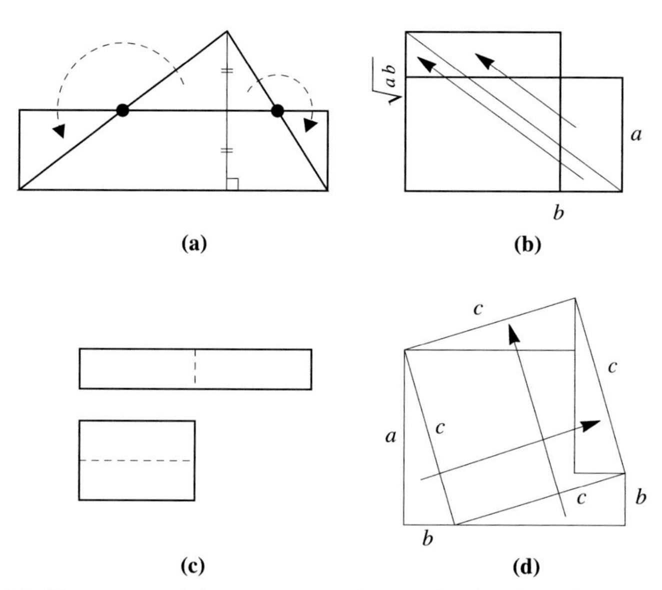

Figure 3.1. The steps needed to get a square from a triangle. The Pythagorean step is done by translations only, a fact that will be important in the circle-squaring work of Chapter 9 (see Lemma 9.24).

rectangle implies that they are, in fact, congruent. The figure does not correctly describe the situation when the length is greater than four times the width, but any such unbalanced rectangle can be transformed to one of the desired type by repeated halving and stacking, as illustrated in Figure 3.1(c). Hence any triangle is congruent by dissection with a square.

The proof is concluded by observing that the Pythagorean Theorem can be proved in a way that can be used to transform two (or more) squares into one by dissection. Consider Figure 3.1(d), which proves that $c^2 = a^2 + b^2$ by showing how the squares on a and b can be transformed by dissection into one on c; this Pythagorean proof is due to physicist George Airy.

Used repeatedly, this construction shows how any finite set of squares can be transformed by dissection to a single square. So the last step is to show that any polygon can be divided into triangles. A simple way to do this is to draw a vertical line through every vertex; this divides the polygon into trapezoids, and each such splits along a diagonal into two triangles. Another approach to triangulation, one that avoids adding any new vertices, can be found in [DR11, p. 233]. Squaring the triangles and combining the squares by the specific construction given earlier yields one square that is congruent by dissection to the original polygon. □

This efficient proof does not lead to particularly efficient or beautiful dissections. For several interesting examples, such as a four-piece squaring of an equilateral triangle, see [Eve63, Chap. 5]. One can ask about restricting the group. The most important result in this area is the Hadwiger–Glur Theorem from 1951

[Bol78, §10]. That gives a criterion for congruence by dissection of polygons using translations only and has this important consequence.

Theorem 3.3 (Hadwiger–Glur). *A convex polygon is congruent by dissection to a square using translations if and only if the polygon is centrally symmetric.*

The theory of geometrical dissections in higher dimensions, or other geometries, is not at all as simple as in the plane. In fact, the third problem on Hilbert's famous list from 1900 asks whether a regular tetrahedron in $\mathbb{R}^3$ is congruent by dissection (into polyhedra) with a cube. All proofs of the volume formula for a tetrahedron were based on a limiting process of one sort or another, such as the devil's staircase [Hea56, p. 390] or Cavalieri's Principle [Moi74]. Hopes for an elegant dissection proof were dashed when Dehn proved, in 1900, that a regular tetrahedron is not congruent by dissection with any cube (see [Bol78] for a proof). But it is possible that for a suitable generalization of dissection where a wider class of pieces is allowed, a regular tetrahedron is piecewise congruent to a cube. Indeed, one consequence of the Banach–Tarski Paradox is that a regular tetrahedron can be cubed if arbitrary sets are allowed as pieces (see Thms. 3.11 and 9.28).

The set-theoretic version of congruence by dissection may be stated in the context of an arbitrary group action.

Definition 3.4. *Suppose G acts on X and $A, B \subseteq X$. Then A and B are* G-equidecomposable *(sometimes called* finitely G-equidecomposable *or* piecewise G-congruent*) if A and B can each be partitioned into the same finite number of respectively G-congruent pieces. Formally,*

$$A = \bigcup_{i=1}^{n} A_i, \qquad B = \bigcup_{i=1}^{n} B_i,$$

$A_i \cap A_j = \varnothing = B_i \cap B_j$ if $i < j \leqslant n$, and there are $g_1, \ldots, g_n \in G$ such that, for each $i \leqslant n$, $g_i(A_i) = B_i$.

The notation $A \sim_G B$ will be used to denote the equidecomposability relation, but the G will be suppressed if X is a metric space and G is the full isometry group, or if it is obvious which group G is meant. Thus, for sets in $\mathbb{R}^n$, equidecomposability means G_n-equidecomposability. We say A is *G-equidecomposable with B using n pieces* (denoted $A \sim_n B$) if the disassembly can be effected with n pieces.

It is straightforward to verify that $\sim_G$ is an equivalence relation. Transitivity of $\sim_G$ is proved in the same way as for congruence by dissection, yielding that if $A \sim_m B$ and $B \sim_n C$, then $A \sim C$, using at most mn pieces. Not surprisingly, then, the relation $\sim_n$ is not transitive. A simple counterexample is $A = \{1, 2, 3, 4\}$, $B = \{1, 2, 5, 6\}$, and $C = \{1, 5, 9, 13\}$; $A \sim_2 B \sim_2 C$, but $A \sim_4 C$ and the 4 cannot be lowered. We may now rephrase more succinctly the notion of a set being G-paradoxical. E is G-paradoxical if and only if E contains disjoint sets

A, B such that $A \sim_G E$ and $B \sim_G E$. One can then obtain immediately the following easy but useful fact, which shows that the property of being G-paradoxical is really a property of the $\sim_G$-equivalence classes in $\mathscr{P}(X)$.

Proposition 3.5. *Suppose G acts on X and E, E' are G-equidecomposable subsets of X. If E is G-paradoxical, so is E'.*

It is not immediately apparent that there is any connection between equidecomposability and congruence by dissection. Indeed, they differ in a most fundamental way. Because there is no restriction on the subsets that may be used to verify that $A \sim B$, there is no guarantee that A and B have the same area (or n-dimensional Lebesgue measure, if $A, B \subseteq \mathbb{R}^n$). For if the pieces are non-Lebesgue measurable, then the straightforward proof that works in the case of congruence by dissection cannot be used, because it involves summing the areas of the pieces. Thus it is conceivable that all polygons in the plane are equidecomposable or that all polyhedra in $\mathbb{R}^3$ are equidecomposable. In fact, the latter assertion is true (see Cor. 3.10), whereas the former is not! The preservation of a given (G-invariant) measure under G-equidecomposability is related to the existence of a finitely additive, G-invariant extension of the measure to all subsets of X, and this, in turn, depends on abstract properties of the group G (and of G's action on X). The second part of this book deals extensively with the question of which groups have the property that invariant measures on an algebra of subsets of X may be extended to invariant, finitely additive measures on all subsets of X.

In a different vein, one can ask whether polygons that are congruent by dissection are necessarily equidecomposable. The problem is that, somehow, the boundaries of the pieces in a geometrical dissection must be accounted for in a precise way. In a typical dissection, such as those in Figure 3.1, the boundaries do double duty and so cannot simply be assigned to one of the pieces. This problem can be solved though, and the main tool is a very important property of the equivalence relation $\sim_G$.

Whenever one has an equivalence relation on the collection of subsets of a set, one may define another relation, $\preccurlyeq$, by $A \preccurlyeq B$ if and only if A is equivalent to a subset of B. Then $\preccurlyeq$ is really a relation on the equivalence classes and, in fact, is reflexive and transitive. The Schröder–Bernstein Theorem of classical set theory states that if the cardinality relation is used—A and B are equivalent if there is a bijection from A to B—then $\preccurlyeq$ is antisymmetric as well; that is, if $A \preccurlyeq B$ and $B \preccurlyeq A$, then A and B have the same cardinality. Thus $\preccurlyeq$ is a partial order on the equivalence classes. (Under the Axiom of Choice, every set is equivalent to an ordinal, so $\preccurlyeq$ is a well-ordering.) Banach realized that the proof of the Schröder–Bernstein Theorem could be applied to any equivalence relation satisfying two abstract properties; in particular, it applies to G-equidecomposability. From now on we use the notation $A \preccurlyeq B$ only in the context of equidecomposability: $A \preccurlyeq B$ means A is G-equidecomposable with a subset of B.

Theorem 3.6 (Banach–Schröder–Bernstein Theorem). *Suppose G acts on X and $A, B \subseteq X$. If $A \preccurlyeq B$ and $B \preccurlyeq A$, then $A \sim_G B$. Thus $\preccurlyeq$ is a partial ordering of the $\sim_G$-classes in $\mathscr{P}(X)$.*

Proof. The relation $\sim_G$ is easily seen to satisfy the following two conditions:

(a) if $A \sim B$, then there is a bijection $g : A \to B$ such that $C \sim g(C)$ whenever $C \subseteq A$, and
(b) if $A_1 \cap A_2 = \varnothing = B_1 \cap B_2$, and if $A_1 \sim B_1$, and $A_2 \sim B_2$, then $A_1 \cup A_2 \sim B_1 \cup B_2$.

The rest of the proof assumes only that $\sim$ is an equivalence relation on $\mathscr{P}(X)$ satisfying (a) and (b).

Let $f : A \to B_1, g : A_1 \to B$, where $B_1 \subseteq B$ and $A_1 \subseteq A$, be bijections as guaranteed by (a). Let $C_0 = A \setminus A_1$ and, by induction, define C_{n+1} to be $g^{-1}f(C_n)$; let $C = \bigcup_{n=0}^{\infty} C_n$. Then it is easy to check that $g(A \setminus C) = B \setminus f(C)$, and hence the choice of g implies that $A \setminus C \sim B \setminus f(C)$. But, by the choice of $f, C \sim f(C)$ and property (b) yields $(A \setminus C) \cup C \sim (B \setminus f(C)) \cup f(C)$, or $A \sim B$ as desired. □

It is clear from the proof that $m + n$ pieces suffice for the final decomposition if m, n, respectively, are used in the hypothesized decompositions. This proof serves as a proof of the classical Schröder–Bernstein Theorem as well, because the cardinality relation satisfies properties (a) and (b).

This theorem eases dramatically the verification of equidecomposability. As an illustration, suppose a subset E of X is G-paradoxical, say, A, B are disjoint subsets of E with $A \sim E \sim B$. Then $E \sim B \subseteq E \setminus A \subseteq E$, so the Banach–Schröder–Bernstein Theorem implies that $E \setminus A \sim E$. This proves the following result.

Corollary 3.7. *A subset E of X is G-paradoxical if and only if there are disjoint sets $A, B \subseteq E$ with $A \cup B = E$ and $A \sim E \sim B$.*

A consequence of Corollary 3.7 is that F_2 is paradoxical in a more complete fashion than shown in Figure 1.1. We can split F_2 into two disjoint sets A and B with each of these sets being equidecomposable to F_2. Figure 3.2 illustrates such a paradox in F_2, where $A = A_1 \cup A_2$ and $B = B_1 \cup B_2$, and $A_1 \cup \sigma A_2 = F_2 = B_1 \cup \tau B_2$.

Another application of the theorem is the following counterintuitive result, the underlying idea of which will appear later (Cor. 3.11 and Thm. 9.3). It yields, for example, that a given disk and a given square may each be split into two Borel sets, such that corresponding pieces are similar; as an exercise, the reader can construct explicitly the four pieces in this special case. Recall that the group of all similarities of $\mathbb{R}^n$ is the group generated by all isometries and all magnifications from the origin, where the latter refers to a function of the form $f(P) = \alpha P$, where $\alpha \neq 0$.

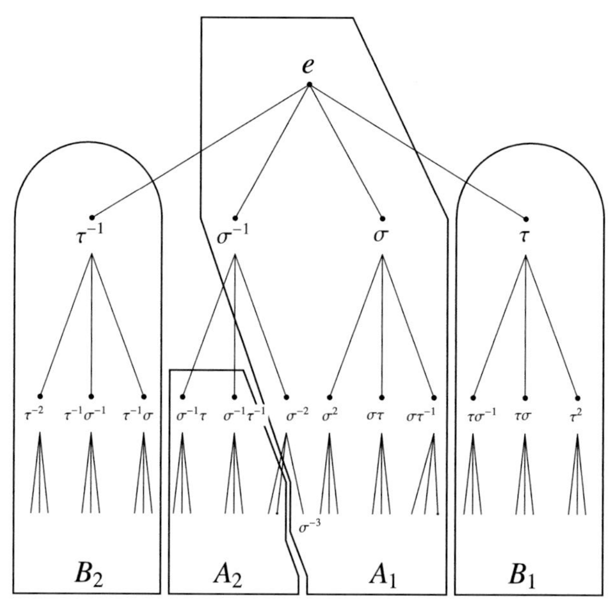

Figure 3.2. An improved paradoxical decomposition of F_2. If $A = A_1 \cup A_2$ and $B = B_1 \cup B_2$, then each of A, B is equidecomposable with F_2.

Corollary 3.8. *Any two subsets X, Y of $\mathbb{R}^n$, each of which is bounded and has nonempty interior, may be partitioned as follows: $X = X_1 \cup X_2$, $Y = Y_1 \cup Y_2$, where X_1 is similar to Y_1, and X_2 is similar to Y_2. If X and Y are Borel sets, so are the four sets in the decomposition.*

Proof. The hypotheses on X and Y guarantee the existence of similarities g_1, g_2 such that $g_1(X) \subseteq Y$ and $g_2(Y) \subseteq X$; simply shrink X so that it fits into Y, and vice versa. Therefore $X \preccurlyeq Y$ and $Y \preccurlyeq X$ with respect to equidecomposability using similarities, and because $m + n = 1 + 1 = 2$, the result now follows from Theorem 3.6. If X and Y are Borel, then the sets introduced by the proof of Theorem 3.6 will also be Borel. □

The following application of Theorem 3.6 is important in that it shows that polygons that are congruent by (geometric) dissection are also equidecomposable, that is, congruent by set-theoretic dissection.

Theorem 3.9. *If the polygons P_1 and P_2 are congruent by dissection, then they are equidecomposable.*

Proof. Let Q_1, Q_2 be the open sets obtained by forming the union of all the interiors of the polygonal subsets of P_1, P_2, respectively, arising from the hypothesized dissection. Then $Q_1 \sim Q_2$, and so the proof will be complete once it is shown that $P_1 \sim Q_1$ and $P_2 \sim Q_2$, that is, that the boundary segments can be absorbed. This follows from the following fact (by setting $A = Q_1$, and $T = P_1 \setminus Q_1$): If A is a bounded set in the plane with nonempty interior and T is a set, disjoint from A, consisting of finitely many (bounded) line segments, then $A \sim A \cup T$.

To prove this fact, let D be a disc contained in A, and let r be its radius. By subdividing the segments in T, we may assume that each one has length less than r. Let θ be any rotation of D about its center having infinite order, let R be any radius of D (excluding the center of D), and let $\overline{R} = R \cup \theta(R) \cup \theta^2(R) \cup \ldots$. Now, if $s \in T$, then $D \cup s \preccurlyeq D$. This is because $\theta(\overline{R})$ is disjoint from R and $D \setminus \overline{R}$, so $D \cup s = (D \setminus \overline{R}) \cup \overline{R} \cup s \sim (D \setminus \overline{R}) \cup \theta(\overline{R}) \cup \sigma(s) \subseteq D$, where σ is any isometry taking s to a subset of R. Because, obviously, $D \preccurlyeq D \cup s$, the Banach–Schröder–Bernstein Theorem implies that $D \sim D \cup s$. Because each of the segments in T may thus be absorbed, one at a time, into D, we have that $D \sim D \cup T$. Adding $A \setminus D$ to both sides and applying Theorem 3.6(b) yields $A \sim A \cup T$, as required. □

Because of the Bolyai–Gerwien Theorem (Thm. 3.2), the preceding theorem implies that any two polygons of the same area are equidecomposable. The converse is true, though much harder to prove, because it follows from the existence of a Banach measure (a finitely additive, G_2 -invariant measure on all subsets of $\mathbb{R}^2$ that extends Lebesgue measure (see Cor. 12.9)).

In 1924, Tarski asked his famous circle-squaring question: Is a disk equidecomposable with a square of the same area? This was resolved in 1990 by Miklos Laczkovich, who showed, surprisingly, that the answer is yes, and only translations were needed. Also, he showed that Theorem 3.9 can be done with translations only (see Thm. 9.22).

The proof of Theorem 3.8 might be called a *proof by absorption*, because it shows how a troublesome set (the boundary segments) can be absorbed in a way that, essentially, renders it irrelevant. Now, we have seen that free groups of rank 2 cause paradoxes when they act without fixed points, and so situations where the fixed points can be absorbed will be especially important. The following proof is typical of the absorption proofs and immediately yields the Banach–Tarski Paradox.

Theorem 3.10. *If D is a countable subset of $\mathbb{S}^2$, then $\mathbb{S}^2$ and $\mathbb{S}^2 \setminus D$ are $SO_3(\mathbb{R})$-equidecomposable (using two pieces).*

Proof. We seek a rotation, ρ, of the sphere such that the sets D, $\rho(D)$, $\rho^2(D), \ldots$ are pairwise disjoint. This suffices, because then $\mathbb{S}^2 = D^* \cup (\mathbb{S}^2 \setminus D^*) \sim \rho(D^*) \cup (\mathbb{S}^2 \setminus D^*) = \mathbb{S}^2 \setminus D$, where $D^* = \bigcup\{\rho^n(D) : n = 0, 1, 2, \ldots\}$. The construction of ρ is similar to the proof of Theorem 2.6. Let ℓ be a line through the origin that misses the countable set D. Let A be the set of angles θ such that for some $n > 0$ and some $P \in D$, $\rho(P)$ is also in D, where ρ is the rotation about ℓ through $n\theta$ radians. Then A is countable, so we may choose an angle θ not in A; let ρ be the corresponding rotation about ℓ. Then $\rho^n(D) \cap D = \varnothing$ if $n > 0$, from which it follows that whenever $0 \leqslant m < n$, $\rho^m(D) \cap \rho^n(D) = \varnothing$ (consider $\rho^{n-m}(D) \cap D$); therefore ρ is as required. □

Corollary 3.11 (The Banach–Tarski Paradox) (AC). *The sphere* $\mathbb{S}^2$ *is* $SO_3(\mathbb{R})$-*paradoxical, as is any sphere centered at the origin. Moreover, any solid ball in* $\mathbb{R}^3$ *is* G_3-*paradoxical and* $\mathbb{R}^3$ *itself is paradoxical.*

Proof. The Hausdorff Paradox (Thm. 2.3) states that $\mathbb{S}^2 \setminus D$ is $SO_3(\mathbb{R})$-paradoxical for some countable set D (of fixed points of rotations). Combining this with the previous theorem and Proposition 3.4 yields that $\mathbb{S}^2$ is $SO_3(\mathbb{R})$-paradoxical. Because none of the previous results depends on the size of the sphere, spheres of any radius admit paradoxical decompositions.

It suffices to consider balls centered at $\mathbf{0}$, because G_3 contains all translations. For definiteness, we consider the unit ball B, but the same proof works for balls of any size. The decomposition of $\mathbb{S}^2$ yields one for $B \setminus \{\mathbf{0}\}$ if we use the radial correspondence: $P \to \{\alpha P : 0 < \alpha \leqslant 1\}$. Hence it suffices to show that B is G_3-equidecomposable with $B \setminus \{\mathbf{0}\}$, that is, that a point can be absorbed. Let $P = (0, 0, \frac{1}{2})$, and let ρ be a rotation of infinite order about the axis that is the horizontal line in the x-z plane containing P. Then, as usual, the set $D = \{\rho^n(\mathbf{0}) : n \geqslant 0\}$ may be used to absorb $\mathbf{0}$: $\rho(D) = D \setminus \{\mathbf{0}\}$, so $B \sim B \setminus \{\mathbf{0}\}$. If, instead, the radial correspondence of $\mathbb{S}^2$ with all of $\mathbb{R}^3 \setminus \{\mathbf{0}\}$ is used, one gets a paradoxical decomposition of $\mathbb{R}^3 \setminus \{\mathbf{0}\}$ using rotations. Because, exactly as for the ball, $\mathbb{R}^3 \setminus \{\mathbf{0}\} \sim_{G_3} \mathbb{R}^3$, $\mathbb{R}^3$ is paradoxical via isometries. □

Because of its use of Theorem 3.9, this proof of the Banach–Tarski Paradox seems to depend on having uncountably many rotations available. But, as a consequence of a more general approach in the next chapter, we shall see (Thm. 13.21) that for subgroups G of $SO_3(\mathbb{R})$, $\mathbb{S}^2$ is G-paradoxical if and only if G has a free subgroup of rank 2. In fact, it follows from Theorem 5.5 that $\mathbb{S}^2$ is paradoxical with only the rotations ϕ and ρ of Theorem 2.1 being used to move the pieces of the decomposition.

The version of the Banach–Tarski Paradox in Corollary 3.11 does not add anything to our knowledge of finitely additive measures not already derivable from the Hausdorff Paradox (see Thm. 2.6), but the result is much more striking, indeed, more bizarre, than Hausdorff's. A ball, which has a definite volume, may be taken apart into finitely many pieces that may be rearranged via rotations of $\mathbb{R}^3$ to form two, or even a million, balls, each identical to the original one! Or, more whimsically, the unit ball may be decomposed into finitely many pieces, forming a three-dimensional jigsaw puzzle with the following property: For each $n \leqslant 1{,}000{,}000$, the pieces of the puzzle may be arranged using rotations to form n disjoint unit balls. Of course, because the Axiom of Choice is used to produce the pieces, the jigsaw would have to be inconceivably sharp!

Rotations preserve volume, and this is why the result has come to be known as a paradox. A resolution is that there may not be a volume for the rotations to preserve; the pieces in the decomposition may be (indeed, will have to be) non-Lebesgue measurable. In fact, we have already seen (Thm. 2.6) that $\mathbb{S}^2$ is SO_3-negligible, that is, there is no rotation-invariant, finitely additive measure defined for all subsets of $\mathbb{S}^2$ (or of the unit ball).

The proof of the Banach–Tarski Paradox, like all proofs of the existence of a nonmeasurable set, is nonconstructive in that it appeals to the Axiom of Choice. It has been argued that the result is so counterintuitive, so patently false in the real world, that one of the underlying assumptions must be incorrect; the Axiom of Choice is usually selected as the culprit. We discuss this argument more fully in Chapter 15, where the role that Choice plays in the foundations of measure theory will be examined in detail. For now, let us mention two points. First, strange results that do not require the Axiom of Choice abound in all of mathematics. In particular, there are several striking paradoxes whose construction does not use AC:

- Sierpiński–Mazurkiewicz Paradox in $\mathbb{R}^2$ (§1.2)
- Mycielski–Wagon Paradox in $\mathbb{H}^2$ (§4.3)
- Satô Paradox in $\mathbb{S}^2 \cap \mathbb{Q}^3$ (Chap. 2)
- Dougherty–Foreman Paradox (§11.2)

These paradoxes are as startling as the duplication of the sphere. Second, the Axiom of Choice is consistent with the other axioms of set theory; as shown by Gödel, the axiom is true in the "constructible universe." Hence, independent of its truth in any individual's view of the set-theoretic universe, the Banach–Tarski Paradox is, at the least, consistent.

Though volume is not one of them, there is one simple property that is preserved by equidecomposability in $\mathbb{R}^3$: If A is bounded, then so is any set equidecomposable with A. Banach and Tarski were able to show that any two bounded sets, each having nonempty interior, are equidecomposable. The condition on interior is used to get started, because the basic Banach–Tarski Paradox is for a solid ball. Thus they generalized their already surprising result so that it applies to solids of any shape. It is a consequence of this strengthening that any bounded subset of $\mathbb{R}^3$ with nonempty interior is paradoxical; in fact, the unit ball is equidecomposable with any other ball, no matter how large or small. Another consequence is that *any* two polyhedra in $\mathbb{R}^3$ are equidecomposable; this should be contrasted with the result (Thm. 3.8 and remarks following) that two polygons are equidecomposable if and only if they have the same area.

Theorem 3.12 (Banach–Tarski Paradox, Strong Form) (AC). *If A and B are any two bounded subsets of $\mathbb{R}^3$, each having nonempty interior, then A and B are equidecomposable.*

Proof. It suffices to show that $A \preccurlyeq B$, for then, by the same argument, $B \preccurlyeq A$, and Theorem 3.6 yields $A \sim B$. Choose solid balls K and L such that $A \subseteq K$ and $L \subseteq B$, and let n be large enough that K may be covered by n (overlapping) copies of L. Now, if S is a set of n disjoint copies of L, then using the Banach–Tarski Paradox to repeatedly duplicate L, and using translations to move the copies so obtained, yields that $L \succcurlyeq S$. Therefore $A \subseteq K \preccurlyeq S \preccurlyeq L \subseteq B$, so $A \preccurlyeq B$. □

This remarkable result usually is viewed negatively because it so forcefully illustrates Hausdorff's Theorem (Thm. 2.6) that certain measures do not exist.

Not all consequences of the Banach–Tarski Paradox are negative, however. Tarski used it to prove a result on finitely additive measures on the algebra of Lebesgue measurable subsets of $\mathbb{R}^3$, and this result was used in work on the uniqueness of Lebesgue measure (see Lemma 11.9 and remarks following Thm. 13.13).

There are some interesting problems of a topological nature regarding the Banach–Tarski Paradox. As pointed out, the pieces into which the sphere $\mathbb{S}^2$ is decomposed in order to be duplicated cannot be Lebesgue measurable and hence cannot be Borel sets. There is another family of subsets of $\mathbb{R}^n$ or $\mathbb{S}^n$ that shares many of the properties of the measurable sets. A set is said to have the *Property of Baire* if it differs from a Borel set by a meager set; that is, the symmetric difference $X \triangle B\ (= (X \setminus B) \cup (B \setminus X))$ is meager for some Borel set B. (Recall that a set is meager—often said to be of first category—if it is a countable union of nowhere dense sets.) In fact (see [Oxt71, p. 20]), the set B in the preceding definition can be taken to be open. Not all sets have the Property of Baire: Using the representation using open sets just mentioned, it is not hard to see that the set M of the proof of Theorem 1.5 (whose construction used the Axiom of Choice) fails to have the Property of Baire. See [Oxt71] for more on this family of sets.

There are many similarities between the σ-algebras of measurable sets and sets with the Property of Baire; for instance, a set is Lebesgue measurable if and only if it differs from a Borel set (in fact an F_σ set) by a null set (meaning a set of measure zero). But the two notions do not coincide. Every subset of $\mathbb{R}^n$ may be split into a null set and a meager set [Oxt71, p. 5], and if a nonmeasurable set is so divided, the meager part cannot by measurable.

While the measurable subsets of $\mathbb{S}^2$ carry a finitely additive (in fact, countably additive) $SO_3(\mathbb{R})$-invariant measure of total measure 1—namely, Lebesgue measure—it was for many years not known whether such a measure exists on the sets having the Property of Baire.

This was known as Marczewski's Problem and was listed as Open Problem 1 in the first edition of this book. It was expected that the result would be yes; that is, that such a measure would exist. But in a stunning piece of work in 1994, Randall Dougherty and Matt Foreman proved [DF94] that the Banach–Tarski Paradox can be constructed in such a way that the pieces have the Property of Baire. Therefore there is no finitely additive, rotation-invariant measure on the Property of Baire subsets of the sphere. A discussion of their work is given in §11.2. The work of Dougherty and Foreman has many surprising consequences. For example (Thm. 11.7), any cube, no matter how small, has pairwise disjoint open subsets $U_1, U_2, \ldots, U_m$ such that, for suitable isometries, $\cup \rho_i U_i$ is dense in the unit cube.

Another old problem on the topology of pieces concerns more general metric spaces. No paradoxical decomposition of the sphere can use only Borel pieces, as previously pointed out. How general is this fact? If $\mathbb{N}$ is given the discrete metric, then any permutation is an isometry, and because $\mathbb{N}$ is countable, all subsets are Borel; hence (see Thm. 1.4), $\mathbb{N}$ is Borel paradoxical. Because of this, and other examples, we restrict our attention to compact metric spaces. Now, any compact metric space bears a countably additive, G-invariant Borel measure of total measure one, where G is the group of isometries of the space. (To prove this, put a metric on G by defining the distance between g_1 and g_2 to be the supremum of the distance between $g_1(x)$ and $g_2(x)$, for $x \in X$. One can check that this turns G into a compact topological group, and therefore G bears a left-invariant Borel measure, ν, of total measure one, namely Haar measure (see [Coh80]). We may use ν to define the desired Borel measure, μ, on X by fixing some $x \in X$ and setting $\mu(A) = \nu(\{g \in G : g(x) \in A\})$.) The existence of such a Borel measure, even if it were only finitely additive, means that a compact metric space is never paradoxical using Borel pieces.

But a rather different problem arises if we relax the definition of isometry by considering partial isometries rather than global isometries. Let us reserve the term *congruent* for two subsets A and B of a metric space to mean that there is a distance-preserving bijection from A to B, that is, $\sigma(A) = B$ for some partial isometry σ. For $\mathbb{R}^n$, this leads to nothing new, because partial isometries extend to isometries of $\mathbb{R}^n$, but in more general spaces, such extensions may not exist. In compact metric spaces, however, it is easy to see that any isometry on A extends uniquely to one on $\overline{A}$, the closure of A (see [Lin26, p. 215]). This congruence relation does not arise from a group action, but the definition of paradoxical decomposition is easily modified to apply. Simply replace the sets g_i, (A_i) and $h_j(B_j)$ by sets congruent to A_i, B_j, respectively. Note that if A is Borel and congruent to B, it is not obvious that B also is Borel; in general, continuous images of even closed sets need not be Borel. But for complete, separable spaces (sometimes called Polish spaces; compact spaces are Polish), a one-to-one, continuous image of a Borel set is Borel (see [Kur66, p. 487]), and hence the Borel sets are, indeed, closed under congruence.

Question 3.13. Is it true that no compact metric space is paradoxical (with respect to congruence) using Borel pieces?

Of course, an affirmative answer would follow from the existence of a finitely additive, congruence-invariant Borel measure on the space, having total measure one. This problem, in the (possibly stronger) measure-theoretic form just given, is an old one; it was posed by Banach and Ulam in 1935 as the second problem in the famous collection of primarily Polish problems known as *The Scottish Book* [Mau81]. There are three interesting partial results. Mycielski [Myc74] showed that any compact metric space admits a countably additive, Borel measure of total measure one that assigns congruent open sets the same measure; hence there are no paradoxical decompositions using just open sets. Because Mycielski's construction yields a countably additive measure, the existence of countable, compact metric spaces (e.g., $\{0, 1, \frac{1}{2}, \frac{1}{3}, \ldots\}$) shows that it cannot be modified to work for closed sets. Nevertheless, Davies and Ostaszewski [DO79] have shown how to construct finitely additive, congruence-invariant measures of total measure one for all countable compact metric spaces. Moreover, Bandt and Baraki [BB86] have shown that under the additional assumption that the metric space is *locally homogeneous* (any two points have congruent neighborhoods), Mycielski's measure does indeed assign congruent Borel sets the same measure.

The strong form of the Banach–Tarski Paradox implies that a regular tetrahedron in $\mathbb{R}^3$ is equidecomposable with a cube. This is not in the spirit of the sort of dissection Hilbert asked for in his third problem, because *any* cube will do; the cube need not have the same volume as the tetrahedron. But this raises the possibility of cubing a regular tetrahedron using pieces restricted to some nice collection of sets more general than polyhedra (the context of Hilbert's Third Problem), but not so general that paradoxes exist and volume is not preserved.

A natural choice is to restrict the pieces to be Lebesgue measurable, for then equidecomposable sets must have the same measure. Recently Grabowski, Máthé, and Pikhurko [GMP∞a, GMP∞b] have proved that this is possible.

Theorem 3.14. *Any tetrahedron in $\mathbb{R}^3$ is equidecomposable with a cube using isometries and pieces that are Lebesgue measurable.*

See Theorem 9.28 for more on this result. R. J. Gardner Gar85 has shown that such a result is not possible if the group is assumed to be discrete. There are related questions and results in $\mathbb{R}^1$ and $\mathbb{R}^2$ (see Chap. 9).

Finally, a different, rather more bizarre question arises if one asks whether the pieces in a paradoxical duplication can be moved physically to form the two new copies of, say, the sphere. Because the pieces do not exist physically, this calls for some clarification. Consider the equidecomposability of a unit ball with two disjoint unit balls. Can the motions of the pieces, assuming the latter somehow to be given, be actually carried out in $\mathbb{R}^3$ in such a way that the pieces never overlap? This was a long-standing open question of de Groot but was settled positively by Trevor Wilson in 2005, while he was an undergraduate at CalTech. Loosely speaking, a ball in $\mathbb{R}^3$ can be partitioned into sets that can be rearranged into two balls of the same radius in such a way that the piece can be simultaneously moved into their new positions so that, at any particular time, the pieces are disjoint. A proof is given in Corollary 10.15.

In the rest of Part I, we show how versions of the Banach–Tarski Paradox can be constructed in other spaces, such as higher-dimensional spheres and non-Euclidean spaces. As pointed out at the end of Chapter 2, the Banach–Tarski Paradox does not exist in $\mathbb{R}^1$ or $\mathbb{R}^2$, but other constructions are possible that have similar measure-theoretic implications. And we also study refinements of the original construction in $\mathbb{R}^3$, showing, for example, how to duplicate a ball using the smallest possible number of pieces.

Notes

The use of dissection to derive area formulas dates at least to Greek geometry, but the converse idea (Theorem 3.2) was not considered until the nineteenth century. F. Bolyai and P. Gerwien discovered Theorem 3.2 independently around 1832 (see [Ger33]), but apparently William Wallace of England had proved the result already in 1807 ([Pla31]; see also [Emc46, p. 225] and [Jac12]). Gerwien also proved [Ger83] that Theorem 3.2 is valid for spherical polyhedra. A complete treatment of the Bolyai–Gerwien Theorem, including a discussion of the smallest possible group of isometries with respect to which it is true, may be found in [Bol78]. For a treatment of the Bolyai–Gerwien Theorem in both Euclidean and hyperbolic geometry, see [MP81, §10.4].

For an exposition of the solution, due to Dehn, of Hilbert's Third Problem, [Bol78]. More advanced topics related to this problem may be found in [Sah79].

The definition of G-equidecomposability is due to Banach and Tarski [BT24, Tar24a], and [BT24] contains many elementary properties of $\sim_G$. The example showing that $\sim_2$ is not transitive is from Sierpiński's monograph [Sie54], which contains a fairly complete discussion of equidecomposability, including the Hausdorff and Banach–Tarski Paradoxes.

Banach's version of the Schröder–Bernstein Theorem (Thm. 3.6) is proved in [Ban24] (see also [Kur66, p. 190]). Theorem 3.9 on the equidecomposability of polygons is due to Tarski [Tar24a] and appears in [BT24] as well. The application of Theorem 3.6 to similarities (Cor. 3.8) is due to Klee [Kle79, p. 139].

Theorem 3.10 and its use in deriving the Banach–Tarski Paradox (Cor. 3.11) from the Hausdorff Paradox are due to Sierpiński [Sie48a, Sie54, pp. 42, 92]. The original derivation that appeared twenty years earlier in [BT24] is slightly different. Theorem 3.12, the strong form of the Banach–Tarski Paradox, appears in the original paper [BT24]. For an exposition of the paradox that includes miscellaneous other dissection results, see [Wap05].

For the reader interested in pursuing the foundations of set theory, modern proofs of Gödel's famous theorem that the Axiom of Choice is consistent with Zermelo–Fraenkel Set Theory may be found in [Kri71] and [Kun80].

Question 3.13, in the form asking about invariant, finitely additive, Borel measures in compact metric spaces, is Problem 2 of *The Scottish Book* [Mau81] and is due to Banach and Ulam. It is also mentioned in [BU48] and [Ula60, p. 43]. De Groot's question about continuous equidecomposability was raised in [Dek58a, p. 25].

4

Hyperbolic Paradoxes

The Banach–Tarski paradox helps us understand the extra richness that arises as we move from two Euclidean dimensions to three. For the line and plane, the isometry group is solvable; this means that certain measures exist and there are no paradoxes of the Banach–Tarski type. But when we leave Flatland for 3-space, we enter a world with a much richer group of geometric transformations. In $\mathbb{R}^3$ there are two independent rotations, and that is the key ingredient for the Hausdorff and Banach–Tarski Paradoxes. Perhaps coincidentally, the move from two to three dimensions is also the point at which geometrical dissections break down: The Bolyai–Gerwien Theorem holds in the plane, but in 3-space, one cannot transform a regular tetrahedron to a cube by geometrical dissection.

In non-Euclidean space, things are different, and we explore the situation in this chapter, focusing on the hyperbolic plane, $\mathbb{H}^2$, where several types of unusual and constructive paradoxes exist.

4.1 The Hyperbolic Plane

The hyperbolic plane can be modeled several ways by Euclidean objects. Simplest is to use the (open) upper half of the complex plane, where hyperbolic lines are semicircles or vertical lines orthogonal to the real axis; this model is denoted $\mathbb{H}^+$. Another useful model is the Poincaré disk, denoted $\mathbb{D}$, where the hyperbolic lines are again arcs arising from circles orthogonal to the boundary (and also the disk's diameters).

Working in $\mathbb{H}^+$, the orientation-preserving isometries are given by linear fractional transformations $z \mapsto (az + b)/(cz + d)$, where $ad - bc > 0$ and $a, b, c, d \in \mathbb{R}$ (see [Gre80, Leh64, MP81]). The composition of two such transformations corresponds to matrix multiplication of the corresponding 2×2 matrices. Because a transformation is unchanged if each entry is divided by the same constant, we may assume $ad - bc = 1$; therefore the group is isomorphic to $PSL_2(\mathbb{R}) = SL_2(\mathbb{R}) \setminus \{\pm I\}$, where I is the identity matrix (SL_2 refers to the special linear group: matrices of determinant 1). The orientation-reversing motions are given by $z \mapsto (a\bar{z} + b) / (c\bar{z} + d)$, where $ad - bc = 1$.

There is a useful classification of the orientation-preserving isometries using the 2×2 matrix of the linear fractional transformation. Let T be a nonidentity element of $PSL_2(\mathbb{R})$, and let tr denote the trace of a matrix (sum of the diagonal elements).

- If $|\mathrm{tr}(T)| < 2$, then T has a fixed point and is called a *hyperbolic rotation*; these are also called *elliptic.*
- If $|\mathrm{tr}(T)| = 2$, then T fixes one point on the boundary of $\mathbb{H}^+$ and can be represented as a conjugation by a linear fractional transformation of the transformation $z \mapsto z + a$, where $a \in \mathbb{R}$. The transformation is called a *Euclidean translation*; these are also called *parabolic.*
- If $|\mathrm{tr}(T)| > 2$, then T has two fixed points on the boundary of $\mathbb{H}^+$ and is called a *hyperbolic translation*; these are also called simply *hyperbolic*.

Simple algebra shows that only the elliptics fix a point in $\mathbb{H}^2$ (see [Leh66, §1.1]). While a free subgroup of $PSL_2(\mathbb{R})$ might contain an elliptic element, such elements can never appear in a free subgroup of $PSL_2(\mathbb{Z})$. This is because any elliptic element of $PSL_2(\mathbb{Z})$ has finite order: If $\mathrm{tr}(\sigma) = 0$, then $\sigma = \left[\begin{smallmatrix} a & b \\ c & -a \end{smallmatrix}\right]$ and $\sigma^2 = \left[\begin{smallmatrix} -1 & 0 \\ 0 & -1 \end{smallmatrix}\right]$, the identity of $PSL_2(\mathbb{Z})$, and if $\mathrm{tr}(\sigma) = \pm 1$, then $\sigma = \left[\begin{smallmatrix} a & b \\ c & \pm 1 - a \end{smallmatrix}\right]$ and σ has order 3. This shows that any independent pair in $SL_2(\mathbb{Z})$, such as $\left[\begin{smallmatrix} 1 & 2 \\ 0 & 1 \end{smallmatrix}\right]$ and its transpose (Prop. 4.4), induces an independent pair of isometries of $\mathbb{H}^2$ such that the group they generate has no nontrivial fixed points. In fact, the lack of elliptic elements in free subgroups holds if $PSL_2(\mathbb{Z})$ is replaced by any discrete subgroup of $PSL_2(\mathbb{R})$; see Theorem 7.7.

A useful fact is that the orientation-preserving transformations are locally commutative on $\mathbb{H}^2$, meaning that the subgroup that fixes a point is commutative.

Proposition 4.1. *The orientation-preserving isometries are locally commutative on* $\mathbb{H}^2$.

Proof. If σ_1 and σ_2 are nonidentity elements that fix z, then, because of the quadratic nature of the fixed-point relation, they fix $\bar{z}$ as well. Choose ρ to be a linear fractional tranformation (complex coefficients allowed) of $\mathbb{C} \cup \infty$ that takes z to 0 and $\bar{z}$ to ∞; then $\tau_i = \rho\,\sigma_i \rho^{-1}$ has $\{0, \infty\}$ as its fixed-point set in $\mathbb{C} \cup \infty$ and so there must be complex a_i such that $\tau_i(z) = a_i z$. But then τ_1 and τ_2 commute, and so σ_1 and σ_2 commute as well. □

The preceding result also follows from the fact that the set of isometries fixing a specified point is isomorphic to the commutative group $SO_2(\mathbb{R})$.

There are two important examples of pairs of hyperbolic isometries that we will use.

The first is $\left[\begin{smallmatrix} 1 & 2 \\ 0 & 1 \end{smallmatrix}\right]$ and $\left[\begin{smallmatrix} 1 & 0 \\ 2 & 1 \end{smallmatrix}\right]$. These are parabolic, and the proof of independence is given in Proposition 4.4; thus they generate the free group F_2 and act without fixed points. A second example is $\sigma = \left[\begin{smallmatrix} 0 & 1 \\ -1 & 0 \end{smallmatrix}\right]$ and $\tau = \left[\begin{smallmatrix} 0 & -1 \\ 1 & 1 \end{smallmatrix}\right]$; here σ^2 and τ^3 are the identity, and they generate the free product $\mathbb{Z}_2 * \mathbb{Z}_3$ (Prop. 4.2).

The parabolic pair just mentioned gives an action of F_2 on $\mathbb{H}^2$ with no nontrivial fixed points. So all the ideas of the Banach–Tarski Paradox can be applied and we can, assuming the Axiom of Choice, get a paradoxical decomposition of the whole plane, $\mathbb{H}^2$. In this chapter we improve this in two ways:

1. The Banach–Tarski paradox based on AC uses nonmeasurabe pieces, but there is no reason that there cannot be a paradox using measurable pieces, so long as the set has measure 0 or ∞. In fact, there is such a thing using very simple pieces built up from squares or triangles (§§4.2, 4.3).
2. There is a paradox for a bounded set in $\mathbb{H}^2$ (§4.6).

Two unexpected bonuses of these investigations are as follows:

- a paradox that mixes the ideas of Banach and Tarski with those of M. C. Escher
- a set in the plane that is congruent to some of its subsets in a way that cannot happen in Euclidean space

4.2 A Hyperbolic Hausdorff Paradox

Recall that whenever G is a discrete subgroup of $PSL_2(\mathbb{R})$ (meaning the set of matrices corresponding to the elements in G contains no convergent sequence of distinct matrices), there is a fundamental polygon for the action of G on $\mathbb{H}^2$. That is, there is a hyperbolic (open) polygon P such that $\{\rho(P) : \rho \in G\}$ are pairwise disjoint and $\bigcup\{\overline{\rho(P)} : \rho \in G\} = \mathbb{H}^2$ (see [Leh66, §1.4]). In other words, $\mathbb{H}^2$ can be tiled by pairwise interior-disjoint copies of the fundamental polygon, one copy for each transformation in G. The classic example of such a tiling is due to Klein and Fricke [KF90] (see [Leh66, p. 29] or [Mag74, p. 174]) and comes from letting G be $PSL_2(\mathbb{Z})$ (known as the *modular group*). This group, whose elements can be viewed as linear fractional transformations, is isomorphic to $\langle \sigma, \tau : \sigma^2 = \tau^3 = e\rangle$, and hence to $\mathbb{Z}_2 * \mathbb{Z}_3$, where $\sigma(z) = -1/z$ and $\tau(z) = -1/(z+1)$; see [Kur56, App. B] or [Leh64 pp. 140, 234] for a proof. Here is a proof that σ and τ generate the free product.

Proposition 4.2. *The matrices* $S = \left[\begin{smallmatrix} 0 & 1 \\ -1 & 0 \end{smallmatrix}\right]$ *and* $T = \left[\begin{smallmatrix} 0 & -1 \\ 1 & 1 \end{smallmatrix}\right]$ *generate the free product* $\mathbb{Z}_2 * \mathbb{Z}_3$.

Proof. First note that S and T have order 2 and 3, respectively. Let $R = T^2$. Suppose w is a nonempty string in S, T, R, with no adjacencies of the form SS, TT, TR, or RT, that equals the identity. Conjugating by S, TS, or RS if necessary, we may assume that $w = Sy \ldots yS\,yS$, where each $y \in \{T, R\}$. We have $ST = \left[\begin{smallmatrix} 1 & 1 \\ 0 & 1 \end{smallmatrix}\right]$ and $SR = \left[\begin{smallmatrix} 1 & 0 \\ 1 & 1 \end{smallmatrix}\right]$. Now start with $(-1, 0)$; the rightmost S in w gives $(0, 1)$, and further applications of ST or SR cannot decrease the entries and so never produce $(-1, 0)$, a contradiction. □

The tiling of $\mathbb{H}^2$ corresponds to the action of the modular group as illustrated in Figure 4.1, where the triangle with vertices at 0 and $\pm\frac{1}{2} + \frac{\sqrt{3}}{2}i$ is the fundamental

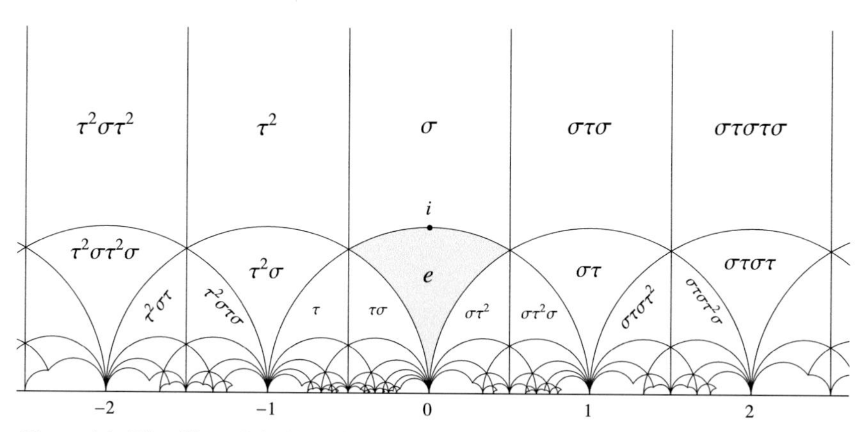

Figure 4.1. The tiling of the hyperbolic plane corresponding to the group generated by σ (order 2) and τ (order 3).

polygon and several copies of this triangle are labeled with the appropriate group elements from the modular group.

Now, as observed by Hausdorff (see Chap. 2 Notes; this was the first example of a paradoxical group), the abstract group $\mathbb{Z}_2 * \mathbb{Z}_3$ is paradoxical ($\mathbb{Z}_2 * \mathbb{Z}_2$ has a cyclic commutator subgroup and so is solvable; such groups are not paradoxical (§12.4)). In fact, there is a partition of the group into $A \cup B \cup C$ such that the Hausdorff relations

$$(*) \qquad \tau(A) = B, \tau^2(A) = C, \sigma(A) = B \cup C$$

hold, where σ and τ are the generators. In short, A is simultaneously a half of the group and a third of the group. This is of course a paradoxical situation: One can use $G = A \cup B \cup C$ and $\sigma A = B \cup C$ to partition G into four sets, each G-congruent to A; the first two such sets then give A and $B \cup C$, as do the last two. So G is a paradoxical group (easy exercise; see [TW14]). Hausdorff constructed the sets inductively in a way that gives preference to B when there is a choice. Here are the placement rules:

Start with $A = \{e\}$, $B = C = \varnothing$ and consider τ^2 as an atom when dealing with the length of a word (so the word τ^2 has length 1). Working inductively by length, any unassigned word w has the form σu, τu, or $\tau^2 u$, where u has been placed.

- If w has τ or τ^2 on the left, place w as forced by the first two relations of $(*)$ (i.e., if $u \in A$, place τu into B and $\tau^2 u$ into C, and similarly moving cyclically if u is in B or C).
- If w has σ on the left, place w into A if $u \in B \cup C$ (forced) and into B if $u \in A$.

Note that the last clause is where the B-preference arises. One could just as easily change this clause so that w is put into C when $u \in A$ to get a C-preferred set of rules. These rules lead to sets that satisfy the Hausdorff relations, but in fact there is a simpler way to get the sets, as we show in the next proof.

Table 4.1. *The three sets of the Hausdorff Paradox in the group* $\mathbb{Z}_2 * \mathbb{Z}_3 = \langle \sigma, \tau : \sigma^2 = \tau^3 = e \rangle$

A	B	C
e	τ	$\tau\tau$
$\sigma\tau$	$\tau\sigma\tau$	$\tau\tau\sigma\tau$
$\tau\sigma$	$\tau\tau\sigma$	σ
$\sigma\tau\tau$	$\tau\sigma\tau\tau$	$\tau\tau\sigma\tau\tau$
$\sigma\tau\sigma\tau$	$\tau\sigma\tau\sigma\tau$	$\tau\tau\sigma\tau\sigma\tau$
$\sigma\tau\tau\sigma$	$\tau\sigma\tau\tau\sigma$	$\tau\tau\sigma\tau\tau\sigma$
$\tau\sigma\tau\sigma$	$\tau\tau\sigma\tau\sigma$	$\sigma\tau\sigma$
$\sigma\tau\sigma\tau\tau$	$\tau\sigma\tau\sigma\tau\tau$	$\sigma\tau\sigma\tau\sigma\tau\sigma$
$\sigma\tau\tau\sigma\tau$	$\tau\sigma\tau\tau\sigma\tau$	$\tau\tau\sigma\tau\sigma\tau\tau$
$\sigma\tau\sigma\tau\sigma\tau$	$\tau\sigma\tau\sigma\tau\sigma\tau$	$\tau\tau\sigma\tau\tau\sigma\tau$
$\sigma\tau\sigma\tau\tau\sigma$	$\tau\sigma\tau\sigma\tau\tau\sigma$	$\tau\tau\sigma\tau\sigma\tau\sigma\tau$
$\sigma\tau\tau\sigma\tau\sigma$	$\tau\sigma\tau\tau\sigma\tau\sigma$	$\tau\tau\sigma\tau\sigma\tau\tau\sigma$
$\sigma\tau\tau\sigma\tau\tau$	$\tau\sigma\tau\tau\sigma\tau\tau$	$\tau\tau\sigma\tau\tau\sigma\tau\sigma$
$\tau\sigma\tau\sigma\tau\sigma$	$\tau\tau\sigma\tau\sigma\tau\sigma$	$\sigma\tau\sigma\tau\sigma$
$\sigma\tau\sigma\tau\sigma\tau\tau$	$\tau\sigma\tau\sigma\tau\sigma\tau\tau$	$\tau\tau\sigma\tau\tau\sigma\tau\tau$
$\sigma\tau\sigma\tau\tau\sigma\tau$	$\tau\sigma\tau\sigma\tau\tau\sigma\tau$	$\tau\tau\sigma\tau\sigma\tau\sigma\tau\tau$
$\sigma\tau\tau\sigma\tau\sigma\tau$	$\tau\sigma\tau\tau\sigma\tau\sigma\tau$	$\tau\tau\sigma\tau\sigma\tau\tau\sigma\tau$
$\sigma\tau\tau\sigma\tau\tau\sigma$	$\tau\sigma\tau\tau\sigma\tau\tau\sigma$	$\tau\tau\sigma\tau\tau\sigma\tau\sigma\tau$
$\sigma\tau\sigma\tau\sigma\tau\sigma\tau$	$\tau\sigma\tau\sigma\tau\sigma\tau\sigma\tau$	$\tau\tau\sigma\tau\tau\sigma\tau\tau\sigma$
$\sigma\tau\sigma\tau\sigma\tau\tau\sigma$	$\tau\sigma\tau\sigma\tau\sigma\tau\tau\sigma$	$\tau\tau\sigma\tau\sigma\tau\sigma\tau\sigma\tau$
$\sigma\tau\sigma\tau\tau\sigma\tau\sigma$	$\tau\sigma\tau\sigma\tau\tau\sigma\tau\sigma$	$\tau\tau\sigma\tau\sigma\tau\sigma\tau\tau\sigma$
$\sigma\tau\sigma\tau\tau\sigma\tau\tau$	$\tau\sigma\tau\sigma\tau\tau\sigma\tau\tau$	$\tau\tau\sigma\tau\sigma\tau\tau\sigma\tau\sigma$
$\sigma\tau\tau\sigma\tau\sigma\tau\sigma$	$\tau\sigma\tau\tau\sigma\tau\sigma\tau\sigma$	$\tau\tau\sigma\tau\sigma\tau\tau\sigma\tau\tau$
$\sigma\tau\tau\sigma\tau\sigma\tau\tau$	$\tau\sigma\tau\tau\sigma\tau\sigma\tau\tau$	$\tau\tau\sigma\tau\tau\sigma\tau\sigma\tau\sigma$
$\sigma\tau\tau\sigma\tau\tau\sigma\tau$	$\tau\sigma\tau\tau\sigma\tau\tau\sigma\tau$	$\tau\tau\sigma\tau\tau\sigma\tau\sigma\tau\tau$
$\tau\sigma\tau\sigma\tau\sigma\tau\sigma$	$\tau\tau\sigma\tau\sigma\tau\sigma\tau\sigma$	$\sigma\tau\sigma\tau\sigma\tau\sigma\tau\sigma$

Note: The assignments of the powers of $\tau\sigma$ and their τ-translates shaded in gray.

Theorem 4.3 (Hausdorff Paradox). *The group $\mathbb{Z}_2 * \mathbb{Z}_3$ is paradoxical. There is a partition of the group into $A \cup B \cup C$ such that $\tau(A) = B$, $\tau^2(A) = C$ and $\sigma(A) = B \cup C$.*

Proof. For $L \in \{\sigma, \tau, \tau^2\}$, let W_L be the set of words having L has the leftmost term (recall that we consider τ^2 as an atom, so it is in W_{τ^2}, not W_τ). One first tries the naive approach $A = W_\sigma$, $B = W_\tau$, and $C = W_{\tau^2}$. This almost works, but e is unassigned. To address this, we absorb e into A as follows, where j runs through $\mathbb{N}$:

$$A = \{\text{all } (\tau\sigma)^j \text{ and all of } W_\sigma, \text{ except } \tau^2(\tau\sigma)^j\};$$

$$(**) \quad B = \{\text{all } \tau(\tau\sigma)^j \text{ and all of } W_\tau, \text{ except } (\tau\sigma)^j\};$$

$$C = \{\text{all } \tau^2(\tau\sigma)^j \text{ and all of } W_{\tau^2}, \text{ except } \tau(\tau\sigma)^j\}.$$

In short, powers of $\tau\sigma$ and their translates by τ or τ^2 are assigned directly, with all other words assigned according to the naive scheme, that is, by their leftmost term. Table 4.1 shows some elements of the three sets.

Verification that the sets satisfy $(*)$ is routine and left as an exercise. A key point is that $\sigma(\tau\sigma)^{j-1} = \tau^2(\tau\sigma)^j$, which means that the step placing τ^2 of the powers of $\tau\sigma$ into C is also placing σ of those powers into C. □

The sets A, B, C of the preceding proof are not the same as Hausdorff's sets. However, they are the same as the sets arising from Hausdorff's method with the preference changed from B to C. Furthermore, if the method of getting the sets in the proof of Theorem 4.1 is changed so that powers of $\tau^2\sigma$ are used in $(**)$ instead of powers of $\tau\sigma$, then the three sets do agree with Hausdorff's original three sets. We leave the proofs as an exercise.

Because of the isomorphism of $\mathbb{Z}_2 * \mathbb{Z}_3$ with $PSL_2(\mathbb{Z})$ and the correspondence between elements of the modular group and tiles in the tiling of Figure 4.1, this immediately gives subsets (also called A, B, C) of $\mathbb{H}^2$ that provide a Hausdorff decomposition of $\mathbb{H}^2 \setminus N$, where N is the measure zero and nowhere dense set consisting of the boundaries of all the polygons. The subset A is just the union of the images of the fundamental triangle using transformations in the subset A of G, and similarly for B and C. One can visualize this in the upper half-plane model: The three sets are shown in Figure 4.2. We have $\tau(A) = B$, $\tau(B) = C$, and $\sigma(A) = B \cup C$. Note (see Fig. 4.1) that τ corresponds to a clockwise rotation.

The use of the half-plane makes it easy to compute the tiling, but the picture is much more pleasing when we transform it to the Poincaré disk. This can be done by a single linear fractional transformation. First we move to the disk model with center corresponding to $-\frac{1}{2} + \frac{\sqrt{3}}{2}i$. In that model, τ is an exact *clockwise* Euclidean rotation through 120°, and τ^2 through 240°. The three sets are shown in Figure 4.3, and it is clear even to a Euclidean observer that each set is a third of the plane.

When we adjust the projection as in Figure 4.3(b) so that the center of the disk is i, then σ is just a 180° rotation, and one sees that A is one-half of the hyperbolic plane: $\sigma(A) = B \cup C$. So this provides a view of the basic idea of the Hausdorff paradox that is completely geometric and uses nice sets; because the fundamental triangle serves as a choice set, the Axiom of Choice is never needed! For an animation showing how sets change when the viewpoint changes, see [Wag07].

The connection to the classical yin-yang motif (Fig. 4.4(a)), with three regions instead of two, is a surprising coincidence of this construction. That motif is not paradoxical in the same way, because moving the viewpoint as done earlier to put i at the center makes the red region take up less than half of the disk (Fig. 4.4(b)); rotation of the red region through 180° will pick up all of blue-and-green, except for a small lens around the origin. The red area in Figure 4.4(b) is 99.7% that of the half-disk. A small surprise, easily explained by reference to Figure 4.2, is that all the cusp points in Figure 4.3(a) lie exactly on the semicircles of Figure 4.4(a).

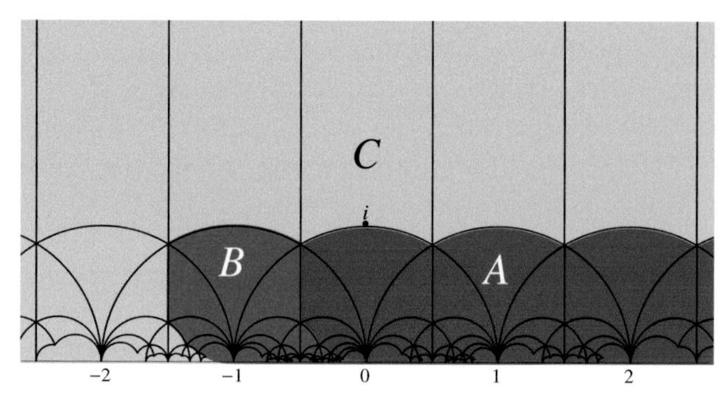

Figure 4.2. The Hausdorff paradox in the upper half-plane.

A nice exercise is to see what the triple yin-yang figure and also the central lens in Figure 4.4(b) look like in the upper half-plane [TW14].

One can also visualize the original Hausdorff paradox which, in the context of Theorem 4.3 and its proof, arises from using powers of $\tau^2\sigma$ to absorb the identity as opposed to powers of $\tau\sigma$. This is shown in Figure 4.5, where we see that the sets are not connected as nicely as they are when $\tau\sigma$ is used. In fact, the proof of Theorem 4.3 works with any word of infinite order used instead of $\tau\sigma$.

4.3 A Banach–Tarski Paradox of the Whole Hyperbolic Plane

The paradox of §4.2 is not of the whole hyperbolic plane, because we have not tried to account for the boundaries of all the triangles. Rather, it is a paradox on $\mathbb{H}^2 \setminus D$, where D is the set consisting of the boundaries of all the tiles. And because σ and τ of §4.2 are elliptic (and hence have fixed points), we cannot use them to get a Hausdorff Paradox that accounts for every last point. To address this point, we consider the nonelliptic subgroup of $PSL_2(\mathbb{Z})$ mentioned in §4.1, namely, let $\sigma(z) = z/(2z+1)$ and $\tau(z) = z+2$ (using the upper half-plane model), and let F be the group they generate. Then, in matrix form, F consists of all matrices in $PSL_2(\mathbb{Z})$ that are congruent to the identity matrix modulo 2 and is called the principal congruence subgroup of the modular group of level 2 (see [Leh66, p. 60]). As already shown (§4.1), F, being a subgroup of $PSL_2(\mathbb{Z})$, has no elliptic elements and hence acts without fixed points. Moreover, σ and τ are independent, and so $F = F_2$. Here is a proof that reduces the issue to the earlier result about $\mathbb{Z}_2 * \mathbb{Z}_3$.

Proposition 4.4. *The two transformations in $SL_2(\mathbb{Z})$ defined by $A = \left[\begin{smallmatrix} 1 & 0 \\ 2 & 1 \end{smallmatrix}\right]$ and $B = \left[\begin{smallmatrix} 1 & 2 \\ 0 & 1 \end{smallmatrix}\right]$ are independent and act on $\mathbb{H}^2$ without fixed points.*

Proof. Let S, T, R be as in Proposition 4.2; then $STST = A$ and $SRSR = B$. Any nontrivial reduced word in $A = SRSR$, $A^{-1} = RSRS$, $B = STST$, and $B^{-1} = TSTS$, viewed as a word in S, T, R, is a nontrivial reduced word in $\mathbb{Z}_2 * \mathbb{Z}_3$. This is because of the eight possible adjacencies AA, BB, AB, BA, AB^{-1}, BA^{-1}, $A^{-1}B$, and $B^{-1}A$, only the last two have any reduction. The first of these is

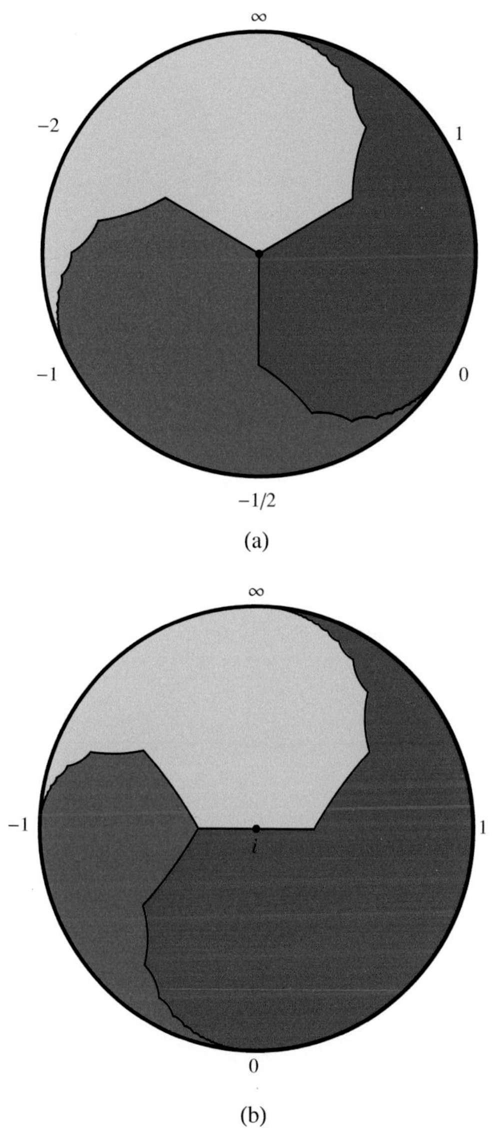

Figure 4.3. The Hausdorff Paradox in the hyperbolic plane. (a) If the sets, reading clockwise from red, are A, B, C, then this shows that A is congruent to B via τ and to C via τ^2. (b) This shows that A is congruent to $B \cup C$ via σ. The labels refer to the corresponding points in the upper half-plane.

$RSRS\,SRSR = RSTSR$; it still ends in R and so leads to no other cancellation. The other case is the same, with T and R switched. Proposition 2 showed that any such word in S, T, R is not the identity. And as shown in §4.1, using only integers means there will be fixed points. □

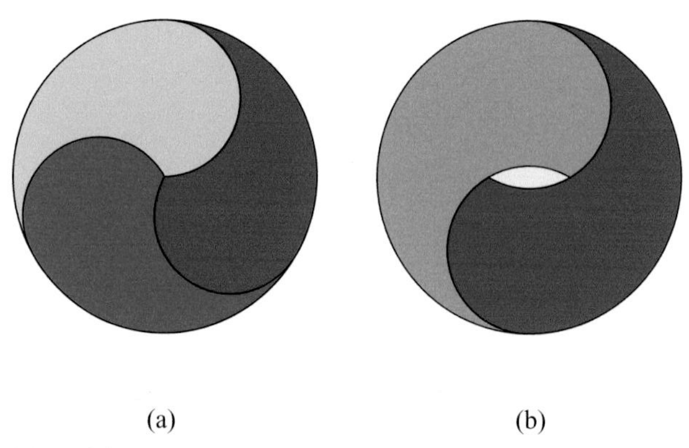

Figure 4.4. (a) A triple yin-yang motif. (b) The transformation of the motif so that *i* is at the center makes the red region larger than half the disk, unlike the situation in Figure 4.3(b), where it is exactly half.

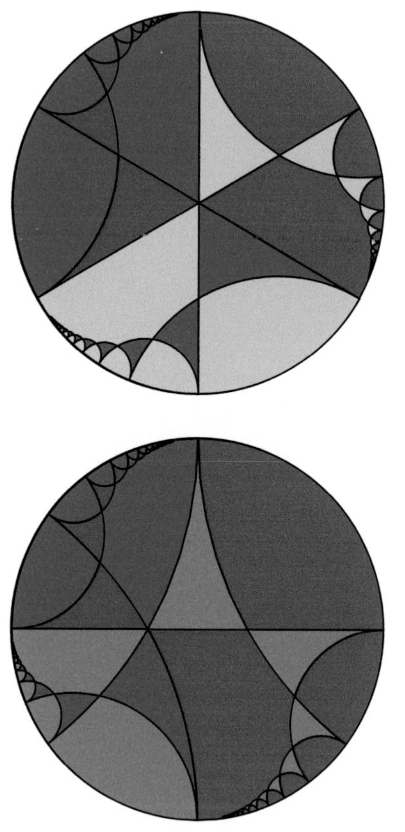

Figure 4.5. A view of Hausdorff's original paradox. The three sets of the upper image are congruent by a rotation. Changing the viewpoint as was done in Figure 4.3 shows that the red set is congruent to the union of the other two.

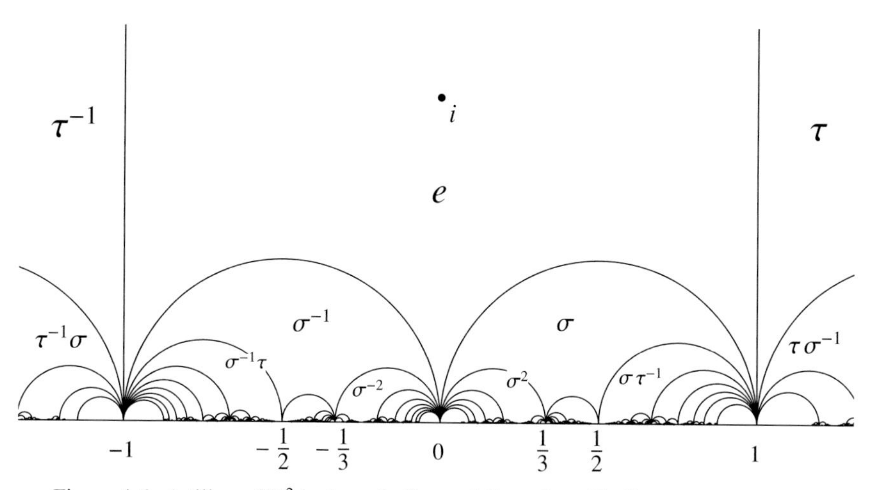

Figure 4.6. A tiling of $\mathbb{H}^2$ by hyperbolic quadrilaterals, with tiles corresponding to elements of a free group of rank 2.

One can also prove this geometrically, working in $\mathbb{H}^+$. Induction on word length can be used to show that $\text{Re}(w(i)) < -1$, $\text{Re}(w(i)) > 1$, $|w(i) + \frac{1}{2}| < \frac{1}{2}$, or $|w(i) - \frac{1}{2}| < \frac{1}{2}$, according as w's leftmost term is τ^{-1}, τ, σ^{-1}, or σ. This implies that $w(i) \neq i$, so $w \neq e$.

The following result does not require the Axiom of Choice.

Theorem 4.5. *There is a paradoxical decomposition of $\mathbb{H}^2$ using isometries in F.*

Proof. Because F is a subgroup of $PSL_2(\mathbb{Z})$, F is discrete, and a fundamental polygon for F's action on $\mathbb{H}^2$ exists. To get a choice set for the orbits of F's action on $\mathbb{H}^2$—without using the Axiom of Choice—we use the fact that the boundary of the fundamental polygon consists of a countable number of sides (open hyperbolic segments) and vertices, and F maps vertices to vertices and sides to sides

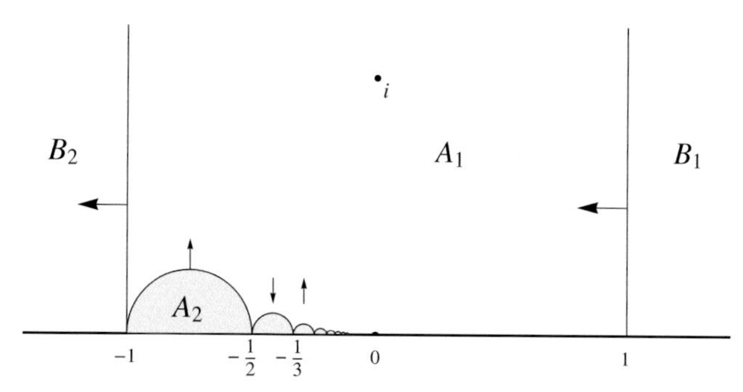

Figure 4.7. A paradoxical decomposition of the hyperbolic plane: $\mathbb{H}^2 = B_1 \cup \tau(B_2) = B_1 \cup (B_2 + 2)$ and $\mathbb{H}^2 = A_1 \cup \sigma(A_2)$. The set A_2 is shown in gray; the sets include boundaries as indicated by arrows.

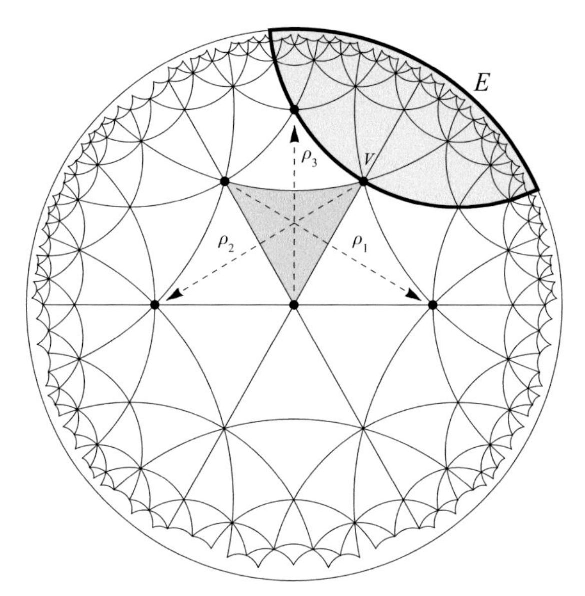

Figure 4.8. The hyperbolic tiling arising from the reflections ρ_i in the sides of the triangle.

(see [Leh66, §§I.4E, F]). It follows that there is a choice set M for the F-orbits that consists of the interior of the fundamental polygon together with some of its vertices and some of the sides. Clearly, M is a Borel set. To summarize, F is a rank 2 free group of hyperbolic isometries acting on $\mathbb{H}^2$ without nontrivial fixed points, and there is a Borel choice set for the orbits of this action. This allows the

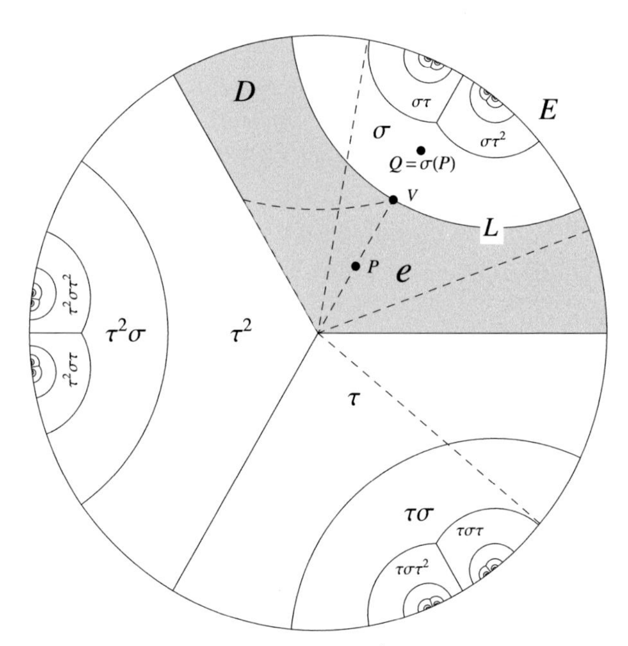

Figure 4.9. The tiling of $\mathbb{H}^2$ using $\langle\sigma, \tau\rangle$.

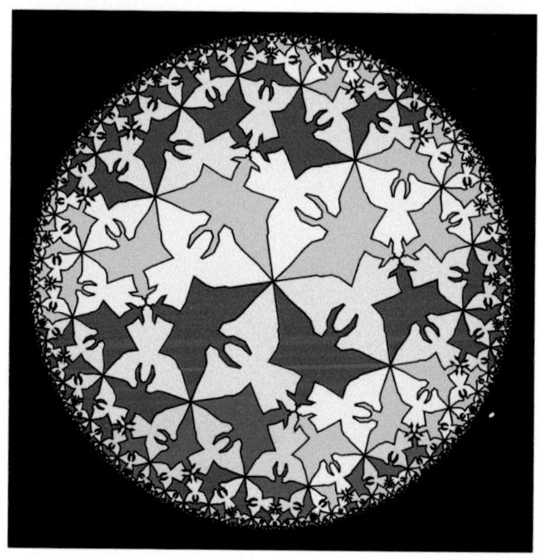

Figure 4.10. A Hausdorff Paradox based on Escher's *Angels and Devils*.

standard paradox for F_2 (Theorem 1.2) to be lifted to all of $\mathbb{H}^2$, taking every single point into account. □

We can give an explicit construction of such a paradox by using a tiling as before. The labeled tiling is shown in Figure 4.6. Note that the geometric arrangement of words is quite similar to the tree representation of F_2 given in Figure 1.1. Now, if the paradoxical decomposition of the group is transferred to $\mathbb{H}^2$ via the labeling of tiles, one gets the four sets A_1, A_2, B_1, B_2 in Figure 4.7. Arrows indicate which sets get the boundary segments. Thus $\mathbb{H}^2$ is divided into four sets such that $\mathbb{H}^2 = B_1 \cup \tau(B_2) = A_1 \cup \sigma(A_2)$; that is, $\mathbb{H}^2$ is paradoxical using four Borel sets. Such Borel sets in $\mathbb{R}^n$ do not exist, although this is not obvious since the fact that $\mathbb{R}^n$ (and $\mathbb{H}^n$) has infinite Lebesgue (or hyperbolic) measure means that a Borel Paradox would yield only the harmless equation $\infty = 2\infty$. The Euclidean cases

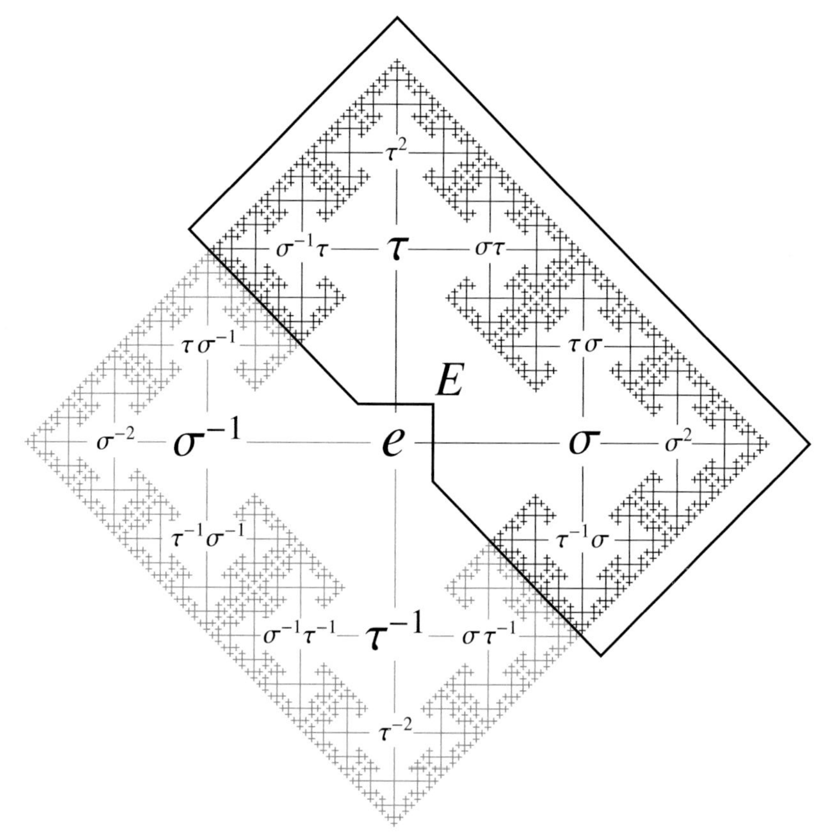

Figure 4.11. A set E such that $\sigma E = E \setminus \{\sigma\}$ and $\tau E = E \setminus \{\tau\}$.

and the measure-theoretic consequences of the hyperbolic paradox are discussed in Chapter 13.

4.4 Paradoxes in an Escher Design

The hyperbolic paradox in §4.2 is based on a direct construction of hyperbolic isometries that generate $\mathbb{Z}_2 * \mathbb{Z}_3$. Curtis Bennett [Ben00] saw how to get similar paradoxes using the group that underlies M. C. Escher's famous woodcut *Circle Limit IV*, also known as *Angels and Devils*. The tiling in Figure 4.8 is the one on which that piece of art is based. It starts with the triangle having angles 60°, 45°, 45°, and then repeatedly reflects that triangle in the edges. In the group theory sense, it arises from the group generated by ρ_1, ρ_2, ρ_3, the three reflections in the sides of the basic triangle. Each of these has the form $z \mapsto \overline{\tau_i(z)}$, where τ_i is the appropriate rotation and the bar is complex conjugation. Because we are in the disk, τ_i will be given by a matrix with complex entries. Moving to the upper half-plane, the representations of τ_1, τ_2, τ_3 are given by the following

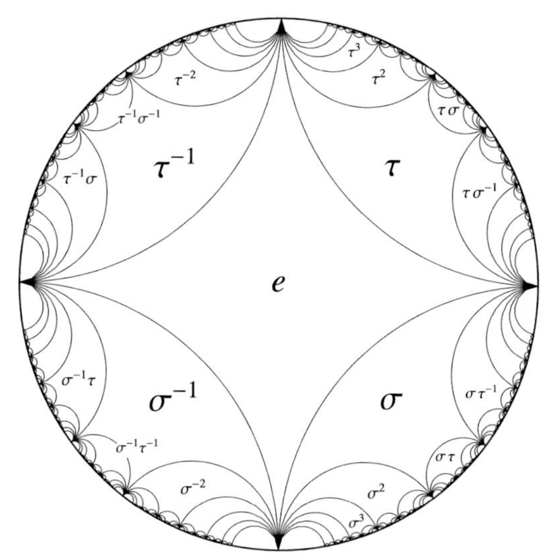

Figure 4.12. A tiling of the hyperbolic plane by squares using the free group generated by σ and τ (see Fig. 4.7 for the upper half-plane version of this tiling). The tiles in the four quadrants are defined by words beginning on the left with τ, τ^{-1}, σ, σ^{-1}, respectively.

Figure 4.13. The set X of red or blue tiles (each of which is a hyperbolic square) contains two tiles whose removal leaves the set geometrically unchanged. If the lower light red square is removed to leave Y, then $\sigma(X) = Y$; the same holds using τ and the upper light blue square.

elements of $SL_2(\mathbb{R})$): $\left[\begin{smallmatrix} 0 & -1 \\ 1 & 1 \end{smallmatrix}\right]$, $\left[\begin{smallmatrix} 1 & 1 \\ -1 & 0 \end{smallmatrix}\right]$, $\left[\begin{smallmatrix} 1-\sqrt{2} & 0 \\ 0 & -\sqrt{2}-1 \end{smallmatrix}\right]$. Note that the first two matrices are inverses, but when viewed in the disk, they have nonreal entries, and so the corresponding orientation-reversing isometries, which use conjugation, are not inverses.

The group for this tiling is known as the triangle group $T(3, 4, 4)$; its generator-and-relation form is $\langle \rho_1, \rho_2, \rho_3 : \rho_i^2 = e, (\rho_1\rho_2)^3 = (\rho_1\rho_3)^4 = (\rho_2\rho_3)^4 = e \rangle$. Let G be the subgroup generated by $\sigma = \rho_3\rho_1\rho_3\rho_1$ and $\tau = \rho_1\rho_2$. We will show that G is the free product $\mathbb{Z}_2 * \mathbb{Z}_3$. Note that σ and τ are orientation preserving and so are in the subgroup known as the von Dyck group $D(3, 4, 4)$; in the upper half-plane, we have the forms $\tau = \left[\begin{smallmatrix} 0 & -1 \\ 1 & 1 \end{smallmatrix}\right]$ and $\sigma = \left[\begin{smallmatrix} -1 & 2(1+\sqrt{2} \\ -\sqrt{2} & 1 \end{smallmatrix}\right]$; τ is the familiar order-3 isometry used in §4.2 and Proposition 4.2; but σ differs from the order-2 rotation used earlier.

Proposition 4.6. *The group $\langle \sigma, \tau \rangle$ is the free product $\mathbb{Z}_2 * \mathbb{Z}_3$.*

Proof. Consider Figure 4.8, where the origin is the fixed point of the order-3 clockwise rotation τ and σ is the order-2 rotation about V. Suppose a nontrivial word w in the abstract free product is the identity, where atoms s and t are used. We may assume, by conjugation, that w has the form $s \dots s$ or $s \dots t$. We will show that $w(0) \neq 0$, where 0 is the origin in Figure 4.8. Because $\tau(0) = 0$, we can remove t or t^2 from the right end of w to get $w = st \dots ts$. Now, the rightmost s leads to σ, which moves 0 into the closed region E. But then each subsequent phrase $\sigma\tau$ or $\sigma\tau^2$ takes the point outside of E by the rotation τ or τ^2 and then, via σ, back into E. So the final location is inside E and is therefore not 0. □

Next we need the fundamental domain for G, which we will obtain by the classic technique of taking the orbit of a point, and then looking at the Voronoi region of the initial point: the set of points closer to the initial point than to any other point of the orbit. We'll use $P = V/2$ as the initial point; see Figure 4.9. Let D be the gray region in Figure 4.9; we will prove that D is the desired fundamental domain. Let L be the line separating D from $\sigma(D)$; observe that $\sigma(L) = L$. An easy induction (using, e.g., the fact that σ maps the complement of E in Fig. 4.8 into E) shows that the orbit points in the first quadrant are P and those of the form $\sigma u(P)$, the ones in the fourth quadrant, are τuP, and the others are $\tau^2 uP$.

In the next discussion, concatenation of points refers to hyperbolic distance, and the term *Voronoi line* refers to the perpendicular bisector of two points. It is easy to check that the Voronoi line for R and τR is the line bisecting the angle at the origin formed by the two points.

Theorem 4.7. *For any $X \in L$ and any word $w \in \langle \sigma, \tau \rangle$, $XP \leq XwP$.*

Proof. By induction on word length; it is clear for $w = e$. Note that $\sigma(L) = L$.

Case 1. $w = \sigma u$. Because $\sigma X \in L$, we have $\sigma X \sigma P = \sigma X\ P \leq \sigma X uP$, the last by the inductive hypothesis. Applying σ gives $XP \leq X\sigma uP$.

Case 2. $\boldsymbol{w = \tau\sigma u}$. We want $XP \le X\tau\sigma uP$. We know that σuP is in region E. Because the line bisecting the angle formed by σuP, 0, and $\tau\sigma uP$ lies outside of E (see dashed lines in Fig. 4.9), $X\sigma uP \le X\tau\sigma uP$; but $XP \le X\sigma uP$ by the inductive hypothesis.

Case 3. $\boldsymbol{w = \tau^2\sigma u}$. Similar to Case 2. □

Corollary 4.8. *The fundamental domain for $\langle\sigma,\tau\rangle$ is the region D in Figure 4.8.*

Proof. For any orbit point σuP, the Voronoi line for P and σuP lies above or equals arc L and so has no impact on the domain. We next show that the Voronoi line for P and τuP (where $u = \sigma v \ldots$) lies completely below the x-axis. Let Y be a point on the positive x-axis. The Voronoi line for τP and τuP is a rotation of the corresponding line for P and uP, which we know is in region E by the theorem. So $Y\tau P \le Y\tau uP$, because Y, on the x-axis, is above this line. But $Y\tau P = YP$. So Y, and hence the entire positive x-axis, is closer to P than τuP, as desired. The case of P and $\tau^2 uP$ is essentially identical to the preceding case. □

Now we can use D to generate a Hausdorff Paradox in $\mathbb{H}^2$ that allows a paradoxical decomposition of the angels and devils. The result is shown in Figure 4.10, where the red set is evidently a third of the plane in one image but a half in the other, where the viewpoint has been moved to the point V of Figure 4.9.

4.5 The Disappearing Hyperbolic Squares

The hyperbolic plane allows the visualization of an interesting phenomenon related to the Banach–Tarski Paradox. The basic absorption technique shows the importance of sets X, whether in a group or a metric space, having a point p so that X is congruent to $X \setminus \{p\}$. Of course, it is easy to find such sets in many infinite objects; for example, in the group of integers $\mathbb{Z}$, we have that $\mathbb{N}$ is congruent to $\mathbb{N} \setminus \{0\}$ by addition of 1. But what if we ask, as Sierpiński did, about sets containing *two* distinct points p, q so that X is congruent to the result of removing either one?

Definition 4.9. *A* weak Mycielski set *in a group or metric space is a set X containing distinct points p, q so that each of $X \setminus \{p\}$ and $X \setminus \{q\}$ is congruent to X.*

These are called weak because of the stronger notion, discussed in §7.3, of sets that are invariant under the removal of *any* finite set.

Theorem 4.10. *There is no weak Mycielski set in $\mathbb{R}^1$ or $\mathbb{R}^2$.*

Proof. The proof for the plane, due to E. G. Straus, is a little complicated, and we refer the reader to [Str57]. For the line, suppose $\sigma(X) = X \setminus \{p\}$ and $\tau(X) = X \setminus \{q\}$, where $\sigma, \tau \in G_1$ and $p \ne q$. Because reflections have order 2, σ and τ must be translations, say, by the nonzero values a, b, respectively.

Then $q \in X \setminus \{p\} = X + a$, so $q - a \in X \setminus \{q\} = X + b$ and $q \in X + a + b = (X \setminus \{p\}) + b \subseteq X + b = X \setminus \{q\}$, a contradiction. □

Finding a weak Mycielski set in a free group is not difficult.

Proposition 4.11. *A free non-Abelian group has a weak Mycielski set.*

Proof. Suppose σ_1 and σ_2 generate F_2. Let S_i be the set of words whose rightmost term is a positive power of σ_i, and let $X = S_1 \cup S_2$. Then $\sigma_i\, S_i = S_i \setminus \{\sigma_i\}$, while $\sigma_i\, S_{3-i} = S_{3-i}$, yielding $\sigma_i\, X = X \setminus \{\sigma_i\}$, as desired. □

The set of the proof is easily visualized in the Cayley diagram of F_2 from Chapter 1, provided we change how the diagram is formed to build up words by multiplication on the left as opposed to the right. That is done in Figure 4.11, which shows the free group on σ and τ: The two lobes form the weak Mycielski set, E.

As usual, the construction in the group yields geometric objects in cases where the group acts without fixed points. So using two independent rotations of the sphere $\mathbb{S}^2$ (Thm. 2.1), we can find a weak Mycielski set on the unit sphere with the set of fixed points deleted. Or we can use the Satô rotations (Thm. 2.1) to get such a set using only rational points on the unit sphere. These sets are not discrete and so are difficult to visualize.

But we can use our knowledge of hyperbolic isometries and tilings to construct a discrete visual representation of a weak Mycielski in the hyperbolic plane. As in §4.3, we will use the free generators $\sigma(z) = z/(2z+1)$ and $\tau(z) = z+2$. There is a subtle point here because the tree of Figure 4.11 forms words by adding letters on the left, while the tree inherent in the tiling for σ and τ adds letters on the right (Fig. 4.12). A consequence is that the ultimate picture is more interesting than the two lobes of a tree!

Now we can interpret the set of Figure 4.11 in the tiling and so visualize a hyperbolic weak Mycielski set. While we could work with single points, we will instead focus on the tiles (viewed as open sets) corresponding to the points, as that is much easier to see. So in Figure 4.13, we have a set X of hyperbolic squares shown in red and blue; the two lighter squares are such that the removal of either leaves a set congruent to X (using σ or τ). For example, σ moves each red square in the path starting at the large red square down to its neighbor. By Theorem 4.10, such a set cannot exist in the Euclidean plane.

In $\mathbb{R}^3$ there do exist Mycielski sets (§7.3), but it seems that such a set cannot be discrete. It is known that there is no discrete free group of isometries in $\mathbb{R}^3$, which means that the obvious route to constructing a discrete weak Mycielski set fails. But we still have the following question.

Question 4.12. Is there is a discrete weak Mycielski set in $\mathbb{R}^3$?

As we will see (§7.3), there is a set X in the hyperbolic plane that is invariant under the deletion of any finite subset. And much more is possible: There is a subset of $\mathbb{S}^2$ or $\mathbb{H}^2$ that is invariant under the deletion or addition of any countable subset.

4.6 A Bounded Hyperbolic Paradox

The earlier work in this chapter concerns paradoxes of all of $\mathbb{H}^2$. But a proper analog of the Banach–Tarski result is a paradox of a bounded set, such as a disk. In 1989, Jan Mycielski devised a way to get such a paradox. The construction in [Myc89] is fundamentally sound, but contained some gaps, which were repaired in [MT13]. We make further modifications here.

The Bolyai–Gerwien Theorem (Thm. 3.2) showing geometric equidecomposability of polygons having the same area is valid in the hyperbolic plane. A nice treatment that proves the hyperbolic and the classic case in a uniform way can be found in [MP81, §10.4]. It is, however, unresolved in $\mathbb{H}^3$ or higher hyperbolic dimensions [Kel00].

4.6.1 Preliminaries

The main idea is to obtain a free non-Abelian group of *piecewise* isometries acting on a bounded set Q in the hyperbolic plane such that the set of fixed points of group elements is countable. We work in $\mathbb{D}$, the Poincaré disk model for $\mathbb{H}^2$: $\{z \in \mathbb{C} : |z| < 1\}$. The hyperbolic distance between two points in $\mathbb{D}$ is as follows, where $\|\cdot\|$ is the standard Euclidean norm:

$$\operatorname{arccosh}\left(1 + 2\frac{\|u - v\|^2}{(1 - \|u\|^2)(1 - \|v\|^2)}\right).$$

Recall the classification of isometries at the beginning of the chapter. In this section the main transformation tool will be the hyperbolic translations.

The Cayley transformation $z \mapsto (z - i)/(z + i)$ maps $\mathbb{H}^+$ to $\mathbb{D}$, and so conjugation by this transformation turns the representations using $PSL_2(\mathbb{R})$ into representations of isometries of $\mathbb{D}$. Working in $\mathbb{D}$, the simplest cases are the rotations $\rho_\theta(z) = e^{i\theta} z$ and the hyperbolic translations $\tau_a(z) = \frac{z+a}{\bar{a}z+1}$, where $|a| < 1$; for our work here we need only the translations τ_a, where a is either real or purely imaginary. Translations do not commute in general, but they do when the parameters are both real or both purely imaginary; in that case the composition formula is $\tau_\alpha \tau_\beta = \tau_{(\alpha+\beta)/(\alpha\beta+1)}$.

We will use the concept of an *equidistant arc*, which will be equidistant from the real or imaginary axis. Such a hyperbolic arc—it is not a geodesic in $\mathbb{H}^2$—is the set of all points whose minimum distance from, say, the imaginary axis is a fixed value b. The arc will usually refer to just one of the pair of arcs defined.

There is a countably additive, isometry-invariant measure defined on the Lebesgue measurable subsets of $\mathbb{D}$: Define it by $\mu(E) = \int_E (1 - x^2 - y^2)^{-2} d\lambda$.

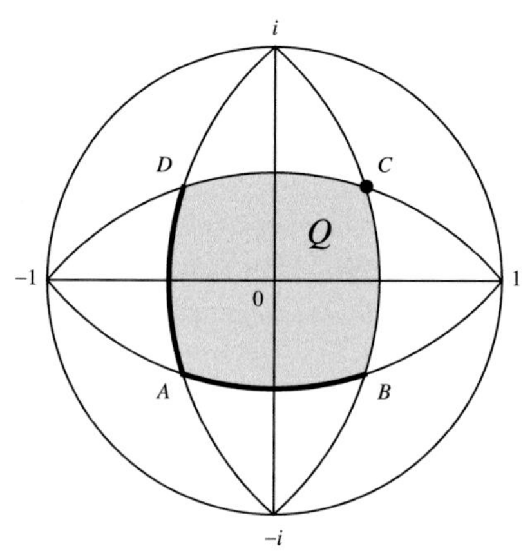

Figure 4.14. The four equidistant lines defining the Mycielski barrel $Q = ABCD$. The exterior labels refer to points in the plane containing the disk.

Therefore, as in the classic case, the Axiom of Choice is necessary, as the paradoxical sets must be nonmeasurable.

4.6.2 The Mycielski Barrel

To start, we need a bounded subset Q of $\mathbb{D}$ with certain nice properties with respect to translations. To get the set, fix α with $0 < \alpha < 1/\sqrt{2}$ and let $b_\alpha = \frac{1}{2}\text{arccosh}(1/(1 - 2\alpha^2))$. Consider the four arcs equidistant at distance b_α from one of the two axes (see Fig. 4.14, where α is chosen to be $\frac{e^2-1}{\sqrt{2(1+e^4)}}$, about 0.606, so that $b_\alpha = 1$). The quadrangle Q_α, called a *Mycielski Barrel*, is defined to be the set of points whose minimum distance to either axis is at most b_α; because we care about individual points when forming decompositions, we define Q_α to include its left and bottom boundary arcs, but only one of the four corners, the intersection of those two boundary arcs (point A in Fig. 4.14). We will usually suppress α and just use Q. Note that if $\alpha \geq 1/\sqrt{2}$, then b_α is infinite or nonreal, and so Q_α does not exist.

We let A, B, C, D be the four corners of Q. One can work out symbolic expressions for the bounding arcs and these points. A key property is that the translation τ_α, where α is real, preserves any horizontal equidistant arc. Therefore such translations carry A to points on AB. The formula for b_α used to define Q was chosen so that the translation that takes A to B is exactly τ_α.

The set Q will play the role of the unit ball in $\mathbb{R}^3$ in the classic paradox. We need two independent transformations (φ and ψ, bijections of Q), and to that end, we divide Q into three pieces, as follows. Let M and M' be the extreme points of the real axis in Q; these values are given by $\pm\tanh(\frac{1}{4}\text{arcsech}(1 - 2\alpha^2))$. Choose a value $\beta < \alpha$ and consider the translation τ_β of Q's left boundary. This defines

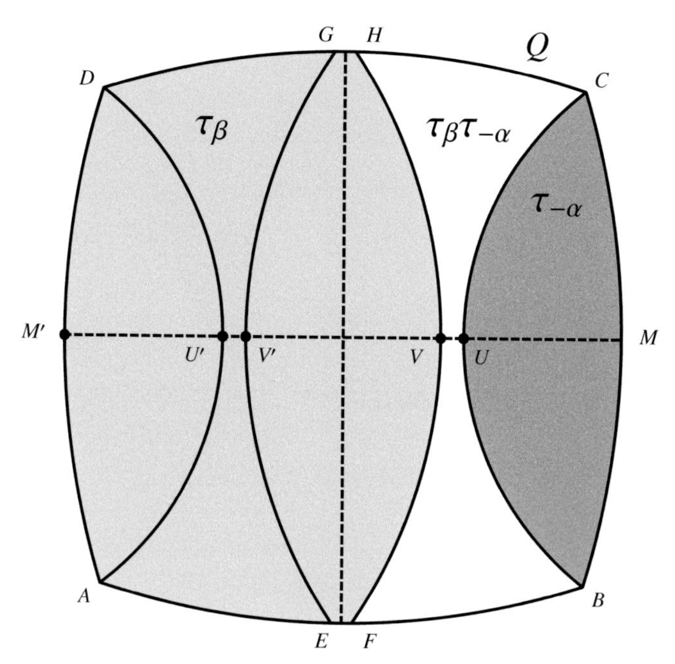

Figure 4.15. The decomposition of a region Q to allow the construction of a bijection using three piecewise translations (the indicated τs) on a quadrangle, a lens, and an hourglass.

arc $EV'G$ in Figure 4.15. Define arc FVH similarly on the positive side. Use τ_α of the left boundary to define arc BUC, and similarly on the negative side to get $AU'D$.

These arcs define a partition of Q into the quadrangle $AEFHGDM'$ (excluding right and top borders), the lens $BMCU$ (including only the left border), and the hourglass $BUCHVF$ (including only the left border). And these three sets allow us to define our first transformation φ as a piecewise translation, as follows:

$$\varphi(z) = \begin{cases} \tau_\beta(z) & \text{if } z \in \text{quadrangle}, \\ \tau_{-\alpha}(z) & \text{if } z \in \text{lens}, \\ \tau_{\beta-\alpha} & \text{if } z \in \text{hourglass}. \end{cases}$$

The second bijection is essentially identical to φ, but working vertically instead of horizontally; thus $\psi = \rho_{\pi/2}^{-1}\varphi\rho_{\pi/2}$. We want arc FVH to be fully left of arc BUC, and that means that β must be larger than $\frac{2M-\alpha-\alpha M^2}{M^2-2\alpha M+1}$, the β-value for which $V = U$ (here M denotes the real coordinate of point M); this lower bound simplifies to $\alpha(1-\sqrt{1-\alpha^2})/(\sqrt{1-\alpha^2}-\alpha^2)$. This lower bound on β is greater than the upper bound $\beta = \alpha$ when $\alpha > \sqrt{2\sqrt{3}-3} = 0.681\ldots$, so we will restrict α to lie under this value when constructing Q_α (see Fig. 4.16). For the strong form of the Banach–Tarski Paradox, we need to know that this construction is sound even

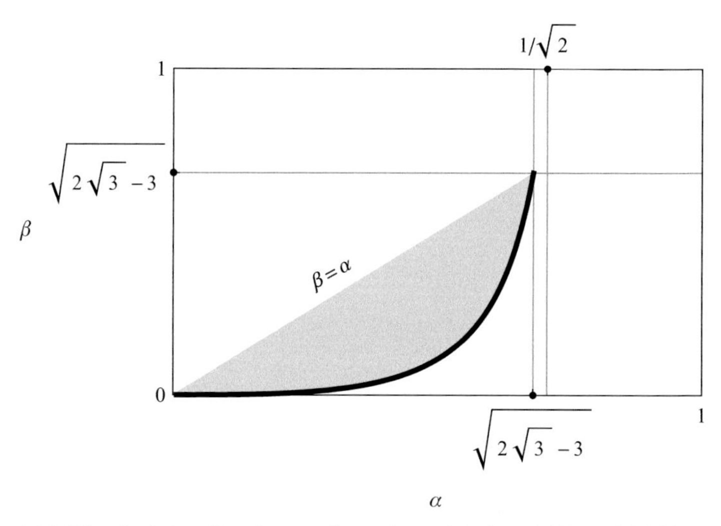

Figure 4.16. The shaded region shows values of α and β that yield Mycielski barrels that might work as the basis of a paradox.

when α is very small. In that case, the quadrangle Q_α looks more like a Euclidean square, but the bijections work in exactly the same way.

4.6.3 A Free Group of Piecewise Isometries

The next mission is to interpret the action of the bijections on Q in a way that relates to integers and so allows an analysis of how φ and ψ interact.

Lemma 4.13. *For every positive integer n, Q decomposes into finitely many regions R_i, symmetric about the real axis, so that for each i, there are nonnegative integers n_1 and n_2 (not both 0) such that, on R_i, $\varphi^n(z) = \tau_\beta^{n_1}\tau_{-\alpha}^{n_2}$. The same is true for ψ, with regions having horizontal symmetry and $\psi^n(z) = \tau_{i\beta}^{n_1}\tau_{-i\alpha}^{n_2}$.*

Proof. Induction on n. The case $n = 1$ uses the regions that define φ (Fig. 4.17, upper left). Each inductive step will yield new regions based on applying φ^{-1} to the regions for the previous case. Precisely, suppose R is a region for φ^n, which equals $\tau_\beta^{n_1}\tau_{-\alpha}^{n_2}$ in that region, and let R' be $\varphi^{-1}(R)$. Subdividing if necessary, assume that R' lies entirely within one of the three sets: quadrangle, hourglass, or lens. Then R' becomes one of the regions for φ^{n+1}, with the pair of integers defined as follows:

$R' \subseteq$ quadrangle: use the pair $(n_1 + 1, n_2)$;

$R' \subseteq$ hourglass: use the pair $(n_1 + 1, n_2 + 1)$;

$R' \subseteq$ lens: use the pair $(n_1, n_2 + 1)$.

For example, in the decomposition for φ^2 in Figure 4.17, the region R' marked $(2, 0)$ is contained in φ^{-1}(quadrangle). Because φ on the quadrangle is τ_β, in

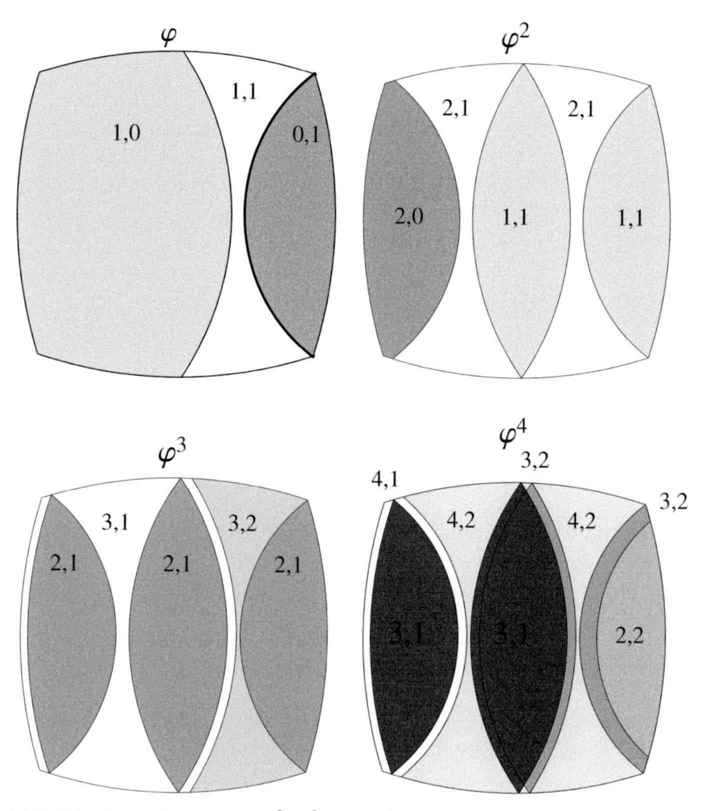

Figure 4.17. The breakdown of φ, φ^2, φ^3, and φ^4, where the integer pairs refer to powers of τ_β and $\tau_{-\alpha}$.

this region, $\varphi^2 = \tau_\beta^2$. In the image for φ^3, the various pairs in the φ^2 image are incremented by 1 in one or both coordinates, and the same again for φ^4. □

Next comes the delicate step: how to choose parameters α and β so that the two bijections φ and ψ of the set Q_α are independent. This is tricky because τ_α and τ_β commute: They are never independent. The key to Mycielski's clever approach is to first look at the case $\beta = i\alpha$ and consider α values that cannot occur geometrically (where $\alpha < 0.6813$ is required) but make sense algebraically.

Lemma 4.14. *If $\alpha = 1/\sqrt{2}$, then any nontrivial word in the translations τ_α and $\tau_{i\alpha}$ does not fix 0 (and so the two translations generate a free group of rank 2).*

Proof. This is a classic geometric argument. Refer to Figure 4.18, where the four hyperbolic lines are drawn so that τ_α maps L_1 to L_2, with similar behavior for $\tau_{i\alpha}$ in the other direction. Let R denote the central region. Then it is easy to prove by induction that $w(R) \subseteq S_1 \cup S_2$ or $w(R) \subseteq T_1 \cup T_2$, depending on the leftmost term of w. Therefore $w(0) \neq 0$. □

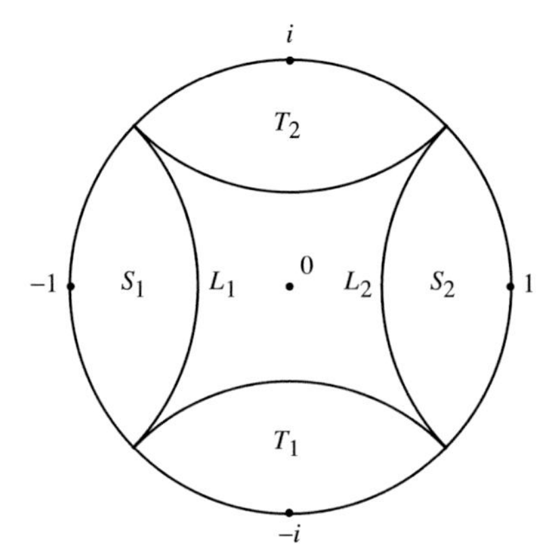

Figure 4.18. A quadruply asymptotic hyperbolic square yields a proof that two orthogonal translations are independent.

The preceding proof, which goes back to Klein and Fricke [KF90], works for $1/\sqrt{2} \le \alpha < 1$; the method can also be used to show more: No nontrivial word has any fixed point. For smaller values of α, the conclusion of the lemma can fail: If $\alpha = \sqrt{\sqrt{2}-1} \approx 0.64\ldots$, then $(\tau_{-i\alpha}\tau_{-\alpha}\tau_{i\alpha}\tau_{\alpha})^2$ is the identity.

Now, the theorem we want is the following.

Theorem 4.15. *There are parameters α, β so that the region Q_α and bijections φ, ψ are properly defined and*

(a) *φ, ψ are independent*
(b) *any nonidentity word in φ and ψ has at most finitely many fixed points in Q_α*

The proof requires this next technical lemma.

Lemma 4.16. *Let X be the set of points (α, β) in the real square $(-1, 1)^2$ such that for some $k \ge 1$ and finite sequences of integers p_j, q_j, r_j, s_j having length $k-1$ or k and with $p_j + q_j \ne 0$ and $r_j + s_j \ne 0$,*

$$(*) \qquad \tau_{-\alpha}^{p_1}\tau_{\beta}^{q_1}\tau_{-i\alpha}^{r_1}\tau_{i\beta}^{s_1}\ldots\tau_{-\alpha}^{p_{k-1}}\tau_{\beta}^{q_{k-1}}\tau_{-i\alpha}^{r_{k-1}}\tau_{i\beta}^{s_{k-1}}\tau_{-\alpha}^{p_k}\tau_{\beta}^{q_k}(0) = 0.$$

Then X is meager.

Proof. Let $f_{\alpha,\beta}$ be the expression on the left in $(*)$. Consider the subset S of the open square $(-1, 1)^2$ defined by

$$S = \{(\alpha, \beta) : \exists k\ \exists \text{ sequences } p_j, q_j, r_j, s_j \text{ such that } f_{\alpha,\beta}(0) = 0\}.$$

Letting $\beta = -\alpha$ in the word defining $f_{\alpha,\beta}$ gives

$$\tau_{-\alpha}^{p_1+q_1}\tau_{-i\alpha}^{r_1+s_1}\ldots\tau_{-\alpha}^{p_{k-1}+q_{k-1}}\tau_{-i\alpha}^{r_{k-1}+s_{k-1}}\tau_{-\alpha}^{p_k+q_k};$$

the hypothesis on the sums guarantees that this is a nontrivial word w in $\tau_{-\alpha}$ and $\tau_{-i\alpha}$. Now if $w(0) = 0$, then when $\alpha = -1/\sqrt{2}$, this contradicts Lemma 2; note that when $k = 1$, w is just $\tau_\alpha^{-p_1-q_1}$, and this cannot fix 0 regardless of α. Therefore the point $(-1/\sqrt{2}, -1/\sqrt{2})$ is not a member of S, and so S is not the full square. But S is a union of countably many algebraic sets, as each collection of sequences p, q, r, s determines an algebraic set in the variables α, β. Each such algebraic set is closed and therefore nowhere dense (for if it contained a nonempty open set, it would have to be all of the square); therefore S is meager. □

A more constructive approach to the case of $\tau_\alpha^p \tau_\beta^q$ comes from the isomorphism tanh from the additive group of reals $\mathbb{R}$ to the interval $(-1, 1)$ under the operation $*$ given by $x * y = \frac{x+y}{xy+1}$. The operation $*$ satisfies $\tau_\alpha \tau_\beta = \tau_{\alpha * \beta}$. Thus we can use this isomorphism to reduce the issue to the additive group of reals. Suppose $\tau_\alpha^p \tau_\beta^q$ is the identity. Then $\alpha' = \operatorname{arctanh} \alpha$ and $\beta' = \operatorname{arctanh} \beta$ are such that $p\alpha' + q\beta' = 0$. The solutions to this form a line in $\mathbb{R}^2$, which is nowhere dense in the plane. Because tanh is continuous, the corresponding set of pairs of translation parameters (α, β) is nowhere dense in the square $(-1, 1)^2$, and the union over all p, q is meager.

Lemma 4.16 yields pairs (α, β) in the gray region of Figure 4.16 that will lead to a free group, as we prove in a moment. It is natural to ask if there is such a pair on the lower boundary of the region, that is, of the form (α, β_α), where β_α is the lower bound of allowable β-values. Possibly there is a transcendental choice of α that leads to such pairs, but that is not known. Note that if $\alpha = \sqrt{2\sqrt{3} - 3}$, the largest possible value, then $\beta_\alpha = \alpha$, and the cube of the commutator of τ_α and τ_{β_α} is the identity.

We can now prove Theorem 4.15; part (a) follows from (b), so we deal only with the fixed point issue.

Proof of Theorem 4.15. Choose (α, β) to avoid the meager set of Lemma 4.16. Therefore we may assume (conjugating by a large power of φ, if necessary) the given word $w = \varphi^{m_1} \psi^{m_2} \ldots \varphi^{m_k}$, with $k \geq 1$ and nonzero exponents. By repeated refinement of the regions provided by Lemma 4.13, get a partition $\{R_i\}$ of Q_α such that every power of φ and ψ in w has the form $\tau_{-\alpha}^{n_1} \tau_\beta^{n_2}$ for some integer exponents when restricted to R_i. Now suppose w fixed infinitely many points in Q_α; then it would fix infinitely many points within one of the regions R_i. But on R_i, w's action is that of the following isometry:

$$\tau_{-\alpha}^{p_1} \tau_\beta^{q_1} \tau_{-i\alpha}^{r_1} \tau_{i\beta}^{s_1} \ldots \tau_{-\alpha}^{p_{k-1}} \tau_\beta^{q_{k-1}} \tau_{-i\alpha}^{r_{k-1}} \tau_{i\beta}^{s_{k-1}} \tau_{-\alpha}^{p_k} \tau_\beta^{q_k}.$$

If $k = 1$, this is simply $\tau_{-\alpha}^{p_1} \tau_\beta^{q_1}$. Because an isometry is determined by three points, this means that this isometry is the identity on all of $\mathbb{D}$, contradicting the choice of parameters α, β. □

These arguments show that, for the choice of α and β as in Theorem 4.15, the group generated by φ and ψ is the free product of two copies of the free Abelian group of rank 2: $(\mathbb{Z} \oplus \mathbb{Z}) * (\mathbb{Z} \oplus \mathbb{Z})$. A natural question is whether one can get the same results by choosing a pair (α, β_α); that is, by choosing parameters on the lower boundary of the legal region in Figure 4.16. Then the hourglass

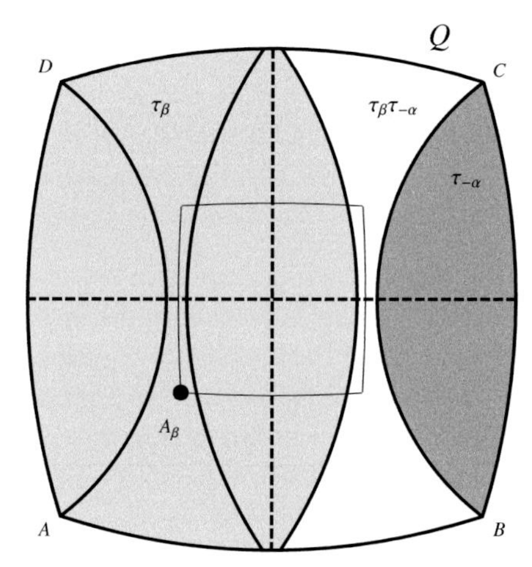

Figure 4.19. Point A_β is a fixed point of the commutator $\tau_{i\beta}^{-1}\tau_\beta^{-1}\tau_{i\beta}\tau_\beta$ and hence a fixed point of $\psi^{-1}\varphi^{-1}\psi\varphi$.

in Figure 4.15 would shrink to a point at its waist. Possibly this works if α is transcendental and not one of the countably many points excluded by the equation $p \operatorname{arctanh} \alpha + q \operatorname{arctanh} \beta_\alpha = 0$.

Theorem 4.17 (AC). *For suitable choice of α, there is a paradoxical decomposition of the corresponding Mycielski barrel Q_α. Any two bounded subsets of $\mathbb{H}^2$ with nonempty interior are equidecomposable.*

Proof. Choose α and β as in Theorem 4.15. Then the set Q' of points fixed by a nontrivial word in the free group $\langle\varphi, \psi\rangle$ is countable. So the free group $\langle\varphi, \psi\rangle$ acts on $Q \setminus Q'$ without fixed points, and so leads to a paradoxical decomposition of $Q \setminus Q'$. The standard method of absorption to handle the fixed points shows that $Q \setminus Q' \sim Q$, so Q is paradoxical. Because there is no problem choosing α to be arbitrarily small, standard techniques using the Banach–Schröder–Bernstein Theorem (as in Thm. 3.11) yield the strong version of the paradox. □

The piecewise isometries used here do not act without fixed points. For consider the smaller Mycielski barrel Q_β inside Q_α, with lower-left corner A_β; then A_β is a fixed point of the commutator in φ and ψ (see Fig. 4.19). In [MT13] it is observed that for small values of α and β, there is a fixed point for the commutator of τ_α and $\tau_{i\beta}$, and therefore this commutator is a rotation. In fact, this is true so long as $\alpha^2 + \beta^2 < 1$: The commutator $\tau_{-i\beta}\tau_{-\alpha}\tau_{i\beta}\tau_\alpha$ has the fixed point $\frac{1}{\alpha^2+\beta^2}(\sqrt{1-(\alpha^2+\beta^2)}-1)(\alpha, \beta)$.

When $\alpha = \beta$, this fixed point is just A, the lower left corner of the Mycielski barrel. So the bound of $1/\sqrt{2}$ is sharp for the fixed-point result, because for any smaller value, the preceding, with $\beta = \alpha$, gives a nontrivial word with a fixed point. As observed by Mycielski, Lemma 2 implies that for any transcendental $\alpha \in (0, 1)$, τ_α and $\tau_{i\alpha}$ are independent; this is because any nontrivial word w yields a formula for $|w(0)|$ that is a polynomial in α, and this polynomial cannot be degenerate because it does not vanish when $\alpha = 1/\sqrt{2}$.

Question 4.18. Is there a subset X of the hyperbolic plane that is bounded and has nonempty interior, and for which two piecewise isometries can be found that are bijections of X and act on X without nontrivial fixed points?

Notes

A good general reference for various properties of isometries in the hyperbolic plane is [Bea83] or [Leh64].

Dekker showed that the group of orientation-preserving isometries of the hyperbolic plane is locally commutative [Dek57]. He considered the hyperbolic plane as points on the hyperboloid $x^2 + y^2 - z^2 = -1$ and showed that the group of orientation-preserving isometries is locally commutative and has two independent elements. He obtained the latter by using cosh and sinh for cos and sin in the rotation ϕ and ψ given in the proof of Theorem 7.4.

The independence of the two matrices in Proposition 4.4 is due to Sanov [San47].

The inductive method of placing words into sets to get a Hausdorff Paradox in $\mathbb{Z}_2 * \mathbb{Z}_3$ is due to Hausdorff [Hau14a, Sie54]. The detailed analysis here showing how to avoid induction and place words in a direct fashion is due to Tomkowicz and Wagon [TW14]. The idea of using the Klein–Fricke tesselation to generate a constructive and visual Hausdorff Paradox is due to Mycielski and Wagon [MW84]. Theorem 4.5 is by Mycielski and Wagon [MW84].

The combining of ideas of Escher and Banach–Tarski is due to Curtis Bennett [Ben00], whose success in obtaining a paradoxical version of *Angels and Devils* solved a problem raised by Wagon.

The use of the hyperbolic plane to visualize a Mycielski set is due to Tomkowicz and Wagon [TW14].

The work in §4.6 on a bounded paradox in $\mathbb{H}^2$ stems from the results of Mycielski [Myc89] in this area, though our approach is quite different. A gap in [Myc89] was fixed in [MT13].

5

Locally Commutative Actions: Minimizing the Number of Pieces in a Paradoxical Decomposition

5.1 A Minimal Decomposition of a Sphere

A careful analysis of the proof presented in Chapter 3 that a sphere may be duplicated using rotations shows that the proof used ten pieces. More precisely, $\mathbb{S}^2 = A \cup B$, where $A \cap B = \varnothing$ and $A \sim_4 \mathbb{S}^2$ and $B \sim_6 \mathbb{S}^2$. By Corollary 3.7, Theorem 2.6, and Proposition 1.10, $\mathbb{S}^2 \setminus D$ splits into A' and B' with $A' \sim_2 \mathbb{S}^2 \setminus D \sim_3 B'$; Theorem 3.9 shows that $\mathbb{S}^2 \sim_2 \mathbb{S}^2 \setminus D$, whence A' and B' yield $A, B \subseteq \mathbb{S}^2$ with the properties claimed. It is easy to see that at least four pieces are necessary whenever a set X, acted upon by a group G, is G-paradoxical. For if X contains disjoint A, B with $A \sim_m X \sim_n B$ and $m + n < 4$, then onc of m or n equals 1. If, say, $m = 1$, then $X = g(A)$ for some $g \in G$, whence $A = g^{-1}(X) = X$ and $B = \varnothing$. It turns out that an interesting feature of the rotation group's action on the sphere allows the minimal number of pieces to be realized: There are disjoint sets $A, B \subseteq \mathbb{S}^2$ such that $A \sim_2 \mathbb{S}^2 \sim_2 B$. Moreover, the techniques used to cut the number of pieces to a minimum lead to significant new ideas on how to deal with the fixed points of an action of a free group, adding to our ability to recognize when a group's action is paradoxical. First, we state precisely the way in which the pieces will be counted.

Definition 5.1. *Suppose G acts on X and $E \subseteq X$. Then E is G*-paradoxical using *r* pieces *if there are disjoint A, B with $E = A \cup B$, $A \sim_m E \sim_n B$, and $m + n = r$.*

Note that this definition is stronger than merely adding to Definition 1.1 the condition that $m + n = 4$, because Definition 5.1 requires $E = A \cup B$ rather than just $E \supseteq A \cup B$. In fact, if r pieces are used in Definition 1.1, then by virtue of Corollary 3.7 and its proof, E is paradoxical using $r + 1$ pieces, in the sense of Definition 5.1. This distinction is strikingly illustrated by the free group, F, of rank 2, which we will here take as being generated by σ, τ. The simple proof of Theorem 1.2 uses four pieces, but because the identity of F does not appear in any of the pieces, it yields only that F is paradoxical using five pieces. To do it with four pieces, one uses the construction of Figure 3.2. The following theorem shows how this construction can be modified to yield an important additional condition that is satisfied by the decomposition.

Theorem 5.2. *The free group F can be partitioned into A_1, A_2, A_3, and A_4 such that $\sigma(A_2) = A_2 \cup A_3 \cup A_4$ and $\tau(A_4) = A_1 \cup A_2 \cup A_4$; therefore F is paradoxical using four pieces. Moreover, for any fixed $w \in F$, the partition can be chosen so that w is in the same piece as the identity of F.*

Proof. The first part was done in Figure 3.2. For the second part, note that the two desired equations are equivalent to four containment relations:

$$\sigma(A_2) \subseteq A_2 \cup A_3 \cup A_4,$$

$$\sigma^{-1}(A_2 \cup A_3 \cup A_4) \subseteq A_2,$$

$$\tau(A_4) \subseteq A_1 \cup A_2 \cup A_4,$$

$$\tau^{-1}(A_1 \cup A_2 \cup A_4) \subseteq A_4.$$

We shall use *domain of σ* to denote A_2, *range of σ* to denote $A_2 \cup A_3 \cup A_4$, and so on.

Suppose that $w = \rho_n \cdots \rho_1$ is given, where each ρ_i, is one of $\sigma^{\pm 1}$, $\tau^{\pm 1}$. Before assigning arbitrary words to the four sets, we show how to successfully place $e, \rho_1, \rho_2\rho_1, \ldots, w$. Place e and w in A_2, A_1, A_4, or A_3 according as ρ_1 is $\sigma, \sigma^{-1}, \tau$, or τ^{-1}. To see how to place the other end segments of w, assume, for definiteness, that ρ_1 is σ. Consider first $u = \rho_{n-1} \cdots \rho_1$, which equals $\rho_n^{-1} w$. If w is in the domain of ρ_n^{-1}, u is placed in the range of ρ_n^{-1}; if w is not in the domain of ρ_n^{-1}, place u in one of the A_i disjoint from ρ_n^{-1}'s range. Then $\rho_{n-2} \ldots \rho_1$ is considered as $\rho_{n-1}^{-1} u$ and placed appropriately, and so on, until only $\rho_1 = \sigma$ remains to be dealt with. Because $\sigma = \sigma e$ and $e \in A_2$, σ must be placed in $A_2 \cup A_3 \cup A_4$, and because $\sigma = \rho_1 = \rho_2^{-1}(\rho_2\rho_1)$, σ must be placed to also satisfy the equation involving ρ_2^{-1} with respect to the location of $\rho_2\rho_1$. But, for each of the three possibilities, σ^{-1}, τ, τ^{-1} for ρ_2^{-1}, $A_2 \cup A_3 \cup A_4$ intersects both the range of ρ_2^{-1} and its complement, so a successful placement of σ is possible. The other cases, $\rho_1 = \sigma^{-1}$ or $\tau^{\pm 1}$, proceed in an identical manner.

Now, an arbitrary word $u \in F$ may be written uniquely as vt, where t is one of $e, \rho_1, \rho_2\rho_1, \ldots, w$. Such words can now be assigned by induction on the length of v. If $v = e$, then u is an end segment of w and has already been assigned. If v does not begin with σ^{-1}, put σvt in $A_2 \cup A_3 \cup A_4$ or A_1 according as v is, or is not, in A_2; this ensures that vt satisfies the first containment. The other noncanceling extensions of vt are dealt with similarly. □

Corollary 5.3. *If F, a free group of rank 2, acts on X without nontrivial fixed points, then X is F-paradoxical using four pieces. Hence any group having a free non-Abelian subgroup is paradoxical using four pieces.*

Proof. Theorem 5.2 gives the four-piece decomposition of F, and it lifts to X by the method of Proposition 1.10. □

Although from a geometrical point of view, it is nicer to have decompositions into as few pieces as possible, such efficient decompositions do not have any new implications for the existence of finitely additive measures. If X is G-paradoxical, then X is G-negligible (Def. 2.4) no matter how many pieces are needed. But Theorem 5.2, in its entirety, can be used to provide a way of dealing with fixed points of action of free groups that is entirely different from the method used in Chapter 3. This, in turn, yields the G-negligibility of X for a wider class of actions than we have yet identified.

If G acts on X and $x \in X$, let Stab(x), the stabilizer of x, denote $\{\sigma \in G : \sigma(x) = x\}$; Stab($x$) is a subgroup of G.

Definition 5.4. *An action of a group G on X is called* locally commutative *if Stab(x) is commutative for every $x \in X$; equivalently, if two elements of G have a common fixed point, then they commute.*

Of course, any action without nontrivial fixed points is locally commutative. A more interesting example is the action of the rotation group $SO_3(\mathbb{R})$ on the sphere $\mathbb{S}^2$. If two rotations of the sphere share a fixed point, they must have the same axis, and therefore they commute; in this case, each Stab(x) is isomorphic to the Abelian group $SO_2(\mathbb{R})$.

The technique used earlier (Prop. 1.10, Cor. 5.3) for transferring a decomposition from a group to a set is based on having a unique representation of each $y \in X$ as $g(x)$, where x is some representative of the orbit of y. If the action has fixed points, then such representations are not unique, but in the case of a locally commutative action, there is an intricate way around this difficulty.

Theorem 5.5 (AC). *If the action of F on X is locally commutative, where F is freely generated by σ and τ, then X is F-paradoxical using four pieces.*

Proof. We shall use Theorem 5.2 to partition X into A_1^*, A_2^*, A_3^*, and A_4^* satisfying $\sigma(A_2^*) = A_2^* \cup A_3^* \cup A_4^*$ and $\tau(A_4^*) = A_1^* \cup A_2^* \cup A_4^*$, which suffices to prove the theorem. However, we shall not use a single partition of F into A_1, A_2, A_3, A_4 but rather many such partitions. More precisely, divide X into orbits with respect to F; then the sets A_i^* will be defined orbit by orbit, appealing to possibly different partitions of F in each orbit.

Each orbit consists entirely of nontrivial fixed points or contains no such points (if $w(x) = x$ and $u \in F$, then $u(x)$ is fixed by uwu^{-1}). For an orbit containing no nontrivial fixed points, the assignment of points to the A_i^* is straightforward, and the same partition of F into A_1, A_2, A_3, A_4 may be used for all such orbits, namely, the one illustrated in Figure 3.2. For each orbit without fixed points, choose any representative point x in the orbit. Then any y in the orbit may be written uniquely as $v(x)$, and y is placed in A_i^* if A_i is the subset of F containing v. Then the equations for the A_i^* (as far as they have yet been defined) follow from the corresponding equations for the A_i.

For the assignment of points in an orbit, $\mathcal{O}$, of fixed points, the choice of x cannot be as arbitrary. Choose a nontrivial word w in F of shortest length fixing an element of $\mathcal{O}$, and let x be one of w's fixed points in $\mathcal{O}$. Let ρ denote whichever of $\sigma^{\pm 1}$, $\tau^{\pm 1}$ is the leftmost term of w. Note that w does not end in ρ^{-1}, for if it did, $\rho^{-1}w\rho$, which fixes $\rho^{-1}(x)$, would be shorter than w. Now, use Theorem 5.2 to partition F into A_1, A_2, A_3, A_4 satisfying the two equations and having the additional property that e and w lie in the same piece of the partition.

We claim that every point y in $\mathcal{O}$ may be written uniquely as $v(x)$, where v does not end in w and does not end in ρ^{-1}. The existence of such a representation is proved by considering a word v of minimal length such that $y = v(x)$. Then v does not end in $w^{\pm 1}$, so in case v does end in ρ^{-1}, we may replace v by vw. The uniqueness of this representation depends on the following important aspect of local commutativity. The only elements of F that fix x are the powers of w. To prove this, note that if u fixes x, then local commutativity implies that $uw = wu$, whence by an elementary property of free groups (see [MKS66, p. 42]), $u = t^j$ and $w = t^k$ for some $t \in F$ and $j, k \in \mathbb{Z}$. But then the minimality of w implies that $|j| > |k|$, so j may be written as $\ell k + r$, where $\ell, r \in \mathbb{Z}$, and $0 \le r < |k|$. Then $x = u(x) = t^r(t^k)^\ell(x) = t^r(x)$, and, again using the minimality of w, this means r must be 0. Therefore k divides j, and u is a power of w, as claimed. Now, if $y = v(x) = u(x)$ are two representations of y in the desired form, then $u^{-1}v(x) = x$, so one of $u^{-1}v$ or $v^{-1}u$ is a positive power of w. Assume it is $u^{-1}v$ (the other case is similar); then either u^{-1} begins with a ρ, contradicting the fact that u does not end in ρ^{-1}, or all of u^{-1} cancels with v, implying that v ends in a w, again a contradiction.

With this representation in hand, we may assign points in $\mathcal{O}$ to the sets A_i^*. Place y in A_i^* if A_i is the subset of F containing v, where $v(x)$ is the unique representation of y defined earlier. To show that this assignment works, consider first the relation $\sigma(A_2^*) \subseteq A_2^* \cup A_3^* \cup A_4^*$. Suppose $y \in A_2^*$—therefore y has the representation $v(x)$, with $v \in A_2$—and consider $\sigma(y)$. If $\sigma v(x)$ is the correct representation of $\sigma(y)$, then because $v \in A_2$ implies $\sigma v \in A_2 \cup A_3 \cup A_4$, $\sigma(y)$ is properly placed in $A_2^* \cup A_3^* \cup A_4^*$. But it is possible that σv ends in w or ρ^{-1}. In the former case, because v does not end in w, σv must equal w, so $\sigma(y) = w(x) = x$. Because σv lies in $A_2 \cup A_3 \cup A_4$, so does w, and by our choice of a partition of F, so does e. And because $e(x)$ is the unique representation of x, this implies that x, and hence $\sigma(y)$, is in $A_2^* \cup A_3^* \cup A_4^*$. If σv ends in ρ^{-1}, then because v does not, v must equal e, and so $y = x$, $\sigma v = \sigma$, $\rho^{-1} = \sigma$, and w begins with σ^{-1}. Now, $e = v \in A_2$, so w is in A_2 and hence $\sigma w \in A_2 \cup A_3 \cup A_4$. But $\sigma w(x)$ is the unique representation of $\sigma(x)$, so $\sigma(y) = \sigma(x) \in A_2^* \cup A_3^* \cup A_4^*$. An identical treatment works for the other three containments involving $\sigma^{-1}(A_2^* \cup A_3^* \cup A_4^*)$, $\tau(A_4^*)$, and $\tau^{-1}(A_1^* \cup A_2^* \cup A_4^*)$, completing the proof. □

Regardless of the number of pieces used, the fact that any locally commutative action of a non-Abelian free group is paradoxical is interesting by itself, because it implies that no invariant, finitely additive measure of total measure one exists; that is, X is F-negligible. There is in fact a simpler proof of this result that, though not efficient in its use of pieces, avoids the rather involved reasoning of Theorem 5.5; see Corollary 10.6.

Corollary 5.6 (AC). *The sphere $\mathbb{S}^2$ is $SO_3(\mathbb{R})$-paradoxical using four pieces, and the four cannot be improved.*

Proof. Because the action of the rotations on the sphere is locally commutative, and because $SO_3(\mathbb{R})$ has a free subgroup of rank 2 (Thm. 2.1), this follows from Theorem 5.5. It was explained at the beginning of this chapter why four is the best possible number of pieces. □

There are other naturally occurring locally commutative actions to which Theorem 5.5 applies. For instance, $SL_2(\mathbb{R})$, the group of area-preserving and orientation-preserving linear transformations of $\mathbb{R}^2$, is locally commutative in its action on $\mathbb{R}^2 \setminus \{\mathbf{0}\}$. If σ_1 and σ_2 fix P, then, choosing a basis that contains P, σ_i may be represented by $\left[\begin{smallmatrix} 1 & a_i \\ 0 & b_i \end{smallmatrix}\right]$. Because $\det(\sigma_i) = 1$, b_i must equal 1, so σ_1 and σ_2 commute. Moreover, $SL_2(\mathbb{Z})$ has two independent elements (Props. 4.2, 4.4). By Theorem 5.5, then, $\mathbb{R}^2 \setminus \{\mathbf{0}\}$ is $SL_2(\mathbb{Z})$-paradoxical using four pieces. Actually, with more work, one can produce two independent elements of $SL_2(\mathbb{Z})$ such that the group they generate has no nontrivial fixed points in $\mathbb{R}^2 \setminus \{\mathbf{0}\}$ (see remarks at beginning of Chap. 8).

A similar situation arises with the isometries of the hyperbolic plane, $\mathbb{H}^2$. Proposition 4.1 showed that the full group is locally commutative. And again, a little more work (Prop. 4.2) yielded a free non-Abelian subgroup that has no fixed points in its action.

5.2 A Minimal Decomposition of a Solid Ball

The preceding work on the sphere may also be applied to decompositions of a solid ball with respect to G_3, the isometry group of $\mathbb{R}^3$. This situation differs from the sphere in two ways: (1) G_3's action is not locally commutative (consider two noncommuting rotations about axes containing the origin), and (2) G_3 does not act on $\mathbb{B}$, the unit ball in $\mathbb{R}^3$. We have already seen (Thm. 3.9) that despite the lack of local commutativity, $\mathbb{B}$ is G_3-paradoxical. But the second point means that it is not clear what the minimal number of pieces is, as the necessity of four pieces depends on having a group of transformations from a set to itself. For instance, in the Sierpiński–Mazurkiewicz Paradox (Thm. 1.7), a set is constructed that is paradoxical using two pieces. Nevertheless, a geometric argument can be used to show that five pieces are necessary for the ball, while the finer analysis available for the sphere allows a five-piece paradoxical decomposition of a ball to be constructed.

Theorem 5.7 (AC). *A solid ball in $\mathbb{R}^3$ does not admit a paradoxical decomposition using fewer than five pieces, and five-piece decompositions of any ball exist.*

Proof. The theorem will be proved for $\mathbb{B}$, the unit ball of $\mathbb{R}^3$ centered at the origin, but the proof applies to any ball. Let $\mathbb{S}$ be the unit sphere that is the surface of $\mathbb{B}$. To see that at least five pieces are necessary, suppose $\mathbb{B} = B_1 \cup B_2 \cup B_3 \cup B_4$, where the B_i are pairwise disjoint and $\sigma_1 B_1 \cup \sigma_2 B_2 = \mathbb{B} = \sigma_3 B \cup \sigma_4 B_4$ for isometries σ_i; this is the only possibility for a decomposition using fewer than five pieces. Not all of the σ_i can fix the origin; otherwise, one copy of $\mathbb{B}$ would be missing the origin, so suppose $\sigma_4(\mathbf{0}) \neq \mathbf{0}$. Then $\sigma_4(\mathbb{B})$ is a unit ball different from $\mathbb{B}$, and it follows that there is a closed hemispherical surface $H \subseteq \mathbb{S}$ that is disjoint from $\sigma_4(\mathbb{B})$. (Let H be the hemisphere of $\mathbb{S}$ symmetric about the point of intersection of $\mathbb{S}$ with the extension of the directed line segment from $\sigma_4(\mathbf{0})$ to $\mathbf{0}$.) Because $\sigma_3(B_3)$ must contain H, B_3 must contain $\sigma_3^{-1}(H)$, a closed hemisphere of $\mathbb{S}$. This means that $(B_1 \cup B_2) \cap \mathbb{S}$ is contained in an open hemisphere of $\mathbb{S}$, namely, the complement of $\sigma_3^{-1}(H)$. Because, therefore, neither B_1 nor B_2 contains a closed hemisphere, the argument used on $\sigma_4(\mathbb{B})$ yields that each of σ_1, σ_2 fix $\mathbf{0}$ and hence map $\mathbb{S}$ to $\mathbb{S}$. But then $(\sigma_1(B_1) \cup \sigma_2(B_2)) \cap \mathbb{S} = \sigma_1(B_1 \cap \mathbb{S}) \cup \sigma_2(B_2 \cap \mathbb{S})$, which is contained in the union of two open hemispheres and hence is a proper subset of $\mathbb{S}$. This contradicts the fact that $\sigma_1(B_1) \cup \sigma_2(B_2) = \mathbb{B}$.

To construct a five-piece decomposition, we shall work separately on each S_r, the sphere of radius r, $0 < r \leq 1$. Use Theorem 2.1 to select two independent rotations σ and τ of $\mathbb{B}$, and let F be the free group they generate. By Theorem 5.5, we may partition each S_r, for $0 < r < 1$, into $A_1^r, A_2^r, A_3^r, A_4^r$ satisfying $\sigma(A_2^r) = A_2^r \cup A_3^r \cup A_4^r$ and $\tau(A_4^r) = A_1^r \cup A_2^r \cup A_4^r$. For S_1, however, we need a partition into five sets, $A_1^1, A_2^1, A_3^1, A_4^1$, and a single point $\{P\}$, satisfying $\sigma(A_2^1) = A_2^1 \cup A_3^1 \cup A_4^1 \cup \{P\}$ and $\tau(A_4^1) = A_1^1 \cup A_2^1 \cup A_4^1 \cup \{P\}$. Such a partition can be constructed by examining the proof of Theorem 5.5 and selecting a single orbit, $\mathcal{O}$, of nonfixed points. There must be such an orbit because F has only countably many fixed points on $\mathbb{S}$, so only countably many points lie in orbits of fixed points. We may now select P to be any point in $\mathcal{O}$ and assign other points $Q \in \mathcal{O}$ to $A_1^1, A_3^1, A_3^1, A_4^1$ according as w begins with a $\sigma, \sigma^{-1}, \tau$, or τ^{-1}, where w is the unique word in $F \setminus \{e\}$ such that $Q = w(P)$. Because $\sigma W(\sigma^{-1}) = F \setminus W(\sigma)$ and $\tau W(\tau^{-1}) = F \setminus W(\tau)$ (see proof of Thm. 1.2), this partition is as desired. Now, $\mathbb{B}$ may be partitioned into B_1, B_2, B_3, B_4 and $\{P\}$, where $B_1 = \{\mathbf{0}\} \cup \bigcup\{A_1^r : 0 < r \leq 1\}$ and for $i = 2, 3, 4$, $B_i = \bigcup\{A_i^r : 0 < r \leq 1\}$. Letting ρ be the translation taking P to the origin, we see that this partition works because $B_1 \cup \sigma(B_2) = \mathbb{B}$ and $B_3 \cup \tau(B_4) \cup \{\rho(P)\} = \mathbb{B}$. □

By partitioning spheres of arbitrary positive radius as in the previous proof, and stealing a point from $\mathbb{S}$ in the same way, we can get a paradoxical decomposition of all $\mathbb{R}^3$ using five pieces. But this decomposition is not the simplest possible. Later we shall construct two independent isometries of $\mathbb{R}^3$ such that the free group that they generate acts on $\mathbb{R}^3$ without nontrivial fixed points (Thm. 6.7). Corollary 5.3 then yields a (best possible) four-piece paradoxical decomposition of $\mathbb{R}^3$.

Although it took more than twenty years (1924–1947), it is remarkable that the original Banach–Tarski duplication of the ball, which was somewhat more complicated than the approach presented in Chapter 3, could be modified and refined to yield a duplication requiring only five pieces, the minimal number possible. In general, optimization along these lines is quite difficult. For instance, if we apply, with minor modification, Theorem 3.8 to the simple two-piece geometric rearrangement of a square into an isosceles right triangle (cut along a diagonal), then five pieces are used to prove these polygons equidecomposable. No lower bound for this problem (except the trivial one, 2) is known.

There is as yet no criterion that can be used to tell, in general, whether an action is paradoxical. But Theorem 5.5 allows such a criterion to be given for the special case of actions of a group of rotations on the sphere. Because the entire action of $SO_3(\mathbb{R})$ on $\mathbb{S}^2$ is locally commutative, it follows from Theorem 5.5 that $\mathbb{S}^2$ is G-paradoxical whenever G is a subgroup of $SO_3(\mathbb{R})$ containing two independent rotations. We shall see in Part II (Thm. 13.23) that these are the only subgroups for which the sphere is paradoxical: If G is a subgroup of $SO_3(\mathbb{R})$ not containing a free subgroup of rank 2, then a finitely additive, G-invariant measure on $\mathcal{P}(\mathbb{S}^2)$ of total measure 1 exists, whence $\mathbb{S}^2$ is not G-paradoxical.

Another situation in which paradoxical actions may be characterized arises if we restrict consideration to decompositions using four pieces. This is because, as we now prove, the converse of Theorem 5.5 is valid, so if G acts on X, then X is G-paradoxical using four pieces if and only if G has a free subgroup of rank 2 whose action on X is locally commutative.

Theorem 5.8. *If G acts on X and X is G-paradoxical using four pieces, then G has two independent elements σ, τ such that the action of F, the group they generate, on X is locally commutative.*

Proof. A four-piece paradoxical decomposition can only arise from a partition of X into A_1, A_2, A_3, A_4 such that for suitable $g_i \in G$, $g_1(A_1)$ and $g_2(A_2)$ partition X, as do $g_3(A_3)$ and $g_4(A_4)$. Let $\sigma = g_1^{-1}g_2$, $\tau = g_3^{-1}g_4$; then the pair A_1, $\sigma(A_2)$ and A_3, $\tau(A_4)$ each partitions X, yielding the four familiar equations:

$$\sigma(A_2) = X \setminus A_1, \sigma^{-1}(A_1) = X \setminus A_2, \tau(A_4) = X \setminus A_3, \tau^{-1}(A_3) = X \setminus A_4.$$

We shall use the terms *domain of* σ, *range of* τ^{-1}, and so on, with respect to these equations as in the proof of Theorem 5.2.

To prove that σ and τ are independent, suppose w is a nontrivial word in σ, σ^{-1}, τ, τ^{-1}; say, $w = \rho_n \ldots \rho_1$, where each ρ_i is one of $\sigma^{\pm 1}$, $\tau^{\pm 1}$. Choose $x \in X$ so that x is not in the domain of ρ_1 but is in the range of ρ_n. These two conditions eliminate at most two of the A_i, so a suitable x exists. But if x is not in the domain of ρ_1, then $\rho_1(x)$ is not in the range of ρ_1, and hence because $\rho_2 \neq \rho_1^{-1}$, $\rho_1(x)$ is not in the domain of ρ_2. Continuing through w in this way yields that $w(x)$ is not in the range of ρ_n. By the choice of x, then, $w(x) \neq x$, so w is not equal to the identity, and the independence of σ, τ is proved.

Suppose the action of F, the free subgroup of G generated by σ and τ, on X is not locally commutative. First we prove the claim that whenever $u, v \in F \setminus \{e\}$ have a common fixed point x in X, then one of the rightmost terms of u or u^{-1}

equals one of the same for v or v^{-1}. For suppose $u = \rho_n \cdots \rho_1$ and $v = \phi_m \cdots \phi_1$ where each ρ_i, ϕ_j is one of $\sigma^{\pm 1}, \tau^{\pm 1}$. If $\rho_1 \neq \phi_1$, then x cannot be in the domains of both ρ_1 and ϕ_1. Suppose x is not in the domain of ρ_1. Then, as proved in the first part of this theorem, $u(x) = x$ is not in the range of ρ_n. If $\rho_n \neq \phi_m$, then because $v(x) = x = u(x)$, $v(x)$ cannot be in the complement of ϕ_m's range too, and so as before, x must be in the domain of ϕ_1. But the domain of ϕ_1 is contained in the range of ρ_n, unless $\rho_n = \phi_1^{-1}$, as sought. Finally, if x is not in the domain of ϕ_1, then the same proof yields $\phi_m = \rho_1^{-1}$.

Now, let u, v be a noncommuting pair in F that shares a fixed point, such that the sum of the lengths of u and v is as small as possible with respect to all other noncommuting pairs that share a fixed point. Suppose $u = \rho_n \cdots \rho_1$ and $v = \phi_m \cdots \phi_1$ ($\rho_i, \phi_j \in \{\sigma^{\pm 1}, \tau^{\pm 1}\}$), and let $x \in X$ be one of the fixed points of both u and v. Then $\rho_n \neq \rho_1^{-1}$; otherwise, by the claim of the previous paragraph, the pair $\rho_1 u \rho_1^{-1}$, $\rho_1 v \rho_1^{-1}$, which fixes $\rho_1(x)$ and does not commute, would have total length smaller, by 2, than that of u, v. Similarly, $\phi_m \neq \phi_1^{-1}$. Now, by the same claim, we may assume $\phi_1 = \rho_1$; otherwise, replace one, or both, of u, v by its inverse. Consider uv^{-1} and $v^{-1}u$, both of which fix x. The minimality of u, v implies that there cannot be so much cancellation in uv^{-1} or $v^{-1}u$ as to affect any of the end terms. For if, say, the u of uv^{-1} was completely absorbed, then the pair uv^{-1}, u^{-1}, which fixes x and does not commute, would have a smaller total length than u, v. But because $\rho_1^{-1} \neq \rho_n$, $\phi_1^{-1} \neq \phi_m$, and $\rho_1 = \phi_1$, the end terms of uv^{-1} and $v^{-1}u$ contradict the claim of the previous paragraph. □

The characterization provided by the theorem just proved and its converse, Theorem 5.5, is especially succinct for actions without nontrivial fixed points: If G acts on X in such a way, then X is G-paradoxical using four pieces if and only if G has a free non-Abelian subgroup. In particular, this applies to the action of a group on itself by left translation, yielding the following corollary.

Corollary 5.9 (AC). *A group G is paradoxical using four pieces if and only if G has a free subgroup of rank 2.*

Proof. By Corollary 5.3 and Theorem 5.8. □

5.3 General Systems of Congruences

Theorem 5.5 and the group theory result upon which its proof is based (Thm. 5.2) are really special cases of a much more general result on partitioning a set or group into subsets satisfying a given system of congruences. Because the derivation of Theorem 5.5 from Theorem 5.2 is independent of the system of congruences, we seek to characterize the sort of systems for which Theorem 5.2, including the assertion about a fixed word w, remains valid. An example for which Theorem 5.2 is false is the single congruence $\sigma(A) = B$, with respect to a partition of F into A and B. For if w is taken to be σ, then if e is in A, w is in B, and if e is not in A, w is not in B, whence e and w cannot be placed together. In fact, this example is essentially the only one for which the sought-after generalization fails, for as we now show, Theorems 5.2 and 5.5 are valid for any system of congruences, except those that, explicitly or implicitly, yield that a set is congruent to its complement. If $D \subseteq \{e, \ldots, r\}$, then D^c denotes $\{1, \ldots, r\} \setminus D$.

Definition 5.10. *Consider an abstract system of m relations involving set-variables $A_1, \ldots, A_r$, each relation having the form*

$$\bigcup\{A_j : j \in L_i\} \cong \bigcup\{A_j : j \in R_i\},$$

where L_i and *R_i are subsets of $\{1, \ldots, r\}$. Because it is intended that the relations be witnessed by isometries, each relation in the system is called a* congruence. *Only congruences where L_i and R_i are nonempty proper subsets of $\{1, \ldots, r\}$ are of interest, and a system where each congruence has that form will be called a* proper system. *Suppose that when we add to a proper system the m complementary congruences, that is, the one referring to L_i^c and R_i^c instead of L_i and R_i, and then consider all congruences obtainable from these 2m ones by transitivity, we do not obtain a congruence of the form $\bigcup\{A_j : j \in D\} \cong \bigcup\{A_j : j \in D^c\}$. Then the original system of congruences is called a* weak system.

The system that we considered in Theorems 5.2 and 5.5 is certainly weak, for if we close $A_2 \cong A_2 \cup A_3 \cup A_4$, $A_4 \cong A_1 \cup A_2 \cup A_4$ under complementation and transitivity (and ignore the ones of the form $A_i \cong A_i$), only the congruences $A_1 \cong A_1 \cup A_3 \cup A_4$ and $A_3 \cong A_1 \cup A_2 \cup A_4$ are added. Another interesting example is the set of $r - 1$ congruences involving $A_1, \ldots, A_r$ given by $A_1 \cong A_2, A_1 \cong A_3, \ldots, A_1 \cong A_r$. If $r = 2$, then $A_1 \cong A_2$ already shows that the system is not weak. But if $r > 2$, then the closure of this system consists of all congruences of the form $A_i \cong A_j$ and its complement,

$$A_1 \cup \cdots \cup A_{i-1} \cup A_{i+1} \cup \cdots \cup A_r \cong A_1 \cup \cdots \cup A_{j-1} \cup A_{j+1} \cup \cdots \cup A_r.$$

Thus, provided $r > 2$, this system, which is satisfied by a partition of a set into r congruent pieces, is a weak system. As a final example, consider the Hausdorff paradoxical partition into three sets satisfying $A_1 \cong A_2$, $A_1 \cong A_3$, $A_1 \cong A_1 \cup A_2$. Clearly this system is not weak, because it implies $A_3 \cong A_1 \cup A_2$. We now generalize Theorem 5.2 by proving a purely group theoretic result about the satisfiability of proper and weak systems of congruences in free groups.

Theorem 5.11. *Suppose $\bigcup\{A_j : j \in L_i\} \cong \bigcup\{A_j : j \in R_i\}$, $i = 1, \ldots, m$, is a proper system of congruences involving $A_1, \ldots, A_r$ and F is the free group generated by $\sigma_1, \ldots, \sigma_m$. Then F may be partitioned into $A_1, \ldots, A_r$ satisfying the system, with σ_i witnessing the ith congruence. If, in addition, the congruences form a weak system, then it can be guaranteed that any fixed word $w \in F$ is placed in the same piece of the partition as e, the identity of F.*

Proof. We shall show how the proof of Theorem 5.2 must be modified to yield this generalization. As in Theorem 5.2, we replace the m congruences by the equivalent set of $2m$ containments:

$$\sigma_i\left(\bigcup\{A_j : j \in L_i\}\right) \subseteq \bigcup\{A_j : j \in R_i\}$$

$$\sigma_i^{-1}\left(\bigcup\{A_j : j \in R_i\}\right) \subseteq \bigcup\{A_j : j \in L_i\},$$

where $i = 1, \ldots, m$. Because each L_i, R_i is nonempty and proper, a straightforward induction exactly as in the last part of Theorem 5.2's proof (taking $w = e$ there) yields the partition of F for an arbitrary system. The stronger result for weak systems is, as in Theorem 5.2, a bit more complicated.

Suppose $w = \rho_n \cdots \rho_1$, with each $\rho_k \in \{\sigma_i^{\pm 1} : i = 1, \ldots, m\}$. As in Theorem 5.2, it is sufficient to assign the end-segments $e, \rho_1, \rho_2\rho_1, \ldots, w$ successfully to sets A_j of the partition, for then the rest of F may be placed by induction. Note that any of the m congruences is equivalent to the one obtained by replacing L_i, R_i by L_i^c, R_i^c, respectively. We use the terms *domain* or *range* of $\sigma_i^{\pm 1}$ as in the proof of Theorem 5.2.

Case 1. For some $k = 1, \ldots, n$, the range of ρ_k (or its complement) intersects both the range of ρ_{k+1}^{-1} and its complement (arithmetic on the index k is assumed to be modulo n; thus $n + 1$ represents 1, etc.).

This case is handled in essentially the same way as the proof of Theorem 5.2. Replacing the congruence for ρ_k by its complementary one if necessary, we may assume that the hypothesis of this case applies to the range of ρ_k rather than its complement. Now, place $\rho_{k-1} \cdots \rho_1$ in the domain of ρ_k, and then place the segments $\rho_{k-2} \cdots \rho_1, \rho_2\rho_1, \ldots, \rho_1, e, w, \rho_{n-1} \cdots \rho_1, \ldots, \rho_{k+1} \cdots \rho_1$ in order to satisfy the appropriate containments, and with w in the same piece as e. This leaves the last end-segment, $\rho_k \cdots \rho_1$, which must be placed in the range of ρ_k and in either the range of ρ_{k+1}^{-1} or its complement. By the hypothesis on k, a successful placement of $\rho_k \cdots \rho_1$ is possible.

Case 2. The hypothesis of Case 1 fails.

The failure of Case 1's hypothesis implies that for each $k = 1, \ldots, n$, the range of p_k equals the range of ρ_{k+1}^{-1} or its complement. But the range of ρ^{-1} equals the domain of ρ, so we have that the range of ρ_k equals the domain of ρ_{k+1} or its complement. Now, place e into the domain of ρ_1, place ρ_1 into the range of ρ_1, place $\rho_2\rho_1$ into the range of ρ_2 or its complement according as the range of ρ_1 equals the domain of ρ_2 or its complement, and continue in this way until w must be placed. Then w is forced to be in either the range of ρ_n or its complement, and hence either in the domain of ρ_1 or its complement. But the latter cannot occur, for otherwise the condition

$$w(\text{domain of } \rho_1) = \text{complement of the domain of } \rho_1$$

is a consequence of the system of congruences, contradicting the weakness of the system. Hence w can be placed in the domain of ρ_1 and in the same set containing the identity, completing the proof of this case, and of the theorem. □

Corollary 5.12 (AC). *Suppose G acts on X and a proper system of m congruences involving $A_1, \ldots, A_r$ is given. If G has m independent elements such that the group F that they generate acts on X without nontrivial fixed points, then X may be partitioned into sets A_i satisfying the system. If the action of F is merely locally commutative, then the partition is possible provided the congruences form a weak system.*

Proof. This follows from the previous theorem in exactly the same way that Corollary 5.3 and Theorem 5.5 follow from Theorem 5.2. □

Now, a free group of rank 2 contains m independent elements for any $m < \aleph_0$; indeed, if σ and τ are independent, then $\{\tau, \sigma\tau\sigma^{-1}, \sigma^2\tau\sigma^{-2}, \ldots\}$ is an infinite set of independent elements, as is $\{\sigma\tau, \sigma^2\tau^2, \sigma^3\tau^3, \ldots\}$ (see [MKS66, p. 43]). Hence, by Theorem 2.1, $SO_3(\mathbb{R})$ contains a free locally commutative subgroup of any finite rank, and therefore by Corollary 5.12, any weak system of congruences is solvable, using rotations, by a partition of $\mathbb{S}^2$. Infinite, even uncountable systems of congruences are discussed in the next chapter.

The general theory of systems of congruences has some interesting, nonparadoxical, geometric consequences concerning partitions of a set into congruent pieces.

Definition 5.13. *A set X is m-*divisible* with respect to a group G acting on X if X splits into m pairwise disjoint, pairwise G-congruent subsets. Each of the m subsets is then called an* mth part *of X. If no group is mentioned, it is understood that the isometry group of X is being used.*

As examples, note that for $m < \infty$ a half-open interval on the real line is m-divisible with respect to translations. Similarly, the circle $\mathbb{S}^1$ is m-divisible. Also, each $\mathbb{R}^n$ is m-divisible: An mth part of $\mathbb{R}^1$ is given by $\bigcup\{[mk, mk+1) : k \in \mathbb{Z}\}$, and this extends easily to $\mathbb{R}^n$. On the other hand, $\mathbb{S}^2$ is not 2-divisible using rotations, because any rotation has a fixed point on the sphere. Of course, the antipodal map ($x \mapsto -x$) yields that $\mathbb{S}^2$ is 2-divisible if arbitrary isometries are allowed: One-half of $\mathbb{S}^2$ is given by the open northern hemisphere together with one-half of the equator.

The m-divisibility of X is equivalent to the existence of a partition of X satisfying the system $\{A_1 \cong A_j : 2 \leq j \leq m\}$. This system is weak if (and only if) $m \geq 3$, so Corollary 5.12 yields divisibility results when locally commutative free groups of rank 2 (and hence rank $\aleph_0$) are present. Applying this to the locally commutative action of $SO_3(\mathbb{R})$ on $\mathbb{S}^2$ yields the following result.

Corollary 5.14 (AC). *For any m with $3 \leq m < \infty$, $\mathbb{S}^2$ is m-divisible.*

Unlike the decompositions of the Banach–Tarski Paradox, it is not clear whether the pieces in a three-piece splitting of the sphere as in Corollary 5.14 can be Lebesgue measurable. This leads to Question 5.15. Because it is consistent with Zermelo–Fraenkel set theory that all sets are Lebesgue measurable (see Thm. 15.1), a negative answer to part (a) implies an affirmative answer to part (b).

Question 5.15. (a) Can $\mathbb{S}^2$ be split into three Lebesgue measurable pieces, any two of which are congruent by a rotation? (b) Is the Axiom of Choice necessary in Corollary 5.14?

In fact, a much stronger result than Corollary 5.14—one that necessitates nonmeasurable pieces—is possible. By considering a more complicated system of congruences, we shall prove (see remarks following Corollary 6.9) that there is a subset of $\mathbb{S}^2$ that is *simultaneously* a third, a quarter, a fifth, and so on, part of $\mathbb{S}^2$; such a set must be nonmeasurable; otherwise, its measure would be equal to both $1/3$ and $1/4$. Chapter 6 contains a discussion of simultaneous m-divisibility for $\mathbb{R}^n$ and for hyperbolic spaces. See also the discussion of Hausdorff decompositions at the end of this chapter.

There are other scattered results on m-divisibility where only one m at a time is considered, but by no means a general theory. Corollary 5.14 can be extended to show that $\mathbb{S}^2$ (and higher-dimensional spheres) is m-divisible using rotations for any m with $3 \leq m \leq 2^{\aleph_0}$ (see Cor. 7.11); indeed, simultaneous m-divisibility is possible here too (see remarks following Cor. 7.11). As pointed out, $\mathbb{S}^1$ is m-divisible for any finite m using half-open arcs; moreover, Vitali's classical example of a non-Lebesgue measurable subset of the circle yields the $\aleph_0$-divisibility of $\mathbb{S}^1$ (see the proof of Thm. 1.5), and this can be extended to all larger cardinals $m \leq 2^{\aleph_0}$ (Cor. 7.11). In short, if arbitrary isometries, rather than just rotations, are allowed, then all spheres are m-divisible for all possibilities for m: $m \leq 2^{\aleph_0}$.

The situation for intervals on the line is a little more complex. The $\aleph_0$-divisibility of $\mathbb{S}^1$ yields in a straightforward manner the existence of $\aleph_0$ pairwise disjoint subsets of a half-open interval, any two of which are equidecomposable (via translations) using two pieces. The extra piece is needed to account for the problem of addition mod 1, if $[0, 1)$ is the interval. Steinhaus [Maz21, p. 8] raised the question whether an interval is, in fact, $\aleph_0$-divisible, and this was settled by von Neumann [Neu28], who proved that all intervals—half-open, open, or closed—are $\aleph_0$-divisible via translations. This result too has been generalized to all m with $\aleph_0 \leq m \leq 2^{\aleph_0}$. Because the finite case for a half-open interval is obvious, such intervals are m-divisible via translations for all possible m. On the other hand, it is known that open or closed intervals are not m-divisible using isometries for any finite $m \geq 2$.

For higher-dimensional balls, it is known that the n-dimensional ball is not m-divisible if $2 \leq m \leq n$; thus it is unresolved whether the disk in the plane is 3-divisible. Kiss and Laczkovich [KL11] proved that the ball in $\mathbb{R}^3$ is m-divisible whenever $m \geq 22$. Thus, in this case, it is unresolved whether the ball is m-divisible for $4 \leq m \leq 21$. They showed also that for every d divisible by 3, the ball is m-divisible when m is sufficiently large. So we have the following open question:

Question 5.16. Is there any $m \geq 2$ so that the unit ball in $\mathbb{R}^4$ is m-divisible?

There are examples of Banach spaces in which the unit ball is 2-divisible; here the definition is widened to allow distance-preserving functions from one subset of the Banach space to another, rather than only bijections from the Banach space to itself that preserve distance. An example is the space of all real sequences converging to 0, with the sup norm. But for a wide class of Banach spaces (reflexive and strictly convex), the unit ball is not 2-divisible (see [Ede00]).

It should be noted that the idea of splitting a set into congruent pieces is, like equidecomposability, a generalization of an older geometric problem: Given a figure (usually in the plane), divide it using finitely many cuts (straight or along curves) into m congruent pieces, where the boundaries of the pieces are ignored. Unlike geometric equidecomposability, for which the Bolyai–Gerwien Theorem (Thm. 3.2) controls what can be done with polygons, this problem is usually attacked on an ad hoc basis. Here is one attractive unsolved problem in this area: If p is an odd prime, is there any way of cutting a square into p congruent pieces other than the obvious way into horizontal (or vertical) rectangles? Note that a polygon is a countable part (in the geometric sense) of $\mathbb{R}^2$ if and only if $\mathbb{R}^2$ can be tiled using $\aleph_0$ copies of the polygon. For a survey of results and questions about such tilings, see [GS86].

Theorem 5.8 and Corollary 5.12 yield necessary and sufficient conditions for the solvability of all weak systems of congruences, namely, the group G acting on X must have a free, locally commutative subgroup of rank 2. The fixed-point condition of the first part of Corollary 5.12, however, is sufficient but not necessary for the solvability of all proper systems. For consider the sphere, $\mathbb{S}^2$, acted upon by its full group, $O_3(\mathbb{R})$, of isometries, including orientation-reversing ones. Any of the latter, when squared, yields an orientation-preserving isometry that can only be a rotation; therefore any subgroup of $O_3(\mathbb{R})$ has nonidentity elements with fixed points on $\mathbb{S}^2$. Nevertheless, a judicious use of the antipodal map shows that all proper systems of congruences are solvable with respect to $O_3(\mathbb{R})$.

Theorem 5.17 (AC). *A proper system $\bigcup\{A_j : j \in L_i\} \cong \bigcup\{A_j : j \in R_i\}$ of congruences involving $A_1, \ldots, A_r$ is solvable via a partition of $\mathbb{S}^2$ provided arbitrary isometries of the sphere may be used to witness the congruences.*

Proof. Let $\zeta : \mathbb{S}^2 \to \mathbb{S}^2$ denote the antipodal map, $\zeta(x) = -x$, and choose $\sigma_1, \ldots, \sigma_m$ to be independent rotations of $\mathbb{S}^2$. Choose any two-coloring of the subsets of $1, \ldots, r$ such that the color of any set differs from the color of its complement. Then define τ_i to be either σ_i or $\sigma_i\zeta$, according as L_i and R_i do, or do not, have the same color.

Note that because each rotation of $\mathbb{S}^2$ is also a linear transformation of $\mathbb{R}^3$, ζ commutes with each σ_i. It follows from this and the fact that $\zeta^2 = e$ that the τ_i are independent, for if a word involving the τ_i equals the identity, then the square of the corresponding word in the σ_i does too, contradicting the independence of the σ_i. Note also that for any rotation θ, $\zeta\theta$ has a fixed point on $\mathbb{S}^2$ if and only if θ is a rotation of order 2. Because there are no such rotations in the free group generated by the σ_i, the local commutativity of that group implies the same for F, the group generated by the τ_i.

We may now proceed as we did for weak systems in Corollary 5.12, that is, use the method of proof of Theorem 5.5, except that a problem will arise when it comes to partitioning an orbit of fixed points. If w is the chosen short word fixing a point in the orbit, then because the given system of congruences is not necessarily weak, Theorem 5.11 as stated does not yield the required partition of F with e and w in the same piece. Nevertheless, the only spot where Theorem 5.11's proof appeals to the weakness of the system is in Case 2, where each R_k, the range of ρ_k, equals either L_{k+1} or its complement (where $n+1$ cycles back to 1). So assume that that is the case and consider the sequence of colors assigned to the sequence of sets:

$$L_1, R_1, L_2, R_2, \ldots, L_n, R_n.$$

Because w has a fixed point, w cannot equal $\theta\zeta$ for a rotation θ, and so w must contain an even number of ζs, that is, an even number of τ_is that equal $\sigma_i\zeta$. This means that there is an even number of color switches in the steps from L to R in the sequence of sets just displayed. The parity of the number, K, of color switches from R-to-L acts like a color operator, transforming the color of L_1 into the color of R_n. That is, if K is odd, then the colors of L_1 and R_n are different, while if K is even, then the colors are the same. Also the parity of K determines whether w is in R_n or its complement.

To be precise, if K is odd, then there is an odd number of switches in the R-to-L steps and the placement of e in $L_1 = \text{domain}(\rho_1)$ forces w to be in the complement of R_n; because the total number of color switches is odd, the complement of R_n must be L_1. Therefore w may be placed in the same piece as e.

If K is even, then w is forced to be in R_n, which has the same color as, and therefore equals, L_1. Again, this allows a successful placement of w. Thus there is a partition of F that can be used to assign the points of the orbit to the appropriate sets. This can be done for each orbit of fixed points, yielding the desired partition of $\mathbb{S}^2$. □

The solution of all proper systems involving $A_1, \ldots, A_r$ is equivalent to the solution of the single, all-encompassing system of $\binom{2^r-2}{2}$ congruences: $\bigcup\{A_j : j \in L_i\} \cong \bigcup\{A_j : j \in R_i\}$, where L, R vary over all pairs of nonempty, proper subsets of $\{1, \ldots, r\}$. A partition of $\mathbb{S}^2$ satisfying this universal system, which exists by the preceding theorem, is surely a striking object. For $\mathbb{S}^2$ is divided into r pieces such that if someone gathers a nonempty, proper collection of these pieces into one pile, and then gathers another such collection into another pile (a piece may appear in both piles), then the union of the pieces in one pile is congruent to the union of the other pile. In particular, each of the r sets is, simultaneously, a half, a third, . . . , an rth part of $\mathbb{S}^2$.

This theorem has one application that is interesting from a historical point of view. Hausdorff's original decomposition of the sphere (§4.2) showed that, except for a countable set, the sphere could be partitioned into A_1, A_2, and A_3, satisfying $A_1 \cong A_2 \cong A_3 \cong A_1 \cup A_2$, where the congruences are realized by rotations. These congruences do not form a weak system, and because rotations have fixed points, it is impossible to eliminate the countable subset of the sphere excluded at the outset. The Banach–Tarski Paradox eliminates this countable set but constructs a partition satisfying a different paradoxical system of congruences. It is a consequence of Theorem 5.16 that if all isometries of the sphere are allowed, then a Hausdorff decomposition of the entire sphere is possible. For a version of the Hausdorff Paradox that retains the restriction to rotations but weakens the assumption that a countable set of points on the sphere must be deleted, see the appendix to [Rob47]. For a constructive geometric realization of the Hausdorff Paradox, that is, one that does not require the Axiom of Choice, see Chapter 4.

Notes

Von Neumann was apparently the first to consider the problem of counting the number of pieces in a paradoxical decomposition; in [Neu29, p. 77] he states, without proof, that nine pieces suffice for the duplication of a solid ball. Sierpiński [Sie45a] was able to improve this to eight, but it was Raphael Robinson [Rob47] who had the idea of first trying to optimize the situation for the sphere and then extending the analysis to the ball.

In fact, most of the main ideas of this chapter stem from Robinson's paper, which contains the proof that five pieces are necessary and sufficient for the ball, and the consideration of arbitrary weak systems of congruences and the application in Corollary 5.14. Question 5.15 was raised by Mycielski [Myc57a, Myc57b].

The generalization of Robinson's work from the special context of a sphere to general locally commutative actions is due to Dekker [Dek56a, Dek58a], who was responsible for the definition of local commutativity and its applications Theorem 5.5 and Corollary 5.12. A special case of these aspects of local commutativity was rediscovered by Akemann [Ake81, Prop. 4]. Dekker also considered the converse problem, proving Theorem 5.8, and investigated higher-dimensional Euclidean and non-Euclidean spaces [Dek56b, Dek57]; these generalizations are treated in Chapters 6 and 7. The part of Theorem 5.8 that yields the independence of σ and τ is similar to a result known as MacBeath's Lemma (see [LU68, LU69]).

Actually, the special case of Dekker's converse that deals solely with groups acting on themselves, Corollary 5.9, had been considered earlier. While a student of Tarski in the 1940s, B. Jónsson proved Corollary 5.9, but because the result did not lead to any progress on the question whether paradoxical groups necessarily had free subgroups of rank 2 (see §12.2), it was never published.

The idea of weaving the antipodal map into strings of rotations to solve all proper systems of congruences on $\mathbb{S}^2$ is due to Adams [Ada54], who proved Theorem 5.16 and derived the Hausdorff Paradox without the exclusion of a countable set.

The m-divisibility of $\mathbb{S}^2$ with respect to rotations for m satisfying $\aleph_0 \leq m \leq 2^{\aleph_0}$ is due to Mycielski [Myc55b], who also proved simultaneous m-divisibility (see remarks following Cor. 7.9). The m-divisibility of $\mathbb{S}^1$ for $\aleph_0 < m < 2^{\aleph_0}$ is due to Ruziewicz [Ruz24]. Von Neumann's proof [Neu28] of the $\aleph_0$-divisibility of intervals is simplified and extended to larger cardinals by Mycielski [Myc57a]. The nondivisibility of an open or closed interval into finitely many congruent pieces is due to Gustin [Gus51] although special cases of this result have been rediscovered by Sierpiński [Sie54, p. 63], Schinzel (see [Myc57a]), and Cater [Cat81]. Van der Waerden [Wae49] asked whether a disk in $\mathbb{R}^2$ is 2-divisible, and a negative answer was provided by Gysin [Gys49]. Puppe (see [HDK64, pp. 27, 81]) extended this to apply to any bounded, closed, convex subset of $\mathbb{R}^2$. R. M. Robinson (see [Wag83]) showed that a ball in $\mathbb{R}^n$ is not m-divisible for any m with $2 \leq m \leq n$. The results on divisibility in Banach spaces are due to Edelstein [Ede88].

6

Higher Dimensions

6.1 Euclidean Spaces

Because paradoxical decompositions depend on free groups, and because the group of rotations of $\mathbb{S}^2$ is contained in higher-dimensional rotation groups, it comes as no surprise that paradoxical decompositions exist for higher-dimensional spaces. This generalization is not completely obvious, though, because the fixed point set of an isometry does expand when the isometry is extended to a higher dimension by fixing additional coordinates. Nevertheless, the basic results on the existence of paradoxical decompositions do extend without requiring any new techniques (see Thm. 6.1). For example, we have already seen that the unit ball in $\mathbb{R}^n$ is G_n-negligible if $n \geq 3$ (Thm. 2.6), and by the theorem of Tarski alluded to just prior to Theorem 2.6, it follows that such balls are paradoxical. But it is useful to see how the decompositions in higher dimensions may be obtained quite directly from the construction on $\mathbb{S}^2$, as is done in Theorem 6.1.

The expansion of the fixed point set is a crucial impasse to generalizing the finer analysis of Chapter 5, however. This is because new fixed points completely destroy the local commutativity of a group when it is viewed as acting on a higher-dimensional space. Nonetheless, locally commutative free groups of isometries (and, where possible, free groups without fixed points) do exist; hence there are minimal paradoxical decompositions in all higher dimensions (Cor. 6.5).

Theorem 6.1 (AC). *Assume $n \geq 3$.*

(a) *Any sphere in $\mathbb{R}^n$ is paradoxical with respect to its group of rotations.*
(b) *Any solid ball in $\mathbb{R}^n$ is G_n-paradoxical, as is $\mathbb{R}^n$ itself.*
(c) *Any two bounded subsets of $\mathbb{R}^n$ with nonempty interior are equidecomposable.*

Proof. (a) The result is true for $n = 3$ (Cor. 3.10), and we proceed from there by induction. We consider $\mathbb{S}^n$, the unit sphere in $\mathbb{R}^{n+1}$ centered at $\mathbf{0}$, but the same proof applies to all spheres. Now, suppose A_i, $B_j \subseteq S^{n-1}$ and $\sigma_i, \tau_j \in SO_n(\mathbb{R})$

witness the fact that $\mathbb{S}^{n-1}$ is paradoxical. Define A_i^*, B_j^* to partition $\mathbb{S}^n$, excluding the two poles $(0, \ldots, 0, \pm 1)$, by putting $(x_1, \ldots, x_n, z)$ in A_i^* or B_j^* according to which of the A_i, B_j contains $(x_1, \ldots, x_n)/|(x_1, \ldots, x_n)|$. Extend σ_i, τ_j to $\sigma_i^*, \tau_j^* \in SO_{n+1}(\mathbb{R})$ by fixing the new axis. In matrix form, this is

$$\sigma_i^* = \begin{bmatrix} & & & 0 \\ & \sigma_i & & \vdots \\ & & & 0 \\ 0 & \cdots & 0 & 1 \end{bmatrix}.$$

Then A_i^*, B_j^*, σ_i^*, τ_j^* provide a paradoxical decomposition of $\mathbb{S}^n \setminus (0, \ldots, 0, \pm 1)$. But any two-dimensional rotation of infinite order, viewed as rotating the last two coordinates and fixing the first $n - 1$ coordinates, can be used as usual (see proof of Cor. 3.10) to show that $\mathbb{S}^n \setminus (0, \ldots, 0, \pm 1) \sim 2\mathbb{S}^n$; by Proposition 3.4 this proves part (a). The rest of the theorem follows from (a) exactly as it does in $\mathbb{R}^3$ (Cor. 3.10, Thm. 3.11). □

As pointed out earlier, we must make a bit of a fresh start to solve arbitrary weak systems of congruences, and hence obtain minimal paradoxical decompositions, for $\mathbb{S}^3$ and beyond. We first consider $\mathbb{S}^3$, where a free non-Abelian group of rotations with no nontrivial fixed points can be constructed directly. Then this group will be combined with the free subgroup of $SO_3(\mathbb{R})$ constructed in Chapter 2 to obtain free subgroups in all higher-dimensional rotation groups, except $SO_5(\mathbb{R})$, which requires special treatment.

Theorem 6.2. *The group $SO_4(\mathbb{R})$ has a free subgroup of rank 2 whose action on the sphere $\mathbb{S}^3$ is without nontrivial fixed points.*

Proof. Choose θ to be an angle whose cosine is transcendental (e.g., $\theta = 1$ radian), and let σ, τ be the rotations in SO_4 given, respectively, by

$$\begin{bmatrix} \cos\theta & -\sin\theta & 0 & 0 \\ \sin\theta & \cos\theta & 0 & 0 \\ 0 & 0 & \cos\theta & -\sin\theta \\ 0 & 0 & \sin\theta & \cos\theta \end{bmatrix} \begin{bmatrix} \cos\theta & 0 & 0 & -\sin\theta \\ 0 & \cos\theta & -\sin\theta & 0 \\ 0 & \sin\theta & \cos\theta & 0 \\ \sin\theta & 0 & 0 & \cos\theta \end{bmatrix}.$$

We shall prove that no nontrivial word in $\sigma^{\pm 1}, \tau^{\pm 1}$ has any fixed points on $\mathbb{S}^3$, which implies both the independence of σ, τ and the lack of fixed points for elements of the group they generate.

A nontrivial word w has one of the four forms $\sigma^{\pm 1} \cdots \tau^{\pm 1}$, $\tau^{\pm 1} \cdots \sigma^{\pm 1}$, $\sigma^{\pm 1} \cdots \sigma^{\pm 1}$, or $\tau^{\pm 1} \cdots \tau^{\pm 1}$. The second form reduces to the first by considering w^{-1}, and repeated conjugation reduces the last two cases to one of the first two, unless w is simply a power of σ or τ. Because $\cos\theta$ is not algebraic, θ is not a rational multiple of π, and therefore powers of the rotation $\left[\begin{smallmatrix} \cos\theta & -\sin\theta \\ \sin\theta & \cos\theta \end{smallmatrix}\right]$, and hence powers of σ or τ, do not have any fixed points. Thus it remains to prove that the same is true for a word, w, of the form $\sigma^{\pm 1} \cdots \tau^{\pm 1}$. This will be done by showing that 1 is not an eigenvalue of the matrix corresponding to such a word w.

Each of $\sigma^{\pm 1}$, $\tau^{\pm 1}$ has the form

$$\begin{bmatrix} P & -Q & R & -S \\ Q & P & -S & R \\ R & S & P & -Q \\ S & -R & Q & P \end{bmatrix},$$

where P and R are polynomials in $\cos\theta$ (with integer coefficients), and Q and S are the products of such polynomials with $\sin\theta$. An obvious induction yields that the same is true for any word w in $\sigma^{\pm 1}$, $\tau^{\pm 1}$. Computing the determinant of $w - \lambda I$ then yields that the characteristic equation for w is

$$\lambda^4 - 4P\lambda^3 + (6P^2 + 2(Q^2 + R^2 + S^2))\lambda^2 - 4P(P^2 + Q^2 + R^2 + S^2)\lambda$$
$$+ (P^2 + Q^2 + R^2 + S^2) = 0.$$

But w is orthogonal, so $P^2 + Q^2 + R^2 + S^2 = 1$ and the characteristic equation is $\lambda^4 - 4P\,\lambda^3 + (4P^2 + 2)\lambda^2 - 4P\,\lambda + 1 = 0$. If 1 is an eigenvalue of w, then $4P^2 - 8P + 4 = 0$. Because $\cos\theta$ is transcendental, this will be a contradiction once it is shown that P, a polynomial in $\cos\theta$, is not just a constant.

Because σ comes from two two-dimensional rotations, for positive integers m, we have that

$$\begin{bmatrix} \cos m\theta & \mp\sin m\theta & 0 & 0 \\ \pm\sin m\theta & \cos m\theta & 0 & 0 \\ 0 & 0 & \cos m\theta & \mp\sin m\theta \\ 0 & 0 & \pm\sin m\theta & \cos m\theta \end{bmatrix}.$$

Because of the identities

$$\cos(m\theta) = 2^{m-1}\cos^m\theta + \text{terms of lower degree in } \cos\theta$$

and

$$\sin(m\theta) = (\sin\theta)(2^{m-1}\cos^m\theta + \text{terms of lower degree in } \cos\theta),$$

$\sigma^{\pm m}$ has the following form, where $\doteq$ denotes that only the term of highest degree in $\cos\theta$ is retained in each entry:

$$\sigma^{\pm m} \doteq 2^{m-1}\cos^{m-1}\theta \begin{bmatrix} \cos\theta & \mp\sin\theta & 0 & 0 \\ \pm\sin\theta & \cos\theta & 0 & 0 \\ 0 & 0 & \cos\theta & \mp\sin\theta \\ 0 & 0 & \pm\sin\theta & \cos\theta \end{bmatrix}.$$

There is a similar representation for $\tau^{\pm k}$. Multiplying the two representations and substituting $1 - \cos^2\theta$ for $\sin^2\theta$ yields the following (where $\epsilon, \delta = \pm 1$):

$$\sigma^{\epsilon m}\tau^{\delta k} \doteq 2^{m+k-2}\cos^{m+k-1}\theta \begin{bmatrix} \cos\theta & -\epsilon\sin\theta & -\epsilon\delta\cos\theta & -\delta\sin\theta \\ \epsilon\sin\theta & \cos\theta & -\delta\sin\theta & \epsilon\delta\cos\theta \\ \epsilon\delta\cos\theta & \delta\sin\theta & \cos\theta & -\epsilon\sin\theta \\ \delta\sin\theta & -\epsilon\delta\cos\theta & \epsilon\sin\theta & \cos\theta \end{bmatrix}.$$

Now, we claim that if $w = \sigma^{\epsilon_n m_n}\tau^{\delta_n k_n}\cdots\sigma^{\epsilon_1 m_1}\tau^{\delta_1 m_1}\}$ and Σ denotes $|k_1| + |m_1| + \cdots + |k_n| + |m_n|$, then

$$w \doteq 2^{\Sigma\mathrm{m}-n-1+}\cos^{\Sigma-1}\theta \begin{bmatrix} \xi\cos\theta & -\mu\sin\theta & -\zeta\cos\theta & -\nu\sin\theta \\ \mu\sin\theta & \xi\cos\theta & -\nu\sin\theta & \zeta\cos\theta \\ \zeta\cos\theta & \nu\sin\theta & \xi\cos\theta & -\mu\sin\theta \\ \nu\sin\theta & -\zeta\cos\theta & \mu\sin\theta & \xi\cos\theta \end{bmatrix},$$

where $\xi, \mu, \nu, \zeta = \pm 1$. To prove this claim, assume inductively that w has the correct form (with ξ_n, μ_n, ζ_n, and ν_n), and consider $\sigma^{\epsilon m}\tau^{\delta k}$. Multiplying yields the desired form, with

$$\xi_{n+1} = \xi_n + \epsilon\mu_n + \delta\nu_n - \epsilon\delta\zeta_n,$$

$$\mu_{n+1} = \mu_n + \epsilon\xi_n + \delta\zeta_n - \epsilon\delta\nu_n,$$

$$\nu_{n+1} = \nu_n - \epsilon\zeta_n + \delta\xi_n + \epsilon\delta\mu_n,$$

$$\zeta_{n+1} = \zeta_n - \epsilon\nu_n + \delta\mu_n + \epsilon\delta\xi_n,$$

but with one less power of 2 than desired. It follows easily from these equations that the equation $\mu\nu = \xi\zeta$, which is true in $\sigma^{\epsilon m}\tau^{\delta k}$, remains true. Hence

$$\xi_{n+1} = \xi_n + \epsilon\mu_n + \delta\nu_n - \epsilon\delta\frac{\mu_n\nu_n}{\xi_n} = \xi_n + \epsilon\mu_n + \delta\nu_n - \xi_n(\epsilon\mu_n)(\delta\nu_n),$$

which, because all terms are ± 1, equals one of ± 2. The same holds true for μ_{n+1}, ν_{n+1}, ζ_{n+1}. Factoring out a 2 yields the correct power of 2 and the correct form, ± 1, for the coefficients, completing the proof of the claim.

This claim shows that P is a polynomial of degree Σ in $\cos\theta$ and, as explained, it follows that 1 is not an eigenvalue of w. □

Corollary 6.3. *If $n \geq 3$ and $n \neq 5$, then $SO_n(\mathbb{R})$ has a free subgroup of rank 2 that is locally commutative in its action on $\mathbb{S}^{n-1}$. If n is a multiple of 4, then $SO_n(\mathbb{R})$ has a free subgroup of rank 2 without any nontrivial fixed points on $\mathbb{S}^{n-1}$.*

Proof. Any $n \geq 3$, except $n = 5$, may be written as $3j + 4k$ for some positive integers j, k. Letting σ_3, τ_3, and σ_4, τ_4 denote, respectively, pairs of independent elements in $SO_3(\mathbb{R})$ and $SO_4(\mathbb{R})$, as constructed in Theorems 2.1 and 6.2, we can define $\sigma, \tau \in SO_n(\mathbb{R})$ by using these pairs on the three- and four-dimensional subspaces of $\mathbb{R}^n$ as specified by the decomposition of n into $3j + 4k$. Thus the matrix of σ is block diagonal, with j 3×3 blocks (σ_3) and k 4×4 blocks (σ_4). Assume the 3×3 blocks precede the 4×4 ones. It is clear that σ and τ are independent. If $j = 0$, then the lack of fixed points in the group generated by σ_4 and τ_4 implies the same for the group generated by σ and τ, proving the assertion of the corollary regarding multiples of 4.

Now, suppose $k > 0$ and w, u are two words in $\sigma^{\pm 1}$, $\tau^{\pm 1}$ that share a fixed point. Note that w and u are ordinary three-dimensional rotations when restricted to a three-dimensional subspace on which they act as if they were the corresponding words in $\sigma_3^{\pm 1}$, $\tau_3^{\pm 1}$. Because they share a fixed point on $\mathbb{S}^{n-1}$, they must share a nonzero fixed point in one of these k three-dimensional subspaces. But then u and w commute when considered as words in $\sigma_3^{\pm 1}$, $\tau_3^{\pm 1}$. By the independence of σ_3 and τ_3, this implies that u and w are commuting words in $\sigma^{\pm 1}$, $\tau^{\pm 1}$. □

K. Satô [Sat97] has shown that some of the results of the preceding corollary can be achieved using orthogonal matrices with rational entries.

This corollary leaves two possibilities outstanding: $SO_{4n+2}(\mathbb{R})$ and $SO_5(\mathbb{R})$. In $SO_m(\mathbb{R})$, m odd, every element has $+1$ as an eigenvalue, and hence has a fixed point on $\mathbb{S}^{n-1}$. Dekker, who discovered the groups of Theorem 6.2 and Corollary 6.3, conjectured [Dek56b] that free groups of rank 2 without nontrivial fixed points exist in the other half of the possible cases, that is, in $SO_{4n+2}(\mathbb{R})$, $n \geq 1$. He also conjectured that $SO_5(\mathbb{R})$, which did not yield to his techniques, has a locally commutative free subgroup of rank 2. Deligne and Sullivan [DS83] settled the $SO_{4n+2}(\mathbb{R})$ case, using relatively deep techniques of algebraic number theory, as well as the theorem of Tits (Thm. 12.6). Their work was significantly extended by A. Borel [Bor83], who showed how to get locally commutative, non-Abelian free subgroups of each $SO_n(\mathbb{R})$, $n \geq 3$; in particular, such subgroups of $SO_5(\mathbb{R})$ exist, as conjectured by Dekker. In fact, Borel's pair of independent rotations of $\mathbb{S}^4$ is similar to the earlier examples in $\mathbb{S}^2$ and $\mathbb{S}^3$. The rotations are defined using angles with a transcendental cosine, and each is built from two 2-dimensional rotations. Borel also showed how fixed-point free, rank-2 free subgroups of $SO_{4n+2}(\mathbb{R})$ could be constructed without using Tits's Theorem. These results, whose proofs use methods from the theory of Lie groups, provide a complete solution to the problem of the existence of these types of free rotation groups: Free non-Abelian subgroups of $SO_n(\mathbb{R})$ that act on $\mathbb{S}^{n-1}$ without nontrivial fixed points, or are locally commutative, exist in all cases, except those excluded for algebraic reasons (solvability or eigenvalues). Moreover, we shall see in Chapter 7 that the set of pairs from $SO_n(\mathbb{R})$ that serve as generators of these groups is dense in $SO_n(\mathbb{R}) \times SO_n(\mathbb{R})$.

Theorem 6.4. *For any even $n \geq 4$, $SO_n(\mathbb{R})$ has a free subgroup of rank 2 having no nontrivial fixed points on $\mathbb{S}^{n-1}$. For any $n \geq 3$, $SO_n(\mathbb{R})$ has a free subgroup of rank 2 whose action on $\mathbb{S}^{n-1}$ is locally commutative.*

Corollary 6.5 (AC). *If $n \geq 2$, then $\mathbb{S}^n$ (and $\mathbb{R}^{n+1} \setminus \{\mathbf{0}\}$) may be partitioned to satisfy any weak system of congruences with respect to the rotation group $SO_{n+1}(\mathbb{R})$. In particular, $\mathbb{S}^n$ is paradoxical using four pieces. If n is odd, $n \geq 3$, then all proper systems of congruences may be solved with respect to the action of $SO_{n+1}(\mathbb{R})$ on $\mathbb{S}^n$. The same is true for even $n \geq 2$ if we enlarge the group witnessing the congruences from $SO_{n+1}(\mathbb{R})$ to $O_{n+1}(\mathbb{R})$.*

Proof. Because a free group of rank 2 contains m independent elements for any m, the parts of this corollary that deal with $SO_{n+1}(\mathbb{R})$ follow from Corollary 5.12 and the free groups of Corollaries 6.3 and 6.4. The result about $O_{n+1}(\mathbb{R})$ follows by closely analyzing the technique of Theorem 5.16, which showed how the presence of the antipodal map could be combined with a locally commutative free group to solve all proper systems of congruences. The essential feature of the locally commutative group used in that proof is that no group element sends any point P on $\mathbb{S}^n$ to its antipode, $-P$; that is, no group element has -1 as an eigenvalue. This is easily seen to be the case for any free subgroup of $SO_3(\mathbb{R})$ and for the subgroup of $SO_4(\mathbb{R})$ constructed in Theorem 6.2; for the latter, note that if $w(P) = -P$, then w^2 would fix P. Hence this property holds for all the locally commutative groups of Corollary 6.3, that is, all relevant cases except $SO_5(\mathbb{R})$.

For $SO_5(\mathbb{R})$ (and this is also true for $SO_3(\mathbb{R})$ and $SO_4(\mathbb{R})$), one can show that, in fact, *any* locally commutative free subgroup of rank 2 has no element with -1 as an eigenvalue. In particular, then, this is true of Borel's group mentioned prior to Theorem 6.4, completing the proof of Corollary 6.5. To prove this fact about $SO_5(\mathbb{R})$, suppose G is a locally commutative subgroup of $SO_5(\mathbb{R})$ freely generated by σ, τ, and some word $u \in G$ has -1 as an eigenvalue. It follows from the orthogonality of u that its real eigenvalues include $+1$, -1, -1, whence u^2 is the identity on a three-dimensional subspace of $\mathbb{R}^5$. Suppose, without loss of generality, that u, and hence u^2, begins with σ. Then u^2 and $w = \tau u^2 \tau^{-1}$ have different leftmost terms and so are not powers of a common word in G. It follows from G's freeness that u^2 and w do not commute (see proof of Thm. 5.5) and hence, by local commutativity, that they do not share a fixed point on $\mathbb{S}^4$. But this contradicts the fact that u^2 and w each fix (pointwise) a three-dimensional subspace of $\mathbb{R}^5$ (because w is a conjugate of u^2), and two such subspaces must have a point in common on $\mathbb{S}^4$. □

The proof in the last paragraph breaks down in $SO_6(\mathbb{R})$ and beyond, and it is not known whether a locally commutative, non-Abelian free subgroup of $SO_n(\mathbb{R})$ ($n \geq 6$) can contain an element having -1 as an eigenvalue. This gap did not affect the proof of Corollary 6.5, because this eigenvalue does not appear among the specific locally commutative groups of Corollary 6.3.

Note that, for n odd, the antipodal map has determinant $+1$ and so lies in $SO_{n+1}(\mathbb{R})$. Hence the technique of the preceding proof, which uses the antipodal map and Theorem 5.16, yields another proof of the part of Corollary 6.5 that deals with n odd, $n \geq 3$. This approach does not require the fixed-point free groups of Theorem 6.4 but uses only the simpler locally commutative groups of Corollary 6.3.

Corollary 6.5 gives a complete answer to the question of m-divisibility of spheres for finite m. But there is a simpler approach that yields a more constructive solution, at least for spheres of odd dimension. For now we consider only finite m, but in the next chapter we shall consider infinite m, showing that any $\mathbb{S}^n$ (except $\mathbb{S}^1$) has a subset that is, simultaneously, a half, a third, ..., a $2^{\aleph_0}$th part of $\mathbb{S}^n$ (see Cor. 7.11 and remarks following).

Theorem 6.6. *Assume m is an integer and $m \geq 2$.*

(a) *If n is odd, then $\mathbb{S}^n$ is m-divisible with respect to $SO_{n+1}(\mathbb{R})$.*

(b) *(AC) If n is even, then $\mathbb{S}^n$ is m-divisible with respect to $SO_{n+1}(\mathbb{R})$ if and only if $m \geq 3$.*

(c) *(AC) For any n, $\mathbb{S}^n$ is m-divisible with respect to $O_{n+1}(\mathbb{R})$.*

Proof. First, observe that the m-divisibility of $\mathbb{S}^n$ can be deduced from that of $\mathbb{S}^j$ and $\mathbb{S}^k$ if $j+1+k+1 = n+1$. For we may partition $\mathbb{S}^n$ by considering a point $P = (x_1, \ldots, x_{n+1}) \in \mathbb{S}^n$ and seeing how the m-division of $\mathbb{S}^k$ treats $(x_1, \ldots, x_{j+1})/\|(x_1, \ldots, x_{j+1})\|$; if $x_1 = \cdots = x_{j+1} = 0$, consider instead $(x_{j+2}, \ldots, x_{j+2+k})/\|(x_{j+2}, \ldots, x_{j+2+k})\|$, which is in $\mathbb{S}^k$. Now, for (a) simply write $n+1$ as $2j$ and use the preceding approach on the j pairs of coordinates, considering a pair as a point on $\mathbb{S}^1$, which is m-divisible for all m. For (b) write $n+1$ as $2j+3$ and use the m-divisibility of $\mathbb{S}^1$ and $\mathbb{S}^2$; recall that the latter uses the Axiom of Choice (see Cor. 5.14). The failure of 2-divisibility arises because every rotation in $SO_{n+1}(\mathbb{R})$ has a fixed point when $n+1$ is odd. Because the antipodal map, which yields 2-divisibility, is in $O_{n+1}(\mathbb{R})$, (c) follows from (a) and (b). □

Because orthogonal maps all fix the origin, the detailed study of such maps leaves unanswered the question of solving congruences by partitions of $\mathbb{R}^n$ and using the isometry group, G_n. For instance, it was shown following Theorem 5.7 how to get a five-piece paradoxical decomposition of $\mathbb{R}^3$ from the four-piece one of $\mathbb{S}^2$ (and the same exists for $\mathbb{R}^n$ if $n \geq 3$ by the four-piece decomposition of the corresponding sphere that exists by Corollary 6.5). But, in fact, four-piece decompositions are possible, because it follows from Theorem 6.7 that any proper system of congruences can be solved by a partition of $\mathbb{R}^n$, provided $n \geq 3$. Note that the particular system asserting that $\mathbb{R}^n$ is m-divisible is trivial to solve for all m, n (see remarks following Def. 5.13).

Theorem 6.7. *There are two independent isometries in G_3 such that the group they generate has no nontrivial fixed points in $\mathbb{R}^3$. The same is true for G_n and $\mathbb{R}^n$ if $n \geq 3$.*

Proof. Let ϕ and ψ be the rotations in $SO_3(\mathbb{R})$ given by

$$\begin{bmatrix} \cos\theta & -\sin\theta & 0 \\ \sin\theta & \cos\theta & 0 \\ 0 & 0 & 1 \end{bmatrix} \text{ and } \begin{bmatrix} 1 & 0 & 0 \\ 0 & \cos\theta & -\sin\theta \\ 0 & \sin\theta & \cos\theta \end{bmatrix},$$

respectively, where the common rotation angle θ is chosen so that $\cos\theta$ is transcendental. Letting $T_{\vec{v}}$ denote the translation of $\mathbb{R}^3$ by the vector $\vec{v}$, and $\vec{i}$, $\vec{k}$ the vectors $(1, 0, 0)$, $(0, 0, 1)$, respectively, define σ to be $T_{\vec{k}}\phi$ and τ to be $T_{\vec{i}}\psi$; σ and τ will be the independent isometries we seek. These two isometries are each screw motions (or glide rotations) of $\mathbb{R}^3$, that is, a rotation followed by a translation in the direction of the rotation's axis. Note that if the translation component of a screw motion is not the identity, then the motion has no fixed points.

We will need to know that the matrix of a word $\phi^{n_1}\psi^{m_1}\cdots\phi^{n_s}\psi^{n_s}$ has the form

$$\begin{bmatrix} P\cos\theta & -\operatorname{sgn}(n_1)\sin\theta & -\operatorname{sgn}(m_1 n_s)\, Q\cos\theta \\ P\sin\theta & Q\cos\theta & -\operatorname{sgn}(m_s)\, Q\sin\theta \\ P\cos\theta & P\sin\theta & P\cos\theta \end{bmatrix},$$

where Q stands for a polynomial in $\cos\theta$ (possibly different in each entry) of degree $d = \Sigma - 1$ ($\Sigma = |n_1| + |m_1| + |n_2| + |m_s|$) with leading coefficient $2^{\Sigma-2s}$, and P represents (possibly different) polynomials in $\cos\theta$ of degree strictly less than d. We omit the details as this may be easily proved by induction in exactly the same manner as the analogous fact proved in Theorem 6.2. (It follows from this representation that ϕ and ψ are independent; see Thm. 2.2.)

Now, it must be shown that no nontrivial word in $\sigma^{\pm 1}$, $\tau^{\pm 1}$ has any fixed points in $\mathbb{R}^3$; the independence of σ, τ follows. Because the existence of a fixed point is invariant under conjugation and inversion, it follows as in the proof of Theorem 6.2 that we need only consider pure powers of σ or τ and words of the form $w = \sigma^{n_1}\tau^{m_1}\cdots\sigma^{n_s}\tau^{n_s}$, with each exponent a nonzero integer. Because $\sigma^n = T_{n\vec{k}}\,\phi^n$ and $\tau^m = T_{m\vec{i}}\,\psi^m$, the pure powers of σ and τ have no fixed points.

It follows easily from the fact that rotations are linear transformations that for any rotation ρ, $\rho T_{\vec{v}} = T_{\rho(\vec{v})}\rho$, and hence $w = T_{\vec{t}}\,\hat{w}$, where $\hat{w}$ is the result of replacing σ, τ in w by ϕ, ψ, respectively, and $\vec{t}$ will be computed in a moment. For $1 \le r \le s$, let $\hat{w}_r$ represent the left segment of $\hat{w}$: $\phi^{n_1}\psi^{m_1}\cdots\phi^{n_r}\psi^{n_r}$. Then using the rule for moving translations left across a rotation, and using the fact that $\psi(\vec{i}) = \vec{i}$, it is easy to see that

$$\vec{t} = n_1\vec{k} + n_2\hat{w}_1(\vec{k}) + \cdots + n_s\hat{w}_{s-1}(\vec{k}) + m_1\hat{w}_1(\vec{i}) + m_2\hat{w}_2(\vec{i}) + \cdots + m_s\hat{w}_s(\vec{i}).$$

Let $\vec{a}$ be a unit vector in the direction of the axis of $\hat{w}$, oriented so that the rotation obeys the right-hand rule with respect to $\vec{a}$. Now, $\vec{t}$ is not necessarily in the same direction as $\vec{a}$, but as long as $\vec{t}$ is not perpendicular to $\hat{w}$'s axis, $w = T_{\vec{t}}\,\hat{w}$ will have no fixed points. Thus the proof will be complete once it is shown that $\vec{t}\cdot\vec{a} \ne 0$; in fact, this condition is equivalent to the nonexistence of a fixed point of a screw motion.

Consider the expansion of $\vec{t}\cdot\vec{a}$ using the formula for $\vec{t}$ just given. There are two sorts of terms in this expansion. The first sort is $n_j\hat{w}_{j-1}(\vec{k})\cdot\vec{a}$. By the invariance of inner product under orthogonal transformations (see App. A), each such term equals $n_j\vec{k}\cdot\hat{w}_{j-1}^{-1}(\vec{a})$. Because the rotation of ρ_1's axis by ρ_2 (ρ_i are rotations) yields the axis of $\rho_2\rho_1\rho_2^{-1}$, this last expression is the dot product of $n_j\vec{k}$ with the appropriately oriented unit vector along the axis of the rotation $\hat{w}_{j-1}^{-1}\hat{w}\hat{w}_{j-1} = \sigma^{n_j}\tau^{m_j}\cdots\tau^{m_s}\sigma^{n_1}\cdots\sigma^{n_{j-1}}\tau^{m_{j-1}}$. Recall that if (A_{ij}) represents a rotation of $\mathbb{R}^3$ with unit axis vector $\vec{b}$ and rotation angle ξ, then $2\vec{b}\sin\xi = (A_{32} - A_{23}, A_{13} - A_{31}, A_{21} - A_{12})$ (see App. A). Also proved in

Appendix A is the fact that the angle of a rotation is the same as the angle of a conjugate of that rotation by another rotation. Hence, if ξ is the angle of rotation of $\hat{w}$, then the term of $\vec{t}\cdot\vec{a}$ being considered equals $n_j(A_{21}-A_{12})/(2\sin\xi)$, where (A_{ij}) is the matrix of $\hat{w}_{j-1}^{-1}\hat{w}\hat{w}_{j-1}$. Using the representation of words in ϕ, ψ given at the beginning of the proof, this equals $(n_j/(2\sin\xi))(P\sin\theta+\operatorname{sgn}(n_j)Q\sin\theta))$. A similar analysis of the other sort of term in $\vec{t}\cdot\vec{a}$, $m_j\hat{w}_j(\vec{i})\cdot\vec{a}$, yields a contribution having the form $(m_j/(2\sin\xi))(P\sin\theta+\operatorname{sgn}(m_j)Q\sin\theta))$. It follows that $(\sin\xi/\sin\theta)(\vec{t}\cdot\vec{a})$ is a polynomial in $\cos\theta$ with leading coefficient $2^{d-2s}(|n_1|+|m_1|+\cdots+|m_s|)$. Therefore, because of the transcendence of $\cos\theta$, $\vec{t}\cdot\vec{a}$ cannot vanish.

Extending the isometries σ and τ to higher dimensions by simply fixing the additional coordinates yields the desired independent isometries of $\mathbb{R}^n$ for any $n\geq 3$. □

Corollary 6.8 (AC). *If $n\geq 3$, then any proper system of m congruences may be solved by a partition of $\mathbb{R}^n$, where isometries are used to realize the congruences. In particular, for $n\geq 3$, $\mathbb{R}^n$ is paradoxical using four pieces.*

Proof. Because a free group of rank 2 contains one of rank m (see remarks following Cor. 5.12), this corollary is an immediate consequence of the preceding theorem and Corollary 5.12. Note that none of the machinery for dealing with weak systems is needed, only the proof of the first part of Theorem 5.11, which is relatively straightforward. □

6.2 Non-Euclidean Spaces

The theory so far has been applied only to spheres and Euclidean spaces and their respective isometry (or rotation) groups, and also to the hyperbolic plane in Chapter 4. Because the matrix computations of this chapter can be applied to elliptic spaces, we can look also at elliptic geometry. If we take $\mathbb{RP}^n$ (real projective n-space: $\mathbb{S}^n$ with antipodal points identified) as our model of $\mathbb{L}^n$, elliptic n-space, with the distance from P to Q defined to be the smaller of the two spherical distances P to Q and $-P$ to Q (equivalently, $\cos^{-1}|P\cdot Q|$), then the isometry group is just the group of rotations of $\mathbb{S}^n$, viewed as acting on $\mathbb{RP}^n$. Because the only orthogonal transformation that collapses to the identity when viewed as acting on $\mathbb{RP}^n$ is the antipodal map, the isometry group of $\mathbb{L}^n$ coincides with $SO_{n+1}(\mathbb{R})$ if n is even and is $SO_{n+1}(\mathbb{R})/\{\pm I\}$ if n is odd. It follows that two independent rotations ϕ, ρ of $\mathbb{S}^n$ remain independent when viewed as rotations of $\mathbb{L}^n$. Moreover, local commutativity (or lack of nontrivial fixed points) of the group generated by ϕ, ρ is preserved, for if u and w both fix P in $\mathbb{L}^n$, then $u(P)=\pm P$ and $w(P)=\pm P$, so u^2 and w^2 share a fixed point on $\mathbb{S}^n$. Hence u^2 and w^2 commute, and this implies the same for u and w (see [MKS66, p. 41]).

Combining these remarks with Theorem 6.4, we see that Corollary 6.5 holds in elliptic space. For each $n \geq 2$, all weak systems of congruences are satisfiable by a partition of $\mathbb{L}^n$, and for odd $n \geq 3$, all proper systems are satisfiable. Note that the antipodal map on $\mathbb{S}^n$ becomes the identity in $\mathbb{L}^n$, so the method of proof of the last part of Corollary 6.5 cannot be used; rather, we use Theorem 6.4's free groups without fixed points directly.

The hyperbolic plane was discussed in depth in Chapter 4, and the results there easily extend to higher dimensions. The work of that chapter allows us to satisfy any collection of congruences because of Corollary 5.12, Proposition 4.4, and the proof method for Proposition 4.5 (which is what allows the next result to hold for Borel sets and without using the Axiom of Choice).

Theorem 6.9. *For any proper system of congruences involving r set-variables, there is a partition of $\mathbb{H}^2$ into Borel sets $A_1, \ldots, A_r$ that satisfy (using isometries) the given system. In particular, $\mathbb{H}^2$ is paradoxical using Borel sets. Moreover, there is a Hausdorff decomposition of $\mathbb{H}^2$ using Borel sets; that is, there is a Borel set in $\mathbb{H}^2$ that is, simultaneously, a half and a third of $\mathbb{H}^2$. These results all hold in higher-dimensional hyperbolic space as well.*

Proof. Only the assertion about $\mathbb{H}^n$ remains to be proved. Using the upper half-space $\{(x_1, \ldots, x_{n-1}, t) : t, x_j \in \mathbb{R}, t > 0\}$ as a model for $\mathbb{H}^n$, an isometry of $\mathbb{H}^2$ may be extended to $\mathbb{H}^n$ by letting it act on (x_1, t), leaving the other coordinates fixed. Moreover, any Borel subset A of $\mathbb{H}^2$ extends to one, A^*, of $\mathbb{H}^n$ by putting a point in A^* if and only if (x_1, t) is in A. It follows that a partition of $\mathbb{H}^2$ that satisfies a given system of congruences induces one of $\mathbb{H}^n$ satisfying the same system. Note that it also follows that the isometry group of $\mathbb{H}^n$ contains a free subgroup of rank 2 that acts without nontrivial fixed points. □

The hyperbolic case differs from the spherical and Euclidean cases in another way. We will see (following Cor. 7.6) that for $\mathbb{S}^n$ ($n \geq 3$, n odd) and $\mathbb{R}^3$, the set of independent pairs of isometries that generate a free group without nontrivial fixed points is dense in $SO_{n+1}^2(\mathbb{R})$ and G_3^2, respectively. But in $\mathbb{H}^2$, this is false. The elliptics, which correspond to matrices whose trace is strictly between -2 and 2, form an open subset of $PSL_2(\mathbb{R})$, and therefore the set of independent pairs generating a fixed-point free group is certainly disjoint from the open set of pairs whose first coordinate is elliptic. Therefore the set of pairs is not dense. As we shall see in Corollary 7.9, $\mathbb{H}^3$ does have this denseness property.

A summary of results about free groups and paradoxical decompositions in spheres and in Euclidean and non-Euclidean spaces appears in Table 6.1. In Chapter 7 these results will be extended still further, to free groups of larger rank and to infinite systems of congruences. With the exception of the hyperbolic plane, these generalizations will resemble closely the results just proved about finite systems of congruences.

Table 6.1. *Decompositions and free groups in various spaces*

X / *G*	Spheres $\mathbb{S}^n$ Orientation-preserving Isometries $SO_{n+1}(\mathbb{R})$	Spheres $\mathbb{S}^n$ Isometries $O_{n+1}(\mathbb{R})$	Elliptic spaces $\mathbb{L}^n$ Isometries	Hyperbolic spaces $\mathbb{H}^n$ Isometries
X is paradoxical using *G* and four pieces ⇔ *G* has a free locally commutative subgroup of rank 2 ⇔ all weak countable systems of congruences are solvable	2, 3, 4,...	2, 3, 4,...	2, 3, 4,...	2, 3, 4,...
G has a free subgroup of rank 2 without fixed points in *X*	3, 5, 7,...	3, 5, 7,...	3, 5, 7,...	2, 3, 4,... (3, 4, 5,...)
All countable proper systems of congruences are solvable	3, 5, 7,...	2, 3, 4,...	3, 5, 7,...	2, 3, 4,...

Note: The table is complete in that the nonappearance of a certain dimension means that the phenomenon is known not to exist. The results of the table hold if "countable" is replaced by "continuum" and "rank 2" by "rank $2^{\aleph_0}$" in all cases except one instance of $\mathbb{H}^2$ (the values in parentheses indicate the situation in this case).

6.3 Tetrahedral Chains

To conclude this chapter, we discuss a geometric problem raised by Steinhaus that has nothing to do with systems of congruences, but its solution uses free groups and the sort of matrix computations that we have been considering. We shall show that the four reflections in the faces of a regular tetrahedron in $\mathbb{R}^3$ are free generators of $\mathbb{Z}_2 * \mathbb{Z}_2 * \mathbb{Z}_2 * \mathbb{Z}_2$; that is, the reflections satisfy no relation except those derivable from the fact that each has order 2. This is false for an equilateral triangle in the plane, because $\phi\psi\phi = \psi\phi\psi$ for two reflections ϕ, ψ in the triangle's sides, but is true for higher-dimensional simplices.

This property of tetrahedral reflections can be used to solve the following geometric problem. Let us call a sequence of regular tetrahedra in $\mathbb{R}^3$ a *Steinhaus chain* if two consecutive tetrahedra share exactly one face, and each tetrahedron is distinct from its predecessor's predecessor. In 1956, Steinhaus asked; Can the last tetrahedron in a Steinhaus chain equal the first one? The answer is clearly yes if cubes are used in place of tetrahedra, but it is also possible using octahedra or dodecahedra or icosahedra [EW15]. A negative answer to the problem for tetrahedra is related to a result on free groups, which we now prove. The problem was originally solved by Świerczkowski [Swi58]; the approach here is due to Dekker [Dek59a].

Theorem 6.10. (a) *Let ϕ_1, ϕ_2, ϕ_3, ϕ_4 be the four reflections in the faces of a regular tetrahedron in $\mathbb{R}^3$. Then no word of the form $\phi_{i_1}\phi_{i_2}\phi_{i_3}\cdots\phi_{i_s}$, where $s \geq 1$ and adjacent terms are distinct, equals the identity; that is, the ϕ_i are generators of $\mathbb{Z}_2 * \mathbb{Z}_2 * \mathbb{Z}_2 * \mathbb{Z}_2$. This same result, with $n+1$ copies of*

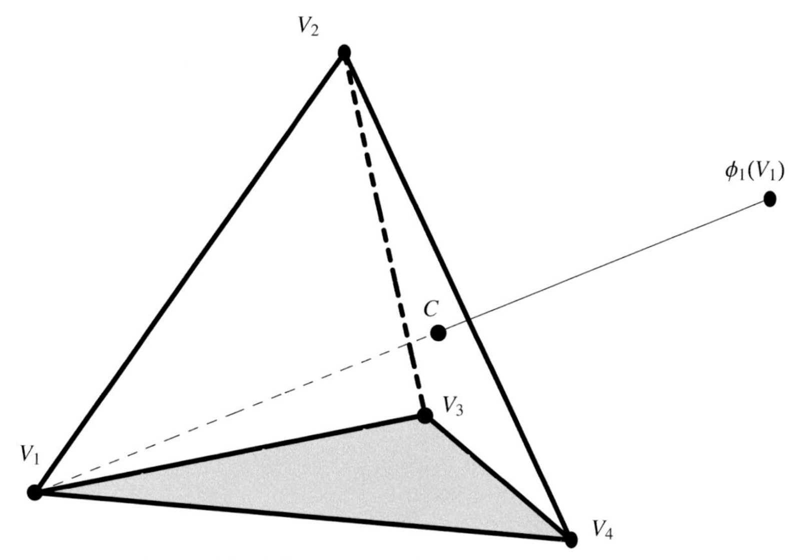

Figure 6.1. Reflection in a side of a regular tetrahedron.

$\mathbb{Z}_2$, *holds for the* $n+1$ *reflections in the faces of a regular n-dimensional simplex, provided* $n \geq 3$.

(b) *The last tetrahedron in any Steinhaus chain cannot equal the first.*

Proof. We prove (b) first, using the notation of (a). Any point in $\mathbb{R}^3$ may be represented uniquely as $x_1V_1 + x_2V_2 + x_3V_3 + x_4V_4$, where the V_j are the vertices of a regular tetrahedron T and $\sum x_i = 1$; these are barycentric coordinates with respect to T. Each ϕ_i may be represented by a 4×4 matrix acting on barycentric coordinates, where the columns of the matrix are the vectors $\phi_i(V_i)$. Then composition corresponds to matrix multiplication.

Because the reflection ϕ_i in the face $x_i = 0$ sends V_i to $C + (C - V_i) = 2C - V_i = 2(\frac{1}{3}\Sigma_{j \neq i} V_j) - V_i$, where C is the centroid of the face opposite V_i (see Fig. 6.1), the matrices for the reflections are

$$M_1 = \begin{bmatrix} -1 & 0 & 0 & 0 \\ \frac{2}{3} & 1 & 0 & 0 \\ \frac{2}{3} & 0 & 1 & 0 \\ \frac{2}{3} & 0 & 0 & 1 \end{bmatrix}, \quad M_2 = \begin{bmatrix} 1 & \frac{2}{3} & 0 & 0 \\ 0 & -1 & 0 & 0 \\ 0 & \frac{2}{3} & 1 & 0 \\ 0 & \frac{2}{3} & 0 & 1 \end{bmatrix},$$

$$M_3 = \begin{bmatrix} 1 & 0 & \frac{2}{3} & 0 \\ 0 & 1 & \frac{2}{3} & 0 \\ 0 & 0 & -1 & 0 \\ 0 & 0 & \frac{2}{3} & 1 \end{bmatrix}, \quad M_4 = \begin{bmatrix} 1 & 0 & 0 & \frac{2}{3} \\ 0 & 1 & 0 & \frac{2}{3} \\ 0 & 0 & 1 & \frac{2}{3} \\ 0 & 0 & 0 & -1 \end{bmatrix}.$$

Figure 6.2. A tetratorus using 174 tetrahedra; if the side length is taken to be 1 cm, then the gap at the end is about the diameter of a proton.

Now suppose the last tetrahedron agrees with the first. Then there is a sequence $\phi_{i_1}, \ldots, \phi_{i_s}$ with no consecutive pair the same so that $M_1 M_{i_2} \cdots M_{i_s}$ is a permutation matrix. We can and will assume that $i_1 = 1$. The next claim shows that one cannot get a 0-1 matrix from such a word. □

Claim. Consider the matrix product $M_1 M_{i_2} \cdots M_{i_s}$ with $2/3$ replaced by x. The polynomials in the second row have x-degree less than s, except for the one in position $(2, i_s)$, which has degree s. And they all have leading coefficient $+1$.

Proof. By induction; it is clear for $s = 1$. Consider what happens when the matrix of a word that ends in M_j, assumed to have the claimed form, is multiplied on the right by M_n, where $n \neq j$. The multiplications by x preserve the claimed property, as the degree rises from n to $n + 1$ in position $(2, n)$ and does not rise at all in the other spots. And the leading coefficient's sign is affected only by the x multipliers.

Now look at the polynomial in the $(2, i_s)$ position; suppose it is $x^s + a_1 x^{s-1} + \cdots + a_s$, where $a_i \in \mathbb{Z}$. Setting $x = 2/3$ and taking a common denominator yields $(2^s + 3a_1 2^{s-1} + \cdots + 3^{s-1} a_s)/3^s$, the numerator of which is not divisible by 3; the fraction is therefore not 0 or 1, as required. The proof for higher dimensions is identical, with $2/3$ replaced by $2/n$.

(a) The result follows from the preceding proof because the identity is one of the permutation matrices excluded by the proof. □

A natural question, something that Świerczkowski also wondered about [Swi07, last chapter], is how small the gap at the end of a Steinhaus chain having no self-intersections can be; such a chain is called *embedded*. Intensive searching by M. Elgersma and S. Wagon led to many examples where the gap at the end is very small. See [EW15] for a complete discusson of the methods. The example of Figure 6.2 has 174 tetrahedra; if the edge length is 1 cm, then the gap at the end is less than $1.4 \cdot 10^{-13}$ cm, about the diameter of a proton. The example was obtained by multiplying the matrices M_i together, where the indices i are defined by the 58-length string $S4\overline{S}p(S4\overline{S})$, where $S = 12342342321423$, $\overline{S}$ is the reversal of S, and p is the 3-cycle permutation (134). The product K of those 58 matrices has eigenvalues $(1, 1, z, \overline{z})$, where z is very close to $e^{2i\pi/3}$, and so K^3 is close to the identity matrix. In fact, it is within about 10^{-13} of the identity matrix and so leads to the nearly closed chain in Figure 6.2. For example, the (1, 1)th entry of the product is the exact rational

$$\frac{12900700781701218214112847414009168077632947067294063741160716994000469754230533339}{12900700781701026662481960358450703949334417416449930858101164413445974926422638 49}.$$

A complication is that the chain corresponding to the word just described has a collision at the end, as opposed to a gap, and so is not embedded. But if one shifts the string 17 characters to the left, then the resulting chain is embedded, with a gap at the end of size about $1.3 \cdot 10^{-13}$. Further work by Elgersma and Wagon led to a chain of 540 tetrahedra for which the gap at the end has size about $6 \cdot 10^{-18}$. The many examples found in [EW15] provide strong evidence that the following conjecture is true.

Conjecture 6.11. For any positive ϵ, there is an embedded Steinhaus chain such that the last tetrahedron is within ϵ of coinciding with the first.

Notes

The first extensive investigation into the situation for higher-dimensional and non-Euclidean spaces was made by Dekker [Dek56b, Dek57, Dek58a], who discovered the free subgroup of $SO_4(\mathbb{R})$ given in Theorem 6.2 and derived Corollary 6.3. He also derived Corollary 6.5 for all cases except $SO_5(\mathbb{R})$ and provided the application to elliptic spaces at the start of §6.2. Dekker [Dek57] also raised the question whether free non-Abelian subgroups of G_n, $n \geq 3$, without nontrivial fixed points in $\mathbb{R}^n$ exist. An affirmative solution was announced by Mycielski [Myc56], but this was premature, and the independent isometries of Theorem 6.7 were discovered shortly thereafter, independently, by Dekker [Dek58b] and by Mycielski and Świerczkowski [MS58]. The two rotations of $\mathbb{R}^3$ at the beginning of Theorem 6.7's proof were first considered by de Groot [Gro56], who proved their independence.

Dekker [Dek56b] conjectured that the cases he could not settle were similar to the ones he could, and this was confirmed, albeit twenty-five years later, by Deligne and Sullivan [DS83], who constructed free non-Abelian groups without fixed points in $SO_n(\mathbb{R})$ for all even $n \geq 4$. And Borel [Bor83] confirmed that

$SO_5(\mathbb{R})$ is not an exception as far as locally commutative free groups are concerned by providing such a free group of rank 2 in $SO_5(\mathbb{R})$. More generally, Borel deals with actions of Lie groups, and he obtains results relating the size of the set of fixed points to the Euler characteristic.

The problem on chains of tetrahedra is due to H. Steinhaus. It was originally solved by Świerczkowski. Dekker [Dek59a] simplified Świerczkowski's approach by proving Theorem 6.10 and deriving from it the solution to the problem in all dimensions greater than or equal to three. These results were later rediscovered by Mason [Mas72]. Elgersma and Wagon [EW15], using different patterns to search for legal chains, found the examples in §6.3 having very small error.

7

Free Groups of Large Rank: Getting a Continuum of Spheres from One

7.1 Large Free Groups of Isometries

The Banach–Tarski Paradox shows how to obtain two spheres, or balls, from one, and it is clear how to get any finite number of balls: Just duplicate repeatedly, lifting the subsets of the new balls back to the original. After all, the joy in owning a duplicating machine is being able to use it more than once. Alternatively, one need only consider the weak system of congruences:

$$A_{2j} \cong \bigcup\{A_i : 1 \leq i \leq 2n, i \neq 2j-1\}, \quad j = 1, \ldots, n.$$

By the remarks following Corollary 5.12, there is a partition of the sphere into sets A_i that satisfy this system with respect to rotations, and therefore $A_{2j} \cup A_{2j-1} \sim \mathbb{S}^2$. Need we stop at just finitely many copies? What about infinitely many, even uncountably many? Using the existence of infinitely many independent rotations $\sigma^i \tau^i$, $i = 0, 1, 2, \ldots$, where σ, τ are independent, it is not hard to see how the results of Chapter 5 on systems of congruences can be made to yield the solvability of any countably infinite weak system by a partition of $\mathbb{S}^2$. Hence $\mathbb{S}^2$ can be partitioned into countably many sets, each of which is $SO_3(\mathbb{R})$-equidecomposable (using just two pieces) with $\mathbb{S}^2$. But even stronger transfinite duplications are possible. One can get a continuum of spheres from one: The sphere can be partitioned into sets B_α, as many sets as there are points on the sphere (i.e., $2^{\aleph_0}$), such that each B_α is $SO_3(\mathbb{R})$-equidecomposable with the sphere.

Although transfinite duplications do not add to our knowledge of finitely additive measures, these results do lead to a deeper understanding of free groups of rotations. Moreover, as in previous chapters, the techniques necessary for transfinite duplications lead to interesting geometrical applications of a different sort.

The first step is to generalize the results of Chapter 5 on general actions of free groups. Because those results do not depend on the finiteness of the system of congruences, this generalization is quite straightforward.

Theorem 7.1 (AC). *Let κ be any cardinal, and suppose F, a group acting on an infinite set X, is free of rank κ, with free generators σ_α, $\alpha < k$. Let*

$\bigcup\{A_\beta : \beta \in L_\alpha\} \cong \bigcup\{A_\beta : \beta \in R_\alpha\}$, $\alpha < \kappa$, *be any proper system of* κ *congruences involving at most* $|X|$ *sets* A_β. *If* F *has no nontrivial fixed points on* X, *then* X *may be partitioned into sets* A_β *satisfying the system, with* σ_α *witnessing the* α*th congruence. If the system is weak, then a solution exists provided only that* F*'s action on* X *is locally commutative.*

Proof. The technique by which Theorem 5.5 was derived from Theorem 5.2, and Corollary 5.12 from Theorem 5.11, makes no use of the finiteness of the system; thus it suffices to show that the generalization of Theorem 5.11 is valid. The proof that F can be partitioned to satisfy the system can be done exactly as in Theorem 5.11, that is, by induction on the length of words in F. To put e and some given word w in the same piece, assuming the system is weak, just consider the group F_0 generated by the finitely many σ_α that appear in w. By Theorem 5.11, F_0 can be partitioned to satisfy the congruences to be witnessed by its generators (the finiteness of r in Thm. 5.11 is unimportant), and e and w will be in the same piece. Then words u in $F \setminus F_0$ can be assigned inductively so that all congruences are satisfied, starting the induction with the longest terminal string of u that is in F_0. □

It should be pointed out that in the proof of Theorem 7.1, as in the proofs of the corresponding results of Chapter 5, there is no guarantee that each of the sets of the constructed partition is nonempty. Indeed, the proof never really uses the fact that there are no more than $|X|$ sets A_β, and if there were more, some would have to be empty. This is not a serious problem though, because most of the particular systems of interest imply that the sets are nonempty (for example, the system asserting m-divisibility or the one asserting the existence of a paradoxical decomposition using four pieces). Moreover, if one weakens the conclusion of Theorem 7.1 slightly by not insisting that the congruences be witnessed by the specific generators σ_α, then a partition can be found that solves the system and contains only nonempty sets. This can be proved by using the local commutativity of F to find a free subgroup, F_0, of the same rank such that there are $|X|$ many F_0-orbits in X, each consisting of nonfixed points of $F_0 \setminus \{e\}$ (for details, see [Dek56b, Thm. 2.9]).

There is a partial converse to Theorem 7.1 in the style of Theorem 5.8, which showed that if one particular weak system is solvable, then the group must have a locally commutative free subgroup of rank 2. The proof of the following theorem is essentially identical to that of Theorem 5.8, with σ_α, $\alpha < \kappa$, defined just as σ, τ are, and replacing σ, τ in the proof.

Theorem 7.2. *If* G *acts on* X *and* X *splits into* κ ($\kappa \geq 2$) *sets, each of which is* G*-equidecomposable with* X *using only two pieces, then* G *has a free subgroup of rank* κ *that is locally commutative on* X.

The rotation groups $SO_n(\mathbb{R})$ all have size $2^{\aleph_0}$, so a free group of rotations cannot have greater rank. In fact, locally commutative free groups of this maximum possible rank exist in $SO_n(\mathbb{R})$ if $n \geq 3$, and combined with Theorem 7.1, this fact will yield the desired generalizations of our earlier results on systems of congruences to continuum-sized systems. In Theorem 7.4(a) we shall construct very explicitly a set of $2^{\aleph_0}$ independent rotations of $\mathbb{S}^2$. Then we shall introduce a different approach, more general but also less direct, that yields large locally commutative free subgroups of $SO_n(\mathbb{R})$ for all $n \geq 3$. We have seen how transcendental numbers can be used to get two independent rotations of $\mathbb{S}^2$ (see first part of Theorem 6.7's proof); to construct $2^{\aleph_0}$ independent rotations, we need a large set of algebraically independent numbers. Recall that a set X of reals (or complex numbers) is *algebraically independent* if $P(x_1, \ldots, x_n) \neq 0$ for any $x_1, \ldots x_n \in X$ and nontrivial polynomial P with rational coefficients.

Theorem 7.3. *There is a set* $\{\upsilon_t : t \in (0, \infty)\}$ *of algebraically independent numbers.*

Proof. If M is an infinite set of real numbers, then the set of reals that are algebraic over M, that is, the set of all $x \in \mathbb{R}$ for which $P(x_1, \ldots, x_n, x) = 0$ for some nontrivial rational polynomial P and $x_i \in M$, has the same size as M. This follows from the countability of the number of possibilities for P. Hence an algebraically independent set that is smaller than the continuum can always be extended to a larger algebraically independent set. This allows a continuum-sized set of algebraically independent reals to be constructed by transfinite induction, although this assumes the existence of a well-ordering of the reals. Alternatively, it follows that a maximal algebraically independent set of reals (which exists by Zorn's Lemma) has cardinality $2^{\aleph_0}$.

These approaches use the Axiom of Choice and, while it is true that that axiom will be used in the application of this result to paradoxical decompositions, it is noteworthy that the desired set can be constructed in a much more effective manner that does not require Choice. Letting $\lfloor x \rfloor$ denote the greatest integer less than or equal to X, define, for any real $t > 0$,

$$\upsilon_t = \sum_{n=1}^{\infty} 2^{(2^{\lfloor tn \rfloor} - 2^{n^2})}.$$

The series is dominated by a geometric series with ratio $1/2$ once $n > t$, so υ_t is a well-defined real; the numbers υ_t are sometimes called *von Neumann numbers*, after their discoverer. To get some feeling for υ_t, consider its binary expansion: When n is large enough, the 2^{n^2} term dominates the exponent, and so the tail of υ_t's binary expansion has its 1s spaced very far apart. This is enough to guarantee, in the same way that the transcendence of the Liouville number, $\sum_{n=1}^{\infty} 10^{-n!}$, is proved (see [Niv67]), that each υ_t is transcendental. Moreover, if $s < t$, then as can easily be seen by choosing two rationals in (s, t), $\lfloor tn \rfloor - \lfloor sn \rfloor \to \infty$ as $n \to \infty$; hence the nth 1 in υ_s's expansion occurs exponentially sooner than the nth 1 in υ_t's expansion. This implies that υ_t is much more closely approximable by rationals than any v_s, $s < t$, and as proved by von Neumann [Neu61], this yields the algebraic independence of the v_t. □

In [Myc64], Mycielski proved a general theorem, without the Axiom of Choice, about relations in separable metric spaces; this yields an alternative, more abstract, proof of the previous result of von Neumann, which is simpler in that the number theory details are avoided. Mycielski's theorem will be discussed further in Theorem 7.5. One application of von Neumann numbers is to answer the question, Can one prove the existence of an uncountable, proper subfield of $\mathbb{R}$ without using the Axiom of Choice? The subfield generated by $\{v_t : 0 < t < 1\}$ is such a field. We now show how algebraically independent numbers can be used to construct a continuum of independent rotations.

Theorem 7.4. (a) *For any $n \geq 3$, there are independent rotations ρ_t, $0 < t < 1$, in $SO_n(\mathbb{R})$ such that the free group they generate is locally commutative on $\mathbb{S}^{n-1}$. If n is even, then the group has no nontrivial fixed points on $\mathbb{S}^{n-1}$.*

(b) *If $n \geq 3$, there are independent isometries ρ_t, $0 < t < 1$, of $\mathbb{R}^n$ such that the subgroup of G_n that they generate has no nontrivial fixed points in $\mathbb{R}^n$.*

(c) *For any $n \geq 2$, there are independent isometries ρ_t, $0 < t < 1$, of hyperbolic n-space, $\mathbb{H}^n$, that generate a locally commutative group.*

Proof. (a) We shall use von Neumann numbers to prove (a) when $n = 3$. Although this approach works for larger n, a different proof for the general case will be given following Theorem 7.5. Let v_t, $0 < t \leq 1$, be the algebraically independent von Neumann numbers of Theorem 7.3. Because $SO_3(\mathbb{R})$ is locally commutative, it is sufficient to construct $2^{\aleph_0}$ independent rotations.

To do this, let $\theta_t = 2 \arctan v_t$; then $\sin \theta_t = 2v_t/(1 + v_t^2)$ and $\cos \theta_t = (1 - v_t^2)/(1 + v_t^2)$. For any θ, let $\sigma(\theta)$, $\tau(\theta)$ denote the two independent rotations in $SO_3(\mathbb{R})$ as defined at the beginning of Theorem 6.7's proof. Now, the desired independent rotations ρ_t may be defined by $\rho_t = \sigma(\theta_t)\tau(\theta_1)\sigma(\theta_t)^{-1}$. To prove the independence of the ρ_t, suppose $w = \rho_{t_1}^{m_1} \cdots \rho_{t_s}^{m_s} = e$. Each entry of w's matrix representation, a_{ij}, is a rational polynomial in $\sin \theta_1$, $\cos \theta_1$ and those $\sin \theta_t$, $\cos \theta_t$ occurring in w, that is, with t equal to one of the t_k. We want to show that at least one a_{ij} or $a_{ii} - 1$ is not the zero polynomial for arbitrary values of t. Suppose otherwise; then the matrix would remain the identity when each of these θ_t is replaced by $k\theta_1$, k being the least index such that $t_k = t$. Because this substitution transforms each $\rho_{t_k}^{m_k}$ in w to $\sigma(k\theta_1)\tau(\theta_1)^{m_k}\sigma(k\theta_1)^{-1}$, which equals $\sigma(\theta_1)^k\tau(\theta_1)^{m_k}\sigma(\theta_1)^{-k}$, w is transformed to a nontrivial word in $\sigma(\theta_1)^{\pm 1}$, $\tau(\theta_1)^{\pm 1}$ that, by the choice of σ, τ, cannot equal the identity. Hence at least one entry a_{ij} of w is a rational polynomial in $2v_1/(1 + v_1^2)$, $(1 - v_1^2)/(1 + v_1^2)$, $2v_t/(1 + v_t^2)$, and $(1 - v_t^2)/(1 + v_t^2)$, where $t \in \{t_1, \ldots t_s\}$, that does not vanish (or equal 1, if $i = j$) for arbitrary values of the v_t. It follows that when the equation $a_{ij} = 0$ (or 1) is turned into $P = 0$ by taking a common denominator, P is a nontrivial polynomial in finitely many of the v_t, contradicting the algebraic independence of the von Neumann numbers.

(b) The proof of this case can be derived from Theorem 6.7 in a way similar to part (a), that is, using von Neumann numbers and conjugation. We omit the details because this will be proved following Theorem 7.5 in a more abstract, but overall simpler, way.

(c) It suffices to consider $\mathbb{H}^2$, for the natural embedding of $\mathbb{H}^2$ into $\mathbb{H}^n$ (see Thm. 6.9) induces an embedding of isometry groups that never adds fixed points and hence preserves local commutativity. Moreover, because the action of $PSL_2(\mathbb{R})$ on the upper half-plane is locally commutative (Prop. 4.1), it suffices to produce a continuum of independent (orientation-preserving) isometries. The construction is simpler if we change the model of the hyperbolic plane from the upper half-plane in $\mathbb{C}$ to points on the upper sheet ($z > 0$) of the hyperboloid $x^2 + y^2 - z^2 = -1$ in $\mathbb{R}^3$. This model has as its metric $d(P, Q) = \operatorname{arccosh}(-g(P, Q))$, where g denotes the inner product $p_1q_1 + p_2q_2 - p_3q_3$ and isometries may be represented by nonsingular 3×3 matrices A such that

$$A^T \begin{bmatrix} 1 & 0 & 0 \\ 0 & 1 & 0 \\ 0 & 0 & -1 \end{bmatrix} A = \begin{bmatrix} 1 & 0 & 0 \\ 0 & 1 & 0 \\ 0 & 0 & -1 \end{bmatrix},$$

and A sends the upper sheet of the hyperboloid to itself rather than to the lower sheet (see [Bea83, §3.7]). Now it may be proved essentially in the same way as in Theorem 6.2 that ϕ and ψ are independent, orientation-preserving isometries of $\mathbb{H}^2$, where

$$\phi = \begin{bmatrix} 1 & 0 & 0 \\ 0 & \cosh\theta & \sinh\theta \\ 0 & \sinh\theta & \cosh\theta \end{bmatrix}, \quad \psi = \begin{bmatrix} \cosh\theta & 0 & \sinh\theta \\ 0 & 1 & 0 \\ \sinh\theta & 0 & \cosh\theta \end{bmatrix},$$

and $\cosh\theta$ is transcendental [Dek57]. We may now use exactly the same technique as in part (a) (i.e., von Neumann numbers and conjugation), using the hyperbolic trigonometric functions and their identities to define a continuum of matrices that correspond to isometries and are independent. □

We now outline an entirely different approach to the construction of large free groups that is more modern and less computational. This approach can be applied to all the cases of Theorem 7.4. The key is the following theorem of Mycielski.

Theorem 7.5. *Let X be a complete, separable metric space with no isolated points, and let $\mathcal{R} = \{R_i : i < \infty\}$ be a set of relations on X, that is, each $R_i \subseteq X^{m_i}$ for some positive integer m_i. Suppose further that each R_i is a meager subset of X^{m_i}. Then there is a subset F of X such that $|F| = 2^{\aleph_0}$ (in fact, F is perfect, i.e., closed and without isolated points), and for each sequence of m_i distinct elements $x_1, \ldots, x_{m_i}$ of F, $x_1, \ldots, x_{m_i} \notin R_i$.*

Proof. Because each R_i is a union of countably many nowhere dense sets, we may reindex and so assume that each R_i is nowhere dense. We shall construct a tree of nonempty open subsets of X having the following form and satisfying items (1)–(4):

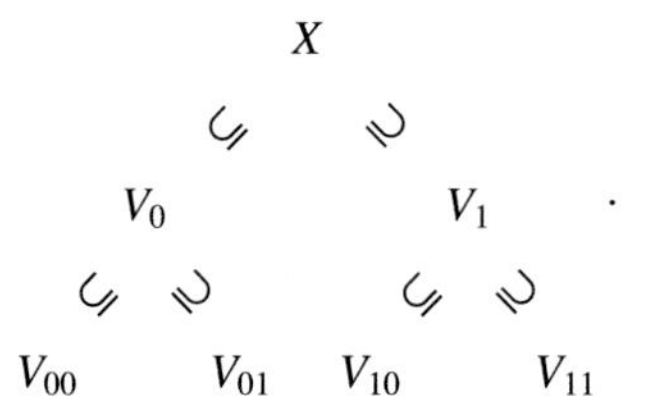

1. $\overline{V}_{si} \subseteq V_s$ for each sequence s and $i = 0, 1$.
2. $V_{s0} \cap V_{s1} = \varnothing$ for each sequence s.
3. If $\ell(s)$ is the length of the sequence s, then the diameter of V_s is at most $1/\ell(s)$.
4. If $s_1, \ldots, s_{m_j}$ are m_j distinct sequences of the same length r, where $r \geq j$, then $V_{s_1} \times V_{s_2} \times \cdots V_{s_{m_j}} \cap R_j = \varnothing$.

The existence of such a tree yields the desired set F as follows. For each infinite binary sequence b, completeness of X implies that $\bigcap\{\overline{V}_s : s \subseteq b\}$ is nonempty and condition (3) yields that this intersection must contain but a single point x_b. Let $F = \{x_b : b \text{ is an infinite binary sequence}\}$; F is perfect and has size $2^{\aleph_0}$. Any m_j-tuple of distinct elements of F comes from distinct nodes at a level past the jth, whence condition (4) implies that the m_j-tuple is not in R_j.

To construct the tree, use induction on levels. Suppose V_s is defined for all s of length n. Because X has no isolated points, V_s contains at least two points, which can be separated by open balls V'_{s0} and V'_{s1}, chosen small enough so that conditions (1)–(3) are satisfied. We shall thin these two balls repeatedly to satisfy (4). Consider each $j \leq n+1$ in turn. If $m_j > 2^{n+1}$, do nothing; (4) is vacuously satisfied with respect to R_j. Otherwise, consider each of the m_j-sized subsets of the 2^{n+1} sequences of length $n+1$ in turn. If A is such a subset, then $\prod\{V'_{si} : si \in A\}$ is open and so must contain an m_j-tuple not in $\overline{R_j}$. There must be an open set U containing this m_j-tuple that remains disjoint from $\overline{R}_j$, and so we may replace each V'_{si} by a smaller nonempty open set so that $\prod\{V'_{si} : si \in A\} \subseteq U$. This guarantees that (4) is satisfied with respect to R_j and A, the set of distinct sequences. Because we may repeat the thinning process to take care of all sets A and all $j \leq n+1$, this yields the next level of the tree as desired. □

The preceding theorem is valid for all complete metric spaces, but this assumes the Axiom of Choice. For separable spaces, AC is avoided because there is a countable basis of open sets that can be well-ordered; hence the choices in the proof of Theorem 7.5 can be made by using this well-ordering and choosing the first basic open set that works. This theorem can be strengthened to show that most (i.e., a comeager set in an appropriate topology) perfect subsets of X satisfy the conclusion (see [Myc73]). Note that Theorem 7.5 yields an alternative proof of Theorem 7.3. For each nonzero polynomial $p(x_1, \ldots, x_n)$ with integer coefficients, let R_p be the subset of $\mathbb{R}^n$ consisting of all zeros of P. Then R_p is closed and contains no nonempty open set (otherwise P is identically zero), so R_p is nowhere dense. Because there are only countably many polynomials with integer coefficients, we may apply Theorem 7.5 to the collection of all the sets R_p to get an algebraically independent set of size $2^{\aleph_0}$.

Now, $SO_n(\mathbb{R})$ is a complete, separable metric space, but to apply Theorem 7.5, we must view $SO_n(\mathbb{R})$ as a real analytic manifold (of dimension $n(n-1)/2$). In fact, $SO_n(\mathbb{R})$ is an analytic submanifold of $\mathbb{R}^{n^2}$ under the embedding induced by considering the entries of the matrix corresponding to an element of $SO_n(\mathbb{R})$. Furthermore, the product $SO_n^m(\mathbb{R})$ is an analytic submanifold of $\mathbb{R}^{mn^2}$. Hence the mn^2 real-valued functions on $SO_n^m(\mathbb{R})$, each of which gives the ijth entry in the kth rotation, are analytic. Moreover, $SO_n(\mathbb{R})$ is connected (and path-connected).

Now, let n be even and consider the problem of getting large free groups in $SO_n(\mathbb{R})$ without fixed points (as in Thm. 6.4(a)). Let w denote a reduced group word ($w \neq e$) in m variables $x_1, \ldots, x_m$. Define $R_w \subseteq SO_n^m(\mathbb{R})$ to consist of those m-tuples $(\sigma_1, \ldots, \sigma_m)$ such that $w\,(\sigma_1, \ldots, \sigma_m)$ has 1 as an eigenvalue (i.e., has a fixed point on $\mathbb{S}^{n-1}$), and let $\mathcal{R}$ be the (countable) collection of all such sets R_w. It remains only to show that each R_w is nowhere dense; for then Theorem 7.5 may be applied to $\mathcal{R}$ to obtain a set F, which clearly will be a set of free generators for a subgroup of $SO_n(\mathbb{R})$ without nontrivial fixed points. We need the following claim:

Claim. For any word w as in the preceding paragraph, there is an analytic $f: SO_n^m(\mathbb{R}) \to \mathbb{R}$ such that $R_w = f^{-1}(\{0\})$.

Proof. First we observe that $(\sigma_1, \ldots, \sigma_m) \in R_w$ if and only if $\det(w(\sigma_1, \ldots, \sigma_m) - I) = 0$. Because the inverse of a matrix in $SO_n(\mathbb{R})$ equals its transpose, $w\,(\sigma_1, \ldots, \sigma_m)$ may be expressed as n^2 polynomials (with integer coefficients) in the mn^2 entries of the σ_i. Because the determinant is also a polynomial function in the entries of a matrix, the vanishing of the determinant is equivalent to the vanishing of a single polynomial in the mn^2 entries of the σ_i. Finally, because the functions giving the entries are analytic, it follows that R_w is the zero-set of a single analytic function on $SO_n^m(\mathbb{R})$. □

This claim immediately yields that R_w is closed, so all that remains is to prove that R_w has empty interior. But it is an easy consequence of the connectedness of $SO_n^m(\mathbb{R})$ that if an analytic function vanishes on a nonempty open set, then it vanishes on all of $SO_n^m(\mathbb{R})$. Hence if R_w contains a nonempty open set, R_w must equal $SO_n^m(\mathbb{R})$. But this is a contradiction to Theorem 6.4, which yields independent σ, ρ such that no word in σ, ρ has a fixed point on $\mathbb{S}^{n-1}$. It follows that $(\rho, \sigma\,\rho\,\sigma^{-1}, \sigma^2\,\rho\,\sigma^{-2}, \ldots, \sigma^{m-1}\,\rho\,\sigma^{-(m-1)})$ lies in $SO_n^m(\mathbb{R}) \setminus R_w$. (Alternatively, one can avoid Thm. 6.4 and obtain a contradiction by using Thm. 1 of [Bor83]; see [MW84].)

The case of odd n, where large locally commutative free groups are sought, can be handled in two ways. First observe that the preceding technique easily solves the $SO_3(\mathbb{R})$ case. Because $SO_3(\mathbb{R})$ is locally commutative on $\mathbb{S}^2$, all that is needed is a perfect set of free generators, and the technique yields such a set if R_w is defined to be the set of m-tuples $(\sigma_1, \ldots, \sigma_m) \in SO_n^m(\mathbb{R})$ such that $w\,(\sigma_1, \ldots, \sigma_m)$ is the identity. Now, one can obtain the higher-dimensional result by appealing to [Bor83, p. 162], where it is shown that $SO_3(\mathbb{R})$ may be represented as a subgroup H of $SO_{n+1}(\mathbb{R})$ $(n \geq 2)$ where H's action on $\mathbb{S}^n$ is locally commutative. The result then follows from the existence of the perfect set in $SO_3(\mathbb{R})$.

Alternatively, one can imitate the proof for even n. Let R_w consist of those m-tuples $(\sigma_1, \ldots, \sigma_m) \in SO_n^m(\mathbb{R})$ such that $w\,(\sigma_1, \ldots, \sigma_m)$ is the identity. Now, let u, v be any two reduced group words in variables $x_1, \ldots, x_r$ that do not commute as abstract words. Then let $R_{u,v}$ consist of all r-tuples $(\sigma_1, \ldots, \sigma_r)$ such that $u\,(\sigma_1, \ldots, \sigma_r)$ and $v\,(\sigma_1, \ldots, \sigma_r)$ have a common fixed point on $\mathbb{S}^{n-1}$. Let $\mathcal{R}$ consist of all these sets R_w and $R_{u,v}$. Then $\mathcal{R}$ is countable, and so as in the previous case, it remains to show that each R_w and $R_{u,v}$ is nowhere dense. For then Theorem 7.5 yields a subset F of $SO_n(\mathbb{R})$ that avoids R_w and $R_{u,v}$. The avoidance of R_w means that F is a set of free generators. Hence noncommuting words in F correspond to abstract noncommuting words u, v used to define $R_{u,v}$. Because F avoids $R_{u,v}$, this means that two noncommuting rotations in the group generated by F cannot share a fixed point; that is, this group is locally commutative. By Theorem 6.4, locally commutative free groups of finite rank exist, and it follows that each R_w and $R_{u,v}$ is not equal to all of $SO_n^m(\mathbb{R})$. Therefore, as in the previous case, the fact that these sets are nowhere dense is an immediate consequence of the following claim and the fact that an analytic function that vanishes on an open set vanishes everywhere.

Claim. The sets R_w and $R_{u,v}$ previously defined are of the form $f^{-1}(\{0\})$ for some analytic F.

Proof. The condition $w(\sigma_1, \ldots, \sigma_m) = 1$ is equivalent to the simultaneous vanishing of n^2 polynomials in the mn^2 entries of the σ_i. This condition is equivalent to the vanishing of a single polynomial by using the trick of summing squares: $\Sigma \mathrm{p}_i^2 = 0$ if and only if each $P_i = 0$. The claim for R_w then follows as in the previous case. Finally, membership of $(\sigma_1, \ldots, \sigma_m)$ in $R_{u,v}$ is equivalent to the existence of a nonzero solution to a homogeneous system of $2n$ equations in n unknowns; the system is the one expressing the fact that u and v have a common eigenvector for the eigenvalue 1. Now, such a system has a nontrivial solution if and only if the determinant of the coefficients in each subsystem of size n vanishes (easy linear algebra exercise). Hence membership in $R_{u,v}$ is equivalent to the simultaneous vanishing of $\binom{2n}{n}$ polynomials in the entries. As before, this is equivalent to the vanishing of a single polynomial, which proves the claim. □

The general technique just discussed can also be used to prove Theorem 7.4(b). We need to work in SG_3, the connected subgroup of G_3 consisting of those isometries whose linear part has determinant $+1$. An element of SG_3 may be viewed as an element of $SL_4(\mathbb{R})$ as follows. If (a_{ij}) represents the linear part of $\sigma \in SG_3$

and (v_1, v_2, v_3) is the vector corresponding to the translational part, then σ may be identified with

$$\begin{bmatrix} a_{11} & a_{12} & a_{13} & v_1 \\ a_{21} & a_{22} & a_{23} & v_2 \\ a_{31} & a_{32} & a_{33} & v_3 \\ 0 & 0 & 0 & 1 \end{bmatrix}.$$

Hence SG_3 may be viewed as an analytic submanifold of $\mathbb{R}^{12}$ and a proof along the same lines as for spheres can be carried out. The key point is that an element σ of SG_3 has a fixed point if and only if a certain polynomial in the entries of the matrix vanishes. To see that such a polynomial exists, recall that σ has a fixed point if and only if the translation vector is perpendicular to the axis of the rotation given by (a_{ij}). But it is shown in Appendix A (Thm. A.6) that the rotation axis of (a_{ij}) is parallel to $(a_{32} - a_{23}, a_{13} - a_{31}, a_{21} - a_{12})$. Hence σ has a fixed point if and only if

$$v_1(a_{32} - a_{23}) + v_2(a_{13} - a_{31}) + v_3(a_{21} - a_{12}) = 0.$$

All the groups of Theorem 7.4 have rank $2^{\aleph_0}$, and using the Axiom of Choice, they all have free subgroups of any smaller rank. Hence, by applying Theorem 7.1, we get the following corollary. The elliptic spaces are handled by viewing the groups of 7.4(a) as groups of isometries of $\mathbb{L}^n$, as in §6.2. Part (a) of Corollary 7.6 for the pair $(\mathbb{S}^n, O_{n+1}(\mathbb{R}))$ is proved by combining the locally commutative group of Theorem 7.4(a) with the availability of the antipodal map as in Theorem 5.16 and Corollary 6.5 That technique requires that no element of the group sends a point P to $-P$; that is, no element has -1 as an eigenvalue. Because this condition is satisfiable for groups of finite rank (see the proof of Cor. 6.5), it may be built into the proof of Theorem 7.4(a) that uses Theorem 7.5. Simply add to $\mathcal{R}$ the relations R'_w, consisting of m-tuples such that $w(\sigma_1, \ldots, \sigma_m)$ has -1 as an eigenvalue. Because membership in R'_w may be expressed as a polynomial, the crucial claim remains valid.

Corollary 7.6 (AC). (a) *Let (X, G) be any of the following pairs: $(\mathbb{S}^n, SO_{n+1}(\mathbb{R}))$, where $n \geq 3$ and n is odd; $(\mathbb{S}^n, O_{n+1}(\mathbb{R}))$, where $n \geq 2$; $(\mathbb{R}^n, G_n)$, where $n \geq 3$; $\mathbb{L}^n$ and its isometry group, where $n \geq 2$ and n is odd. Then any proper system of at most $2^{\aleph_0}$ congruences involving at most $2^{\aleph_0}$ sets is solvable, using G, by a partition of X.*

(b) *The same is true when restricted to weak systems in the following cases: $(\mathbb{S}^n, O_{n+1}(\mathbb{R}))$, $\mathbb{L}^n$ and its isometry group, and $\mathbb{H}^n$ and its isometry group, where $n \geq 2$.*

(c) *In all the cases of (a) and (b), X may be partitioned into $\{A_\alpha : \alpha < 2^{\aleph_0}\}$ such that $A_\alpha \sim_G X$ using two pieces.*

For spheres and Euclidean, hyperbolic, and elliptic spaces, this corollary shows that in all cases where the set or space is paradoxical, one can get a continuum of copies of the set from one.

A consequence of the analysis of the sets R_w in the preceding approach to large free groups is finer information about free groups of finite rank. Consider the case of $SO_n(\mathbb{R})$, $n \geq 4$ and n even. Theorem 6.4 shows that free non-Abelian subgroups without nontrivial fixed points exist, but in fact the set of pairs that generate such subgroups is dense in the space of all possible pairs. This is because each R_w (see comments after claim following Thm 7.5), where w is a word in two variables, is nowhere dense. Hence the union of these R_w is meager in $SO_n^2(\mathbb{R})$ and the complement of this union is comeager and, by the Baire Category Theorem, dense. Similarly, the fact that R_w and $R_{u,v}$ are nowhere dense yields that for any $n \geq 3$, the set of pairs (σ, τ) in $SO_n^2(\mathbb{R})$ such that σ and τ freely generate a locally commutative subgroup of $SO_n(\mathbb{R})$ is comeager. So, while the specific examples of independent pairs of isometries constructed by Hausdorff et al. are necessary to establish these results, we see that such pairs are quite abundant.

While the remarks of the previous paragraph are valid in $\mathbb{R}^3$ as well, the situation in $\mathbb{R}^4$ and beyond is less clear. The proof of Theorem 7.4(b) uses the fact that the existence of a fixed point of an element of SG_3 is a polynomial condition; it is not clear that this is true for SG_4. Thus Theorem 7.4(b) is valid for all $n \geq 3$, but it is not known that the set of pairs from G_4 that freely generate a group without nontrivial fixed points is comeager. Moreover, the technique used by Borel [Bor83] does not apply, because his results are for semisimple groups.

The hyperbolic case provides an interesting counterpoint as far as large free groups without nontrivial fixed points are concerned, as well as for the related question on the density of independent pairs. Returning to the upper half-plane model of hyperbolic 2-space, $\mathbb{H}^2$, recall that the orientation-preserving isometries correspond to linear fractional transformations that we identify with $PSL_2(\mathbb{R})$ and that may be classified as parabolic, hyperbolic, or elliptic according as the trace is < 2, $= 2$, or > 2; the latter are the only ones with fixed points in $\mathbb{H}^2$. Because matrices A and $-A$ are identified in $PSL_2(\mathbb{R})$, we can ignore negative traces. Another useful classification is the discrete/nondiscrete classification of subgroups of $PSL_2(\mathbb{R})$. A subgroup G of $PSL_2(\mathbb{R})$ is discrete if the collection of matrices corresponding to elements of G contains no convergent sequence of distinct matrices. An easy fact [Leh66, p. 12] is that a cyclic subgroup of $PSL_2(\mathbb{R})$ is discrete if and only if it contains no elliptic element of infinite order. Also, note that a discrete subgroup must be countable.

If a free subgroup of $PSL_2(\mathbb{R})$ is discrete, it can have no elliptic elements, for such an element would be of infinite order and so would generate a nondiscrete subgroup. It is essentially this fact that was used in §4.1 to show that any free subgroup of $PSL_2(\mathbb{Z})$ has no elliptic elements. It follows from work of Siegel [Sie50] that the converse of this fact holds too.

Theorem 7.7. *A free subgroup of $PSL_2(\mathbb{R})$ is discrete if and only if it has no elliptic elements. Hence any group of isometries of $\mathbb{H}^2$ generated by an uncountable set of independent elements contains an isometry fixing a point in $\mathbb{H}^2$.*

Proof. The forward direction was already proved and the reverse direction for cyclic free groups is a consequence of the fact about cyclic subgroups of $PSL_2(\mathbb{R})$ previously stated. We prove the remaining case, noncyclic free groups, guided by some algebraic computation (we use *Mathematica*). Suppose G is nondiscrete, has no elliptic elements, and has independent elements σ and τ. We use $[x, y]$ for the commutator of two group elements. In this proof, the norm of a matrix is the standard induced norm: $||A|| = \max\{||A \cdot x|| : ||x|| = 1\}$; we will use the well-known inequality (for the 2×2 case)

$$||A||_{\text{max}} \leq ||A||_{\text{Euclidean}} \leq \sqrt{2}\,||A||,$$

where the first norm is the maximum absolute value of the entries and the second is the Euclidean length of the 4-vector. We also use the fact that $||A|| = ||A^{-1}||$ when $\det A = 1$; this holds because (a) the norm is the largest singular value of A (the square root of the largest eigenvalue of $A^T A$) and (b) the eigenvalues of a 2×2 matrix of determinant 1 are λ and $1/\lambda$. Note also that $||\sigma|| \geq 1$ because $|\sigma_{11} + \sigma_{22}| \geq 2$ so that $\sigma \cdot (1, 0)$ or $\sigma \cdot (0, 1)$ is not inside the unit circle. Because G is nondiscrete, there are elements in G that are arbitrarily close to the identity matrix.

Claim 1. For any $\epsilon > 0$, G has two independent elements ρ and ϕ that are conjugate and within ϵ of the identity.

Proof of claim. Choose ρ within $\epsilon||\sigma||^{-2}||\tau||^{-2}$ of the identity. Conjugate at most once to get $\rho = \sigma^i \ldots \sigma^j$ where $i, j \in \mathbb{Z}$; this leaves ρ within $\epsilon \big/ ||\tau||^2$ of the identity, because

$$\frac{\|\sigma\rho\sigma^{-1}x\|}{\|x\|} = \frac{\|\sigma\rho\sigma^{-1}x\|}{\|\rho\sigma^{-1}x\|}\,\frac{\|\rho\sigma^{-1}x\|}{\|\sigma^{-1}x\|}\,\frac{\|\sigma^{-1}x\|}{\|x\|} \leq \|\sigma\|\|\rho\|\|\sigma^{-1}\| = \|\sigma\|^2\|\rho\|.$$

Then let $\phi = \tau\rho\tau^{-1}$, which is within ϵ of the identity. Then ρ and ϕ are independent because there is no nontrivial cancellation when words in $\rho^{\pm 1}$ and $\phi^{\pm 1}$ are formed. □

Claim 2. For any $\epsilon > 0$, G has a hyperbolic element within ϵ of the identity.

Proof of claim. Assume $\epsilon < 1$ and take ρ and ϕ from Claim 1 using $\epsilon/\sqrt{2}$. Assume both are parabolic, because otherwise either satisfies the claim. Then $\rho\,\phi$ is hyperbolic. If not, then $\text{tr}(\rho) = \text{tr}(\rho\,\phi) = 2$. Conjugate ρ to $\rho^* = \left[\begin{smallmatrix} 1 & b \\ 0 & 1 \end{smallmatrix}\right]$ (with $b \neq 0$) and conjugate all of G the same way. Then $\phi^* = \left[\begin{smallmatrix} a & d \\ c & 2-a \end{smallmatrix}\right]$. The trace of $\rho\,\phi$ (same as $\text{tr}(\rho^*\phi^*)$) is then $2 + bc$, and so we have $c = 0$, which means $a = 1$ and $\phi\,\rho = \rho\,\phi$, a contradiction. Because the induced norm is submultiplicative $(||A||\;||B|| \leq ||AB||)$, $\rho\,\phi$ satisfies the ϵ condition.

To finish, use Claim 2 to get a hyperbolic u close enough to the identity so that the largest eigenvalue is $1 + \epsilon$, where $0 < \epsilon < 1/7$; the explicit eigenvalue formula shows that being within $1/115$ of the identity in each coordinate suffices. Diagonalize u, inverting if necessary, to get $\left[\begin{smallmatrix} 1+\epsilon & 0 \\ 0 & (1+\epsilon)^{-1} \end{smallmatrix}\right]$ (which we still call u), and similarly conjugate all of G so that we now work in G^*, the conjugated group, which has all the same properties as G. Choose $\rho \in G^*$ as in Claim 1, with each entry within 1 of the identity matrix entry, but also so that ρ has no zero elements. This can be done by using Claim 1 to get ρ and ϕ within $1/2$ of the identity in each coordinate and observing that one of ρ, ϕ, $\rho\phi$ has the desired no-zero property; the last will be within 1 of the identity. The details are an exercise, a key point being that if ρ and ϕ are both upper triangular, then $[[\rho, \phi], [\rho, \phi^{-1}]]$ is the identity, contradicting independence.

So we have $\rho_{11}\rho_{22} \neq 1$. Let $w = \rho u \rho^{-1}$ and $\delta = w_{11} - 1$. The goal is to show that one of γ, ζ is elliptic, where $\gamma = [u, w]$ and $\zeta = [u, \gamma]$. To prove that neither is parabolic, we will need two conditions: $\delta \neq \epsilon$ and $\delta \neq -\epsilon/(1+\epsilon)$. It turns out that $\delta - \epsilon = (\rho_{11}\rho_{22} - 1)\epsilon(\epsilon + 2)/(\epsilon + 1)$, which is nonzero. If the second condition fails, redefine w using ρ^2 instead of ρ. This changes w_{11} by the nonzero quantity $\epsilon(\epsilon + 2)(\rho_{11}\rho_{22} - 1)(\rho_{11}^2 + \rho_{22}^2 + 1)$ and so changes δ. The two bounds, $1/7$ on ϵ and 1 on $||\rho - I||_{\max}$, imply that $|\delta| < 1$. The two inequations for δ show that $\mathrm{tr}(\gamma) \neq 2$ and, using also $\delta < 1$, that $\mathrm{tr}(\zeta) \neq 2$. The proof then concludes by showing that the ratio $(\mathrm{tr}(\zeta) - 2)/(\mathrm{tr}(\gamma) - 2)$ must be negative. The product of this ratio with $(1+\delta)^{-1}(1+\epsilon)^3(\epsilon\,(2+\epsilon))^{-2}$ is $-(1-\delta)(\epsilon+1) - \epsilon^2$, which is negative when $\epsilon > 0$ and $\delta < 1$. Thus one of the two commutators is the elliptic we seek. □

If M is a set of independent isometries of $\mathbb{H}^2$, then so is $\{\sigma^2 : \sigma \in M\}$, and each σ^2 is orientation preserving. This reduces the second assertion to a statement about $PSL_2(\mathbb{R})$, which follows from the first part of the theorem because an uncountable subgroup of $PSL_2(\mathbb{R})$ is necessarily nondiscrete. □

With more work one can improve Theorem 7.7: A nonsolvable, nondiscrete subgroup of $SL_2(\mathbb{R})$ is dense in $SL_2(\mathbb{R})$. This result yields (the reverse direction of) Theorem 7.7 as a corollary because the elliptics ($|\mathrm{trace}| < 2$) form a nonempty open subset of $SL_2(\mathbb{R})$ and because a free group of rank 2 is not solvable. The stronger theorem, which was pointed out to the authors by A. Borel and D. Sullivan, is proved as follows. Let F be the subgroup in question, let $\overline{F}$ denote the closure of F in $SL_2(\mathbb{R})$, and let H be the component of the identity in $\overline{F}$. Assume, to get a contradiction, that $\overline{F} \neq SL_2(\mathbb{R})$. Because F is nondiscrete, $H \neq \{e\}$, whence H is a one- or two-dimensional Lie subgroup of $SL_2(\mathbb{R})$. Using Lie algebras, it can be shown that H is conjugate to either $SO_2(\mathbb{R})$ or a subgroup of the group of upper triangular matrices. In either case, it is easy to check that the normalizer of H in $SL_2(\mathbb{R})$ is solvable, contradicting the fact that the nonsolvable group $\overline{F}$ normalizes H.

Theorem 7.7 is in direct contrast to the cases of Theorem 7.4(a) and (b), where uncountable free groups without fixed points are constructed for other spaces. The general technique, using nowhere dense sets and analytic functions, used to deduce Theorem 7.4(a) and (b) from the corresponding results about free groups of rank 2 might lead one to expect that whenever there is a pair of independent elements generating a group without fixed points, there is in fact continuum many such elements. But this is false in the case of $\mathbb{H}^2$. Where does the general proof break down? The elliptic transformations in $PSL_2(\mathbb{R})$ form an open set (trace$^2 < 4$) and are therefore not nowhere dense. Unlike the spherical and Euclidean cases, the (orientation-preserving) isometries of $\mathbb{H}^2$ with a fixed point are not the zero set of an analytic function. The statement that $\left[\begin{smallmatrix} a & b \\ c & d \end{smallmatrix}\right]$ corresponds to an isometry with a fixed point is expressed by a polynomial inequality ($(a+d)^2 < 4$), not a polynomial equality, as happens in $SO_n(\mathbb{R})$ and in the group of orientation-preserving isometries of $\mathbb{R}^3$. As pointed out following Theorem 6.9, this difference is also reflected in the fact that the set of pairs from $PSL_2(\mathbb{R})$ that generate a free group without nontrivial fixed points in $\mathbb{H}^2$ is not dense.

Concerning higher-dimensional hyperbolic spaces, it is easy to see that in the isometry group of $\mathbb{H}^n$, $n \geq 4$, there do exist free groups of continuum rank without nontrivial fixed points in $\mathbb{H}^n$. This is because the action of the Euclidean isometry group G_{n-1} on $\mathbb{R}^{n-1}$ can be mimicked in $\mathbb{H}^n$. If $\sigma \in G_{n-1}$, then send the point $(x_1, \ldots, x_{n-1}, t)$ of $\mathbb{H}^n$ (using the upper half-space model) to $(\sigma(x_1, \ldots, x_{n-1}), t)$. This yields an isometry of $\mathbb{H}^n$ that, assuming σ fixes no point of $\mathbb{R}^{n-1}$, has no fixed point. Hence the assertion about $\mathbb{H}^n$, $n \geq 4$, follows from Theorem 7.4(b), the corresponding result about $\mathbb{R}^n$, $n \geq 3$.

The case of $\mathbb{H}^3$, however, requires a more subtle approach. The action of $PSL_2(\mathbb{C})$ on $\mathbb{C} \cup \{\infty\}$ as linear fractional transformations can be extended to $\mathbb{R}^3 \cup \{\infty\}$ by identifying $\mathbb{C} \cup \{\infty\}$ with $\mathbb{R}^2 \cup \{\infty\}$ and using some geometry related to the fact that each linear fractional transformation is a Möbius transformation of $\mathbb{R}^2 \cup \{\infty\}$. Such an extension preserves $\mathbb{H}^3$ and corresponds to the group of orientation-preserving isometries of $\mathbb{H}^3$. Alternatively, one can use an approach based on quaternions; see [Bea83] for both approaches. It turns out that the orientation-preserving isometries of $\mathbb{H}^3$ with a fixed point correspond (except for the identity) to matrices whose trace is real and lies in the interval $(-2, 2)$. Such elements of $PSL_2(\mathbb{C})$ are called *elliptic*.

Theorem 7.8. *There is a free subgroup of $PSL_2(\mathbb{C})$ of rank $2^{\aleph_0}$ that contains no elliptic element. Hence there is a rank $2^{\aleph_0}$ free group of isometries of $\mathbb{H}^3$ whose action on $\mathbb{H}^3$ is without nontrivial fixed points.*

Proof. There is no loss in working in $SL_2(\mathbb{C})$ rather than $PSL_2(\mathbb{C})$. For a nonidentity reduced word w, let

$$R_w = \{(\sigma_1, \ldots, \sigma_m) \in SL_2(\mathbb{C})^m : w(\sigma_1, \ldots, \sigma_m) \text{ is elliptic}\}.$$

As in the previous proofs using Theorem 7.5, it is sufficient to prove that R_w is nowhere dense. The complication is that R_w is not the zero-set of a polynomial function of the $8m$ real numbers giving the entries of the σ_i. Rather, if $a_1, \ldots, a_{8m}$ are these reals, then $(\sigma_1, \ldots, \sigma_m) \in R_w$ if and only if $p(a_1, \ldots, a_{8m}) = 0$ and

$q(a_1, \ldots, a_{8m}) < 4$, where the polynomial p is an expression for the imaginary part of the trace, while q equals the square of the real part of the trace.

Now, suppose R_w fails to be nowhere dense. Then the closed set $\{(\sigma_1, \ldots, \sigma_m) : p(a_1, \ldots, a_{8m}) = 0\}$, which contains R_w, must contain a nonempty open set. Because $SL_2(\mathbb{C})^m$ is a connected, real analytic submanifold of $\mathbb{R}^{8m}$, this means that p must be identically zero on $SL_2(\mathbb{C})^m$. But this means that the trace of $w(\sigma_1, \ldots, \sigma_m)$ is real for all choices of $\sigma_1, \ldots, \sigma_m$, which leads to a contradiction, as follows.

We shall need the fact, due to Magnus and Neumann (see Thm. 8.1(a)), that the two matrices $\rho = \left[\begin{smallmatrix} 1 & 1 \\ 1 & 2 \end{smallmatrix}\right]$ and $\tau = \left[\begin{smallmatrix} 5 & 2 \\ 2 & 1 \end{smallmatrix}\right]$ are free generators of a subgroup of $SL_2(\mathbb{R})$ that, except for the identity, consists only of hyperbolic elements; in other words, the absolute value of the trace of any nontrivial word in ρ, τ is greater than 2. Returning to the problem at hand, define for each $z \in \mathbb{C}$

$$\rho_z = \frac{1}{1+z-z^2}\begin{bmatrix} 1 & z \\ z & 1+z \end{bmatrix} \quad \text{and} \quad \tau_z = \frac{1}{1+4z-4z^2}\begin{bmatrix} 1+4z & 2z \\ 2z & 1 \end{bmatrix}.$$

Then define the complex function

$$f(z) = \operatorname{tr}(w(\rho_z, \tau_z\rho_z\tau^{-1}, \ldots, \tau_z^{m-1}\rho_z\tau^{-(m-1)})).$$

Then $f(z)$ is a rational function, whose denominator has four zeros. Let Ω be a region containing 0 and 1 but excluding these four zeros. Then f is analytic in Ω, so by the Open Mapping Theorem, $f(\Omega)$ either is just a single point or contains a nonempty open set. If the trace of $w(\sigma_1, \ldots, \sigma_m)$ is always real, only the first possibility can hold. But this is the desired contradiction because $f(0) = \operatorname{trace}(I) = 2$, while $f(1)$ is the trace of a reduced nonidentity word in ρ and τ, whence $f(1) \neq 2$. □

Corollary 7.9. *The set of pairs of isometries of $\mathbb{H}^3$ that generate a free group of elliptic elements is dense in the isometry group $PSL_2^2(\mathbb{C})$.*

Proof. The preceding proof shows that each R_w is nowhere dense. And the argument following Corollary 7.6 then shows that the set of pairs in the assertion is comeager. □

Tomkowicz [Tom∞] generalized Theorem 5.17 and applied this generalization to $\mathbb{H}^2$, obtaining the following result.

Theorem 7.10 (AC). *In $\mathbb{H}^2$, any proper system of at most $2^{\aleph_0}$ congruences using at most $2^{\aleph_0}$ sets is solvable.*

His proof uses Theorem 7.5 and the aforementioned generalization of Theorem 5.17 to get a free locally commutative subgroup of orientation-preserving hyperbolic isometries with some additional properties. Table 6.1 summarizes this sort of result about large free groups.

Corollary 7.6(b) yields the κ-divisibility of $\mathbb{S}^2$ with respect to $SO_3(\mathbb{R})$ whenever $3 \leq \kappa \leq 2^{\aleph_0}$, and this can be combined with splittings of $\mathbb{S}^1$ to obtain the

κ-divisibility of higher-dimensional spheres, without having to use free locally commutative groups in the higher-dimensional rotation groups.

Corollary 7.11 (AC). *For any cardinal κ satisfying $3 \le \kappa \le 2^{\aleph_0}$ and any $n \ge 1$, $\mathbb{S}^n$ is κ-divisible with respect to $SO_{n+1}(\mathbb{R})$.*

Proof. For $\mathbb{S}^2$, this follows by applying Corollary 7.6(b) to the weak system $A_0 \cong A_\alpha, \alpha < \kappa$. Once the case of $\mathbb{S}^1$ is handled, the divisibility of $\mathbb{S}^n$, $n > 2$, follows by combining the partitions of $\mathbb{S}^1$ and $\mathbb{S}^2$, as was done in Theorem 6.6 (or, apply Cor. 7.6(b) for larger n). For $\mathbb{S}^1$, the result is obvious if $\kappa < \aleph_0$, and the $\aleph_0$-divisibility of $\mathbb{S}^1$ is implicit in the classical construction of a non-Lebesgue measurable subset of the circle: The sets M_i in the proof of Theorem 1.5 are pairwise rotationally congruent. For $\aleph_0 < \kappa < 2^{\aleph_0}$, the same proof as for $\kappa = \aleph_0$ works, except that the countable subgroup of rotations through rational multiples of π must be replaced by a κ-sized subgroup of $SO_2(\mathbb{R})$. Such a subgroup is easily obtainable by taking any collection of κ rotations of the circle and considering the group they generate. □

This result seems to indicate that all spheres have the same divisibility properties (into three or more pieces), but in fact each $\mathbb{S}^n$, $n \ge 2$, has a divisibility property that $\mathbb{S}^1$ does not. It is a consequence of Corollary 7.6(a) that if $n \ge 2$, then $\mathbb{S}^n$ has a subset E such that, for each cardinal κ with $2 \le \kappa \le 2^{\aleph_0}$, $\mathbb{S}^n$ is divisible into κ pieces, each of which is $O_{n+1}(\mathbb{R})$-congruent to E. In other words, E is a half of $\mathbb{S}^n$ and a third of $\mathbb{S}^n$ and ... and a $2^{\aleph_0}$th part of $\mathbb{S}^n$! To obtain E, simply choose a partition of $\mathbb{S}^n$ into sets A_α, $1 \le \alpha \le 2^{\aleph_0}$, satisfying the following proper system of congruences:

$$A_1 \cong A_\alpha,\ 1 \le \alpha < 2^{\aleph_0}$$

$$A_1 \cong \bigcup\{A_\beta : \kappa \le \beta \le 2^{\aleph_0}\},\ \kappa \text{ a cardinal and } 2 \le \kappa \le 2^{\aleph_0}.$$

Then let $E = A_1$. For any κ in $[2, 2^{\aleph_0}]$, $\mathbb{S}^n$ splits into the sets A_α, where $1 \le \alpha < \kappa$, and the set $\bigcup\{A_\beta : \kappa \le \beta \le 2^{\aleph_0}\}$, all of which are congruent to E. If one deletes the single congruence $A_1 \cong A_2 \cup A_3 \cup \cdots$, the resulting system is weak (see [Myc55b]) and is therefore solvable in all the cases of Theorem 7.4. Hence there is a subset of $\mathbb{S}^2$, for example, that is simultaneously a third, a quarter, ..., a $2^{\aleph_0}$th part of $\mathbb{S}^2$ with respect to the group of rotations. Because a finitely additive, $O_2(\mathbb{R})$-invariant measure on $\mathbb{S}^1$ of total measure one exists (Cor. 12.9), such a subset of the circle cannot exist: For $n < \infty$, an nth part of $\mathbb{S}^1$ would have to have measure $1/n$, and for $\kappa \ge \aleph_0$ a κth part would have measure zero. Hence $\mathbb{S}^1$ does not have this simultaneous divisibility property that the higher-dimensional spheres do.

7.2 Large Free Semigroups of Isometries

The isometry group of the Euclidean plane, being solvable, does not contain any free non-Abelian subgroup, but we have seen that free subsemigroups of rank 2 exist. In the Sierpiński–Mazurkiewicz Paradox (Thms. 1.7–1.9), such a subsemigroup was used to construct a nonempty paradoxical subset of the plane. The von Neumann numbers (or any set of algebraically independent numbers) allow us to define a free semigroup of isometries of much larger rank, obtaining a refinement of the Sierpiński–Mazurkiewicz Paradox that still avoids the Axiom of Choice. First, we give the straightforward generalization of Proposition 1.9.

Proposition 7.12. *Suppose a group G acts on X and G contains a (necessarily free) semigroup S generated by σ_α, $\alpha \in I$, with the following property: for some $x \in X$, any two elements of S with different leftmost terms yield different points when applied to X. Then some nonempty subset E of X contains pairwise disjoint sets A_α, $\alpha \in I$, such that $\sigma_\alpha^{-1}(A_\alpha) = E$.*

Proof. Let E be the S-orbit of x. Then the hypothesis guarantees that the sets $\sigma_\alpha(E)$ are pairwise disjoint subsets of E, and $\sigma_\alpha^{-1}(\sigma_\alpha(E)) = E$. The fact that the hypothesis on the σ_α implies freeness is proved in the same way as in Theorem 1.8. □

Theorem 7.13. (a) *There are isometries of $\mathbb{R}^2$, $\{\sigma_t : 0 < t \leq 1\}$, such that the subsemigroup of G_2 that they generate satisfies the hypothesis of Proposition 7.12, with $x = (0, 0)$.*

(b) *There is a nonempty subset E of $\mathbb{R}^2$ that may be partitioned into a continuum of sets, each of which is congruent to E.*

Proof. (a) As usual, let $\theta_t = 2 \arctan v_t$, $0 < t \leq 1$, where the v_t are the von Neumann numbers. Let u_t be the complex number $e^{i\theta_t}$; we identify $\mathbb{R}^2$ with $\mathbb{C}$. Note that the u_t are algebraically independent. For if $P = 0$, where P is a rational polynomial in finitely many of the u_t, then because $u_t = (1 - v_t^2 + 2v_t i)/(1 + v_t^2)$, we may take a common denominator to get $Q = 0$, where Q is a polynomial over $\mathbb{Q}(i)$ in the corresponding v_t. Then Q cannot be identically 0, for otherwise P would vanish when each of its variables is set equal to u_1, contradicting the transcendence of u_1 (which follows from the transcendence of its real part). Hence the v_t satisfy an algebraic relation over $\mathbb{Q}(i)$ and, because i is algebraic, over $\mathbb{Q}$ as well, a contradiction.

For $0 < t < 1$, let $\sigma_t(z) = u_t z + u_t$, and let $\sigma_1(z) = u_1 z$; these complex functions correspond to isometries of $\mathbb{R}^2$. Suppose $w_1 = \sigma_{t_1} \cdots \sigma_{t_r}$ and $w_2 = \sigma_{s_1} \cdots \sigma_{s_m}$ are nontrivial words in σ_t, $0 < t \leq 1$, with $t_1 \neq s_1$; hence we may assume $t_1 \neq 1$. Then $w_1(0)$ and $w_2(0)$ are polynomials in the u_{t_i}, u_{s_i}, respectively, and $w_1(0)$ has but a single term of degree 1, u_{t_1}, while $w_2(0)$ has no such term, if $s_1 = 1$, or also a single such term, u_{s_1}. Therefore if $w_1(0) = w_2(0)$, subtraction would yield a polynomial relation among finitely many of the algebraically independent numbers u_t, a contradiction.

(b) Apply Proposition 7.12 to the semigroup of part (a). The sets $\sigma_\alpha(E)$ form, in general, a partition of $E \setminus \{x\}$. But in this particular case, $\sigma_1(0, 0) = (0, 0)$, so $(0, 0) \in \sigma_1(E)$, and we really do have a partition of E. □

Another question raised regarding the Sierpiński–Mazurkiewicz Paradox was whether there could be an uncountable subset of the plane that is paradoxical using two pieces; the example in Theorem 1.7 is countable. Here too von Neumann numbers allow the desired extension to be obtained without the Axiom of Choice.

Theorem 7.14. *There is a subset of the plane, with the cardinality of the continuum, that is paradoxical using two pieces. Assuming the Axiom of Choice, such sets exist having any infinite cardinality not exceeding* $2^{\aleph_0}$.

Proof. Let u_t, $0 < t \leq 1$, be the algebraically independent complex numbers of modulus 1 defined from the von Neumann numbers in Theorem 7.13(a); let u denote u_1. Let $\sigma(z) = uz$ and $\tau(z) = z + 1$, and let S be the subsemigroup of G_2 (again, $\mathbb{R}^2$ is identified with $\mathbb{C}$) generated by σ and τ. Then the desired set E may be defined as

$$E = \{w(z) : w \in S \text{ and } z = \sigma^{-k}(u_t), 0 < t < 1, k = 0, 1, 2, \ldots\}.$$

Clearly $E \supseteq \sigma(E) \cup \tau(E)$ and $\sigma^{-1}(\sigma(E)) = E = \tau^{-1}(\tau(E))$. Moreover, because each $\sigma^{-k}(u_t) = \sigma(\sigma^{-(k+1)}(u_t)) \in \sigma(E)$, $E = \sigma(E) \cup \tau(E)$. It remains to prove that $\sigma(E)$ and $\tau(E)$ are disjoint, but it is easy to see, because words w correspond to polynomials with coefficients of the form mu^n, $m, n \in \mathbb{N}$, that an equality $\sigma w_1 \sigma^{-j}(u_t) = \tau w_2 \sigma^{-k}(u_s)$ implies a nontrivial polynomial relation among u, u_t, and u_s (or just u, u_t if $s = t$), in violation of their algebraic independence. Assuming Choice, $(0, 1)$ may be well ordered in type $2^{\aleph_0}$, and so we need only replace $\{u_t : 0 < t < 1\}$ by any κ-sized subset (κ infinite) to get an example of size κ. □

7.3 Sets Congruent to Proper Subsets

To conclude this chapter, we discuss some geometric problems whose solutions are closely connected with free groups of isometries. These problems arise from the possibility of a set being congruent to a proper subset. To summarize some background about this notion, recall that we have already seen, and made much use of, the fact that a subset of the circle can be congruent to a proper subset: $D = \{\rho^n(1, 0) : n \in \mathbb{N}\}$ is congruent to $D \setminus \{(1, 0)\}$ if ρ is a rotation about the origin of infinite order. In general, call a set in any metric space *compressible* if it is congruent to a proper subset. Recall that congruence may be witnessed by partial isometries: A is congruent to B if there is a distance-preserving bijection from A to B. In any compact metric space, an isometry from A to B can always be extended uniquely to one from $\overline{A}$ to $\overline{B}$.

Now, it is easy to see that no bounded subset of the real line is compressible although, of course, $\mathbb{N}$ is congruent to $\mathbb{N} \setminus \{0\}$. As for the plane, we have the bounded compressible set D. An interesting result of Lindenbaum shows that topologically, the example D is as simple as possible. Because D is countable, it is an F_σ set. But it is not a G_δ, because if $D^c = \mathbb{S}^1 \setminus D$ were an F_σ then, because D is dense, any closed set in the union would be nowhere dense; so D^c would be meager and therefore $D \cup D^c$ would be meager, a contradiction. In fact, no bounded compressible set can be both an F_σ and a G_δ. Following the notation of descriptive set theory for the Borel hierarchy, let $\mathbf{\Delta}_2^0$ denote sets that are in both F_σ and G_δ.

Theorem 7.15. *Any $\mathbf{\Delta}_2^0$ subset of a compact metric space is noncompressible; hence any bounded $\mathbf{\Delta}_2^0$ subset of $\mathbb{R}^n$ is noncompressible.*

Proof. First we prove that closed sets in a compact space X are not compressible. Suppose $\sigma(A) \subseteq A$, where σ is a distance-preserving function on A. For $a \in A$, let $\overline{\sigma}(a) = \{\sigma^n(a) : n \in \mathbb{Z}$ and $\sigma^n(a)$ is defined$\}$. If $\overline{\sigma}(a)$ is finite, then $a = \sigma^n(a)$ for some $n > 0$, whence $a \in \sigma(A)$. If $\overline{\sigma}(A)$ is infinite, then by compactness of X, $\overline{\sigma}(a)$ has a limit point. It follows that for any $\epsilon > 0$, there are distinct $m, n \in \mathbb{Z}$ with $d(\sigma^m(a), \sigma^n(a)) < \epsilon$, whence, assuming $m < n$, $d(a, \sigma^{n-m}(a)) < \epsilon$. This yields that a is a limit point of $\{\sigma^n(a) : n > 0\}$, and because $\sigma(A)$, which is congruent to a closed set, must itself be closed, $a \in \sigma(A)$. We have proved that $A \subseteq \sigma(A)$, and therefore that A is not compressible.

For any subset A of X define the *residue* of A, A_R, to be $A \cap \overline{\overline{B} \setminus A}$. If $\sigma(A) = B \subseteq A$, where σ is a distance-preserving function on A (which, by compactness, is assumed to be defined on $\overline{A}$), then we claim that $\sigma(A_R) \subseteq A_R$ and $A \setminus B \subseteq A_R \setminus \sigma(A_R)$. By the result on closed sets just proved, $\sigma(\overline{A}) = \overline{B}$ cannot be a proper subset of $\overline{A}$; therefore $\overline{A} = \overline{B}$ and $\overline{B} \setminus B = \overline{A} \setminus B \supseteq \overline{A} \setminus A$, whence $\overline{\overline{B} \setminus B} \supseteq \overline{\overline{A} \setminus A}$. But σ takes $\overline{A} \setminus A$ to $\overline{B} \setminus B$, and hence also takes $\overline{\overline{A} \setminus A}$ to $\overline{\overline{B} \setminus B}$; because these sets are closed, this means $\overline{\overline{B} \setminus B} = \overline{\overline{A} \setminus A}$. The fact that σ is a homeomorphism from $\overline{A}$ to $\overline{B}$ yields that $\sigma(A_R) = (\sigma(A))_R$, whence $\sigma(A_R) = B_R = B \cap \overline{\overline{B} \setminus B} \subseteq A \cap \overline{\overline{A} \setminus A} = A_R$. For the second part of the claim, $A \setminus B \subseteq \overline{A} \setminus B = \overline{B} \setminus B \subseteq \overline{\overline{B} \setminus B} = \overline{\overline{A} \setminus A}$, so $A \setminus B \subseteq A_R$. This proves the claim because $A \setminus B$ is disjoint from $\sigma(A)$, and hence from $\sigma(A_R)$. (In fact, it can be shown that $A \setminus B = A_R \setminus \sigma(A_R)$.)

Now, for any set A, we may define a sequence of residues: $A_0 = A$, $A_{\alpha+1} = (A_\alpha)_R$, and $A_\gamma = \bigcap_{\alpha<\gamma} A_\alpha$ for limit ordinals γ. There must be some ordinal β such that $A_\beta = A_{\beta+1} = \ldots$. The Baire Category Theorem, which is valid in complete, and hence in compact, metric spaces, can be used to show that if A is $\mathbf{\Delta}_1^0$, then $A_\beta = \varnothing$ (see [Hau57, §30] or [Kur66, §§12, 34]). (The converse is true in all metric spaces. If A_β vanishes, then $A \in \mathbf{\Delta}_1^0$ (see [Hau57, §30]). The claim shows that a compressible set A can be carried through the successor stages of the sequence of residues, and it is easy to cross the limit stages as well. Because the empty set is obviously incompressible, A must be too. □

The compressible sets $\mathbb{N}$ and D led Sierpiński to wonder if a set could contain two or more points, each of whose deletion, separately, leaves a set congruent to the original. He proved the following result for the line, and a proof was given in Theorem 4.10; the case of $\mathbb{R}^2$, which is somewhat more complicated and whose proof we omit, was proved by Straus [Str57].

Theorem 7.16. *There is no subset E of either $\mathbb{R}^1$ or $\mathbb{R}^2$ such that for two distinct points P and Q in E, E is congruent to $E \setminus \{P\}$ and to $E \setminus \{Q\}$.*

The proof for $\mathbb{R}^1$ (§4.6) uses the commutativity of translations or, more precisely, the fact that the equation $x^2y^2x^{-2}y^{-2} = e$ is universally satisfied in G_1. The proof for $\mathbb{R}^2$ likewise uses a universal equation, namely, $[[x^2, y^2], [x^2, y^2]] = e$ (see App. A), but also uses the geometric form of elements of G_2. That the existence of a universal equation does not suffice is shown by the following example. Let G be the group generated by $\left[\begin{smallmatrix} 2 & 0 \\ 0 & 1 \end{smallmatrix}\right]$ and $\left[\begin{smallmatrix} 1 & 0 \\ 0 & 2 \end{smallmatrix}\right]$, and let $E = \{(0, 2^n), (2^n, 0) : n \in \mathbb{N}\}$. Then $E \setminus \{(0, 1)\}$ and $E \setminus \{(1, 0)\}$ are G-congruent to E despite the fact that G is Abelian.

Nonetheless, in free groups of rank 2 or more, where of course no nontrivial equation is universally satisfied, sets congruent to any maximal proper subset do exist. This fact was used by Mycielski to construct a subset of $\mathbb{R}^3$ that is congruent to the remainder after the deletion of any finite set of points. This construction is similar to that of paradoxical decompositions, although the Axiom of Choice is not needed; first show that the set exists in a free group, and then transfer it to a set on which the group acts without fixed points.

Definition 7.17. *If G acts on X, then a nonempty subset E of X is a* Mycielski set *if, for any $p \in E$, there is $\sigma \in G$ such that $\sigma(E) = E \setminus \{p\}$.*

If E is a Mycielski set, then for any finite subset D, there is σ so that $\sigma(E) = E \setminus D$: Just deal with the points in D in sequence. A weaker notion, where E is G-congruent only to two specified point-deleted sets $E \setminus \{p\}$ and $E \setminus \{q\}$, was discussed in §4.5.

Theorem 7.18. (a) *A free group F of rank 2 has a Mycielski set (with respect to its action on itself by left multiplication), as does any group with a subgroup isomorphic to F.*

(b) *If F, as in (a), acts on X without nontrivial fixed points, then X has a Mycielski set.*

(c) *Any sphere in $\mathbb{R}^3$ contains a Mycielski set with respect to its group of rotations.*

Proof. (a) If F is freely generated by τ and ρ, let $\sigma_i = \rho^i \tau^i$ for $i \in \mathbb{N}$; the σ_i are independent. Let F_ω be the subgroup of F generated by the σ_i, and let $\{d_n : n \in \mathbb{N}\}$ enumerate the elements of F_ω. Do it by taking more generators and increasing length, as follows:

$$e,$$
$$\sigma_0^{-1}, \sigma_0,$$
$$\sigma_1^{-1}, \sigma_1,$$
$$\sigma_0^{-1}\sigma_1^{-1}, \sigma_0^{-1}\sigma_1, \sigma_0\sigma_1^{-1}, \sigma_0\sigma_1, \sigma_1^{-1}\sigma_0^{-1}, \sigma_1^{-1}\sigma_0, \sigma_1\sigma_0^{-1}, \sigma_1\sigma_0,$$
$$\sigma_2^{-1}, \sigma_2,$$

words of length 2 in $\sigma_0^{\pm 1}, \sigma_1^{\pm 1}, \sigma_2^{\pm 1}$ (and not in any previous level)

words of length 3 in $\sigma_0^{\pm 1}, \sigma_1^{\pm 1}, \sigma_2^{\pm 1}$ (and not in any previous level)

$\sigma_3^{-1}, \sigma_3,$ etc.

Now let

$$V_n = \{\phi\, \sigma_n^{-m}\, d_n : n = 0, 1, 2, \ldots; m = 1, 2, \ldots;$$
$$\phi \in F_\omega \text{ and } \phi \text{ does not end in } \sigma_n \text{ on the right}\}$$

Because no d_n has σ_n on its left, the form $\phi\, \sigma_n^{-m}\, d_n$ is always a reduced word.

Then $\sigma_n V_n = V_n \cup \{d_n\}$, while $\sigma_k V_n = V_n$ if $k \neq n$. It follows that $E = F_\omega \setminus \bigcup_n V_n$ is the desired Mycielski set, as it is easy to verify that for any n, $\sigma_n(E) = E \setminus \{d_n\}$.

(b) Let E be a Mycielski set in F and let x be any element of X. Then the set $\{w(x) : w \in E\}$ is a Mycielski set in X.

(c) Let σ, τ be the Satô rotations of the unit sphere, $\mathbb{S}^2$ (Thm. 2.1). They act without fixed points on $\mathbb{S}^2$ less a countable set, so by (b), there is a subset of the sphere that is a Mycielski set. To get the set on some other sphere, use the appropriate affine transformation of the set in the unit sphere. □

The preceding proof works just as well if we use finite sets directly instead of singletons. See §4.6 for a concrete visualization of a weak Mycielski set in the hyperbolic plane. And the preceding proof also extends easily to larger cardinals: If F is a free group of rank κ, and λ any cardinal satisfying $\kappa^\lambda = \kappa$, then F has a subset E such that for any $\leq\lambda$-sized subset D of E, there is some $\sigma \in F$ with $\sigma(E) = E \setminus D$. But Mycielski [Myc58b] showed that much more is possible. First of all, the action of the free group need only be assumed to be locally commutative. And, more strikingly, one can arrange things so that small sets can be added as well as deleted. We omit the proof (see [Myc58b]), which involves an inductive construction more intricate than in Theorem 7.18.

Theorem 7.19 (AC). *Suppose F, a free group of rank κ, is locally commutative in its action on X and λ is a cardinal satisfying $\kappa^\lambda = \kappa$. Then there is a subset E of X such that for any two sets, $D_1, D_2 \subseteq X$ with $|D_i| \leq \lambda$, there is some $\sigma \in F$ with $\sigma(E) = (E \setminus D_1) \cup D_2$.*

Corollary 7.20 (AC). *Each of* $\mathbb{S}^n$, $\mathbb{L}^n$, $\mathbb{H}^n$ $(n \geq 2)$, *or* $\mathbb{R}^n$ $(n \geq 3)$ *has a subset that, from the point of view of isometries, is unchanged by adding or deleting countably many points.*

The corollary is an immediate consequence of Theorems 7.19 and 7.4.

The techniques of the preceding results are limited to groups with free non-Abelian subgroups, because such groups are the only ones containing Mycielski sets (compare this with Thm. 12.5).

Theorem 7.21. *If G has a Mycielski set E with respect to its left action on itself, then G has a free subgroup of rank 2.*

Proof. If $\sigma \in E$, then $E\sigma^{-1}$ is also a Mycielski set; thus we may assume that E contains the identity e. Let $\sigma \in G$ map E to $E \setminus \{e\}$; then $\sigma = \sigma e \in E$, so there is some $\tau \in G$ $(\tau \neq \sigma)$ with $\tau E = E \setminus \{\sigma\}$. We claim that σ and τ are independent. Because σ and τ cannot have finite order, any nontrivial vanishing word in $\sigma^{\pm 1}$, $\tau^{\pm 1}$ can be transformed by inversion and conjugation into one of the form $w = \sigma^{m_1}\tau^{n_1}\sigma^{m_2}\cdots\tau^{n_k}$, where $k \geq 1$, $m_1 > 0$, and all exponents are nonzero integers. Choose such a w such that k is as small as possible and, given k, $\sum_{i=1}^{k} m_i + n_i$, is minimal as well.

We shall prove by induction that each right-hand end-segment of w belongs to E. The first such segment is simply τ or τ^{-1}, both of which belong to E by $\tau E = E \setminus \{\sigma\}$. Assume that the proper end-segment u of w belongs to E. Then, clearly, σu and τu lie in E. Moreover, by the minimality of w, $u \neq e$; hence $\sigma^{-1} u \in E$. Finally, if $\tau^{-1} u$ is the next subword and does not lie in E, then $u = \sigma$, and it may easily be checked that σu^{-1}, which equals the identity, is shorter than w with respect to one of the two minimality conditions on w. This covers all four cases, completing the induction. But this means $\sigma^{-1} = \sigma^{m_1 - 1}\tau^{n_1}\cdots\tau^{n_k} \in E$, violating $\sigma E = E \setminus \{e\}$. □

The preceding proof assumed only that left translates of E give the point-deleted subsets; in fact, it is sufficient to assume that left or right translates are used in each case [Str59]. The proof leaves open the question of what happens when E is a weak Mycielski set: has two points so that $\sigma E = E \setminus \{p\}$ and $\tau E = E \setminus \{q\}$.

Question 7.22. If a group G has a weak Mycielski set with respect to its left action on itself, must G contain a free subgroup of rank 2?

Notes

Infinite sets of congruences were first investigated for partitions of $\mathbb{S}^2$, independently by Dekker and de Groot [DG54, DG56] and by Mycielski [Myc55b]. Theorem 7.1, in its generality, was formulated and proved by Dekker [Dek56]. The extension of Theorem 7.1 ensuring that the sets A_α are all nonempty is due to Dekker [Dek56b], as is its converse, Theorem 7.2 [Dek56a].

Lebesgue and Steinitz had realized that $2^{\aleph_0}$ algebraically independent reals exist, but Mazurkiewicz [Maz20] asked if the existence of a proper uncountable subfield of the reals could be proved without using the Axiom of Choice. Von Neumann [Neu61] answered this question affirmatively by giving a constructive proof of Theorem 7.3. The alternative, less computational approach using Theorem 7.5 is due to Mycielski [Myc64].

Sierpiński [Sie45b] was the first to construct an uncountable free group of isometries; he used algebraically independent numbers to prove Theorem 7.4 for $SO_3(\mathbb{R})$. Earlier, Nisnewitsch [Nis40] had proved that any free group is isomorphic to a subgroup of $GL_2(K)$ for some field K. Mycielski [Myc55b] used Sierpiński's large free group to solve large sets of congruences on $\mathbb{S}^2$ (Cor. 7.6 for $\mathbb{S}^2$). Simultaneously, Dekker and de Groot obtained these results for $\mathbb{S}^2$ [Gro54, Gro56, DG54, DG56], and Dekker [Dek56b, Dek57] carried out the extension to the other spaces of Corollary 7.6, in those cases where he had solved the problem for finite systems of congruences (see Notes to Chap. 6). The idea of using Theorem 7.5 to obtain large free groups (thus generalizing the results of Deligne and Sullivan for $SO_n(\mathbb{R})$, n even, and Borel for $SO_5(\mathbb{R})$ given in Chapter 6) is due to Mycielski [MW84]. The fact, stated after Corollary 7.6, that the set of pairs generating certain free subgroups of $SO_n(\mathbb{R})$ is comeager follows from Borel's work [Bor83]; he used a different technique to show that the sets R_w are nowhere dense.

Theorem 7.7 on uncountable free subgroups of $PSL_2(\mathbb{R})$ is due to Siegel [Sie50]; the proof presented here uses some ideas of R. Riley. For extensions of Siegel's result, see [Jor77]. Theorem 7.8 is due to Mycielski and Wagon [MW84]. The observation that free groups in G_3 can be used to prove Theorem 7.8 in $\mathbb{H}^4$ and beyond is due to A. Borel.

Steinhaus first raised the question of whether the set E of the Sierpiński–Mazurkiewicz Paradox (Thm. 1.7) could be taken to be uncountable. This was answered affirmatively by Ruziewicz [Ruz21], but he needed the Axiom of Choice. Still using Choice, Lindenbaum [LT26, p. 327] proved that there is a nonempty subset of the plane that is equidecomposable with κ many copies of itself, for any $\kappa \leq 2^{\aleph_0}$. This was all prior to the discovery of von Neumann numbers. Von Neumann [Neu61] deduced from these numbers a result that eliminated the Axiom of Choice from Ruziewicz's work, and Sierpiński [Sie47] refined Ruziewicz's result even more, giving the proof of Theorem 7.14. Moreover, Sierpiński [Sie47] saw how von Neumann numbers could be used to eliminate choice from Lindenbaum's theorem (Theorem 7.13).

The definition of incompressible sets (also known as *monomorphic sets*) is due to Lindenbaum [Lin26], who proved Theorem 7.15. The example of a subset of the circle that is congruent to a proper subset of itself is due to Tarski [Tar24a]; see [Lin26, p. 217]. *The Scottish Book* [Mau81, p. 67] contains a weaker version of Theorem 7.15 due to Banach and Ulam.

In [Sie50a] and [Sie54, pp. 7–10] (see also [Sie50b]), Sierpiński proved Theorem 7.16 for $\mathbb{R}^1$ and gave an erroneous construction of a set E in the plane congruent to $E \setminus \{P\}$ and $E \setminus \{Q\}$ for distinct $P, Q \in E$. The error was discovered by

Mycielski (see [Sie55, Sie54, p. 116]), who went on to show [Myc54, Myc55a] that Mycielski sets exists in $\mathbb{R}^3$ (Thm. 7.18). The proof of Theorem 7.18 given here is from [Str57], where Straus settled the question for $\mathbb{R}^2$ (Thm. 7.16). The extensions of these results contained in Theorem 7.19 and Corollary 7.20 are due to Mycielski [Myc58b], although the existence of a Mycielski set in the hyperbolic plane was proved earlier by Viola [Vio56]. Mycielski [Myc56, Myc58a] also investigated these types of questions in more general analytic manifolds. Theorem 7.21, which shows that the collection of groups that do not contain Mycielski sets coincides with groups not having a free non-Abelian subgroup, is due to Straus [Str59].

8

Paradoxes in Low Dimensions

The isometry group of the plane does not contain a free noncommutative subgroup, and so the construction of the Banach–Tarski Paradox does not work in the plane, or in the line. Moreover, G_1 and G_2 are solvable, so we can construct Banach measures (isometry-invariant, finitely additive measures on all subsets of $\mathbb{R}^1$ or $\mathbb{R}^2$, Cor. 12.9) showing that no bounded set with interior is paradoxical (an unbounded paradoxical subset of the plane was constructed in Thm. 1.7). But an enlargement of the group, to the area-preserving affine transformations $SA_2(\mathbb{R})$, does yield free groups and the resulting paradoxes. In this chapter we study in detail the sorts of paradoxes that arise from $SA_2(\mathbb{R})$ and $SA_2(\mathbb{Z})$, as well as the related linear groups acting on the punctured plane. In addition, we show how a slightly different type of paradox can be given for the line using contractions.

8.1 Paradoxes in the Plane

8.1.1 Paradoxes of $\mathbb{R}^2$ and $\mathbb{Z}^2$

Recall that the group $A_n(\mathbb{R})$ of affine transformations of $\mathbb{R}^n$ consists of transformations of the form $\sigma = \tau L$, where τ is a translation and L is a linear transformation. Moreover, σ magnifies area by the factor $|\det \sigma|$, where the determinant of σ is defined to be $\det L$. Thus $SA_n(\mathbb{R})$, the group of affine transformations of determinant $+1$, consists of the affine transformations that preserve area and orientation. The group $SA_2(\mathbb{R})$ is quite a bit larger than the subgroup of orientation-preserving isometries, as we shall see in a moment, but in dimension 1, nothing new appears. Any affine transformation, $ax + b$, of determinant ± 1 is, in fact, an isometry. We use $SA_2(\mathbb{Z})$ for the analogous group using integers for all the entries, including the translations. Here we identify the group of integer translations with the additive $\mathbb{Z}^2$.

We know that the two transformations in $SL_2(\mathbb{Z})$ given by $\left[\begin{smallmatrix} 1 & 2 \\ 0 & 1 \end{smallmatrix}\right]$ and $\left[\begin{smallmatrix} 1 & 0 \\ 2 & 1 \end{smallmatrix}\right]$ are independent (Prop. 4.4). This holds also when 2 is replaced by any number larger than 1. In fact, independence holds for a wider class of matrices, as follows. Say that a matrix entry is *dominant* if it is strictly greater in absolute value than all other entries. Then $A = (a_{ij})$ and $B = (b_{ij})$, two members of $SL_2(\mathbb{Z})$, are independent if a_{12} dominates A and b_{21} dominates B. This result is due to K. Goldberg and M. Newman [GN57].

The next theorem describes some useful free groups in two dimensions. The group of the previous paragraph (in fact, all of $SL_2(\mathbb{R})$) is locally commutative on $\mathbb{R}^2 \setminus \{\mathbf{0}\}$ (proved following Cor. 5.6), so part (a) improves that. Part (d) shows that we avoid the solvability issue of (c) when we restrict the action to integer points.

Theorem 8.1. (a) *$SL_2(\mathbb{Z})$ has a rank-2 free subgroup with no fixed points in $\mathbb{R}^2 \setminus \{\mathbf{0}\}$.*

(b) *$SA_2(\mathbb{R})$ has a subgroup that is locally commutative on $\mathbb{R}^2$ and is free on $2^{\aleph_0}$ generators.*

(c) *A subgroup of $SA_2(\mathbb{R})$ whose action has no fixed points is solvable; in fact, such a subgroup is nilpotent.*

(d) *$SA_2(\mathbb{Z})$ has an F_2-subgroup that acts on $\mathbb{Z}^2$ without fixed points.*

Proof. (a) Magnus and Neumann [Mag73, Neu33] showed that the two matrices $\left[\begin{smallmatrix} 1 & 1 \\ 1 & 2 \end{smallmatrix}\right], \left[\begin{smallmatrix} 5 & 2 \\ 2 & 1 \end{smallmatrix}\right]$ work.

(b) Satô [Sat03] proved that the following two transformations are as required: $\left[\begin{smallmatrix} x \\ y \end{smallmatrix}\right] \mapsto \left[\begin{smallmatrix} 2 & \theta \\ 0 & 1/2 \end{smallmatrix}\right]\left[\begin{smallmatrix} x \\ y \end{smallmatrix}\right] - \left[\begin{smallmatrix} 1 \\ 0 \end{smallmatrix}\right]$ and $\left[\begin{smallmatrix} x \\ y \end{smallmatrix}\right] \mapsto \left[\begin{smallmatrix} 2 & 0 \\ \theta & 1/2 \end{smallmatrix}\right]\left[\begin{smallmatrix} x \\ y \end{smallmatrix}\right] - \left[\begin{smallmatrix} 0 \\ 1/2 \end{smallmatrix}\right]$, where θ is any transcendental number. Then he was able to extend this to get a group of continuum rank. His construction relies on the fact that the set $\{\overline{\sigma} = (\sigma_1, \ldots, \sigma_n) \in SA_2(\mathbb{R}) : \overline{\sigma}$ fails to generate a free locally commutative group$\}$ is contained in a countable union of proper algebraic subsets of $SA_2(\mathbb{R})$. Because such a union is a meager set and $SA_2(\mathbb{R})$ is a complete separable metric space, we can apply a theorem of Mycielski (see remarks after Thm. 7.5).

(c) Suppose G is a group as hypothesized; using τ to denote a translation and L a linear transformation, if $\tau\, L \in G$, then $\det(L - I) = 0$, where I is the identity matrix, and therefore L has $+1$ as eigenvalue. Now suppose $\sigma = \tau L$ and $\rho = \tau' L'$ are in G. Then (see App. A) $\sigma\tau = \tau^* L\, L'$, and each of L, L', LL' has $+1$ as an eigenvalue, and so all three matrices have trace 2. Hence, choosing an appropriate basis, $L = \left[\begin{smallmatrix} 1 & b \\ 0 & 1 \end{smallmatrix}\right]$, $L' = \left[\begin{smallmatrix} \alpha & \beta \\ \gamma & 2-\alpha \end{smallmatrix}\right]$, and $LL' = \left[\begin{smallmatrix} \alpha+b\gamma & * \\ * & 2-\alpha \end{smallmatrix}\right]$. The trace of the last being 2 implies that $b\gamma = 0$. Now, if $b \neq 0$, then $\gamma = 0$, and because determinants are 1, $\alpha = 1$ and $L' = \left[\begin{smallmatrix} 1 & \beta \\ 0 & 1 \end{smallmatrix}\right]$. If instead $b = 0$, then L is the identity, and we can use a basis change to get L' into the form $\left[\begin{smallmatrix} 1 & \beta \\ 0 & 1 \end{smallmatrix}\right]$. So, in either case, σ and ρ, viewed as 3×3 matrices (App. A), have the form $\left[\begin{smallmatrix} 1 & B & t_1 \\ 0 & 1 & t_2 \\ 0 & 0 & 1 \end{smallmatrix}\right]$. Matrix algebra on these matrices shows that $\sigma\rho\sigma^{-1}\rho^{-1}$ is a pure translation; therefore any commutator of commutators is the identity, and the group is solvable.

To go further, let $\tau'' L''$ be a third element of G. Case 1: One of L, L', L'' is the identity. Then use a change of basis to get the three matrices to be, in some order $\left[\begin{smallmatrix} 1 & 0 \\ 0 & 1 \end{smallmatrix}\right], \left[\begin{smallmatrix} 1 & b \\ 0 & 1 \end{smallmatrix}\right], \left[\begin{smallmatrix} \alpha & \beta \\ \gamma & 2-\alpha \end{smallmatrix}\right]$. As before, $b\gamma = 0$, and the matrices extend in the same way to upper diagonal 3×3 forms with 1s on the main diagonal. Matrix multiplication then shows that $xyzyx = yxzxy$ holds. Case 2: None of the three linear transformations is the identity. Then $L = \left[\begin{smallmatrix} 1 & b \\ 0 & 1 \end{smallmatrix}\right]$, $L' = \left[\begin{smallmatrix} 1 & \beta \\ 0 & 1 \end{smallmatrix}\right]$, $L'' = \left[\begin{smallmatrix} 1 & \delta \\ 0 & 1 \end{smallmatrix}\right]$, and the corresponding 3×3 forms are as before and the identity holds. So the group satisfies $xyzyx = yxzxy$, which means the group is nilpotent [NT63].

(d) Satô [Sat99] found two transformations that work: $\left[\begin{smallmatrix} x \\ y \end{smallmatrix}\right] \mapsto \left[\begin{smallmatrix} 7 & 3 \\ 9 & 4 \end{smallmatrix}\right]\left[\begin{smallmatrix} x \\ y \end{smallmatrix}\right] + \left[\begin{smallmatrix} 1 \\ -1 \end{smallmatrix}\right]$ and $\left[\begin{smallmatrix} x \\ y \end{smallmatrix}\right] \mapsto \left[\begin{smallmatrix} 94 & 39 \\ 147 & 61 \end{smallmatrix}\right]\left[\begin{smallmatrix} x \\ y \end{smallmatrix}\right] + \left[\begin{smallmatrix} 3 \\ 2 \end{smallmatrix}\right]$. □

Theorem 8.1(c) implies that there is no free subsemigroup in this special affine group, and therefore (by Thm. 14.30) the existence of $SA_2(\mathbb{R})$-paradoxical sets in the plane requires transformations having a fixed point. Theorem 5.12 yields the following application of the groups of Theorem 8.1. For the second assertion of (c), use Theorem 8.1(a) to get a paradox of the punctured plane, and then absorb the origin by a translation of one unit in the x-direction.

Corollary 8.2 (AC). (a) *All countable, proper systems of congruences are solvable by a partition of* $\mathbb{R}^2 \setminus \{\mathbf{0}\}$ *using* $SL_2(\mathbb{Z})$.

(b) *All weak systems of congruences are solvable by a partition of* $\mathbb{R}^2$ *using* $SA_2(\mathbb{R})$. *So* $\mathbb{R}^2$ *is* $SA_2(\mathbb{R})$*-paradoxical using four pieces.*

(c) *All countable, weak systems of congruences are solvable by a partition of* $\mathbb{Z}^2$ *using* $SA_2(\mathbb{Z})$. *Also* $\mathbb{R}^2$ *is* $SA_2(\mathbb{Z})$*-paradoxical.*

Even though Theorem 8.1(b) cannot be improved to avoid all fixed points, that does not settle the question of solving all congruences. Recall that a similar situation held for $\mathbb{S}^2$: Theorem 5.16 showed how to solve all systems despite the lack of an F_2 without fixed points.

Question 8.3. Are all (possibly uncountable) proper systems of congruences solvable by a partition of $\mathbb{R}^2$ using area-preserving affine transformations, and not necessarily restricted to the orientation-preserving ones?

8.1.2 Paradoxes of Bounded Sets in the Plane

Our next goal is to obtain a paradoxical decomposition of a set of positive finite area, such as the unit square. We shall show (Thm. 8.5) that this can be done using any independent pair in $SL_2(\mathbb{Z})$ and translations, thus staying within $SA_2(\mathbb{R})$.

Let J be the half-open unit square $[0, 1) \times [0, 1)$. The general results of Chapter 5 apply to a group acting on a set, but $SL_2(\mathbb{Z})$ does not act on the square. This difficulty can be handled by using a planar version of arithmetic modulo 1; because all translations are affine, this will not take us out of $SA_2(\mathbb{R})$. More precisely, let $\approx$ denote the equivalence relation on $\mathbb{R}^2$ determined by the lattice $\mathbb{Z}^2$: $P \approx Q$ if and only if $Q = P + (m, n)$ for integers m, n. For any $P \in \mathbb{R}^2$, let $\hat{P}$ be the unique point in J such that $\hat{P} \approx P$. Let H denote the group of affine transformations τL, where τ is any real translation and $L \in SL_2(\mathbb{Z})$; $H = \pi^{-1}(SL_2(\mathbb{Z}))$ where $\pi : A_2 \to GL_2(\mathbb{R})$ is the canonical homomorphism. For any $\sigma \in H$, let $\hat{\sigma}$ be the function with domain J defined by $\hat{\sigma}(P) = \widehat{\sigma(P)}$.

Proposition 8.4. *The mapping $\sigma \mapsto \hat{\sigma}$ is a homomorphism from H to the group of area-preserving, piecewise affine bijections from J to J. When restricted to $SL_2(\mathbb{Z})$, this homomorphism is an isomorphism onto its image.*

Proof. Clearly $\hat{\sigma}: J \to J$. The following property of $\approx$, which follows easily from the fact that $L(\mathbb{Z}^2) \subseteq \mathbb{Z}^2$ for $L \in H$, is the crux of the proof: If $P \approx Q$ and $\sigma \in SL_2(\mathbb{Z})$, then $\sigma(P) \approx \sigma(Q)$. It follows that $\hat{\sigma}$ is one-to-one on J. Moreover, any $P \in J$ equals $\hat{\sigma}(\hat{Q})$ where $Q = \sigma^{-1}(P)$, so $\hat{\sigma}(J) = J$. Because $\sigma(J)$ is a parallelogram, $\hat{\sigma}$ decomposes into finitely many functions $\tau_i\sigma$, where τ_i is a translation by a point in $\mathbb{Z}^2$; hence $\hat{\sigma}$ is piecewise in H and so preserves area.

Now, if $P \in J$, then again using the crucial property of $\approx$,

$$\hat{\sigma}_1\hat{\sigma}_2(P) = \hat{\sigma}_1\widehat{\sigma_2(P)} = \widehat{\sigma_1(\widehat{\sigma_2(P)})} = \widehat{\sigma_1(\sigma_2(P))} = \widehat{\sigma_1\sigma_2(P)}.$$

This shows that $\sigma \mapsto \hat{\sigma}$ is a homomorphism. To see that the transformation $L \mapsto \hat{L}$, $\lambda \in SL_2(\mathbb{Z})$, is an isomorphism, it is sufficient to check that the only transformation of $SL_2(\mathbb{Z})$ that gets taken to e_J, the identity on J, is e, the identity in $SL_2(\mathbb{Z})$. Suppose $\hat{L} = e_J$. There is some translation τ such that on a polygon, $\hat{L} = \tau L$. Hence τL is the identity on three noncollinear points; because τL is affine, this means that $\tau L = e$. But then $L = \tau^{-1}$, which is impossible unless L is the identity, as required. □

If F is any free subgroup of $SL_2(\mathbb{Z})$, then by this proposition, $\hat{F} = \{\hat{L} : L \in F\}$ is a free group of the same rank, acting on J. Each element of F fixes the origin, and may fix other points as well. But because the fixed point sets are not too large, they may be absorbed analogously to the way a similar problem was handled in the Banach–Tarski Paradox (Thm. 3.9, Cor. 3.10). To do this, we need to bring in translations.

Theorem 8.5 (The von Neumann Paradox for the Plane) (AC). *If σ_1, σ_2 are any two independent elements of $SL_2(\mathbb{Z})$ and G is the subgroup of $SA_2(\mathbb{Z})$ generated by σ_1, σ_2, and T, the group of all translations, then J is G-paradoxical. Moreover, any two bounded subsets of $\mathbb{R}^2$ with nonempty interior are G-equidecomposable.*

Proof. Let F be the group generated by σ_1 and σ_2, and let $\hat{F}$ and $\hat{T}$ be the images of F and T under the reduction modulo $\mathbb{Z}^2$. If D is the set of points in J that are fixed by some nonidentity element of $\hat{F}$, then $\hat{F}$ acts without nontrivial fixed points on $J \setminus D$. Because $\hat{F}$ is freely generated by $\hat{\sigma}_1$, $\hat{\sigma}_2$ (Prop. 8.4), it follows from Proposition 1.10 that $J \setminus D$ is $\hat{F}$-paradoxical. Each element of $\hat{F}$ is piecewise in G (because $T \subseteq G$), which implies that $J \setminus D$ is G-paradoxical. By Proposition 3.4, then, the proof will be complete once we show that J and $J \setminus D$ are T-equidecomposable.

Because every point in D is a fixed point of an affine map, and hence is contained in an affine subspace, and because F is countable, D splits into $D_0 \cup D_1$, where D_0 is a countable set of points in J and D_1 is a countable set of line segments. Now, the technique of Theorem 3.9 shows that for any countable subset

C of J, $J \sim_T J \setminus C$ using two pieces. For it suffices to find a translation τ such that $C \cap \hat{\tau}^n(C) = \varnothing$ for any $n \geq 1$, and this may easily be done once C is enumerated as $\{P_0, P_1, P_2, \ldots\}$. For each i, discard the countably many translations τ such that $\tau(P_i) \approx P_j$ for some $j \geq i$; a nondiscarded translation is as required. We shall reduce the general case, $J \setminus D$, to the case of a deleted countable set by finding a countable set C such that $J \setminus D \sim_T J \setminus C$. It will then follow that $J \setminus D \sim_T J$, whence $J \setminus D \sim_T J$, completing the proof that J is G-paradoxical.

We claim that there is some translation τ such that $D \cap \hat{\tau}^n(D)$ is countable for any integer $n \neq 0$. Enumerate the segments in D_1 as $S_0, S_1, S_2, \ldots$. Suppose n, as well as i, j with $0 \leq i, j$ are given integers. If S_i and S_j are not parallel, we do nothing. For then, no matter what τ is, $\hat{\tau}^n(S_j)$ intersects S_i in at most a single point. But if they are parallel, choose a point P on S_j and discard all translations τ such that for some $(m_1, m_2) \in \mathbb{Z}^2$, $\tau^m(P)$ lies on the straight line containing the segment $S_i + (m_1, m_2)$. This guarantees that if τ is not discarded, then $\hat{\tau}^n(S_j) \cap S_i = \varnothing$. For each of the countably many triples n, i, j, at most countably many lines of translations (where T is identified with $\mathbb{R}^2$) are discarded, and hence altogether only countably many lines of translations are discarded. Therefore, on any nondiscarded line, only countably many translations are discarded, and so there must be one, call it τ, left over. Now, τ is as desired; for if $D \cap \hat{\tau}^n(D)$ is uncountable, then some segment in D_1 must overlap with one in $\hat{\tau}^n(D_1)$, and any translation for which this could happen was discarded.

Let $C = \bigcup\{D \cap \hat{\tau}^n(D) : n = \pm 1, \pm 2, \ldots\}$. Then C is a countable subset of D and it follows from the claim about τ that the sets $\hat{\tau}^n(D \setminus C)$, $n \geq 0$, are pairwise disjoint, and disjoint from C. If A is the complement, in $J \setminus C$, of the union of this sequence of sets, then

$$J \setminus C = \left(\bigcup_{n=0}^{\infty} \hat{\tau}^n(D \setminus C) \right) \cup A \sim_T \left(\bigcup_{n=1}^{\infty} \hat{\tau}^n(D \setminus C) \right) \cup A = J \setminus D,$$

as required.

The G-equidecomposability of any two bounded sets with nonempty interior follows from the Banach–Schröder–Bernstein Theorem exactly as in the proof of Theorem 3.11, using squares instead of balls. Note that for any transformation s of the form $s(x, y) = (\alpha x, \alpha y)$, where $\alpha > 0$, $sGs^{-1} \in G$. It follows that a square of any size is G-paradoxical if the unit square is, and this allows the method of Theorem 3.11 to be used. $\square$

The proof of Theorem 8.5 is much simpler if the free generators σ_1 and σ_2 are such that F has no nontrivial fixed points in $\mathbb{R}^2 \setminus \{\mathbf{0}\}$ (e.g., the matrices of Thm. 8.1(a)). For then D, the set of fixed points of $\hat{F} \setminus \{e\}$, consists only of a countable set of points, and the last half of the proof is unnecessary. Furthermore, if one restricts σ_i to be the two Magnus–Neumann matrices, then there is a simple alternate route to the conclusion of Theorem 8.5. One needs only the group generated by the two Magnus–Neuman matrices, the translations $\mathbb{Z}^2$, and a single real translation τ such that $\tau(\mathbb{Q}^2) \cap \mathbb{Q}^2 = \varnothing$; see [Myc98].

The von Neumann Paradox shows how paradoxical decompositions in the plane completely analogous to the Banach–Tarski Paradox can be obtained if we are willing to step outside the isometry group. In fact, adjoining a certain single linear transformation to the isometry group G_2 is sufficient. More precisely, let σ be the shear of the plane defined by $\sigma(x, y) = (x + y, y)$; $\sigma = \left[\begin{smallmatrix} 1 & 1 \\ 0 & 1 \end{smallmatrix}\right] \in SL_2(\mathbb{Z})$. Let G_2^* be the group generated by G_2 and σ; then Theorem 8.5 holds with respect to G_2^*. Because $\left[\begin{smallmatrix} 1 & 2 \\ 0 & 1 \end{smallmatrix}\right]$ and its transpose are both in G_2^* (the former is σ^2, the latter is $\rho\sigma^{-2}\rho^{-1}$, where ρ is the rotation $\left[\begin{smallmatrix} 0 & -1 \\ 1 & 0 \end{smallmatrix}\right]$), and because $G_2^* \supseteq T$, the hypothesis and hence the conclusion of Theorem 8.5 are valid.

Next we focus on the situation where only linear transformations are used; we first sketch a proof (Cor. 8.10) that any square not containing the origin is $SL_2(\mathbb{R})$-paradoxical and then provide all details of Laczkovich's Theorem (Thm. 8.15) that any two bounded sets that have nonempty interior and positive distance from the origin are $SL_2(\mathbb{R})$-equidecomposable.

In §4.6.3 we proved that the strong form of the Banach–Tarski Paradox holds in the hyperbolic plane. The construction used a free group of piecewise isometries: The group acts on a bounded subset of $\mathbb{H}^2$ and has a countable set of fixed points. Here we present a similar technique for the action of $SL_2(\mathbb{R})$ on $\mathbb{R}^2 \setminus \{\mathbf{0}\}$. Namely, developing an idea suggested by Mycielski [Myc98], Tomkowicz [Tom11] showed that there is a free non-Abelian group of piecewise linear, orientation- and area-preserving transformations that acts on the punctured disk $D = \{(x, y) \in \mathbb{R}^2 : 0 < x^2 + y^2 < 1\}$ without fixed points. Here we sketch the main idea, giving an idea how the piecewise isometries are found, and derive a paradox.

We start with an easy lemma. Let ρ_θ denote counterclockwise rotation by θ, with ρ reserved for the 90° case. Let us say that an element of $SL_2(\mathbb{R})$ has *orthogonal eigenvectors* if it has two real eigenvalues and their eigenvectors are orthogonal.

Lemma 8.6. *For any $\phi \in SL_2(\mathbb{R})$ having orthogonal eigenvectors, $\rho\phi^{-1}(D) = \phi(D)$.*

Proof. We have that, for real λ, $\phi = A^{-1} \left[\begin{smallmatrix} \lambda & 0 \\ 0 & 1/\lambda \end{smallmatrix}\right] A$, where A is orthogonal. The result is then a consequence of the fact that $\phi(D)$ and $\phi^{-1}(D)$ are ellipses centered at the origin and congruent by ρ (Fig. 8.1). □

For any $\phi \in SL_2(\mathbb{R})$, define a piecewise linear transformation $\hat{\phi}: D \to D$ by

$$\hat{\phi}(X) = \begin{cases} \phi(X) & \text{if } \; X \in D \cap \phi^{-1}(D) \\ \rho(X) & \text{if } \; X \in D \setminus \phi^{-1}(D) \end{cases}.$$

Then $\hat{\phi}$ is a bijection of D.

Proof. If $Y \in \phi(D)$, then $\hat{\phi}(\phi^{-1}Y) = \phi(\phi^{-1}(Y)) = Y$, while if $Y \notin \phi(D)$, then let $X = \rho^{-1}(Y)$; then $\rho X \notin \phi(D)$, and so by Lemma 8.6, $X \notin \phi^{-1}(D)$, and so $\hat{\phi}(X) = \rho(X) = Y$. The function is one-one because it is easy to define an inverse using inverses of ϕ and ρ.

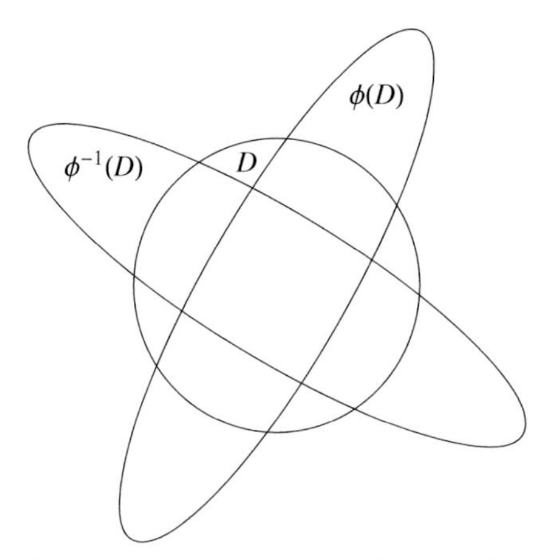

Figure 8.1. If ϕ has orthogonal eigenvectors, the two ellipses are congruent by a 90° rotation.

Let $\phi_1 = \left[\begin{smallmatrix} 3/2 & 0 \\ 0 & 2/3 \end{smallmatrix}\right]$ and $\theta_1 = 0$, $\theta_2 = \pi/3$, and $\theta_3 = 2\pi/3$. Consider the transformations $\phi_2 = \rho_{\theta_2}\phi\rho_{-\theta_2}$ and $\phi_3 = \rho_{\theta_3}\phi\rho_{-\theta_3}$. Let $\Lambda = \hat{\phi}_1\hat{\phi}_2\hat{\phi}_3$. Then, because $3/2 < \sqrt{3}$, we have $D \subset \phi^{-1}(D) \cup \phi_2^{-1}(D) \cup \phi_3^{-1}(D)$ (Fig. 8.2), and this gives property $(*)$.

$(*)$ For every $X \in D$, $\Lambda(X) = f_1 f_2 f_3(X)$, where f_i is one of $\{\phi_i, \rho\}$ and not all the f_i are ρ.

An eigenvalue computation shows that this particular Λ fixes no nonorigin point in D and is of infinite order. So this is the starting point: It is shown in [Tom11] how to define two piecewise bijections Φ and Ψ, each using a triple

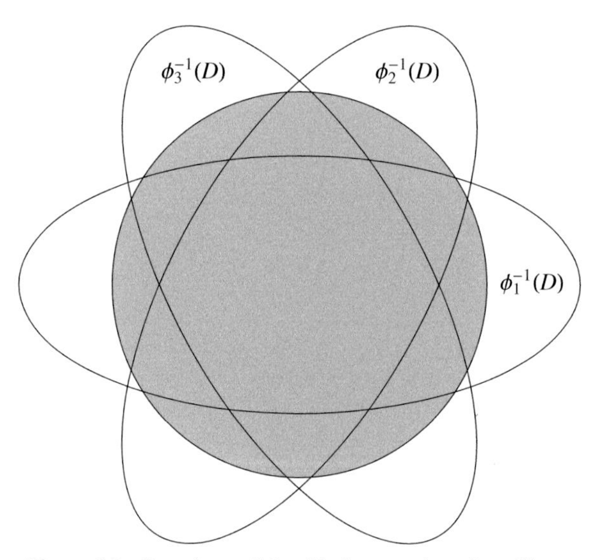

Figure 8.2. Covering a disk with three copies of an ellipse.

(ϕ_1, ϕ_2, ϕ_3) as earlier, but with a particular choice of matrix $\left[\begin{smallmatrix} \kappa & 0 \\ 0 & 1/\kappa \end{smallmatrix}\right]$ and rotation angles θ_i, so that they each satisfy $(*)$ and yield the next theorem. □

Theorem 8.7. *There are two bijections Φ and Ψ of D, each of which is a piecewise linear transformation in the style of the definition of Λ, that generate a free group that acts on D without fixed points. The basic functions ϕ_i can all be defined using the same value of κ, which can be taken arbitrarily close to 1.*

This theorem means that any punctured disk is paradoxical under the action of $SL_2(\mathbb{R})$. One wonders about a version of the strong Banach–Tarski Paradox. A proof would follow from a version of Theorem 8.7 for bijections of an annulus centered on the origin, but the methods of [Tom11] do not easily lead to such an extension. However, applying some results from graph theory and the fact that the action of $SL_2(\mathbb{R})$ on the punctured plane is locally commutative, Miklos Laczkovich [Lac99] proved that a version of the strong Banach–Tarski Paradox is possible. Namely, he proved that any two bounded subsets of the plane that have nonempty interior and are bounded away from the origin are $SL_2(\mathbb{R})$-equidecomposable. We present his proof at the end of this section.

As a simple application of Theorem 8.7, we show that the unit square less the origin is $SL_2(\mathbb{R})$-paradoxical. For the rest of this section, G denotes $SL_2(\mathbb{R})$. The next results will use one result from Chapter 10.

Lemma 8.8 (AC). *Let T be an open triangle with a vertex at the origin. Then T is G-equidecomposable with any punctured disk centered at the origin.*

Proof. Theorem 8.7 implies that any two punctured disks centered at the origin are equidecomposable. For suppose D_1 and D_2 are such disks, with $D_1 \subseteq D_2$. Then for some $\phi \in G$ and rotation ρ, $D_2 \subseteq \bigcup \rho^k \phi(D_1)$. But D_1 is paradoxical, so the union is G-equidecomposable to a subset of D_1, and the Banach–Schröder–Bernstein theorem applies.

Now let T_0 be a sector of a punctured disk lying inside T; let D_0 be the punctured disk with the same radius as T_0. Because D_0 is paradoxical and $T_0 \subseteq D_0$, Corollary 10.22 yields $T \succeq T_0 \sim_G D_0$. But $D_0 \sim_G D_1 \succeq T$. So T is G-equidecomposable with D_0 and hence with any punctured disk around the origin. □

Theorem 8.9 (AC). *Let A, B be bounded subsets of the punctured plane such that each contains an open triangle as in the preceding lemma. Then $A \sim_G B$.*

Proof. Choose punctured disks D_1 and D_2 containing A and B, respectively. Let T_1 and T_2 be the triangles contained in A and B, respectively. By Lemma 8.8, we have $T_1 \sim_G D_2$ and $T_2 \sim_G D_1$. Then $A \subseteq D_1 \sim_G T_2 \subseteq B \subseteq D_2 \sim_G T_1 \subseteq A$, so the Banach–Schröder–Bernstein theorem yields $A \sim_G B$. □

Corollary 8.10 (AC). *Any square, open or closed, that does not contain the origin is G-paradoxical.*

Laczkovich [Lac99] strengthened the preceding work, at least for sets that are bounded away from the origin, and we present his proof now. The key is relating decompositions to certain results of graph theory, so we begin with two lemmas that do that. For a family H of functions mapping a set X to itself and sets $A, B \subseteq X$, define the bipartite graph $\Gamma_H(A, B)$ on parts A and B by including edge $x \leftrightarrow y$ when $f(x) = y$ for some $f \in H$.

Lemma 8.11. (a) *Let G act on X and let $A, B \subseteq X$. Then $A \sim_G B$ iff there is a finite subset H of G such that $\Gamma_H(A, B)$ admits a perfect matching.*

(b) *Let Γ be a connected bipartite graph that has at most one cycle and is such that every vertex has finite degree that is at least 2. Then Γ has a perfect matching.*

Proof. (a) Immediate from the definitions.

(b) The cycle (if it exists) has even length and so admits a matching. Delete the cycle and all incident edges and view the remainder as a forest whose roots are vertices that were neighbors of a cycle-vertex. Each branch in the forest must be infinite, and an infinite branch has a perfect matching via alternate edges. So proceed by matching one maximal branch at a time, removing it and dealing with the residue in the same way. Strictly this is a countable induction always using the branch containing the first unused vertex. □

Using Lemma 8.11, Laczkovich showed that any two bounded subsets of $\mathbb{R}^2$ with nonempty interior and each of positive distance from the origin are $SL_2(\mathbb{R})$-equidecomposable. In particular, any annulus in the plane that is centered at the origin is $SL_2(\mathbb{R})$-paradoxical. We first need a general lemma that relates equidecomposability to local commutativity.

Lemma 8.12 (AC). *Suppose F, a free group generated by the elements $\{f_1, \ldots, f_n\}$, acts in a locally commutative way on a set X. Let $A, B \subseteq X$ be such that*

(i) *for any $x \in A$, $f_i(x) \in B$ for at least two is*
(ii) *for any $y \in B$, $f_i^{-1}(y) \in A$ for at least two is*

Then A and B are F-equidecomposable.

Proof. Define $A_i = A \cap f_i^{-1}(B)$ and let $H = \{f_i \restriction A_i : i = 1, \ldots, n\}$ and $\Gamma = \Gamma_H(A, B)$. By Lemma 8.11(a), we have to show that Γ admits a perfect matching. It is enough to show that any connected component Γ_1 of Γ satisfies the hypothesis of Lemma 8.11(b). The maximum degree of vertices in Γ is n, and conditions (i) and (ii) imply that every vertex has degree at least 2. It remains to show that Γ_1 contains at most one cycle.

For every edge $x \leftrightarrow y$ of Γ_1 with $x \in A$ and $y \in B$, choose i so that $f_i(x) = y$, and then define $\phi(x \leftrightarrow y) = f_i \in F$. Note that $\phi(x \leftrightarrow y)(x) = y$. If $W = (p_0, \ldots p_n)$ is a simple path or cycle in Γ that starts in A, then we define a word $\alpha \in F$, which we associate to W, by

$$\alpha = \phi(p_{n-1} \leftrightarrow p_n)^{-1} \ldots \phi(p_2 \leftrightarrow p_3)\phi(p_2 \leftrightarrow p_1)^{-1}\phi(p_0 \leftrightarrow p_1).$$

Then $\alpha(p_0) = p_n$. For W as before, we have that for each $i = 1, \ldots, n-1$, $p_{i-1} \neq p_{i+1}$ and hence $\phi(p_{i-1} \leftrightarrow p_i) \neq \phi(p_i \leftrightarrow p_{i+1})$; this implies that α is a reduced word in the group.

Suppose now that there are two distinct simple cycles in Γ_1: $C_1 = (p_0, \ldots, p_{n-1}, p_n = p_0)$ and $C_2 = (q_0, q_1, \ldots, q_{k-1}, q_k = q_0)$; shifting, if necessary, assume that if the cycles share a vertex, then $p_0 = q_0$ and $p_1 \neq q_1$. Because Γ_1 is connected, there is a path from C_1 to C_2; choose P to be the shortest such path, and assume that it connects p_0 to q_0 (if the cycles are not disjoint, take $P = \varnothing$). Then $P = p_0 \leftrightarrow t \leftrightarrow \ldots \leftrightarrow q_0$, where t, if it exists, is not in $C_1 \cup C_2$. Let α, β, γ be the group words associated to C_1, C_2, P, respectively. Then p_0 is a common fixed point of α and $\gamma^{-1}\beta\gamma$, and so the two words commute. Because we are in a free group, the words must be powers of a common word, and so there are nonzero integers j and m with $\alpha^m = (\gamma^{-1}\beta\gamma)^j$. Now, in the disjoint case, $\phi(p_0 \leftrightarrow p_1)$ differs from ϕ(first edge in P), while in the other case, it differs from ϕ(first edge in C_2). Therefore the rightmost term of α is not the rightmost term of $\beta\gamma$, which, because the group is free, contradicts the preceding equality of powers. □

We need two more lemmas that will also be useful in the one-dimensional work in §8.2.

Lemma 8.13. *Suppose a_i, b_i, c_i, and d_i are algebraically independent and consider matrices $f_i = \frac{1}{\sqrt{a_i d_i - b_i c_i}} \begin{bmatrix} a_i & b_i \\ c_i & d_i \end{bmatrix} \in SL_2(\mathbb{R})$. The matrices are independent and so are free generators of a free subgroup of $SL_2(\mathbb{R})$.*

Proof. Suppose a nontrivial reduced word $w = f_1^{n_1} \cdots f_k^{n_k}$ is the identity. Then $w = \begin{bmatrix} A & B \\ C & D \end{bmatrix}$, where the entries are polynomials in a_i, b_i, c_i, d_i having integer coefficients. The hypothesis then implies that $A = D = 1$ and $B = C = 0$, as polynomial identities. This means that substituting any matrices h_i in place of f_i into w yields the identity. Now let f and g be $\begin{bmatrix} 1 & 0 \\ 2 & 1 \end{bmatrix}$ and $\begin{bmatrix} 1 & 2 \\ 0 & 1 \end{bmatrix}$, respectively. They are independent (Prop. 4.4), and therefore $\{f^k g^k\}_{k \geq 1}$ are independent too. Let $h_i = f^i g^i$; then $h_1^{n_1} \cdots h_k^{n_k}$ is not the identity, a contradiction. □

Lemma 8.14. *There is a dense set of real numbers consisting of algebraically independent numbers.*

Proof. By induction on the set of rational intervals, using the fact that the set of numbers that are algebraic over a given finite set is countable, and so cannot exhaust an interval. □

Theorem 8.15 (AC). *If A, B are bounded subsets of $\mathbb{R}^2 \setminus \{\mathbf{0}\}$ with nonempty interior and having distance from the origin that is positive, then A and B are $SL_2(\mathbb{R})$-equidecomposable.*

Proof. Assume A and B are closed. This suffices, because the Banach–Schröder–Bernstein Theorem then yields the general result (for general A, B, use the special case on a closed superset of A and a closed subset of B, and vice versa). Clearly any $x \in \mathbb{R}^2 \setminus \{\mathbf{0}\}$ is mapped by some $\alpha_x \in SL_2(\mathbb{R})$ into the interior of B. Because any linear function and the entire action are continuous, there is $\epsilon_x > 0$ such that, for any number z and any $\alpha \in SL_2(\mathbb{R})$, $|z - x| < \epsilon_x$ and $\|\alpha - \alpha_x\| < \epsilon_x$ imply that $\alpha(z)$ is in the interior of B, where $\|\cdot\|$ denotes the norm that is the maximum absolute value of the four entries.

Because A is compact, there is a finite $K \subset A$ such that the open disks $B(x, \epsilon_x)$, for $x \in K$, cover A. Working in the other direction is the same: There are transformations β_y for each $y \in B$ and positive numbers η_y so that B is covered by the disks $B(y, \eta_y)$, for y in some finite subset L of B, and for any w and β so that $|w - y| < \eta_y$ and $\|\beta - \beta_y\| < \eta_y$, we have that $\beta(w)$ is in the interior of A.

Lemma 8.14 gives algebraically independent reals $a_{ij}^x, b_{ij}^x, c_{ij}^y, d_{ij}^y$ $(i, j = 1, 2, x \in K, y \in L)$ such that the as and bs are within ϵ_x of the corresponding coefficients in α_x, and similarly for the cs and ds with respect to β_y. Consider the elements of $SL_2(\mathbb{R})$ obtained by normalizing $\left[\begin{smallmatrix} a_{11}^x & a_{12}^x \\ a_{21}^x & a_{22}^x \end{smallmatrix}\right]$, and similarly using b, c, and d. By Lemma 8.13, these transformations generate a free subgroup F of $SL_2(\mathbb{R})$, whose action on the punctured plane is locally commutative (proved following Cor. 5.6). The hypotheses of Lemma 8.12 are satisfied, and so A and B are $SL_2(\mathbb{R})$-equidecomposable. □

Theorem 8.15 is a variation on the strong form of the Banach–Tarski paradox; a similar result holds in hyperbolic space (Thm. 4.17). So we have the general problem of when a metric space is "strongly paradoxical" in the sense that all bounded sets with interior are equidecomposable using isometries (or, more generally, area-preserving transformations). Are there conditions on a metric space and its isometry group that yield such a result about strong paradoxes?

8.2 Paradoxes of the Real Line

The idea of allowing affine maps, so fruitful in the plane, does not help with the line because the measure-preserving affine transformations are just the isometries. How much does the isometry group have to be expanded before a paradoxical decomposition of an interval arises? We care only about transformations that preserve Lebesgue measure, as there is no surprise in the fact that the function $2x$ can be used to duplicate an interval. Let $G(\lambda)$ be the group of all bijections f from $\mathbb{R}$ to $\mathbb{R}$ such that both f and f^{-1} are Lebesgue measurable and preserve Lebesgue measure: $\lambda(f^{\pm 1}(A)) = \lambda(A)$ if A is measurable. This group, a rather large extension of the isometry group, is rich enough to produce paradoxical decompositions on the line. The two results together are quite analogous to what happens in the hyperbolic plane (§§4.3, 4.6.3).

The next lemma was proved in [MT∞a], and it generalizes and simplifies the proof of Theorem 16 in [Sie54]. The lemma can be seen as a one-dimensional version of Theorem 3.9.

Lemma 8.16. *If C is a bounded subset of $\mathbb{R}^1$ with cardinality less than $2^{\aleph_0}$, then $[0, 1)$ and $[0, 1) \setminus C$ are equidecomposable using translations.*

Proof. Because C is bounded, there is $C_1 \subseteq [0, 1)$ with $C_1 \sim C$. View C_1 as a subset of the unit circle $\mathbb{S}^1$ in the usual way. For each $x \in D$, there are fewer than $2^{\aleph_0}$ rotations of the circle taking x to another point in D, so there is a rotation ρ such that the sets $\rho^n(D)$ are disjoint sets. Then using ρ and the identity gives a two-piece decomposition of $\mathbb{S}^1$ with $\mathbb{S}^1 \setminus D$. Unwrapping the points back to the interval gives the absorption of D. □

This lemma tells us that $[0, 1) \sim_{G(\lambda)} (0, 1)$, from which it follows that $[0, 1] \sim_{G(\lambda)} (0, 1] \sim_{G(\lambda)} [0, 1) \sim_{G(\lambda)} (0, 1)$.

Theorem 8.17 (AC). (a) *Any interval on the line is $G(\lambda)$-paradoxical. Any two bounded subsets of $\mathbb{R}$ with nonempty interior are $G(\lambda)$-equidecomposable.*
(b) *The real line is $G(\lambda)$-paradoxical using Borel sets.*

Proof. (a) It is a general fact of measure theory that there is a bijection $f : [0, 1) \to \mathbb{S}^2$ such that both f and f^{-1} take measurable sets to measurable sets and preserve measure (the measure on $\mathbb{S}^2$ is $\lambda/(4\pi)$, normalized surface Lebesgue measure). This follows from [Roy68, Thm. 9, p. 327], for example. These properties of f imply that if $\sigma \in SO_3(\mathbb{R})$, then $f^{-1}\sigma f$ is a measure-preserving bijection of $[0, 1)$ to itself (which, by periodic extension, may be considered as an element of $G(\lambda)$). Now, the Banach–Tarski Paradox states that $\mathbb{S}^2$ is $SO_3(\mathbb{R})$-paradoxical, via σ_i and A_i, say. It follows that $[0, 1)$ is $G(\lambda)$-paradoxical, using pieces $f^{-1}(A_i)$ and transformations $f^{-1}\sigma f$. This technique applies to any half-open interval, and the case of the other intervals follows from Lemma 8.16 and the remark following it. The second part follows from the usual technique using the Banach–Schröder–Bernstein Theorem (see Thm. 3.11).

(b) First we construct a rank-2 free subgroup of $G(\lambda)$ having no fixed points in its action on $\mathbb{R}$. This can be done using functions that are piecewise linear of slope 1. For every permutation π of $\mathbb{Z}$, define $f_\pi : \mathbb{R} \to \mathbb{R}$ by $f_\pi(x) = \pi(\lfloor x \rfloor) + \text{frac}(x)$. The graph of f_π is a collection of diagonals in the unit squares arising in the integer lattice $\mathbb{Z}^2$. It is easy to see that $f_{\pi \circ \rho} = f_\pi \circ f_\rho$.

Now, consider the free group $F = \langle \sigma, \tau \rangle$ and enumerate F as $\{w_i : i \in \mathbb{Z}\}$ in any constructive way. Define a permutation π of $\mathbb{Z}$ by $\pi(i) = j$ when $w_i\sigma = w_j$, and define ρ similarly using τ. Then the two permutations π, ρ generate a free group, and the corresponding functions f_π, f_ρ generate a free subgroup H of $G(\lambda)$ whose action on the real line has no fixed points. For this last, observe that if a word $w = \phi_1 \cdots \phi_n \in H$ fixed $x \in \mathbb{R}$, then it would fix $m = \lfloor x \rfloor$. But the sequence $m, \phi_n(m), \phi_{n-1}\phi_n(m), \ldots, w(m)$ cannot terminate in m because this sequence corresponds to a sequence of reduced words in F.

It is not hard to see that $[0, 1)$ is a choice set for the orbits of H. So the usual lifting of a paradox in F_2 (Cor. 3.7, Fig. 1.5) yields the desired paradox. Because $[0, 1)$ is a union of closed intervals, this paradox uses only F_σ sets. □

Von Neumann found an entirely different sort of paradox on the line, based on using linear fractional transformations as a way of bringing free groups of 2×2 matrices to bear. Let L be the group of linear fractional transformations of $\mathbb{R} \cup \{\infty\}$ of the form $x \mapsto (ax + b)/(cx + d)$, $ad - bc = 1$, where the usual arithmetic of ∞ is used when x is ∞ or $-d/c$. This is a group action of L on $\mathbb{R} \cup \{\infty\}$ and any $\sigma \in L$ is a strictly monotonic continuous function when restricted to an interval of $\mathbb{R}$ not containing $-d/c$. Then each $\sigma \in L$ can be identified with the coefficient matrix $\left[\begin{smallmatrix} a & b \\ c & d \end{smallmatrix}\right] \in SL_2(\mathbb{R})$, and L is isomorphic to $PSL_2(\mathbb{R})$.

Definition 8.18. *A transformation $\sigma \in L$ will be called a contraction with respect to an interval of $\mathbb{R}$ if, for some $\epsilon < 1$, $|\sigma(x) - \sigma(y)| \leq \epsilon|x - y|$ for all x, y in the interval.*

The linear fractional transformations do not contain any measure-preserving maps except the isometries, but Lebesgue measure λ behaves nicely with respect to contractions in the sense that it shrinks.

Proposition 8.19. *If $\sigma \in L$ is a contraction on $[a, b]$ and A is a measurable subset of $[a, b]$, then $\lambda(\sigma(A)) < \lambda(A)$.*

Proof. Let σ contract by ϵ. Then consider outer Lebesgue measure λ^*. Assume $A \subseteq (a, b)$. If A is covered by open subintervals (a_i, b_i) of (a, b), then the intervals $(\sigma(a_i), \sigma(b_i))$ cover $\sigma(A)$, and each $\sigma(b_i) - \sigma(a_i) < \epsilon(b_i - a_i)$. Hence $\lambda^*(\sigma(A)) \leq \epsilon\lambda(A)$, and because σ^{-1} is a measurable function, $\sigma(A)$ is a measurable set and $\lambda(\sigma(A)) = \lambda^*(\sigma(A))$. □

Now, if all sets were Lebesgue measurable, by the preceding result an interval could not be paradoxical using contractions. Indeed, the image of $[0, 1]$ under a piecewise contraction with respect to $[0, 1]$, using measurable pieces, has measure at most 1. But the von Neumann paradox on the line (Thm. 8.20) gives such a construction for contractions using arbitrarily small ϵ; of course, nonmeasurable sets are used. An elegant approach to this paradox is due to M. Laczkovich [Lac91b], who used ideas of graph theory. A map $f: A \to \mathbb{R}$ is a *piecewise contraction* if there is a finite partition $\{A_i : 1 \leq i \leq n\}$ of A such that each $f \restriction A_i$ is a contraction. We have the following theorem.

Theorem 8.20 (The von Neumann Paradox for the line) (AC). *Let I and J be two bounded intervals of $\mathbb{R}$. Then, for any positive ϵ, there is a bijection from I onto J that is a piecewise contraction with coefficient ϵ.*

The proof requires three lemmas. Let's use $\|f\|$ to denote the supremum of the absolute value of a bounded real function f on an interval.

Lemma 8.21. *Suppose $g(x) = \alpha x + \beta$, $[-M, M]$ is an interval, $\alpha > 0$, and ϵ is some positive number. Then there is a linear fractional transformation $f(x) = \frac{ax+b}{cx+d}$ such that a, b, c, d are algebraically independent, $ad - bc = 1$, and, on the interval $[-M, M]$, $\|g - f\| < \epsilon$ and $\|f'\|$ is within ϵ of $|\alpha|$.*

Proof. Take a, b, c, d to be in the algebraically independent set of Lemma 8.14 and lying within δ of $\alpha, \beta, 0, 1$, respectively, where the positive number δ will be determined separately for each of the two conclusions but is always assumed smaller than α, so that $a > 0$. Then the minimum of the two δ-values will guarantee that both conclusions hold.

Suppose the exact differences between a, b, c, d and $\alpha, \beta, 0, 1$ are given by δ_i, respectively. Then, after some simplification using the triangle inequality and $|\delta_i| < \delta$, the difference between α and $f'(x)$, in absolute value, is bounded by

$$\frac{\delta(3 + |\beta| + (M^2 + 4M + 2)|\alpha|)}{(1 + x\delta_3 + \delta_4)^2}.$$

So it is easy to choose δ so that the numerator is less than $\epsilon/4$. To deal with the denominator, make sure that $\delta < \min[1/6, 1/(3M)]$. It follows that $|\delta_3 x + \delta_4| < 1/2$ for all x in the interval. Then $(1 + x\delta_3 + \delta_4)^2 > 1/4$ and the reciprocal is under 4, which means the overall absolute difference from α is under ϵ.

For the other condition, compute the difference $|f(x) - g(x)|$, which is bounded by

$$\frac{\delta(M + 1)(M|\alpha| + |\beta| + 1)}{|1 + \delta_3 x + \delta_4|}.$$

As in the other case, it is easy to choose δ so that this quantity is under ϵ. By choosing c, d to be sufficiently close to 0, 1 respectively, the determinant is close to a, which is positive. A rescaling then gets $ad - bc$ equal to 1 exactly without changing the function. □

Lemma 8.22. *Let J_1 and J_2 be bounded open intervals. Then there are real numbers $c_1, \ldots, c_{k_1}, c_{k_1+1}, \ldots, c_{k_1+k_2}$ such that $\{c_i + \frac{1}{2}J_1 : i \leq k_1\}$ and $\{c_i + \frac{1}{2}J_1 : k_1 + 1 \leq i \leq k_1 + k_2\}$ each cover J_2.*

Proof. Easy exercise. □

Lemma 8.23. *Let $\{(a_i, b_i) : i = 1, \ldots, n\}$ be a covering of an interval $J = [a, b]$. Then there is $\epsilon > 0$ such that $\{(a_i + \epsilon, b_i - \epsilon) : i = 1, \ldots, n\}$ also covers J.*

Proof. Let ϵ be less than half of the smallest difference between any two distinct values chosen from the set of all a_i, b_i and also a, b. □

Proof of 8.20. The proof here will work for $\epsilon = 1/2$, but all the lemmas, and the proof, hold just as well when some positive integer N replaces 2 and ϵ is $1/N$. We may assume that I and $J = (a_0, b_0)$ are open and that the length of I is not greater than $b_0 - a_0$, because otherwise there is a simple contraction from I to J. Let I_1 and I_2 be two disjoint open subintervals of I. We need linear fractional transformations $f_1, \ldots, f_n$, with algebraically independent coefficients, that satisfy the following:

(i) $f_i \restriction I_1$ is a contraction
(ii) for any $x \in I_1$, $f_i(x) \in J$ for at least two distinct is
(iii) for any $y \in J$, $f_i^{-1}(y) \in I_1$ for at least two distinct is

Apply Lemma 8.22 to get $\{c_i\}$, yielding two covers, $\mathcal{C}_1, \mathcal{C}_2$, of J using transformations of I_1 by linear functions by g_i having linear coefficient $1/2$. We will show how to define f_i by Lemma 8.21 to get each of the three conditions. Then to get all three, one can use the minimum of the three values of ϵ.

(i) Use any ϵ less than $1/2$ in Lemma 8.21. Because of the derivative condition, $\|f_i'\| < 1$ on I_1, yielding (i).

(ii) Let ϵ be smaller than any distance from an endpoint of the interval in either cover $\mathcal{C}_j$ to either a_0 or b_0. Use this ϵ to define the family $\{f_i\}$ by Lemma 8.21. Let y be the midpoint of J. It lies in some interval of each cover $\mathcal{C}_j$, and each of these two intervals must be entirely contained within J. Let $x \in I_1$. Because $f_i(x)$ is within ϵ of $g_i(x)$, it follows that $f_i(x) \in J$.

(iii) Choose ϵ as in Lemma 8.23 so that it works for both covers $\mathcal{C}_j$, and choose f_i by Lemma 8.21 for this ϵ. Let $\mathcal{D}_1$ and $\mathcal{D}_2$ be the perturbed covers: Make each interval in $\mathcal{C}_j$ smaller by ϵ at each end. Any $y \in J$ lies in an interval from each of $\mathcal{D}_j$. Let f_m be the function corresponding to the interval in $\mathcal{D}_1$. Because $f_m(I_1)$ is an interval whose endpoints are no more than ϵ from the endpoints of $g_m(I_1)$, it follows that $f_m^{-1}(y) \in I_1$, and the same reasoning applies to the second interval.

Let $H = \{f_1, \ldots, f_n\}$ and $\Gamma = \Gamma_H(I_1, J)$; by (ii) and (iii), the degree of each vertex of Γ is at least 2. Let F be the group generated by the f_i; by Lemma 8.13 and because composition of these functions corresponds to matrix multiplication, F is a free group on these generators. Let C be the union of those connected components of Γ that contain at least one fixed point of a nonidentity element of F; each function has at most one fixed point, and so C is countable. Now consider $y \in \Gamma \setminus C$ and observe that the component containing y does not contain a cycle, because such would lead to a word in H that, by freeness, cannot be the identity; and nontrivial fixed points were taken care of by the exclusion of C. Therefore, by Lemma 8.11(b), the component containing y has a perfect matching. This is true for any component in $\Gamma \setminus C$, and so $\Gamma_H(I_1 \setminus C, J \setminus C)$ has a perfect matching. By Lemma 8.11(a), there is a bijection f from $I_1 \setminus C$ onto $J \setminus C$ that is a piecewise contraction.

Because $J \cap C$ is countable, there is an injection $g: J \cap C \to I_2$ such that g^{-1} is a piecewise contraction. To define g, cover the interval J by the intervals $A_1, \ldots, A_m$ of length less than half the length of I_2. Determine a translation t_1 such that $t_1(A_1) \subseteq \frac{1}{2}I_2$. Now let $C_2 = (C \setminus A_1) \cap A_2$. Because there are countably many points in $t_1(A_1 \cap C)$ and also in C_2,there are countably many translations that take a point in C_2 to a point in $t_1(A_1 \cap C)$. Hence we can pick a translation t_2 such that $t_2(A_2) \subseteq \frac{1}{2}I_2$ and $t_2(C_2)$ is disjoint from $t_1(A_1 \cap C)$. Then we continue in the same way until we get a translation t_m. We have the simple expansion $F(x) = 2x$, a bijection from $\frac{1}{2}I_2$ to I_2. Hence the function g defined from the functions $f \circ t_i$, for $i = 1, \ldots, n$, is the desired injection.

Define the map g_1 by $g_1(x) = f^{-1}(x)$, if $x \in J \setminus C$ and $g_1(x) = g(x)$ if $x \in J \cap C$. Then g_1 is a bijection from J onto a subset of I. Now taking a contraction h

from I into J, we can apply the Banach–Schröder–Bernstein Theorem regarding the piecewise contractions g_1^{-1} and h. □

The theorem can be extended to more general sets as follows.

Corollary 8.24 (AC). *Let A and B be bounded subsets of $\mathbb{R}$ with nonempty interior. Then there is a piecewise contraction from A onto B.*

Proof. Take intervals I, J so that $I \subseteq A$ and $B \subseteq J$. Let g be an injective contraction from A into the interior of B. By Theorem 8.20, there is a piecewise contraction h that is a bijection from I to J. Let h_0 be the restriction of h to $h^{-1}(B)$. Then h_0^{-1} is an injection from B into A. Now it is enough to apply a variant of the Banach–Schröder–Bernstein Theorem to get partitions $A = A_1 \cup A_2$ and $B = B_1 \cup B_2$ such that $g(A_1) = B_1$ and $h_0(A_2) = B_2$. Then the map f defined to be g on A_1 and h_0 on A_2 is the claimed piecewise contraction. □

Corollary 8.25 (AC). *Suppose $\mu: \mathcal{P}(\mathbb{R}) \to [0, \infty)$ is a finitely additive measure with $\mu([0, 1]) = 1$. Then for any $\epsilon, K > 0$, there is a set $A \subseteq [0, 1]$ with $\mu(A) > 0$ and σ, a contraction of $[0, 1]$ with factor ϵ, such that $\mu(\sigma(A)) \geq K\mu(A)$. Furthermore, there is a set $B \subseteq [0, 1]$ and a Lebesgue measure-preserving bijection τ of $[0, 1]$ to itself such that $\mu(\tau(B)) \neq \mu(B)$.*

Because of the strong version of the Banach–Tarski Paradox, it is clear how to get a version of Theorem 8.20 in $\mathbb{R}^3$: shrink a ball radially as much as desired and then use isometries to get as large a ball as desired. Sierpiński [Sie46, Sie48b] has shown how a contraction-type paradox in the plane can be derived directly from the Banach–Tarski Paradox of the sphere. He proved that for any $r > 0$, there is a bijection f from the unit disk to the disk of radius r that, piecewise, contracts distances.

Some further results in this area are of interest. Assume that there is a piecewise contraction, with n pieces, mapping A onto B. An outer measure argument as in Lemma 8.14 shows that $\lambda^*(B) \leq n\lambda^*(A)$, where λ^* is outer Lebesgue measure. Thus if A and B are Lebesgue measurable, we have $\lambda(B) \leq n\lambda(A)$. But Laczkovich [Lac92c] showed even more: $\lambda(B) \leq \frac{n}{2}\lambda(A)$, and further, $\frac{n}{2}$ is best possible. To be precise, he proved the following theorem. We omit the proof but note that a main step is Lemma 8.12.

Theorem 8.26 (AC). *(a) Suppose A and B are Lebesgue measurable subsets of $\mathbb{R}$ and f is a piecewise contraction from A onto B using n pieces. Then $\lambda(B) < \frac{n}{2}\lambda(A)$.*

(b) Suppose A is a Lebesgue measurable subset of $\mathbb{R}$ and J is an interval of length less than $\frac{n}{2}\lambda(A)$ for some integer n. Then there is a surjective piecewise contraction $f: A \to J$. If A is an interval, then f can be taken to be a bijection.

This theorem implies that the assertion about $n/2$ is best possible, and it also tells us that there is a von Neumann paradox using just three pieces. Take

$I = [0, 1]$ and $J = [0, 5/4]$. There is no such paradox using two pieces, because part (i) of the theorem would imply that $\lambda(J)$ is smaller than $\lambda(I)$, so the situation is not paradoxical.

Let I and J be intervals on the line. Laczkovich showed in [Lac88a] that if the length of J is less than twice the length of I, then the von Neumann paradox can be realized using three translations and one contraction. And he showed that this is impossible if the length condition is false.

Notes

The extension of Proposition 4.4 to larger off-diagonal numbers is due to Brenner [Bre55] (see also [MK66, p. 100]). Brenner also characterized the matrices that appear in the group generated by $\left[\begin{smallmatrix} 1 & m \\ 0 & 1 \end{smallmatrix}\right]$ and its transpose ($m \geq 2$). Theorem 8.1(c), presented here for the first time, solves Problem 3 in [MW84]. Further results on independent pairs of 2×2 matrices may be found in [CJR58, GN57, LU68, LU69, Mag73, Mag75, Ree61]. Independent pairs in $PSL_2(\mathbb{Z})$ that generate a group consisting of only hyperbolic elements were considered by Magnus [Mag73], who used work of Neumann [Neu33]; see also [Myc77b].

The idea of expanding the isometry groups in a way that produces a generalization of the Banach–Tarski Paradox in the plane and on the line is due to von Neumann [Neu29]. He showed that the unit square was $SA_2(\mathbb{R})$-paradoxical; although he made use of the fact that $SL_2(\mathbb{Z})$ has pairs of independent elements, his final proof used pairs in $SL_2(\mathbb{R}) \setminus SL_2(\mathbb{Z})$, defined from small, algebraically independent numbers. The version of Theorem 8.5 presented here and the observation that the addition of a single shear to the planar isometry group is sufficient to produce paradoxes are due to Wagon [Wag82]. This latter result was motivated by work of Rosenblatt [Ros81] on the uniqueness of Lebesgue measure as a shear-invariant measure, which is discussed in Chapter 13.

The existence of paradoxes on the line, using linear fractional transformations, was proved by von Neumann [Neu29]. Similar results for the plane were obtained by Sierpiński [Sie48b]. Another result related to pathology involving contractions of nonmeasurable sets can be found in [Juz82]. The ideas used in Theorem 8.17(b) were proposed independently by Kandola and Vandervelde [KV15], and by Tomkowicz.

9

Squaring the Circle

The study of equidecomposability and measures led Tarski to formulate a problem in 1925 that was one of the most perplexing in the field, and especially intriguing because of its kinship to the ancient geometrical problem of squaring the circle. The strong Banach–Tarski Paradox implies that, using arbitrary sets, a solid ball in $\mathbb{R}^3$ is equidecomposable to a cube—indeed, to a cube of *any* size. But in the plane, a Banach measure (Cor. 12.9) means that such a strong result is not possible: If a disk is equidecomposable with a square, then the two regions must have the same area. This leads naturally to Tarski's Circle-Squaring Problem: Is a disk in the plane equidecomposable with a square of the same area?

For sixty-five years, there was little progress, with various partial results that provided no clue to the answer to Tarski's problem. Things seemed not so different from the situation faced by Greek geometers: Polygons presented no great difficulty (the Bolyai–Gerwien Theorem squares any polygon; Thm. 3.8), but the passage to the circle was a mystery. Then, in a tour de force in 1990, Miklós Laczkovich showed that not only can one square the circle in this sense but it can be done using translations alone. In this chapter, we give an essentially complete proof of this surprising result. The proof calls upon several fields of mathematics (analysis, number theory, geometry, set theory), and it is amazing that one person was able to put it all together. It is surely the single most impressive result in this whole area.

9.1 Changing the Group

When one makes a strong geometrical restriction, the circle-squaring problem can be resolved negatively. We use the term *Jordan domain* to refer to a Jordan curve (a simple closed curve) together with its interior. Call two Jordan domains *scissors-congruent* if one can be decomposed into finitely many pairwise interior-disjoint Jordan domains that, ignoring boundaries, can be rearranged by isometries to form the other. This generalizes the notion of congruence by dissection (Def. 3.1) in which the domains are assumed to be polygons (i.e., the scissors can cut only on straight lines). Thus any polygon is scissors-congruent to a square of the same area. Dubins, Hirsch, and Karush [DHK63] investigated whether the ability to cut along arbitrary Jordan curves allows a circle to be squared. They proved that the answer is NO.

Theorem 9.1. *A square is not scissors-congruent to a disk.*

In fact, Dubins, Hirsch, and Karush established that a disk (more generally, a solid ellipse) is scissors-congruent to no other convex region. Theorem 9.1 has the following corollary for Tarski's Problem.

Corollary 9.2. *A disk is not equidecomposable to a square if the pieces of the decomposition are restricted to interiors of Jordan curves or arcs of such curves.*

The result of the corollary is not known when one expands the allowable pieces to the Borel sets. For an example of two Jordan domains that are not scissors-congruent but are Borel equidecomposable, see [Gar85b].

There is a similarity between Tarski's question and Question 3.13, because both deal with equidecomposability of elementary figures in a context that precludes paradoxical decompositions. Returning to scissors-congruence, we give a short proof of Theorem 9.1 when the pieces are assumed to have rectifiable boundaries. The full proof is an intricate argument in plane topology.

Proof of Theorem 9.1, Rectifiable Case. Let C be a fixed circle. If E is a union of finitely many pairwise interior-disjoint regions, each of which is the interior and boundary of a rectifiable Jordan curve, then the boundary of E, ∂E, is rectifiable as well. Hence we may define a number $\mu(E)$ as follows. If A is an arc of ∂E, then A is called *convex relative to* E if the convex hull of A is contained in E; A is called *concave relative to* E if the convex hull of A is disjoint from the interior of E. For any point P on ∂E, let $f_E(P) = +1$ (resp., -1) if P is contained in the interior of an arc A of ∂E such that A is convex (resp., concave) relative to E, and A is congruent to an arc of C. Then let $\mu(E) = \int_{\partial E} f_E ds$, the integral of f_E on ∂E with respect to an arc-length parametrization of ∂E. Now, μ is clearly isometry-invariant and, because of the cancellation that arises when a concave piece of arc is matched with a convex piece, $\mu(E_1 \cup E_2) = \mu(E_1) + \mu(E_2)$ if the two sets are interior disjoint. It follows that μ is invariant under scissors-congruence. Because μ of a disk is the circumference of the bounding circle, while μ assigns any square the value 0, this proves the rectifiable case of the theorem. □

Some positive results can be easily derived for variations of the problem. If, in addition to isometries, we allow the shear $\sigma = \left[\begin{smallmatrix} 1 & 1 \\ 0 & 1 \end{smallmatrix}\right]$, then Theorem 8.5 can be applied to turn a disk to a square (see the remarks following the proof of Thm. 8.5). But this has little relevance to the spirit of Tarski's problem, because the group then allows the disk to be turned into a square of any size.

Another extension of the isometry group yields a different sort of positive result. Let G be the group of all similarities of the plane, that is, the group generated by all isometries and all magnifications from a point. Because isometries are affine, G consists of all transformations $d\sigma$ where $\sigma \in G_2$ and $d = \left[\begin{smallmatrix} \alpha & 0 \\ 0 & \alpha \end{smallmatrix}\right]$, $\alpha > 0$, is a magnification from the origin. A similarity $\left[\begin{smallmatrix} \alpha & 0 \\ 0 & \alpha \end{smallmatrix}\right]\sigma$, with $\sigma \in G_2$, is called *ϵ-magnifying* if $1 - \epsilon \leq \alpha \leq 1 + \epsilon$.

Theorem 9.3. *For any $\epsilon > 0$, a disk is G-equidecomposable with a square of the same area using similarities that are ϵ-magnifying. Moreover, the pieces in the decomposition are Borel sets.*

Proof. Let C be the unit disk and S a square of side-length $\sqrt{\pi}$. By the Banach–Schröder–Bernstein Theorem and its constructive proof, it is sufficient to find a piecewise ϵ-magnifying similarity from C to a subset of S, and the same from C to a superset of S. For then $C \preceq S \preceq C$, and because the proof of Theorem 3.6 introduces no new similarities, this implies that $C \sim_G S$ using ϵ-magnifications.

Choose n so large that the disk, when shrunk radially by $1 - \epsilon$, fits inside a regular n-gon inscribed in C. Because, by Tarski's version of the Bolyai–Gerwien Theorem (Thm. 3.8), the polygon is equidecomposable (using isometries) with a square smaller than S, C can be packed into S too, provided it is preshrunk by $1 - \epsilon$. To get a superset of S, choose n so large that the disk, when expanded radially by $1 + \epsilon$, contains the regular n-gon circumscribed about C, and proceed in the same way. It is easy to see that the results used in this proof—Theorems 3.2, 3.6, and 3.8—do not introduce any non-Borel sets. □

Tarski's Problem is equivalent to asking for Theorem 9.3 to hold with $\epsilon = 0$ (but without the Borel restriction). Among the many measures constructed in Part II will be a finitely additive extension μ of Lebesgue measure in the plane with the property that $\mu(s(A)) = \alpha^2\mu(A)$ for each similarity s, where α is the amount by which s magnifies distances (Cor. 13.5). It follows that Theorem 9.3 is false if the square's area does not equal that of the circle.

9.2 The Squaring of the Circle

In this section we give a complete proof of Laczkovich's positive answer [Lac90, Lac92b] to Tarski's Circle-Squaring Problem from 1925. The presentation is self-contained, except for its use of the classic, not difficult, Hall–Rado Theorem of graph theory and some results of number theory that are variations of the Erdős–Turán Theorem. The level of complexity and intricacy in this work is high, but the result is so beautiful and unexpected that it is worthy of such a detailed study here. The discussion in this section is heavy on notation, so we list here the important items for convenience:

- $J = [0, 1) \times [0, 1)$.
- E is the standard basis of two vectors in $\mathbb{R}^2$.
- For a Jordan curve K, K^* is the Jordan domain made up of the region inside or on the curve.
- For a polygon P, $p(P)$ is its perimeter.

- A *unit square* is a half-open square $[a, a+1) \times [b, b+1)$, where $a, b \in \mathbb{Z}$.
- For a lattice polygon P, $\hat{P}$ is the union of all the unit squares contained in P^*.
- $\langle x \rangle$ is the distance from a real number x to the nearest integer; frac(x) is $x - \lfloor x \rfloor$, the fractional part of x.
- For A a finite subset of J and H a measurable subset of the plane, the *discrepancy* of A with respect to H is $D(A, H) = \left| \frac{|A \cap H|}{|A|} - \lambda(H) \right|$.
- For a discrete subset S and a measurable subset H of the plane, $\Delta(S, H) = ||S \cap H| - \lambda(H)|$.
- The *Z-neighborhood* of a set X is $U(X, Z) = \{y : \exists \vec{x} \in X$ such that $\|\vec{x} - \vec{y}\| \leq z\}$, where $\|\cdot\|$ is Euclidean distance.
- For $\vec{u} \in \mathbb{R}^2$, X a finite subset of J, and a positive integer N, F_N is defined by $F_N(\vec{u}, X) = \{\text{frac}(\vec{u} + n_1\vec{x}_1 + n_2\vec{x}_2) : n_i = 0, \ldots, N-1\}$.
- $X \sim_T Y$ means that X and Y are equidecomposable by translations.
- $\ell(x) = \max(2, \ln x)$.
- Ψ is used for a function from $\mathbb{N}$ to $[0, \infty)$ such that $\sum_{k=0}^{\infty} \Psi(2^k)/2^k$ is convergent.
- For positive C, $N(C)$ is the least positive integer N such that $16(N+1)^2 < (1 + \frac{1}{4C})^N$.
- Given vectors $\vec{u}$ and $X = \{\vec{x}_1, \vec{x}_2\}$ and a plane set H, $S(\vec{u}, H) = \{\vec{n} \in \mathbb{Z}^2 : \text{frac}(\vec{u} + \vec{n} \cdot X) \in H\}$.

Our presentation focuses on the essentials. For a deeper discussion of the motivation underlying the methods, see [Lac89, GW89].

9.2.1 Preliminaries: Six Lemmas

To prove the powerful Equidecomposability Criterion (§9.2.2), we need several lemmas. First, some notation. We use p for the perimeter of a polygon (and also for the perimeter of the complement of a polygon); a *lattice point* is any element of $\mathbb{Z}^2$. In this section, a *unit square* is a half-open square of the form $[a, a+1) \times [b, b+1)$ where (a, b) is a lattice point; J denotes the fundamental unit square $[0, 1)^2$. Let $\mathcal{H}$ be the family of all nonempty unions of finitely many unit squares. A *lattice polygon* P is an orthogonal polygon with lattice points as vertices. If it is a square, it is called a *lattice square*; for such a square P, $s(P)$ denotes its side-length. For a Jordan curve K, let K^* refer to the closed set that is K together with its interior. When P is a lattice polygon, let $\hat{P}$ be the union of all the unit squares contained in P^*; $\hat{P}$ is just P^* less some bounding segments and is an element of $\mathcal{H}$ (Fig. 9.1).

Lemma 9.4. *If $H \in \mathcal{H}$, one (perhaps more) of the following assertions is true:*

(i) *$H = \hat{P}$ for some lattice polygon P.*
(ii) *There are disjoint sets $H_1, H_2 \in \mathcal{H}$ with $H = H_1 \cup H_2$ and $p(H) = p(H_1) + p(H_2)$.*
(iii) *There are sets $H_1, H_2 \in \mathcal{H}$ such that $H_1 \subset H_2$, $H = H_2 \setminus H_1$, and $p(H) = p(H_1) + p(H_2)$.*

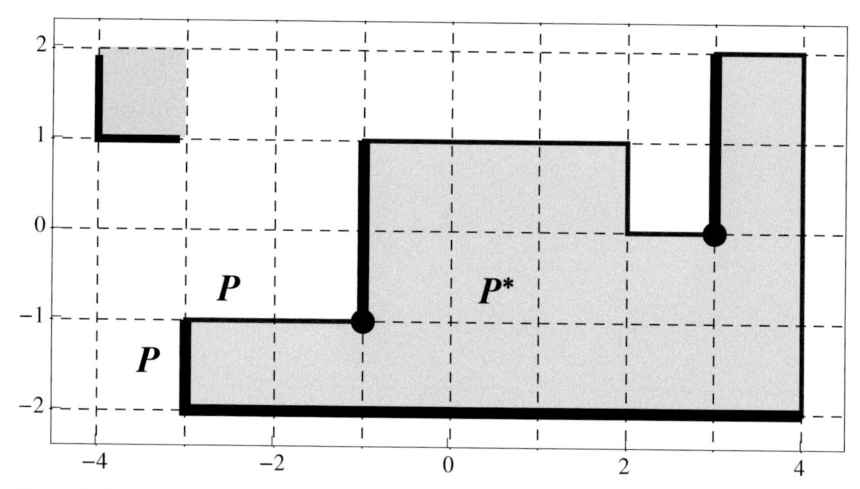

Figure 9.1. A unit square is at upper left; the boundary of the large gray region is a lattice polygon P; the domain P^* is the gray region together with the full boundary; $\hat{P}$ is the open gray region together with the thick black border and the two large points.

Proof. Let $B = \partial H$, the set of segments forming the boundary of H. If B is just one simple polygon, we have (i) with $H = \hat{B}$ (Fig. 9.2(a)). If B is not a single simple polygon, then break any nonsimple polygon in B into a set of simple polygons and discard any that lie inside any other. If the resulting collection (three simple polygons in Fig. 9.2(b)) is not just a single simple polygon, we have (ii), where H_1 and H_2 might share finitely many boundary points or have disjoint boundaries.

There remains the case that B consists of a simple polygon P with simple or nonsimple polygons inside it (Fig. 9.2(c)). Let H_1 consist of all the squares inside P, except those in H. Let $H_2 = H \cup H_1$, a disjoint union. □

Call two distinct unit squares *adjacent* if their boundaries have a common segment. The *derived set* H' of some $H \in \mathcal{H}$ is the union of H and unit squares adjacent to at least one square in H. Starting with $H^{(0)} = H$, iteration of the

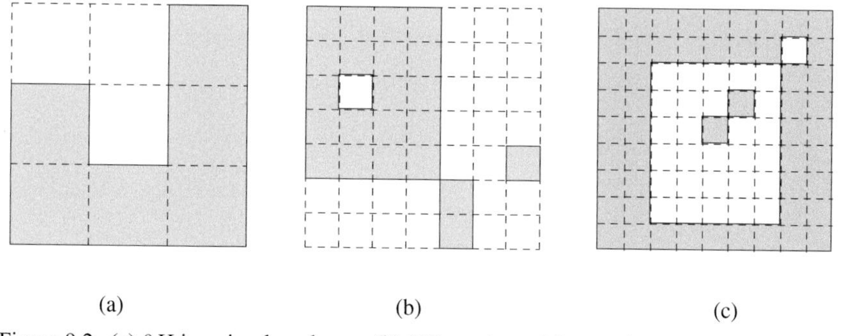

Figure 9.2. (a) ∂H is a simple polygon. (b) ∂H consists of three polygons, one of which is not simple. (c) ∂H includes one all-encompassing simple polygon.

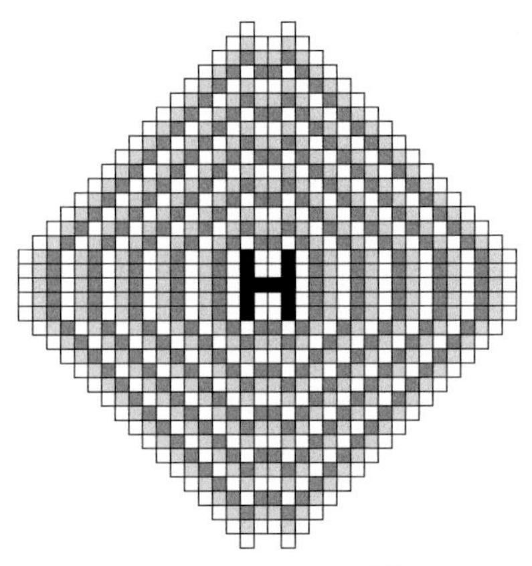

Figure 9.3. The derived sets up to $H^{(16)}$ starting from H.

process leads to the *derived sequence*: $H^{(n+1)}$ is the derived set of $H^{(n)}$ (Fig. 9.3). This set approaches a diamond as n increases.

The diamond shape of derived sets means that, very roughly, the area of $H^{(n)}$ is quadratic in n, which is ultimately greater than any constant times the perimeter, which is linear in n. The next lemma makes this precise.

Definition 9.5. *For any positive C, let $N(C)$ be the least positive integer N such that* $16(N+1)^2 < (1+\frac{1}{4C})^N$.

Lemma 9.6. *Suppose $C > 0$ and $H \in \mathcal{H}$. Then there is a positive integer $n \leq N(C)$ so that $\lambda(H^{(n)}) \geq Cp(H^{(n)}) + \lambda(H)$. The same is true if H is instead the complement of an element of $\mathcal{H}$.*

Proof. Abbreviate $N(C)$ to N. Let $p_n = p(H^{(n)})$, a positive integer; then $\partial H^{(n)}$ consists of p_n unit-length segments. Each such segment is part of the boundary of a square in $H^{(n+1)} \setminus H^{(n)}$. But each such square contains at most four such segments. Therefore

(1) $p_n \leq 4\lambda(H^{(n+1)} \setminus H^{(n)})$.

Suppose the lemma is false, so that for all positive integers $n \leq N$,

$$\lambda(H^{(n)} \setminus H) < Cp_n \leq 4C\lambda(H^{(n+1)} \setminus H^{(n)}) = 4C\lambda(H^{(n+1)} \setminus H) - 4C\lambda(H^{(n)} \setminus H).$$

This implies $\lambda(H^{(n+1)} \setminus H) > (1+\frac{1}{4C})\lambda(H^{(n)} \setminus H)$ and therefore, using (1) for p_0,

(2) $\lambda(H^{(N+1)} \setminus H) > (1+\frac{1}{4C})^N \lambda(H' \setminus H) \geq (1+\frac{1}{4C})^N \frac{p_0}{4}$.

We next prove $\lambda(H^{(N+1)} \setminus H) \leq 4(N+1)^2 p_0$, which, with (2), contradicts the definition of N. Let Q be a unit square contained in $H^{(N+1)} \setminus H$. Then there is a sequence $Q_0, Q_1, \ldots$ of at least two and at most $N+2$ adjacent unit squares so that $Q_0 \subseteq H$ and the last one is Q. Because the closures of Q_1 and H intersect, we can choose $\vec{z}$ to be a lattice point in $\partial Q_1 \cap \partial H$. Let T be the lattice square with center $\vec{z}$ and side-length $2N$; then $Q \subset \hat{T}$. Because ∂H contains at most p_0 lattice points, this method covers $H^{(N+1)} \setminus H$ with p_0 squares, each having area $4(N+1)^2$; this gives the last inequality needed to conclude the proof. The complementary case is identical. □

Next comes a short technical lemma that makes use of the preceding one.

Lemma 9.7. *Let N denote $N(C)$. Then for any $H \in \mathcal{H}$ and $C > 0$, there is $K \in \mathcal{H}$ such that*

(a) $H \subseteq K \subseteq H^{(N)}$
(b) $\lambda(H^{(N)}) \geq Cp(K) + \lambda(K)$

Proof. Apply Lemma 9.6 to $A = \mathbb{R}^2 \setminus H^{(N)}$ to get a positive integer $n \leq N$ such that

(1) $\lambda(A^{(n)} \setminus A) \geq Cp(A^{(n)})$.

Define K to be $\mathbb{R}^2 \setminus A^{(n)}$; then $K \subset \mathbb{R}^2 \setminus A = H^{(N)}$ (see Fig. 9.4). Suppose (a) fails. Then there is a unit square Q contained in H but not in K. Then $Q \subset A^{(n)}$, and so there is a sequence of at most $n+1$ adjacent unit squares going from Q to a square $Q_0 \subseteq A$. Because $Q \subseteq H$, this implies $Q_0 \subseteq H^{(n)} \subset H^{(N)} = \mathbb{R}^2 \setminus A$, which

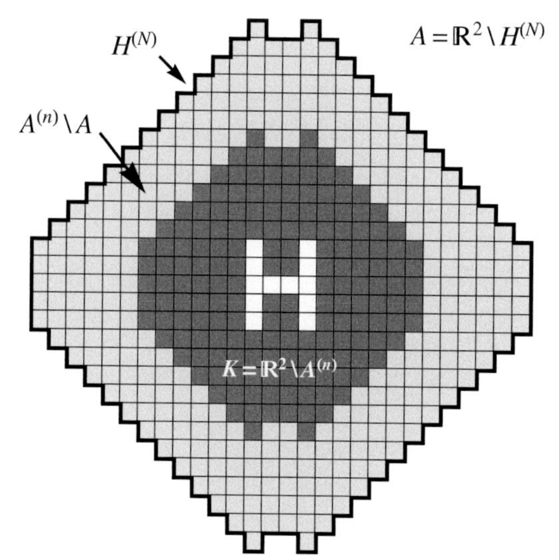

Figure 9.4. The central dozen white squares are H; the collection of all squares shown is $H^{(12)}$; the exterior of that set is A, the complement of $H^{(12)}$. The light gray squares are $A^{(6)} \setminus A$, while the dark gray squares, together with H, form the complement of $A^{(6)}$. In this example, suppose $C = 4$. Then the area of $H^{(12)}$ is 496, greater than area$(K) + Cp(K) = 184 + 4 \cdot 70$.

is impossible. For (b), $A^{(n)} \setminus A = H^{(N)} \setminus K$ and $p(K) = p(A^{(n)})$, and the inequality follows from (1). □

The key circle-squaring idea is the construction of several types of bijections involving lattice points. The next lemma is the first of several such constructions. For a bijection f between two sets in metric spaces, define the *spread* of the bijection to be $\sup_{x \in \text{dom}F} \|f(x) - x\|$. A bijection from S, a subset of $\mathbb{R}^2$, to $\mathbb{Z}^2$ is called *bounded* if its spread is finite.

It is useful to define, for measurable plane sets A and H, the difference between the counting measure and Lebesgue measure: $\Delta(A, H) = ||A \cap H| - \lambda(H)|$. Also neighborhoods are described this way, where X and Z are subsets of any metric space: $U(X, Z) = \{y : \exists \vec{x} \in X \text{ such that } \|\vec{x} - \vec{y}\| \le Z\}$.

Lemma 9.8 (The Bijection Lemma). *Let S be a discrete subset of $\mathbb{R}^2$ and suppose that there is $C > 0$ so that, for any lattice polygon P, $\Delta(S, \hat{P}) \le Cp(P)$. Then there is a bijection $\Phi : S \to \mathbb{Z}^2$ with spread at most $N(C) + \sqrt{2}$.*

Proof. We start with a claim to be proved by induction on the perimeter of H.

Claim. For any $H \in \mathcal{H}$, $\Delta(S, H) \le Cp(H)$.

Proof of claim. Given H, one of Lemma 9.4 (i)–(iii) must hold. If (i) holds, the claim follows immediately from $H = \hat{P}$; this includes the base case where $p(H) = 4$. Now assume the claim is valid for any $H_0 \in \mathcal{H}$ with $p(H_0) < p(H)$. If (ii) holds, then $p(H_i) < p(H)$ for $i = 1$ and 2 and the claim is true for the H_i by the inductive assumption. Therefore $\Delta(S, H) \le \Delta(S, H_1) + \Delta(S, H_2) \le C(p(H_1) + p(H_2)) = Cp(H)$. And if (iii) of Lemma 9.4 holds, then we have $p(H_i) < p(H)$ for $i = 1$ and 2 and therefore $\Delta(S, H) = ||S \cap H_2| - \lambda(H_2) - (|S \cap H_1| - \lambda(H_1))| \le \Delta(S, H_1) + \Delta(S, H_2) \le C(p(H_1) + p(H_2)) = Cp(H)$. This proves the claim.

Returning to the lemma, let $N = N(C)$ and $M = N + \sqrt{2}$, and note that the conclusion is the assertion that the bipartite graph Γ has a perfect matching, where the two parts of the graph are S and $\mathbb{Z}^2$ and edges are $\vec{x} \leftrightarrow \vec{y}$ where $\|\vec{x} - \vec{y}\| \le M$. The degree of any vertex is finite, so the Hall–Rado Theorem (see App. C) yields such a matching provided their condition holds: Any set of k vertices in one part has at least k neighbors in the other part.

Consider a k-element set $A \subseteq \mathbb{Z}^2$, and let H be the union of the unit squares whose lower-left corner is in A. Then $H \in \mathcal{H}$ and $\lambda(H) = k$. By Lemma 9.6, there is a positive integer $n \le N$ such that $\lambda(H^{(n)}) - Cp(H^{(n)}) \ge \lambda(H) = k$. Then, by the claim, $|S \cap H^{(n)}| \ge \lambda(H^{(n)}) - Cp(H^{(n)}) \ge k$.

So we have $H^{(n)} \subset U(H, n) \subset U(A, n + \sqrt{2}) \subseteq U(A, M)$ and therefore $|S \cap U(A, M)| \ge k$. This shows that A is adjacent to at least k vertices in S.

For the remaining case, consider a set $B \subseteq S$ having k elements. Let H be the union of the unit squares that meet B. Then $H \in \mathcal{H}$, and by Lemma 9.7, there is $K \in \mathcal{H}$ such that $H \subseteq K \subseteq H^{(N)}$ and $\lambda(H^{(N)}) \ge \lambda(K) + Cp(K)$.

Combining the claim and the preceding inequality gives $k \le |S \cap H| \le |S \cap K| \le \lambda(K) + Cp(K) \le \lambda(H^{(N)})$ and therefore $|\mathbb{Z}^2 \cap H^{(N)}| = \lambda(H^{(N)}) \ge k$.

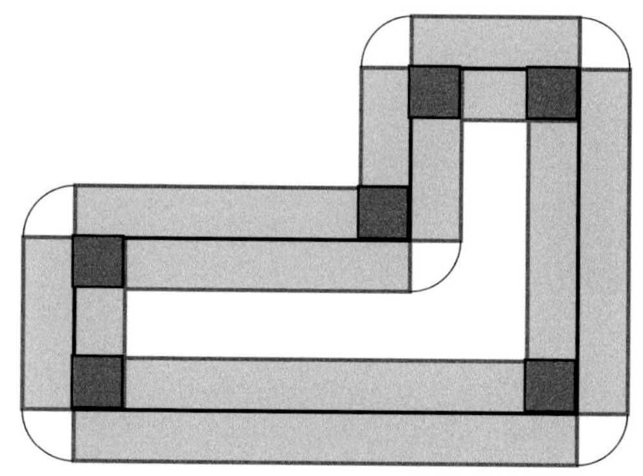

Figure 9.5. The overlap at a corner (dark gray) can be used to cover the missing quarter-disk opposite the overlap, thus capturing the entire δ-neighborhood of P.

Because $H^{(N)} \subset U(H, N) \subset U(B, N) \subset U(B, M)$, we have $|\mathbb{Z}^2 \cap U(B, M)| \geq k$, yielding the desired k neighbors of B in $\mathbb{Z}^2$. The Hall–Rado condition is therefore satisfied, as desired. □

Next we prove a simple result that relates the area of a boundary neighborhood of a polygon to its perimeter.

Lemma 9.9 (Fat Perimeter Lemma). *If P is an orthogonal polygon, then, for any nonnegative δ, $\lambda(U(P, \delta)) \leq 2\delta p(P)$.*

Proof. Expand each edge of the polygon δ units on both sides to get a region of area at most $2\delta p(P)$ (Fig. 9.5). There will be overlap on the inside of each convex vertex; these overlaps can be used to cover the quarter-circle in $U(P, \delta)$ at the outside corner. Similarly, exterior overlap at a concave vertex can be used to cover the needed part just inside the vertex. □

The next lemma involves lattice squares whose side-length is a power of 2. Two such squares never overlap unless one is contained in another. So for any polygon one can get the set of maximal lattice squares whose side-length is a power of 2 and interior lies inside the polygon by just discarding any square contained in a larger one that lies within the polygon. This gives a collection of lattice squares whose interiors approximate the polygon (Fig. 9.6). We define the distance between two sets as follows: $\text{dist}(X, Y) = \inf_{\vec{x} \in X, \vec{y} \in Y} \|\vec{y} - \vec{x}\|$.

Lemma 9.10. *Suppose P is an orthogonal polygon, P^* is the polygon and its interior, and $\{R_1, \ldots, R_m\}$ is the set of all maximal lattice squares with side-length a power of 2 and interior lying inside P^*. Let $Q_i = \hat{R}_i$. Then*

(1) $P^* \setminus U(J, \sqrt{2}) \subseteq \bigcup_{j=1}^{m} Q_j$.
(2) *if $\Psi : \mathbb{N} \to [0, \infty)$ is such that $\sum_{n=0}^{\infty} \Psi(2^n)/2^n$ converges to C, then $\sum_{j=1}^{m} \Psi(s(R_j)) < 6Cp(P)$.*

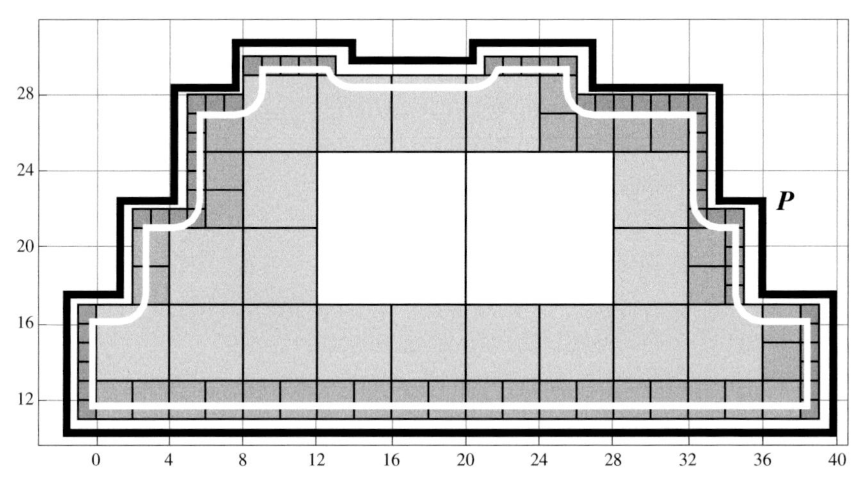

Figure 9.6. For the polygon P, the smallest gray squares form $\mathcal{D}_0$, the next largest are $\mathcal{D}_1$, the next largest are $\mathcal{D}_2$, and the two large white ones are $\mathcal{D}_3$. The white curve encloses $P^* \setminus U(P, \sqrt{2})$.

Proof. Any point in $P^* \setminus U(P, \sqrt{2})$ lies in a unique unit square inside P and so in a square Q_j, proving (1). Let $\mathcal{D} = \{Q_1, \ldots, Q_m\}$, $\mathcal{D}_k = \{Q \in \mathcal{D} : s(Q) = 2^k\}$, and $n_k = |\mathcal{D}_k|$; so, in Figure 9.6, $\mathcal{D}_0$ consists of the smallest gray squares. For any $Q \in \mathcal{D}_k$, there is, by maximality, a lattice square S that is not in $\mathcal{D}$ but contains Q and has $s(S) = 2^{k+1}$. Therefore $\text{dist}(Q, P) \le 2^k\sqrt{2}$ and $Q \subseteq P^* \cap U(P, 2^k 2\sqrt{2})$, proving

(3) $\bigcup\{Q : Q \in \mathcal{D}_k\} \subseteq P^* \cap U(P, 2^k 2\sqrt{2})$.

By the Fat Perimeter Lemma, $\lambda(U(P, \sqrt{2}2^{k+1})) \le \sqrt{2}2^{k+2}p(P)$. On the other hand, $\lambda(\bigcup \mathcal{D}_k) = n_k 2^{2k}$, and so (3) gives $n_k \le 4\sqrt{2}2^{-k}p(P)$. Now, $\sum_{j=1}^{m} \Psi(s(Q_j)) = \sum_{k=0}^{\infty} n_k \Psi(2^k) \le 4\sqrt{2}p(P) \sum_{k=0}^{\infty} \Psi(2^k)/2^k < 6Cp(P)$. □

This concludes the preliminary work; we next relate the concept of translation equidecomposability to the constructions and estimates of the preceding lemmas.

9.2.2 A Condition for Equidecomposability

The main theorem of this section is a powerful tool for equidecomposability. So it is not surprising that the proof is complicated. The central concept is the notion of discrepancy: how well a finite set approximates an infinite set. For a finite subset A of J and a measurable subset H of the plane, the *discrepancy of A* with respect to H is $D(A, H) = \left|\frac{|A \cap H|}{|A|} - \lambda(H)\right|$. And much of the work involves a detailed study of certain sets of lattice points, using the two concepts in the next definition.

Definition 9.11. (a) *Given vectors $\vec{u}$ and $X = \{\vec{x_1}, \vec{x_2}\}$ and a plane set H, let $S(\vec{u}, H)$ be the set of lattice points defined by $S(\vec{u}, H) = \{\vec{n} \in \mathbb{Z}^2 : frac(\vec{u} + \vec{n} \cdot X) \in H\}$.*

(b) *For vectors $\vec{u}$ and $X = \{\vec{x_1}, \vec{x_2}\}$ and a positive integer N, the set $F_N(\vec{u}, X)$ is the subset of J of size at most N^2 defined to be $\{frac(\vec{u} + n_1\vec{x_1} + n_2\vec{x_2}) : n_i = 0, \ldots, N-1\}$.*

The next result is central as it shows how certain special pairs of vectors can be used to deduce translation equidecomposability of two given sets. The existence of such vectors for various cases, like the disk and square, will be proved in the following subsections. We use E for the standard basis $\{(1,0), (0,1)\}$.

Theorem 9.12 (The Equidecomposability Criterion) (AC). *Suppose H_1 and H_2 are measurable subsets of J with $\lambda(H_1) = \lambda(H_2) = \alpha^2 > 0$. Let $X = \{\vec{x_1}, \vec{x_2}\}$ be vectors in J such that*

(a) *$E \cup X$ is linearly independent over $\mathbb{Q}$.*

(b) *There is $\Psi : \mathbb{N} \to [0, \infty)$ so that $\sum_{k=0}^{\infty} \Psi(2^k)/2^k$ converges to a finite sum C and, for any $\vec{u} \in \mathbb{R}^2$, positive integer N, and $j \in \{1, 2\}$, $D(F_N(\vec{u}, X), H_j) \le \Psi(N)/N^2$.*

Then $H_1 \sim_T H_2$.

Proof. Fix $\vec{u} \in J$ and let S_i, or sometimes $S_i(\vec{u})$, denote $S(\vec{u}, H_i)$. The key is the construction of a bounded bijection $\Phi_{\vec{u}} : S_1 \to S_2$. We first show that α^2 roughly determines the proportion of lattice points in a lattice square that lie in S_j (note that $\alpha \le 1$).

Claim 1. For every lattice square R and $j \in \{1, 2\}$, $||Q \cap S_j| - \alpha^2\lambda(Q)| \le \Psi(s(Q))$.

Proof of claim 1. Suppose $s(Q) = N$ and Q's lower left corner is $\vec{a} \in \mathbb{Z}^2$. Then, where inequalities with vectors represent the inequality in all components,

$$
\begin{aligned}
|Q \cap S_j| &= |\{\vec{n} \in \mathbb{Z}^2 : \text{frac}(\vec{u} + \vec{n} \cdot X) \in H_j \text{ and } \vec{a} \le \vec{n} < \vec{a} + (N, N)\}| \\
&= |\{\vec{m} \in \mathbb{Z}^2 : \text{frac}(\vec{u} + \vec{a} \cdot X + \vec{m} \cdot X) \in H_j \text{ and } 0 \le \vec{m} < N\}| \\
&= |F_N(\vec{u} + \vec{a} \cdot X, X) \cap H_j|.
\end{aligned}
$$

Using (b) with $\vec{u} + \vec{a} \cdot X$ as the first argument to F_N, and using linear independence of $X \cup E$ to get $|F_N(\vec{u} + \vec{a} \cdot X, X)| = N^2$, we prove the claim as follows:

$$
\Psi(N) \ge N^2 D(F_N(\vec{u} + \vec{a} \cdot X, X), H_j) = N^2 \left| \frac{|F_N(\vec{u} + \vec{a} \cdot X, X) \cap H_j|}{|F_N(\vec{u} + \vec{a} \cdot X, X)|} - \alpha^2 \right|
$$

$$
= N^2 \left| \frac{|Q \cap S_j|}{|F_N(\vec{u} + \vec{a} \cdot X, X)|} - \alpha^2 \right| = \left| \frac{N^2}{|F_N(\vec{u} + \vec{a} \cdot X, X)|} |Q \cap S_j| - N^2\alpha^2 \right|
$$

$$
= ||Q \cap S_j| - \lambda(Q)\alpha^2|.
$$

Now we start the construction of the key bijection. The next claim is that αS_j satisfies the hypothesis of the bijection lemma (Lemma 9.8).

Claim 2. For any $\vec{u} \in \mathbb{R}^2$, $j \in \{1, 2\}$, and lattice polygon P, $\Delta(\alpha S_j, \hat{P}) \leq C_1 p(P)$, where $C_1 = (4\sqrt{2} + 6C)/\alpha$.

Proof of claim 2. Fix j and P. Let P_α be the scaled polygon $\alpha^{-1}P$. Apply Lemma 9.10 to P_α to get lattice squares R_i such that the lemma's conclusion holds for P_α; then $\bigcup_{i=1}^m Q_i \subseteq \hat{P}_\alpha$. Therefore

(1) $\bigcup_{i=1}^m Q_i \subseteq \hat{P}_\alpha \subseteq U(P_\alpha, \sqrt{2}) \cup \bigcup_{i=1}^m Q_i$.

Let s_i denote $s(Q_i)$. Define positive reals V, W, Y as follows, where all sums and unions in this proof run from 1 to m:

- $V = \alpha^2 \sum s_i^2$
- $W = \sum \Psi(s_j)$
- $Y = |S_j \cap U(P_\alpha, \sqrt{2})|$

We need the following inequality:

(2) $V - W - Y \leq |S_j \cap \hat{P}_\alpha| \leq V + W + Y$.

This inequality bounds the error when V is used to approximate the count of lattice points in $\hat{P}_\alpha$ that are also in S_j; the error is bounded by two terms: Y, which is a boundary error, and W, which is the total error within the approximating squares. We now make this precise. For the first inequality, apply the left half of claim 1 to each Q_i and sum, getting $V = \Sigma\alpha^2 s_i^2 \leq \sum \Psi(s_i) + \sum |Q_i \cap S_j| \leq |\hat{P}_\alpha \cap S_j| + W$.

For the second inequality, start with the right half of (1) and intersect the sets with S_j to get

$$|\hat{P}_\alpha \cap S_j| \leq |S_j \cap U(P_\alpha, \sqrt{2})| + |\bigcup (Q_j \cap S_j)| = Y + |\cup (Q_j \cap S_j)|.$$

Now use the right half of claim 1 in the form $|Q \cap S_j| \leq \alpha^2 \lambda(Q) + \Psi(s(Q))$. This gives $|\hat{P}_\alpha \cap S_j| \leq Y + \Psi(s_j) + \Sigma\alpha^2\lambda(Q_j) = Y + W + V$, which establishes (2).

Next we want

(3) $||S_j(\vec{u}) \cap \hat{P}_\alpha| - \lambda(\hat{P})| \leq (4\sqrt{2} + 6C)\frac{1}{\alpha}p(P)$.

We have three bounds:

- $|V - \lambda(\hat{P})| \leq \lambda(\alpha U(P_\alpha, \sqrt{2})) \leq \lambda(U(P_\alpha, \sqrt{2})) \leq 2\sqrt{2}\ p(P_\alpha) \leq 2\sqrt{2}\frac{1}{\alpha}p(P)$ (by (2) and Lemma 9.9)
- $Y \leq |\mathbb{Z}^2 \cap U(P_\alpha, \sqrt{2})| \leq 2\sqrt{2}\ p(P_\alpha) = 2\sqrt{2}\frac{1}{\alpha}\ p(P)$ (by Lemma 9.9)
- $W \leq 6C\frac{1}{\alpha}p(P)$ (by Lemma 9.10 applied to P_α)

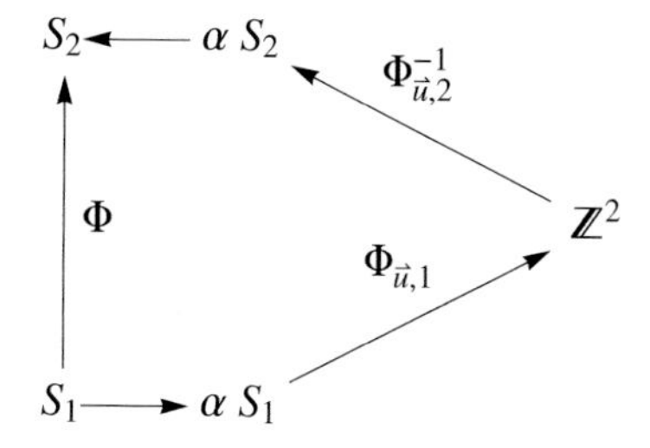

Figure 9.7. The two bijections into $\mathbb{Z}^2$ combine with α-scaling to get a bijection Φ from S_1 to S_2.

Using the triangle inequality,

$$\begin{aligned}||S_j(\vec{u}) \cap \hat{P}_\alpha| - \lambda(\hat{P})| &\leq ||S_j(\vec{u}) \cap \hat{P}_\alpha| - V| + |V - \lambda(\hat{P})| \\ &\leq W + Y + 2\sqrt{2}\frac{1}{\alpha}p(P) \\ &\leq 6C\frac{1}{\alpha}p(P) + 2\sqrt{2}\frac{1}{\alpha}p(P) + 2\sqrt{2}\frac{1}{\alpha}p(P)\end{aligned}$$

as desired.

Because a rescaling shows that $|S_j(\vec{u}) \cap \hat{P}_\alpha| = |\alpha S_j(\vec{u}) \cap \hat{P}|$, (3) yields the conclusion of claim 2.

Claim 3. For any $\vec{u} \in \mathbb{R}^2$, there is a bijection $\Phi: S_1 \to S_2$ having spread at most $C_2 = 2\alpha^{-1}(N(C_1) + \sqrt{2})$.

Proof of claim 3. Because claim 2 gives us the hypothesis of the bijection lemma, we get two bijections $\Phi_{\vec{u},j}: \alpha S_j \to \mathbb{Z}^2$ with spread at most $N(C_1) + \sqrt{2}$. Define $\Phi(\vec{n})$ to be $\alpha^{-1}\Phi^{-1}_{\vec{u},2}\Phi_{\vec{u},1}(\alpha\vec{n})$; then Φ is a bijection from S_1 onto S_2 (Fig. 9.7). Note that the spread of $\Phi^{-1}_{\vec{u},2}$ is the same as the spread of $\Phi_{\vec{u},2}$. Using $\vec{m}_1$ for $\Phi_{\vec{u},1}(\alpha\vec{n}) - \alpha\vec{n}$ and $\vec{m}_2$ for $\Phi_{\vec{u},2}(\Phi_{\vec{u},1}(\alpha\vec{n})) - \Phi_{\vec{u},1}(\alpha\vec{n})$, we get

$$\alpha^{-1}\Phi_{\vec{u},2}(\Phi_{\vec{u},1}(\alpha\vec{n})) = \alpha^{-1}(\alpha\vec{n} + \vec{m}_1 + \vec{m}_2) = \vec{n} + \alpha^{-1}(\vec{m}_1 + \vec{m}_2).$$

Because each $\|\vec{m}_i\| \leq N(C_1) + \sqrt{2}$, this proves the claim.

Now we can conclude the proof of the theorem. Let G be the additive subgroup of $\mathbb{R}^2$ generated by E and X. We partition the additive group $\mathbb{R}^2$ into cosets of G. Choose a coset $\mathcal{E}$ in $\mathbb{R}^2/G$ and pick some $\vec{u} \in \mathcal{E}$. Then, by (a), every $\vec{z} \in \mathcal{E}$ has a unique representation $\vec{z} = \vec{u} + \vec{n} \cdot X + \vec{m} \cdot E$, where $\vec{n}, \vec{m} \in \mathbb{Z}^2$.

Suppose $\vec{z} \in H_1$; then, using the representation just mentioned, $\text{frac}(\vec{u} + \vec{n} \cdot X) \in H_1$ and therefore $\vec{n} \in S_1(\vec{u})$. Let Φ be the bijection of claim 3 for the chosen $\vec{u}$. Let $\vec{n}'$ be $\Phi(\vec{n})$. Because $\vec{n}' \in S_2(\vec{u})$, we have $\text{frac}(\vec{u} + \vec{n}' \cdot X) \in H_2$, and so there is $\vec{m}' \in \mathbb{Z}^2$ such that $\vec{u} + \vec{n}' \cdot X + \vec{m}' \cdot E \in H_2$. Define the map $\chi_{\vec{u}}(\vec{z}) = \vec{u} + \vec{n}' \cdot X + \vec{m}' \cdot E$. This map is well defined from $H_1 \cap \mathcal{E}$ into $H_2 \cap \mathcal{E}$ because the vectors in X and E are linearly independent. Because Φ is a bijection from $S_1(\vec{u})$ onto $S_2(\vec{u})$, $\chi_{\vec{u}}$ is a bijection from $H_1 \cap \mathcal{E}$ onto $H_2 \cap \mathcal{E}$.

The spread bound for Φ tells us that $\|\vec{n}' - \vec{n}\| \leq C_2$ and, because $\vec{z}$ and $\chi_{\vec{u}}(\vec{z})$ are in J, we have $\|\chi_{\vec{u}}(\vec{z}) - \vec{z}\| \leq \sqrt{2}$ and so

$$|(\vec{n}' - \vec{n}) \cdot X + (\vec{m}' - \vec{m}) \cdot E| \leq |\chi_{\vec{u}}(\vec{z}) - \vec{z}| \leq \sqrt{2}.$$

This implies $\|\vec{m}' - \vec{m}\| \leq \|(\vec{m}' - \vec{m}) \cdot E\| \leq \sqrt{2} + \|(\vec{n}' - \vec{n}) \cdot X\| \leq \sqrt{2} + C_2 \max(|\vec{x}_1|, |\vec{x}_2|)$. So, letting $C_3 = \sqrt{2} + C_2 \max(\|\vec{x}_1\|, \|\vec{x}_2\|)$, we have proved that for any $\vec{z} \in H_1$, there are vectors $\vec{q}, \vec{r} \in \mathbb{Z}^2$ such that

(7) $\|\vec{q}\| \leq C_2$ and $\|\vec{r}\| \leq C_3$

and $\chi_{\vec{u}}(\vec{z}) = \vec{u} + \vec{q} \cdot X + \vec{r} \cdot E$.

Let $\{\vec{d}_t\}_{t=1}^{L}$ enumerate the set of vectors $\vec{q} \cdot X + \vec{r} \cdot E$ appearing in (7). This set is finite with size at most $(2C_3 + 1)^2$; this bound depends only on α and Ψ. We have therefore proved that for any $\vec{z} \in H_1 \cap \mathcal{E}$, there is $t \leq L$ such that $\chi_u(\vec{z}) = \vec{z} + d_t$.

Because the proof does not depend on the choice of coset $\mathcal{E}$ and because each $\vec{d}_t \in G$, we have a bijection $\chi : H_1 \to H_2$ such that for any $\vec{z}$ there is a t such that $\chi(\vec{z}) = \vec{z} + \vec{d}_t$. Define the sets $A_t = \{\vec{z} \in H_1 : \exists t \leq L \text{ such that } \chi(\vec{z}) = \vec{z} + \vec{d}_t\}$. Then the two families $\{A_t\}_{t=1}^{L}$, $\{A_t + \vec{d}_t\}_{t=1}^{L}$ witness the translation equidecomposability of H_1 and H_2. □

9.2.3 Some Estimates from Number Theory

In this subsection, we estimate some sums involving the distance of a point in the plane from the nearest lattice point. Estimates of this kind were obtained by W. Schmidt [Sch64] and refined by Laczkovich to meet his needs.The distance of a real number x from the nearest integer—$\min\{\text{frac}(x), 1 - \text{frac}(x)\}$—will be denoted by $\langle x \rangle$. Furthermore, let $\ell(x)$ be 2 when $x < e^2$ and $\ln x$ otherwise: $\ell(x) = \max(2, \ln x)$. The use of ℓ is a minor technicality, and the reader can just think of it as being the natural logarithm.

We start with a fairly simple lemma that is needed in the more complex estimates. It contains two well-known results of Diophantine approximation [Cas57, p. 121].

Lemma 9.13. (a) *For almost every $x \in [0, 1)$, there are only finitely many positive integers k for which $\langle kx \rangle \geq k^{-2}$ is false.*

(b) *For almost all pairs $x, y \in [0, 1)$, there are only finitely many pairs of positive integers h, k for which $\langle hx + ky \rangle \geq (hk)^{-2}$ is false.*

Proof. (a) First, $\langle kx \rangle < k^{-2}$ means that there is $m \in \mathbb{N}$ so that $|kx - m| < k^{-2}$; this in turn is equivalent to $\left|x - \frac{m}{k}\right| < k^{-3}$. For any positive integer k, look at the set of reals $x \in [0, 1)$ so that, for some m, $\left|x - \frac{m}{k}\right| < k^{-3}$. This defines a set

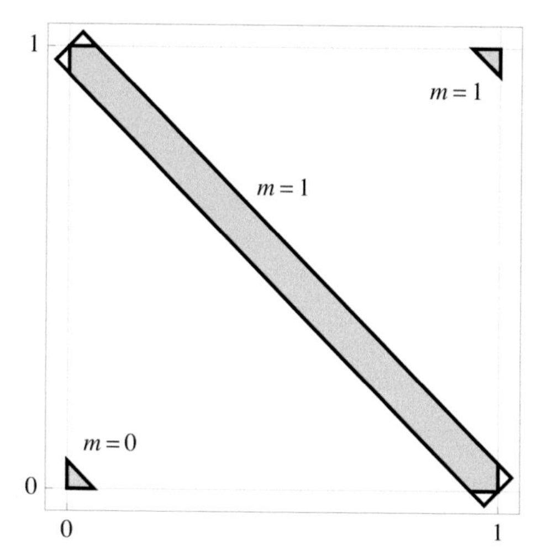

Figure 9.8. The shaded region is $D \cap J$, in the case $h = k = 2$; the pieces can be rearranged into a rectangle of length $\sqrt{2}$.

of intervals of diameter $2k^{-3}$ with centers exactly $1/k$ apart. So the measure of the set of $x \in [0, 1)$ for which there is a k with $\langle kx \rangle < k^{-2}$ is at most $2\sum k^{-2}$. Furthermore, the set of such x satisfying the inequality with $k \geq K$ has measure $2\sum_{k \geq K} k^{-2}$, which is a tail of a convergent series and so can be made arbitrarily small. Therefore the set of x for which the inequality has infinitely many solutions has measure under ϵ for any positive ϵ, and hence has measure 0.

(b) Let $D_{h,k}$ be the set of $(x, y) \in J$ such that $\langle hx + ky \rangle < (hk)^{-2}$; let $M = \left[\begin{smallmatrix} h & 0 \\ 0 & k \end{smallmatrix}\right]$. Then

$$\begin{aligned} D_{h,k} &= \bigcup_m \{(x, y) \in J : |M \cdot (x, y) - m| \leq (hk)^{-2}\} \\ &= \bigcup_m M^{-1} \cdot \{(x, y) \in M \cdot J : |x + y - m| \leq (hk)^{-2}\} \\ &= M^{-1} \cdot \left(\bigcup_m \{(x, y) \in M \cdot J : |x + y - m| \leq (hk)^{-2}\} \right). \end{aligned}$$

Let D be this last union; $D \subseteq M \cdot J = [0, h) \times [0, k)$. Because the condition defining D is invariant modulo 1, we can focus on $D \cap J$; there only the values $m = 0, 1, 2$ are relevant (Fig. 9.8). The long borders of the central strip are given by $y = 1 - x \pm (hk)^2$, from which the width of the strip is $\sqrt{2}(hk)^{-2}$. Moving the small corner pieces shows that $D \cap J$ is congruent by dissection to a rectangle containing the central strip: Its dimensions are $\sqrt{2} \times \sqrt{2}(hk)^{-2}$. So the area of $D \cap J$ is $2(hk)^{-2}$ (the case $h = k = 1$ is a trivial exception: the area is 1 not 2). Extending to $M \cdot J$ tells us that D's total area is $2(hk)^{-1}$. But $D_{h,k}$ is a scaling of D by M^{-1}, so a factor of $\det(M^{-1}) = (hk)^{-1}$ arises, giving $\lambda(D_{h,k}) = 2(hk)^{-2}$.

Let D_K be the set of $(x, y) \in J$ for which there are h, k, with at least one of h, k not less than K, so that $\langle hx + ky \rangle < (hk)^{-2}$. Now, suppose that for (x, y) there are infinitely many pairs h, k so that $(x, y) \in D_{h,k}$. Then, for any K, $(x, y) \in D_K$. But the area of D_K is at most $2\sum_{h\geq K}\sum_{k=1}^{\infty} 2(hk)^{-2}$, where the leading 2 is due to the symmetry in h and k. This is $4\big(\sum_{k=1}^{\infty} k^{-2}\big)\big(\sum_{h\geq K} h^{-2}\big) = \frac{2}{3}\pi^2 \sum_{h\geq K} h^{-2}$. But this last tail of a convergent series is, for large enough K, under any positive ϵ, and so the set of points for which the lemma's inequality is false has measure 0. □

And one more simple lemma.

Lemma 9.14. *If f is a measurable real function on* $[0, 1]$ *with a finite integral, $a \in \mathbb{R}$, and $k \in \mathbb{N}$, then* $\int_0^1 f(\langle a + kx\rangle)\,dx = 2\int_0^{1/2} f(x)\,dx$.

Proof. Periodicity with respect to $[0, 1]$ and a substitution yield $\int_0^1 f(\langle a + kx\rangle)\,dx = \int_{-a/k}^{1-a/k} f(\langle a + kx\rangle)\,dx = \int_0^1 f(\langle a + kx\rangle)\,dx$. Then periodicity with respect to $[0, 1/k]$ and another substitution finishes it as follows:

$$\int_0^1 f(\langle kx\rangle)\,dx = k\int_0^{\frac{1}{k}} f(kx)\,dx = 2\int_0^{\frac{1}{2k}} f(kx)k\,dx = 2\int_0^{\frac{1}{2}} f(x)\,dx.$$

□

Now we begin a series of estimates. The first two obtain bounds on some reciprocal sums involving the nearest integer function.

Theorem 9.15 (AC) (First Estimation Theorem). *For almost every $(x, y) \in J$ and every $\epsilon > 0$, there is $C > 0$ such that for every positive integer n,*

$$\sum_{k=1}^{n} \frac{1}{k\langle kx\rangle\langle ky\rangle} \leq C\ell^{3+\epsilon}(n).$$

Proof. Use the functions $Q_k(w) = \frac{1}{\langle kw\rangle|\ln\langle kw\rangle|^{1+\epsilon}}$ and $K_k(w) = \frac{1}{k\ell^{1+\epsilon}(k)}Q_k(w)$. Consider the following family of functions $f_k(x, y) = K_k(y)Q_k(x) = \frac{1}{k\ell^{1+\epsilon}(k)}Q_k(x)Q_k(y)$ with domain J, where singularities (x or $y = 1/k$ or 0) are irrelevant because we are ignoring sets of measure 0; let f be their sum: $f(x, y) = \sum_{k=1}^{\infty} f_k(x, y)$.

Claim 1. For almost all x and y, $f(x, y)$ is finite.

We simplify the problem by integrating to reduce the dimension. The basic idea is to reduce the convergence question for $f = \sum K_k(y)Q_k(x)$ to the same for $\sum K_k(y)$.

Claim 2. For almost all y, $\sum_{k=1}^{\infty}\int_0^1 f_k(x, y)\,dx$ is finite.

Proof of claim 2. For any y, Lemma 9.14 gives

$$\int_0^1 f_k(x, y)\,dx = K_k(y)\int_0^1 Q_k(x)\,dx = 2K_k(y)\int_0^{\frac{1}{2}} \frac{1}{x(-\ln x)^{1+\epsilon}}\,dx = \frac{2K_k(y)}{(\ln 2)^{\epsilon}\epsilon}.$$

So claim 2 follows from the next claim.

Claim 3. For almost all y, $\sum_{k=1}^{\infty} K_k(y)$ converges.

As before, the lemma gives $\int_0^1 Q_k(y)dy = \frac{2}{(\ln 2)^\epsilon \epsilon}$. The integral test yields convergence of $\sum_{k=1}^{\infty} \frac{1}{k\ell^{1+\epsilon}(k)}$, and so the following series converges:

$$\sum_{k=1}^{\infty} \left(\frac{1}{k\ell^{1+\epsilon}(k)} \int_0^1 Q_k(y)\, dy \right).$$

The monotone convergence theorem (which requires countably additivity and hence requires some form of the Axiom of Choice; §15.3) allows the interchange of the sum and integral, yielding

$$\int_0^1 \left(\sum_{k=1}^{\infty} \frac{1}{k\ell^{1+\epsilon}(k)} Q_k(y) \right) dy = \int_0^1 \sum_{k=1}^{\infty} K_k(y)\, dy < \infty.$$

But this means that the integrand is finite for almost all y. This proves claim 3, and therefore claim 2.

Now, interchanging the sum and limit in claim 2 gives, for almost all y, the finiteness of $\int_0^1 \sum_{k=1}^{\infty} f_k(x, y)\, dx < \infty$. The integrand must be finite for almost all x. Because Fubini's Theorem shows that the two almost everywhere conditions lead to a measure-zero failure set in two dimensions, this proves claim 1.

To finish, we use Lemma 9.13(a), which implies that there is a positive C' (depending on x, and assumed smaller than 1) such that $\langle kx \rangle \geq C' k^{-2}$ for every k. For such x, taking logarithms and some simple algebra yields

$$\frac{1}{\ell(k)} \leq \frac{2 + |\ln C'|}{|\ln \langle kx \rangle|^{1+\epsilon}}.$$

This implies that for almost all x, y,

$$\frac{1}{k\ell^{3+\epsilon}(k)\langle kx \rangle \langle ky \rangle} \leq \frac{(2 + |\ln C'|)^2}{k\ell^{1+\epsilon}(k)\langle kx \rangle \langle ky \rangle |\ln \langle kx \rangle \ln \langle ky \rangle|^{1+\epsilon}}$$
$$= (2 + |\ln C'|)^2 K_k(y)\, Q_k(x).$$

The right side sums (almost always) to a finite value by claim 1. Therefore for almost all x, y,

$$C = \sum_{k=1}^{\infty} \frac{1}{k\ell^{3+\epsilon}(k)\langle kx \rangle \langle ky \rangle} < \infty.$$

Because ℓ is monotonic, $\sum_{k=1}^{n} \frac{1}{k\langle kx \rangle \langle ky \rangle} \leq \sum_{k=1}^{n} \frac{\ell^{3+\epsilon}(n)}{\ell^{3+\epsilon}(k)} \frac{1}{k\langle kx \rangle \langle ky \rangle} < C\ell^{3+\epsilon}(n)$, as desired. □

Next we derive a similar result, but in four dimensions.

Theorem 9.16 (AC) (Second Estimation Theorem). *For almost every $(x_1, y_1, x_2, y_2) \in J^2$ and every positive ϵ and C, there is $C' > 0$ such that for every*

$n \in \mathbb{N}$,

$$\sum_{h=1}^{n}\sum_{k=1}^{n} \frac{1}{hk\langle hx_1 + ky_1\rangle\langle hx_2 + ky_2\rangle} \le C' l^{6+\epsilon}(n).$$

Proof. The proof is similar to the preceding proof. Let $\vec{X} = (x_1, y_1)$ and $\vec{Y} = (x_2, y_2)$. Let $Q_{h,k}(\vec{V}) = \frac{1}{\langle hv_1+kv_2\rangle|\ln(\langle hv_1+kv_2\rangle)|^{1+\epsilon}}$ and $K_{h,k}(\vec{V}) = \frac{1}{hk\ell^{1+\epsilon}(h)\ell^{1+\epsilon}(k)} Q_k(\vec{V})$. Define the functions $f_{h,k}(\vec{X}, \vec{Y}) = K_{h,k}(\vec{Y})Q_{h,k}(\vec{X})$, and let f be the sum $f(\vec{X}, \vec{Y}) = \sum_{h=1}^{\infty}\sum_{k=1}^{\infty} f_{h,k}(\vec{X}, \vec{Y})$.

The claims of the preceding proof become the following, and claim 2 implies claim 1 exactly as before.

Claim 1. For almost all $\vec{X}$ and $\vec{Y}$, $f(\vec{X}, \vec{Y})$ is finite.

Claim 2. For almost all $\vec{Y}$, $\sum_{h=1}^{\infty}\sum_{k=1}^{\infty}\int_J f_{h,k}(\vec{X}, \vec{Y})\, d\vec{X}$ is finite.

Claim 3. For almost all $\vec{Y}$, $\sum_{h=1}^{\infty}\sum_{k=1}^{\infty} K_{h,k}(\vec{Y})$ converges.

Proof of claim 2. For any $\vec{Y}$, Lemma 9.14 gives

$$\begin{aligned}
\int_J f_{h,k}(\vec{X}, \vec{Y})\, d\vec{X} &= K_{h,k}(\vec{Y})\int_0^1\int_0^1 Q_{h,k}(x_1, y_1)\, dx_1 dy_1 \\
&= K_{h,k}(\vec{Y})\int_0^1\int_0^1 \frac{1}{\langle hx_1 + ky_1\rangle|\ln(\langle hx_1 + ky_1\rangle)|^{1+\epsilon}}\, dx_1 dy_1 \\
&= K_{h,k}(\vec{Y})\int_0^1 2\int_0^{1/2} \frac{1}{x_1(-\ln x_1)^{1+\epsilon}}\, dx_1 dy_1 \\
&= K_{h,k}(\vec{Y})\int_0^1 \frac{2}{(\ln 2)^{\epsilon}\epsilon}\, dy_1 = K_{h,k}(\vec{Y})\frac{2}{(\ln 2)^{\epsilon}\epsilon}.
\end{aligned}$$

And now the claim follows from claim 3.

Proof of claim 3. As in claim 2, the lemma gives $\int_0^1\int_0^1 Q_{h,k}(x_1, y_1)\, dx_1 dy_1 = \frac{2}{(\ln 2)^{\epsilon}\epsilon}$. The integral test (twice) yields convergence of $\sum_{h=1}^{\infty}\sum_{k=1}^{\infty}\frac{1}{hk\ell^{1+\epsilon}(h)\ell^{1+\epsilon}(k)} = \sum_{h=1}^{\infty}\frac{1}{h\ell^{1+\epsilon}(h)}\left(\sum_{k=1}^{\infty}\frac{1}{k\ell^{1+\epsilon}(k)}\right)$; in fact, it converges to the square of the innermost series. Therefore the following series converges:

$$\sum_{h=1}^{\infty}\sum_{k=1}^{\infty}\left[\left(\frac{1}{hk\ell^{1+\epsilon}(h)\ell^{1+\epsilon}(k)}\right)\int_0^1\int_0^1 Q_{h,k}(x_1, y_1)\, dx_1 dy_1\right)\right].$$

The monotone convergence theorem then yields the result as before.

Next we use Lemma 9.13(b), which gives, for almost every $(x_1, y_1, x_2, y_2) \in J^2$, a number C_1 such that $C_1 h^{-2}k^{-2} \le \langle hx_1 + ky_1\rangle$ and $C_1 h^{-2}k^{-2} \le \langle hx_2 + ky_2\rangle$ for all positive integers h and k. For these quadruples we get, taking logarithms and with j either 1 or 2, $|\ln\langle hx_j + ky_j\rangle| \le (2 + |\ln C_1|)(\ell(h) + \ell(k))$. But ℓ was defined to be no less than 2, so a product of two ℓs is not less than their sum, and the preceding inequality becomes

$$\frac{1}{\ell(h)\ell(k)} \leq \frac{1}{\ell(h)+\ell(k)} \leq \frac{2+|\ln C_1|}{|\ln\langle hx_j+ky_j\rangle|} \leq \frac{2+|\ln C_1|}{|\ln\langle hx_j+ky_j\rangle|^{1+\epsilon}}.$$

This implies

$$\sum_{h=1}^{\infty}\sum_{k=1}^{\infty}\frac{1}{hk\ell^{3+\epsilon}(h)\ell^{3+\epsilon}(k)\langle hx_1+ky_1\rangle\langle hx_2+ky_2\rangle} < \infty$$

for almost every $(x_1, x_2), (y_1, y_2) \in \mathbb{R}^2$ and every $\epsilon > 0$. And this yields the result in the same way as at the end of the proof of Theorem 9.15. □

We conclude by combining the preceding theorems with a classic result of discrepancy theory—the Erdős–Turán–Koksma formula—to get a strong bound on a global notion of discrepancy. Basic discrepancy is defined as follows, where A is a finite subset of J and H is measurable: $D(A, H) = \left|\frac{|A\cap H|}{|A|} - \lambda(H)\right|$.

Definition 9.17. *For A, a finite subset of J, define the* discrepancy $D(A)$ *to be the supremum of $D(A, H)$ over all half-open subrectangles H contained in J.*

Theorem 9.18 (AC) (Third Estimation Theorem). *For almost every pair of vectors $\vec{x}, \vec{y} \in J$ and every $\epsilon > 0$, there is a number C such that, for every $\vec{u}$ and positive integer N, and with $X = \{\vec{x}, \vec{y}\}$,*

(1) $D(F_N(\vec{u}, X)) \leq C\frac{\ell^{6+\epsilon}(N)}{N^2}$.

Proof. We will show that if $(x_1, x_2, y_1, y_2) \in J^2$ is such that the conclusions of Theorems 9.15 and 9.16 hold for each of the pairs (x_1, y_1) and (x_2, y_2) and for each of the quadruples $(\pm x_1, \pm x_2, \pm y_1, \pm y_2)$, then (1) is valid. Because the two theorems assert that almost every 4-vector has this property, this will suffice.

To estimate the discrepancy of $F_N(\vec{u}, X)$, we apply the Erdős–Turán–Koksma formula [KN, p. 116]. That formula gives an absolute constant C such that, for any positive integer m,

(2) $D(F_N(\vec{u}, X)) \leq C\left(\frac{1}{m} + \frac{1}{N^2}\Lambda\right)$,

where $\|\vec{h}\| = \max(|h_1|, |h_2|)$, $r(\vec{h}) = \max(|h_1|, 1)\max(|h_2|, 1)$, and

$$\Lambda = \sum_{0<\|\vec{h}\|\leq m}\left(\frac{1}{r(\vec{h})}\left|\sum_{0\leq n,k<N} e^{2\pi i(h_1(u_1+nx_1+ky_1)+h_2(u_2+nx_2+ky_2))}\right|\right).$$

Let Λ_1 be the sum of those terms in Λ for which $h_1 > 0$ and $h_2 = 0$. We will need this next identity, which can be derived by summing the geometric series, converting the exponentials to trigonometric form, and removing terms on the unit circle:

$$\left|\sum_{n=0}^{N-1} e^{2\pi i n\beta}\right| = |\csc(\pi\beta)\sin(\pi\beta N)| \le |\csc(\pi\beta)|.$$

Using this identity then gives, for any positive integer h,

$$\left|\sum_{0\le n,k\le N-1} e^{2\pi i h(u_1+nx_1+ky_1)}\right| = |e^{2\pi i h u_1}| \left|\sum_{0\le n,k\le N-1} e^{2\pi i h(nx_1+ky_1)}\right|$$

$$= \left|\sum_{0\le n,k\le N-1} e^{2\pi i h n x_1} e^{2\pi i h k y_1}\right| = \left|\sum_{n=0}^{N-1} e^{2\pi i h n x_1} \sum_{k=0}^{N-1} e^{2\pi i h k y_1}\right|$$

$$\le \frac{1}{|\sin(\pi h x_1)\sin(\pi h y_1)|} \le \frac{1}{4\langle h x_1\rangle\langle h y_1\rangle}.$$

Summing and applying the Second Estimation Theorem then gives

$$\Lambda_1 = \sum_{h=1}^{m} \frac{1}{h}\left|\sum_{0\le n,k\le N} e^{(2\pi i h(u_1+nx_1+ky_1))}\right| \le \sum_{h=1}^{m} \frac{1}{h\langle h x_1\rangle\langle h y_1\rangle} \le C\ell^{3+\epsilon}(m).$$

Let Λ_2 be similar, using terms for which $h_1 < 0$ and $h_2 = 0$. Then

$$\Lambda_2 \le \sum_{h=1}^{m} \frac{1}{h\langle -hx_1\rangle\langle -hy_1\rangle} \le \sum_{h=1}^{m^2} \frac{1}{h\langle h x_1\rangle\langle h y_1\rangle} \le C\ell^{3+\epsilon}(m).$$

And we get the same estimates for the symmetric case where $h_1 = 0$ and $h_2 \ne 0$.

Next, let Λ_3 be the sum of terms in which both h_1, h_2 are positive. Then Theorem 9.16 for (x_1, x_2, y_1, y_2) gives

$$\Lambda_3 \le \sum_{h_1,h_2=1}^{m} \frac{1}{h_1 h_2\langle h x_1 + h_2 x_2\rangle\langle h_1 y_1 + h_2 y_2\rangle} \le C\ell^{6+\epsilon}(m).$$

We get the same estimates for the sums of those terms in which $h_1 > 0, h_2 < 0$, and similar cases. So in total we have $\Lambda \le C_2\ell^{6+\epsilon}(m)$, where C_2 is a constant depending only on $\vec{x}$, $\vec{y}$, and ϵ. Applying (2) concludes the proof. □

And two final estimates are needed: The first is known as the Erdős–Turán Theorem, though the version here is a variation due to Laczkovich, and the second theorem is yet another variation on Erdős–Turán. These results concern plane regions bounded by graphs of functions. For a function $f: [0, 1) \to [0, 1)$, define $\mathrm{gr}(f)$ to be the filled graph: $\mathrm{gr}(f) = \{(x, y) : 0 \le x < 1, 0 \le y \le f(x)\}$. The next theorem connects $D(S, \mathrm{gr}(f))$, the discrepancy with respect to the graph of a function (the graph could be part of a circle), to $D(S)$, a supremum of discrepancies for rectangles. This is a glimpse of the application to circle-squaring as it relates curves to polygons. However, even when f is the identity function, the result is important; in that case, the graph is a right isosceles triangle, and this theorem is the key to showing that such triangles can be squared using translations. We present an outline of the proof of the next theorem; the full details are in [Lac90, pp. 97–100].

Theorem 9.19 (Erdős–Turán Theorem). *Let $f\colon [0, 1) \to [0, 1)$ be strictly monotonic, and suppose there is a constant C, $0 < C \le 1$, such that, for any distinct reals $x, y \in [0, 1)$, $|f(x) - f(y)| \ge C|x - y|$.*

Then, for any finite set $S = \{(x_n, y_n)\}$, a subset of J having size N, and m is a positive integer,

$$D(S, \operatorname{gr}(f)) \le \max\left(8D(S), \frac{216}{C}E_m\right),$$

where E_m is given by

$$E_m = \frac{1}{m} + \sum_{h=1}^{m} \frac{1}{h}\left(\frac{1}{N}\left|\sum_{n=1}^{N} e^{2\pi hi(f(x_n)-y_n)}dx\right|\right.$$
$$\left. + \left|\int_0^1 e^{2\pi hi f(x)}\,dx - \frac{1}{N}\sum_{n=1}^{N} e^{2\pi hi\, f(x_n)}\right|\right).$$

Proof Outline. For a real t, let f_t be the upward translation of f by t: $f_t(x) = f(x) + t$; let $\overline{f}_t$ be the fractional part of f_t. Define $R(t)$, a periodic function on the reals with period 1, to be the signed discrepancy $D(S, \operatorname{gr}(\overline{f}_t))$, that is, the usual discrepancy less the absolute value:

$$R(t) = \frac{1}{N}|S \cap \operatorname{gr}(\overline{f}_t)| - \lambda(\operatorname{gr}(\overline{f}_t)).$$

When $t = 0$, the translation and fractional parts are irrelevant, so $|R(0)| = D(S, \operatorname{gr}(f)))$, the object of study in the theorem. Now, $R(t)$ is a periodic function with only finitely many simple discontinuities. Let $M = \max_t |R(t)|$. The theorem will follow from the result that $M \le \max\left(8D(S), \frac{216}{C}E_m\right)$. The key to this bound on M is a detailed study of the Fourier series of $R(t)$. If $A(t)$ denotes the number of points in S for which $y_n \le \overline{f}_t(x_n)$, then $R(t) = \frac{A(t)}{N} - \int_0^1 f_t(x)\,dx$, and this form is amenable to Fourier analysis. In the simple but important case that $f(x) = x$, $R(t)$ becomes just $\frac{A(t)}{N} - \frac{1}{2}$.

Suppose the Fourier series of the $R(t)$ is $\sum_{h=0}^{\infty} c_h e^{2\pi iht}$. Then, by linearity, $c_h = \frac{1}{N}a_h - b_h$. Because M is invariant under upward translation and reduction mod 1 of S, we may assume that S has the property that the average of its y-values is $1/2$.

First compute a_h and b_h. To start, $b_h = \int_0^1 (\int_0^1 \overline{f}_t(x)\,dx)e^{-2\pi iht}dt$. Some easy work based on switching the integration order and then using integration by parts gives $b_h = \frac{-1}{2\pi ih}\int_0^1 e^{2\pi ihf(x)}\,dx$ if $h \ne 0$, while $b_0 = 1/2$.

Next consider $a_h = \int_0^1 A(t)e^{-2\pi i\,ht}dt$. Again, a relatively easy analysis turns a_h into $\sum_{n=1}^{N}(1 - y_n) = N - \sum y_n = N/2$ when $h = 0$, and $\sum_{n=1}^{N}\frac{1}{2\pi ih}(e^{2\pi ih(f(x_n)-y_n)} - e^{2\pi ihf(x_n)})$ otherwise.

Now, $c_0 = \frac{1}{N}a_0 - b_0 = \frac{1}{N}\frac{N}{2} - \frac{1}{2} = 0$ and

$$c_h = \frac{1}{2\pi ihN}\sum_{n=1}^{N}(e^{2\pi ih(f(x_n)-y_n)} - e^{2\pi ihf(x_n)}) + \frac{1}{2\pi ih}\int_0^1 e^{2\pi ihf(x)}\,dx.$$

We need to bound the absolute values; rearranging gives

$$|c_h| \le \frac{1}{2\pi hN}\left|\sum_{n=1}^{N} e^{2\pi ih(f(x_n)-y_n)}\right| + \left|\frac{1}{2\pi h}\int_0^1 e^{2\pi ihf(x)}\,dx - \frac{1}{2\pi hN}\sum_{n=1}^{N} e^{2\pi ihf(x_n)}\right|.$$

The proof concludes with some computations that make use of the Fejér kernel (see [Lac90, p. 100]). The max is dealt with by assuming $M > 8D(S)$ and showing, under this assumption, that $M \le \frac{216}{mC} + 16\sum_{h=1}^{m}|c_h|$. Substituting the upper bound on $|c_h|$ turns this into the desired bound on M, and hence on $R(0)$. □

And now comes the final result about discrepancy needed to square the circle. This is the theorem that provides the vectors X that are needed to define the decomposition of a disk.

Theorem 9.20 (Erdős–Turán Variation). *Let f be twice differentiable on* $[0, 1]$ *with $f(0) = 0$ and $f(1) = 1$. Suppose that $f'(x)$ is always positive and $f''(x)$ is bounded away from 0. Then, for almost every set X of two vectors in $\mathbb{R}^2$, there is a constant C such that for any $\vec{u}$ and N,*

$$D(F_N(\vec{u}, X), gr(f)) \le CN^{-4/3}\ell^7(N).$$

We refer to [Lac90, pp. 105–110] for the proof. It uses all the previous results in this section—the three estimation theorems and the Erdős–Turán Theorem—and the overall technique is similar to the proof techniques exhibited in the preceding proofs. It calls on one additional lemma—a bound on $\int_0^1 e^{2\pi i(hf(x)-kx)}\,dx$ with $h, k \in \mathbb{Z}$—which is derived from some well-known bounds on oscillatory integrals known as the Van der Corput inequalities.

9.2.4 Behold! The Grand Problem No Longer Unsolved: The Circle Squared, beyond Refutation

The section title comes from one of the many books published by misguided circle-squarers. The only true circle-squarer is Miklós Laczkovich. In this section we conclude his proof that, from a modern set-theoretic point of view, circle-squaring is both a grand problem and no longer unsolved.

In the context of classic straightedge-and-compass Euclidean constructions, it is not difficult to construct a square equal in area to a given polygon. And when using geometric equidecomposability, again the "squaring" of a polygon is not difficult: Any polygon is congruent by dissection to a square, as shown by the Bolyai–Gerwien Theorem (Thm. 3.2). And extending the toolbox to general equidecomposability leads to nothing new (Thm 3.8). But if we restrict to translations, the situation is quite different; see Theorem 3.3.

The first consequence of all the machinery and estimates in the preceding sections is that, working with general equidecomposability, the situation is very different: Laczkovich proved that any polygon is equidecomposable to a square using translations. The most difficult step is handling an isosceles right triangle; we do that first, making use of many of the complicated estimates we have discussed.

Theorem 9.21 (AC). *Any isosceles right triangle is translation equidecomposable to a square.*

Proof. Of course, throughout this section, the triangle and square are considered to be two-dimensional objects. Consider the isosceles right triangle $A = \mathrm{gr}(f)$, where f is the identity function. We will show that $A \sim_T Q$, where Q is the square $[0, 1/\sqrt{2})^2$. Let $\Psi(x) = \ell^7(x)$; $\sum_{k=0}^{\infty} \Psi(2^k)/2^k$ is finite. By the Equidecomposability Criterion (Thm. 9.12), it suffices to show that there are two vectors $X = \{\vec{x}, \vec{y}\}$ and a constant C such that $X \cup E$ is linearly independent over $\mathbb{Q}$ and, for any $\vec{u} \in \mathbb{R}^2$ and positive integer N,

(1) $D(F_N(\vec{u}, X), A) \leq C\frac{\Psi(N)}{N^2}$
(2) $D(F_N(\vec{u}, X), Q) \leq C\frac{\Psi(N)}{N^2}$

Because Q is a subrectangle of J, the Third Estimation Theorem (Thm 9.18) with $\epsilon = 1$ yields (2) for almost every $\vec{x}, \vec{y} \in \mathbb{R}^2$. Because a measure-1 set has pairs of vectors that satisfy the linear independence condition, it is enough to show that (1) is true for almost every $\vec{x}, \vec{y}$. We will prove that if $(x_1, x_2, y_1, y_2) \in J^2$, it is such that

(a) the conclusion of the Third Estimation Theorem holds with $X = \{(x_1, x_2), (y_1, y_2)\}$, $\epsilon = 1$, and for any $\vec{u} \in \mathbb{R}^2$ and positive integer N
(b) the conclusion of the First Estimation Theorem (Thm 9.15) holds with $\epsilon = 1$ and with (x, y) being either one of (x_1, y_1) or $(x_1 - x_2, y_1 - y_2)$

then (1) holds for $X = \{(x_1, x_2), (y_1, y_2)\}$ and any $\vec{u}$ and N.

Apply Theorem 9.19 with S being the N^2-element set $F_N(\vec{u}, X)$. Because f is the identity, the constant C in that theorem is just 1. This gives $D(F_N(\vec{u}, X), A)) \leq \max(8D(F_N(\vec{u}, X)), 216E_m)$, where

$$E_m = \frac{1}{m} + \sum_{h=1}^{\infty} \frac{1}{h} \left(\frac{1}{N^2} \left| \sum_{0 \leq n,k \leq N-1} e^{2\pi hi(u_1 + nx_1 + ky_1 - u_2 - nx_2 - ky_2)} \right| \right.$$
$$\left. + \left| \int_0^1 e^{2\pi hi\, x}\, dx - \frac{1}{N^2} \sum_{0 \leq n,k \leq N-1} e^{2\pi hi(u_1 + nx_1 + ky_1)} \right| \right)$$

By (a), the first argument to max in Theorem 9.19 is under $C\Psi(N)/N^2$, for some constant C. For the second argument, the integral in E_m vanishes, and so, using the same argument as in the proof of the Third Estimation Theorem, and using (b) twice for the final inequality, we have

$$E_m \le \frac{1}{m} + \frac{1}{N^2} \sum_{h=1}^{m} \left(\frac{1}{h\langle h(x_1 - x_2)\rangle \langle h(y_1 - y_2)\rangle} + \frac{1}{h\langle hx_1\rangle \langle hy_1\rangle} \right)$$

$$\le \frac{1}{m} + \frac{C'}{N^2} \ell^4(m).$$

Setting $m = N^2$ yields (1). So it remains to show that the preceding argument applies to almost all (x_1, x_2, y_1, y_2) in J^2.

By the First Estimation Theorem, there is a measure zero set $S \subset J$ such that, for any $(x, y) \in J \setminus S$, there is a constant C such that for $\epsilon = 1$ and any positive integer n, the conclusion of that theorem holds. We claim that for almost every $(x_1, x_2, y_1, y_2) \in J^2$, we have $(x_1 - x_2, y_1 - y_2) \notin S$. Indeed, $L(x_1, x_2, y_1, y_2) = (x_1 - x_2, y_1 - y_2, x_1, y_1)$ is a nonsingular linear transformation and $(x_1 - x_2, y_1 - y_2) \in S$ if and only if $L(x_1, x_2, y_1, y_2) \in S \times J$. Because the preimage of a measure 0 set under L has measure 0, this proves that for almost all pairs of vectors $\vec{x}, \vec{y} \in J$, (a) and (b) hold, and this yields (2), as desired. This completes the proof for one specific right triangle. But because, for any similarity σ and translation τ, $\sigma^{-1}\tau\sigma$ is a translation, the method works on any triangle. □

Next, we build on earlier geometric work and extend the preceding result to all polygons.

Theorem 9.22 (Polygon Equidecomposability Theorem) (AC). *Any two polygons of the same area are translation equidecomposable.*

We need two lemmas: the first uses standard techniques and allows us to ignore one-dimensional sets when studying plane equidecomposability.

Lemma 9.23 (Absorption Lemma). *Let K be the interior of a Jordan domain. If H is any union of finitely many line segments, then $K \sim_T K \cup H$.*

Proof. Deal with the segments one at a time. Suppose first that H is a line segment that is very short relative to the size of K. Translate H into K so that a small rectangle can be erected in K using H as base. It is easy to see (conjugating by a rotation to get into an orthogonal orientation) that any rectangle is translation equidecomposable with the rectangle less any subset of a side: We can drop down one dimension by using Lemma 8.16, which says that a closed interval is translation equidecomposable with the interval less an endpoint. This can be applied to the closed interval orthogonal to the side in question for those base points lying in the set to be absorbed. Because any segment can be broken up into finitely many segments shorter than a given length, this takes care of all of H. □

A consequence of the lemma (either by modifying the proof or using Thm. 3.6) is that if P is the interior of a polygon, then any two sets S_i with $P \subseteq S_i \subseteq \overline{P}$ are translation equidecomposable.

A parallelogram is centrally symmetric, so the next lemma follows from the Hadwiger–Glur Theorem of 1951 (Thm. 3.3; [Bol78, §10]), which states that a convex polygon is congruent by dissection to a square using translations if and only if the polygon is centrally symmetric. But there is a very simple self-contained proof in the case we need here. The main idea—the use of the reverse Pythagorean theorem—is elegant and was pointed out to us by Laczkovich.

Lemma 9.24. *Any two parallelograms of the same area are translation equidecomposable.*

Proof. The previous lemma allows us to ignore the issue of boundary lines. A parallelogram (call its area α) is congruent by translation to a rectangle by the standard method of cutting off and sliding the corners. The rectangle can be turned into a square, in a skew orientation, by translation as in Theorem 3.2. Now the single skew square can be broken into two axis-aligned squares by the *reverse* of the Pythagorean theorem, using the method of Airy (see Fig. 3.1(d)). By the reverse of the rectangle-squaring step already used, each of the two squares can be turned into an axis-aligned rectangle having base of length $\sqrt{\alpha}$; the two rectangles can then be stacked to get an axis-aligned square. Doing this for both parallelograms yields the same square, so the result follows by transitivity. □

Proof of the Polygon Equidecomposability Theorem. By transitivity, it suffices to work with a given polygon P and an axis-aligned square Q of the same area. Decompose P into interior-disjoint triangles T_i (see proof of Thm. 3.2). Use horizontal lines to divide Q into rectangles R_i so that area$(R_i) =$ area(T_i). It is sufficient to prove that $T_i \sim_T R_i$. There is the issue of assigning the sides of the triangles to particular triangles, but Lemma 9.23 shows that any boundary pieces can be absorbed, using only translations, into a triangle or rectangle. So the borders of the triangles and rectangles are irrelevant to the construction and can be assigned arbitrarily.

For each i, there is a linear transformation L_i such that $L_i(T_i)$ is an isosceles right triangle. Assume by translation that the triangle has one vertex at the origin. If the other two vertices are V_1 and V_2, then let L_i be the inverse of the transformation determined by the matrix whose columns are V_1 and V_2. Theorem 9.21 then yields $L_i(T_i) \sim_T S_i$, where S_i is a square. Lemma 9.24 shows that S_i is translation equidecomposable to the parallelogram $L_i(R_i)$. Therefore $L_i(T_i) \sim_T L_i(R_i)$. But for any translation τ, $L_i^{-1}\tau L_i$ is a translation, so we have the desired $T_i \sim_T R_i$. □

Now we can take the final step toward the squaring of the circle. First we apply the Equidecomposability Criterion and the Erdős–Turán variation.

Theorem 9.25 (Graph Equidecomposability) (AC). *Let K be a Jordan curve made up of three subarcs OA, AB, and BO, where the first two are line segments and the third is a twice differentiable curve that lies inside P, the parallelogram determined by O, A, and B, is never tangent to a side of P, and has curvature bounded away from 0. Then K^* is translation equidecomposable to Q, an axis-aligned square of the same area.*

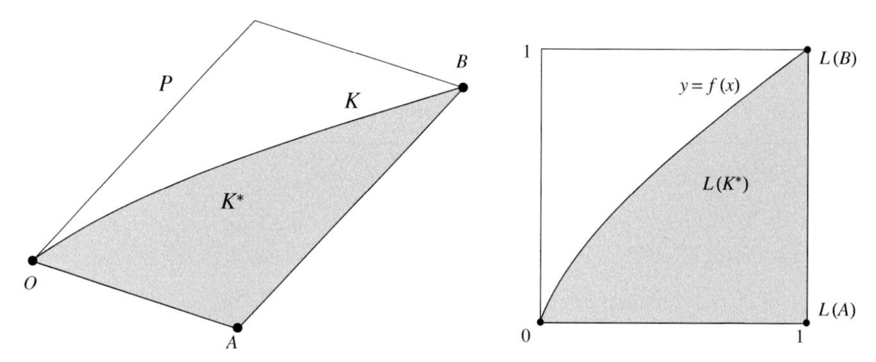

Figure 9.9. At left is a Jordan domain K^* as in Theorem 9.25. At right is the transformed region, the graph of an inequality $0 \leq y \leq f(x)$.

Proof. The setup is as in Figure 9.9. Assume, by translating, that $O = (0, 0)$. Let L be the linear transformation of $\mathbb{R}^2$ such that $L(A) = (1, 0)$ and $L(B) = (1, 1)$. Then $L(OB)$ is contained in $[0, 1]^2$. The conditions on differentiability and curvature imply that the curve $L(OB)$ is convex or concave with respect to the x-axis; it follows that there is an increasing function $f : [0, 1] \to [0, 1]$ whose graph is $L(OB)$. It then follows from the differentiability and tangency conditions that $f'(x)$ is always positive.

Because the curvature of f at $(x, f(x))$ is $f''(x)/(1 + [f'(x)]^2)^{3/2}$, it follows from the curvature condition that f satisfies the hypothesis of Theorem 9.21. So for almost every pair X of vectors of $\mathbb{R}^2$, there is a constant C' such that, for any $\vec{u}$ and N,

(1) $D(F_N(\vec{u}, X), \mathrm{gr}(f)) \leq C' N^{-4/3} \ell^7(N)$.

Let $Q_1 \subseteq J$ be an axis-aligned square with lower-left corner at the origin and with area equal to that of $\mathrm{gr}(f)$. Let Q_2 be a square congruent to Q_1 so that both Q_2 and $L^{-1}(Q_2)$ are far away from the origin; of course, $Q \sim_T Q_1 \sim_T Q_2$. Let $P = L^{-1}(Q_2)$, a closed parallelogram. By the Third Estimation Theorem, for almost all pairs of vectors X, there is a constant C'' such that

(2) $D(F_N(\vec{u}, X), Q_1) \leq C'' N^{-2} \ell^7(N)$.

Therefore we can choose a pair of vectors X and a single constant C so that, with this C, (1) and (2) hold for any $\vec{u}$ and N. Let $\Psi(N) = C N^{2/3} \ell^7(N)$. Then $\Sigma \Psi(2^k)/2^k$ converges. Because $N^2 D(F_N(\vec{u}, X), \mathrm{gr}(f)) \leq \Psi(N)$ and $N^2 D(F_N(\vec{u}, X), Q_1) \leq \Psi(N)$ hold for any $\vec{u}$ and N, the Equidecomposability Criterion yields that $\mathrm{gr}(f) \sim_T Q_1$, and so $\mathrm{gr}(f) \sim_T Q_2$. Simple conjugation yields $L^{-1}(\mathrm{gr}(f)) \sim_T P$.

Now we have $L(K^*) = \mathrm{gr}(f) \cup (\{1\} \times [0, 1])$, and hence K^* differs from $L^{-1}(\mathrm{gr}(f))$ by only the segment AB. Because P is a parallelogram, the absorption lemma yields $K^* \sim_T P$. Also the areas of P and Q each equal that of K^*, so the Polygon Equidecomposability Theorem yields $P \sim_T Q$. Transitivity concludes the proof. □

The preceding result is enough to solve the classic problem.

Corollary 9.26 (AC). *Any disk can be squared with translations.*

Proof. Work with the unit disk; radii from the origin to $(-1, 0)$ and $(0, 1)$ define two subsets, one convex and one not, that meet the conditions of the preceding theorem. Take one of the subsets as closed and the other as partially open. By the theorem, the closure of each set is translation equidecomposable to an axis-aligned square. And the Absorption Lemma means this is true for the partially open set too. Translation then allows the squares to share a border line, and the resulting polygon is translation equidecomposable to a single square by the Polygon Equidecomposability Theorem. □

But Laczkovich's work goes beyond circle-squaring, as it applies to a large variety of plane sets.

Theorem 9.27 (AC). *Let K be a Jordan curve and Q a square whose area is the area of K^*. Suppose K is composed of subarcs $K_1, \ldots, K_n$ such that*

- *each K_i is either the graph of a function or a vertical line segment*
- *for each K_i that is the graph of a nonlinear function, the function is twice differentiable with curvature bounded away from 0*
- *K has no cusps; that is, at the common endpoint of two adjacent subarcs, the half-tangents to the two curves do not coincide*

Then $K^* \sim_T Q$.

Proof. The curvature condition means that the second derivative for each of the graphs is never 0 (except for the linear case). This means that each graph splits into two pieces on which the function is monotonic. So assume that all nonlinear subarcs consist of graphs of monotone functions. Let an end of a subarc be called a *knot*.

Claim. There are points A_i on K (indexed in order around K) and corresponding points P_i in K^* (not necessarily distinct) such that each knot is one of the A_i, and the following holds, where indices are interpreted mod m.

- The line segments $p_i^- = P_iA_i$ and $p_i^+ = P_iA_{i+1}$ are in K^* and hence p_i^-, p_i^+, and the subarc A_iA_{i+1} of K form a Jordan curve T_i for each i.
- The sets T_i^* are interior disjoint.
- For each i, either T_i is a triangle or it satisfies the conditions of Theorem 9.25.

Proof of claim. Throughout this proof a *square* is an open, axis-aligned square. And we will use several times the fact that the interior of K is an open set (by the Jordan curve theorem).

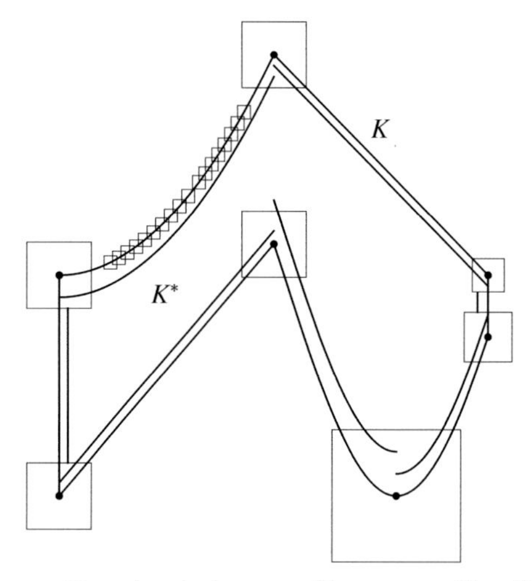

Figure 9.10. The curve K can be nicely covered by squares. Here there are seven large squares at the knots. Then the subarcs are all translated so as to thicken K by a small amount, and the curve outside the knot-squares is covered by small squares that do not break the inside boundary of the fattened curve.

Step 1. For each knot A, let D_A be a square centered at A so that $\partial D_A \cap K$ consists of two points and the closures of the knot-squares are pairwise dijsoint. Also choose the square small enough so that the angle bisector of the two tangents at the point does not intersect either subarc inside the square. See Figure 9.10.

Step 2. Translate each subarc up or down into the interior of K (left or right for the vertical lines) so that the translated arc intersects the knot-squares at each end in the same edge as the untranslated subarc. Also choose the translation so small that the translated arcs do not, outside of the the knot-squares at their ends, strike any of the other translated or untranslated subarcs or squares. For the graphs, this can be accomplished by using $f(x) \pm \epsilon$; the vertical line segments can just be translated horizontally. This gives a fattened version of K (see Fig. 9.10).

Step 3. For each point x on K and not in the (open) knot squares, define a small square D_x centered at x and so that $D_x \cap K^*$ lies between the arc forming K and the translated arc from step 2. A sampling of such squares is shown in Figure 9.10.

Step 4. The collection of squares from steps 1 and 2 forms an open cover of the compact set K, so there is a finite subcover. Each square in the selection, being open, intersects other squares (Fig. 9.11).

Step 5. Let the points A_i be all the intersection points of K and the boundaries of the squares from step 4, together with all the knots.

Step 6. Define the points P_i using some cases. If A_i is a knot, then P_i and P_{i-1} can each be taken to be the point where the tangent angle bisector at the

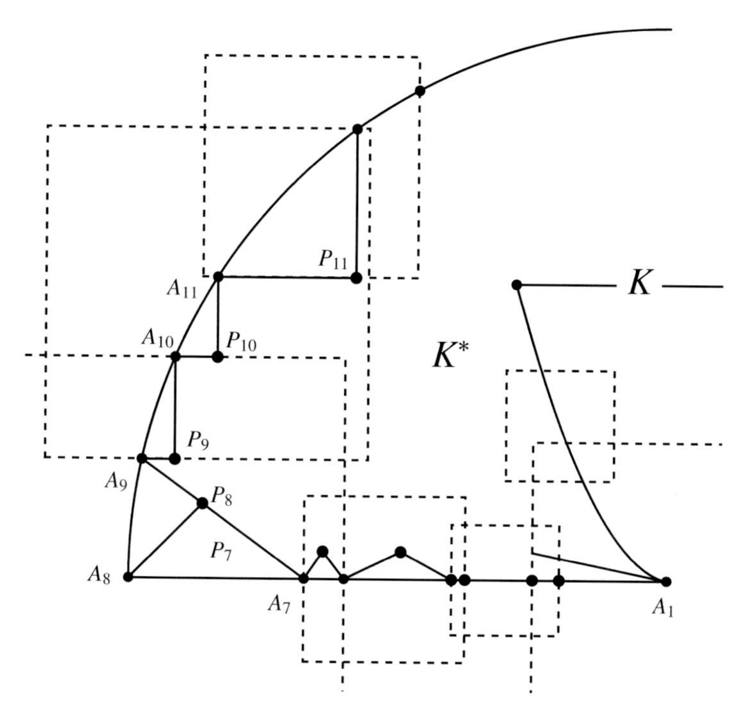

Figure 9.11. Once K is covered by finitely many squares, it is easy to find points A_i and P_i that decompose K^* into one polygon and many three-sided regions. The lines into K^* from A_1 and A_8 bisect the angles of the tangents.

knot meets the segment $A_{i-1}A_{i+1}$ (see Fig. 9.11); this works because of the angle bisector condition in step 1. If the subarc is a horizontal or vertical line, it is easy to find a small V-shape to define P_i. Now suppose the arc is the graph of $y = f(x)$, $A_i = (x_i, y_i)$, and $A_{i+1} = (x_{i+1}, y_{i+1})$. If f is increasing and K^* is below the graph or f is decreasing and K^* is above the graph, then let $P_i = (x_{i+1}, y_i)$, a point that lies inside any of the squares containing A_i or A_{i+1}. In the remaining cases, let $P_i = (x_i, y_{i+1})$. This proves the claim.

Because there are no cusps, we can apply Theorem 9.25 to get disjoint squares $Q_1, \ldots, Q_m$ such that $T_i \sim_T Q_i$. The Polygon Equidecomposability Theorem gives a square Q_0 disjoint from the Q_i such that the polygon $K^* \setminus \bigcup T_i^*$ is translation equidecomposable to a square Q_0. The Absorption Lemma allows this last polygon to be taken as open. Therefore $K^* \sim_T \bigcup_{i=0}^{m} Q_i$, and a final application of Polygon Equidecomposability gives $K^* \sim Q$. □

In fact, the theorem holds even if the cusp assumption is removed; this is shown in [Lac90] using some ideas from the type semigroup.

9.3 Generalizations and Open Problems

In later work, Laczkovich improved his initial results from [Lac90]. The improvements are in two directions. Consider first the proof of the Equidecomposability Criterion. In claim 3 we constructed a bijection with finite spread, transforming one discrete subset of $\mathbb{R}^2$ into another one. Laczkovich [Lac92a] showed how one can construct such a bijection transforming a discrete set to another when both sets are in $\mathbb{R}^n$. This step allowed him [Lac92b] to extend his Equidecomposability Criterion to all Euclidean spaces $\mathbb{R}^n$. Second, Laczkovich was able to connect the Equidecomposability Criterion to a description of boundaries of sets in terms of box dimension. The upper box dimension of a bounded subset E of $\mathbb{R}^n$ (same as the upper entropy index of Kolmogorov) is defined as follows. Let Q_m^n be the set of cubes $\left[\frac{a_1-1}{m}, \frac{a_1}{m}\right] \times \cdots \times \left[\frac{a_n-1}{m}, \frac{a_n}{m}\right]$, where $a_i \in \mathbb{Z}$, and let $N(n, E)$ be the number of such cubes Q having nonempty intersection with E. Then the *upper box dimension* is $\limsup_{m\to\infty} \frac{\ln N(m,E)}{\ln m}$. Laczkovich [Lac92b] showed that if two bounded measurable subsets of $\mathbb{R}^n$ of the same measure have boundaries whose upper box dimension is less than n, then the sets are translation equidecomposable.

As an application of this last result, consider bounded convex sets in $\mathbb{R}^n$. Namely, for any convex subset A of $\mathbb{R}^n$, the upper box dimension of ∂A is at most $n-1$ (see [Egg63, Thms. 41, 42]). Therefore every ball is translation equidecomposable to a cube of the same volume. Moreover, this settles the following one-dimensional problem raised by C. A. Rogers: Is the set

$$A = \left(\frac{1}{3}, \frac{2}{3}\right) \bigcup \left(\frac{7}{9}, \frac{8}{9}\right) \bigcup \left(\frac{25}{27}, \frac{26}{27}\right) \bigcup \cdots$$

translation equidecomposable to the interval $(0, 1/2)$? Because the upper box dimension of $\partial A = 0$, we obtain an affirmative answer to this question (but see §10.3 for an open question about A).

Another direction concerns the existence of measurable decompositions: The pieces are to be Lebesgue measurable. Because a disk and square are measurable, one can hope for a decomposition using measurable pieces. Because the Equidecomposability Criterion uses the Axiom of Choice and there is no description of the nature of the pieces, the work in §9.2 yields no information about measurability. But several spectacular recent results of Grabowski, Máthé, and Pikhurko [GMP∞a, GMP∞b] show that it can be done with measurable pieces. Let $\mathcal{LB}$ denote the family of sets in $\mathbb{R}^n$ that are measurable and have the Property of Baire; that is, $\mathcal{LB} = \mathcal{L} \cap \mathcal{B}$. So a set is in $\mathcal{LB}$ iff it differs from an open set by a measurable meager set.

Theorem 9.28 (AC). (a) *In $\mathbb{R}^n$ $(n \geq 3)$ any two sets of $\mathcal{L}_b$ that have nonempty interior and the same Lebesgue measure are equidecomposable with pieces in $\mathcal{LB}$.*

(b) *If A and B are sets in $\mathbb{R}^n$ of the same nonzero finite Lebesgue measure and if the upper box dimension of the boundary of each of A and B is less than* *n, then A and B are equidecomposable using translations and with pieces in $\mathcal{LB}$.*

In particular, a disk is equidecomposable to a square of the same area using translation and pieces in $\mathcal{LB}$. And the same is true for a regular tetrahedron and a cube of the same volume, and so their work sheds light on Hilbert's Third Problem. So now the notable question in this area is whether one can get decompositions using Borel sets.

Question 9.29. Is a disk in the plane equidecomposable to a square of the same area using pieces that are Borel sets?

Theorem 9.28(a) is false in $\mathbb{R}^2$ by a result of Laczkovich (see Thm. 10.34) that shows that there are Jordan domains that have the same measure but are not equidecomposable even using arbitrary pieces. And it fails in $\mathbb{R}^1$ as well (see the claim in the proof of Thm. 10.38).

The work of Grabowski, Mathé, and Pikhurko disproved one conjecture of R. Gardner—that if a polytope and convex body are equidecomposable using pieces in $\mathcal{L}$ and isometries from an amenable group, then they are equidecomposable using convex pieces. But another conjecture of Gardner [Gar89] remains open. Amenable groups are discussed in Chapter 12 (all solvable groups are amenable); for amenable groups of isometries G, two Lebesgue measurable sets that are G-equidecomposable must have the same measure (Thm. 12.9).

Conjecture 9.30. Suppose $A, B \subset \mathbb{R}^n$ are bounded Lebesgue measurable sets that are equidecomposable with respect to an amenable group of isometries (and therefore have the same measure). Then they are equidecomposable using measurable pieces (and all isometries).

The conclusion of this conjecture uses all isometries, not just those in the amenable group. Now, if one were to formulate the conjecture so that the conclusion used the amenable group of the hypothesis, it would be more powerful. But, as observed by Laczkovich [Lac88b], that version is false. He showed that there are four one-dimensional isometries $S = \{g_i\}$ so that the unit interval $[0, 1]$ is equidecomposable to itself using those isometries (which come from the solvable group G_1), but a measurable equidecomposition using S does not exist.

Let u be an irrational number between 0 and $1/2$ and use it to define the rectangle with vertices p_i as in Figure 9.12. Let R be its perimeter, let σ_i be an isometry of the line that extends the linear function defined by the segment $p_i p_{i+1}$ (where index 5 becomes 1), and let $S = \{\sigma_i\}$. Note that the σ_i are, in order, $x + u$, $-x + u$, $x - u$, $-x + 2 - u$.

Theorem 9.31. *The unit interval is equidecomposable to itself using functions in S, but such a decomposition using S and measurable pieces does not exist.*

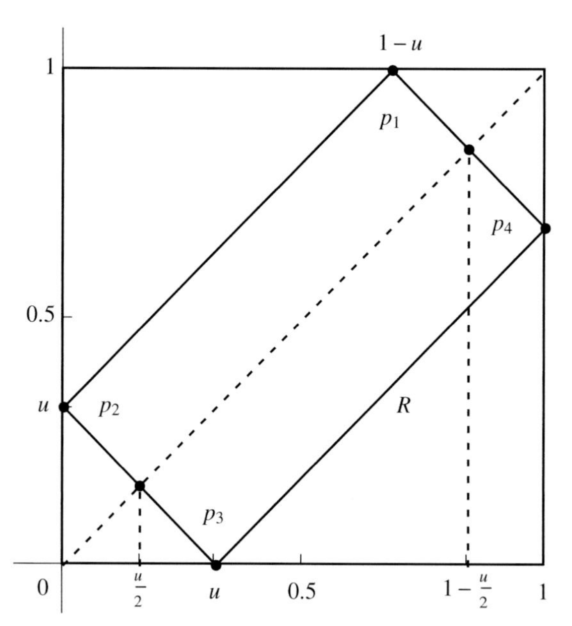

Figure 9.12. The perimeter of the rectangle defined using $u = 1/\pi$ contains a graph of a bijection from the unit interval to itself.

Proof. We will relate decompositions to matchings. By a *matching* (in the context of the unit interval) we mean a set of ordered pairs $M \subset [0, 1]^2$ that defines a bijection. Laczkovich then proved the following claims about R to get his example.

Claim 1. The set R contains a matching.

Such a matching shows that $[0, 1]$ is equidecomposable to itself using S. The proof uses the bipartite graph whose vertices consist of two copies of $[0, 1]$, with an edge from x in one part to y in the other whenever $(x, y) \in R$. All vertices have degree 2, except for four degree-1 vertices: 0 or 1 in either part; therefore the graph decomposes into paths and cycles. The graph has a perfect matching, and that yields the claimed matching. Any infinite path, finite even path, or finite even cycle has such a matching. Any bipartite graph has no odd cycle. To see that there is no path with an odd number of vertices, observe that such a path must go from 0 to 1 in the same part. Then one proves by induction that for any path $(x_1, \ldots, x_n)$, $x_n = \pm x_1 + 2a_n + b_n u$, where $a_n, b_n \in \mathbb{Z}$. Now, if x_1 is 0 and x_n is 1, then $b_n u$ is an odd integer, which contradicts u's irrationality.

For the rest we need to review some measure theory (for more detail, see [Mat95]). Suppose (X, d) is a metric space. Then μ is called an *outer measure* if it is a $[0, \infty]$-valued function defined on $\mathcal{P}(X)$ that is subadditive and such that $\mu(\varnothing) = 0$. Furthermore, we say that μ is a *metric outer measure* if $\mu(A \cup B) = \mu(A) + \mu(B)$ for any two sets A, B having positive distance from each other.

The most important examples of metric outer measures are Lebesgue outer measure in $\mathbb{R}^n$ and d-dimensional outer Hausdorff measure in $\mathbb{R}^n$, which we will define shortly. The restriction of outer measure to the sets A (called *μ-measurable sets*) satisfying the condition $\mu(X \cap A) + \mu(X \setminus A) = \mu(X)$ for any set X is a measure, that is, a countably additive function defined on a σ-algebra and with $\mu(\varnothing) = 0$. In the case of metric outer measures, every Borel set is μ-measurable.

Now, we need the notion of one-dimensional Hausdorff measure H defined on certain subsets of $\mathbb{R}^2$. Such a measure allows one to measure sets such as rectifiable plane curves; it is also related to the notion of Hausdorff dimension, useful in the study of fractals. It is defined as follows: for any $A \subseteq \mathbb{R}^2$ and $\delta > 0$, define

$$H_\delta = \inf\left\{\sum_{i=1}^{\infty} \text{diam}\,(U_i) : A \subseteq \bigcup U_i,\ \text{diam}\,(U_i) < \delta\right\}.$$

Then $H(A) = \lim_{\delta \to 0} H_\delta$; the limit exists because $\mathcal{H}_\delta$ decreases as δ approaches 0. This defines a metric outer measure, and the corresponding H-measurable sets are denoted $\mathcal{H}$.

For example, a disk in the plane has infinite H-measure, while H agrees with one-dimensional Lebesgue on any measurable (with respect to λ_1) subset of a line. In what follows, we will redefine H to be the Hausdorff measure on the H-measurable subsets of R, normalized so that $H(R) = 1$; and $\mathcal{H}$ will refer to the H-measurable subsets of R. An H-measurable and measure-preserving function T from R to R is said to be an *ergodic transformation* if, for $A \in \mathcal{H}$ with $T(A) = A$, we have that $H(A) = 0$ or $H(A) = 1$.

Claim 2. There is no matching contained in R that is in $\mathcal{H}$.

Define $f : R \to R$ to be the piecewise vertical translation that takes $(x, y) \in R$ to the other point in R with the same x-coordinate, with the exceptions that f fixes p_1 and p_4; g is defined similarly in the other direction.

The two functions f and g just defined are homeomorphisms of R onto itself and $g \circ f$ is ergodic on R; the proof of this is based on constructing a measure-preserving homeomorphism $h : R \to \mathbb{S}^1$ and then showing that $hgfh^{-1}$ is just an irrational rotation of the circle (see [Lac88b] for more details). Figure 9.13 shows how the composition acts on R. Moreover, if $M \subset R$ is matching, then $M \cup f(M) = R$ and $M \cap f(M)$ has two points; the same holds for g. Now suppose that $M \in \mathcal{H}$. Because $H(R) = 1$, H vanishes on single points, and $H(M) = H(f(M))$, it must be that $H(M) = 1/2$.

On the other hand, the properties of the transformations f and g imply that $g(f(M))$ and M differ in a finite set.

Proof. $M \cap f(M)$ and $M \cap g(M)$ are each finite (two points). So $g(M) \cap g(f(M))$ is finite. Also $M \cup g(M) = R = M \cup f(M)$ and so $M \cup g(M) = g(M) \cup g(f(M))$. Subtracting $g(M)$ from both sides of this last, we have that

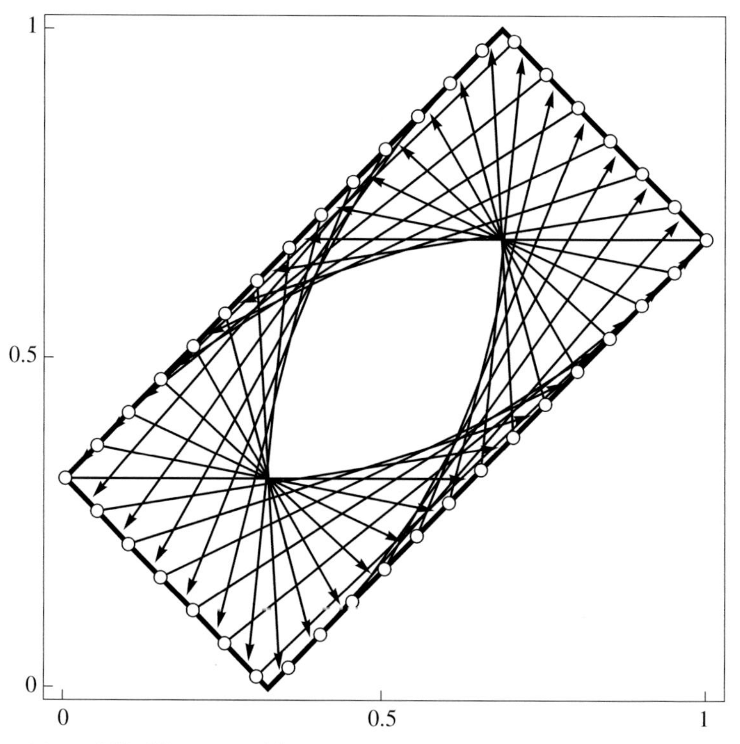

Figure 9.13. The composition $f \circ g$ wraps the perimeter R around itself.

$M \setminus B = g(f(M)) \setminus A$, for finite sets A, B, which implies that the symmetric difference of $g(f(M))$ and M is finite. □

Let $M_1 = \bigcap_{i=-\infty}^{\infty} (gf)^i(M)$. This equals M up to a countable set, and such has H-measure 0. But M_1 is $g \circ f$ invariant and so has measure 0 or 1, a contradiction.

Now we can conclude with an argument that travels from the line to the plane.

Claim 3. The interval $[0, 1]$ is not equidecomposable to itself using S and Lebesgue measurable pieces.

Proof of claim 3. Let A be a Lebesgue measurable subset of $[0, 1]$. It is clear that in $\mathbb{R}$, Lebesgue measure and one-dimensional Hausdorff measure (call it H_1) coincide; so $\lambda(A) = H_1(A)$. Now consider $\sigma_i \in S$; its form is $\pm x + b$ where we take the domain to be $[0, 1]$. Define Π to be the projection $\Pi : \text{graph}(\sigma_i) \to [0, 1]$. From now on we work in $\mathbb{R}^2$ and consider $[0, 1]$ as being a subset of $\mathbb{R}^2$. Clearly, for any Lebesgue measurable subset A of $\mathbb{R}^1$, $H_1(A)$ is the same as $H(A)$ when A is viewed in $\mathbb{R}^2$. Observe now that the transformation $\Pi^{-1} : [0, 1] \to \text{graph}(\sigma_i)$ is a similarity, with factor $\sqrt{2}$. So for any Lebesgue measurable subset of $[0, 1]$, $\Pi^{-1}(A) \in \mathcal{H}$. Now, suppose that $[0, 1]$ is equidecomposable to itself using Lebesgue measurable sets and isometries from S. The inverse projections of the sets lead to a matching in R that is also in $\mathcal{H}$, contradicting claim 2. □

Notes

The circle-squaring problem of Tarski appears in [Tar25] and has been mentioned often in the literature [Kle79, Sie50b, Wag81b]. Theorem 9.1 is due to Dubins, Hirsch, and Karush [DHK63] (see also [Sal69]). For various generalizations of the notion of scissors-congruence, see Sah [Sah79]. The idea of using the Banach–Schröder–Bernstein Theorem to attack an approximate form of the circle-squaring problem using ϵ-magnifications is due to Klee; Theorem 9.3 as stated here is due to Henle and Wagon.

The presentation of the circle-squaring work in §9.2 follows that of Laczkovich [Lac90].

In recent work, Grabowski, Mathé, and Pikhurko [GMP∞a] proved that circle-squaring is possible using Lebesgue measurable pieces. A general criterion for equidecomposability using measurable pieces of any two bounded measurable subsets of $\mathbb{R}^n (n \geq 3)$ is given in [GMP∞b].

10

The Semigroup of Equidecomposability Types

The concept of equidecomposability is much easier to work with if one can add sets, so that a paradoxical decomposition becomes just $X = 2X$. Tarski developed a way to do this in 1938 [Tar38b], and his idea has many applications. The most striking is the theorem of Tarski (Cor. 11.2) relating the existence of measures to the nonexistence of paradoxes. Here we will present the method of turning sets into an abstract object with addition (an additive semigroup), as well as several applications. Noteworthy ones are a Cancellation Law for equidecomposability and also a proof that any two subsets of $\mathbb{S}^2$ with nonempty interior are equidecomposable using rotations. This expanded context for equidecomposability will also yield a simpler proof that a locally commutative action of a free non-Abelian group is paradoxical (Thm. 5.5). And a certain variation on this setup will yield the result that the Banach–Tarski Paradox can be done with pieces that are moved so that they never collide.

10.1 The Semigroup of Equidecomposability Types

10.1.1 The Addition of Sets

We begin, appropriately enough, by setting up a space where a set is expanded into infinitely many copies of itself.

Definition 10.1. *Suppose the group G acts on X. Define an enlarged action as follows. Let $X^* = X \times \mathbb{N}$, let $G^* = \{(g, \pi) : g \in G$ and π is a permutation of $\mathbb{N}\}$, and let the group G^* act on X^* by $(g, \pi)(x, n) = (g(x), \pi(n))$. If $A \subseteq X^*$, then those $n \in \mathbb{N}$ such that A has at least one element with second coordinate n are called the* levels *of A.*

The action of G^* extends that of G and treats all the levels in the same way; for example, if $E \subseteq X$, then $E \times \{n\}$ is G^*-congruent to $E \times \{m\}$. Moreover, G-equidecomposability and G^*-equidecomposability are closely related: If $E_1, E_2 \subseteq X$, then $E_1 \sim_G E_2$ if and only if $E_1 \times \{m\} \sim_{G^*} E_2 \times \{n\}$ for all $m, n \in \mathbb{N}$.

Now, the equidecomposability class of some $E \subseteq X$ in $\mathcal{P}(X^*)$ with respect to G^* is much more valuable than the $\sim_G$-equivalence class of E in $\mathcal{P}(X)$. Because copies of E at different levels can be identified with E, the set $E \times \{0, 1\}$ serves as a representative of what we intuitively would like to call $2E$, and if E is G-paradoxical, then $E \times \{0\} \sim_{G^*} E \times \{0, 1\}$, that is, $E = 2E$. We make this more precise by defining an addition operation for those subsets of X^* having only finitely many levels.

Definition 10.2. *Let G, X, G^*, X^* be as in Definition 10.1.*

(a) A subset A of X^* is called bounded *if it has only finitely many levels. The equivalence class with respect to G^*-equidecomposability of a bounded $A \subseteq X^*$ is called the* type *of A and is denoted $[A]$. The collection of types of bounded sets will be denoted by $\mathcal{S}$.*

(b) For $[A], [B] \in \mathcal{S}$, define $[A] + [B]$ to be $[A \cup B']$, where B' is an upward shift of B so that the levels of B' are disjoint from those of A; that is, $B' = \{(b, m + k) : (b, m) \in B\}$, where k is sufficiently large.

It is a simple matter to check that $+$ is well defined: $[A] + [B]$ is independent of the choice of representatives A, B and of the integer k used to shift B upward. Moreover, $+$ is commutative and associative, whence $(\mathcal{S}, +)$ is a commutative semigroup, called the *type semigroup*. Note that $[\varnothing] = e$ serves as an identity for $+$. If $E \subseteq X$, then $[E]$ is used to denote $[E \times \{0\}]$.

For any commutative semigroup with identity, there is a natural way of multiplying elements by natural numbers: $n\alpha = \alpha + \alpha + \cdots + \alpha$ with n summands. Also, there is a natural ordering given by $\alpha \leq \beta$ if and only if $\alpha + \gamma = \beta$ for some γ in the semigroup. Note that $[A] \leq [B]$ if and only if $A \preceq B$, that is, A is G^*-equidecomposable with a subset of B. These semigroup operations satisfy many familiar axioms: $n(m\alpha) = (nm)\alpha$, $(n + m)\alpha = n\alpha + m\alpha$, $n(\alpha + \beta) = n\alpha + n\beta$, $n\alpha \leq n\beta$, if $\alpha \leq \beta$, $n\alpha \leq m\alpha$ if $n \leq m$, and $\alpha + \gamma \leq \beta + \gamma$ if $\alpha \leq \beta$. The order will satisfy antisymmetry (if $\alpha, \beta \in \mathcal{S}$, then $\alpha \leq \beta$ and $\beta \leq \alpha$ imply $\alpha = \beta$), and this means that it gives the type semigroup the structure of an ordered semigroup. The fact that only bounded subsets of X^* were considered when types were formed means that the type semigroup satisfies an Archimedean condition with respect to $[X]$: For each $\alpha \in \mathcal{S}$, there is some $n \in \mathbb{N}$ such that $\alpha \leq n[X]$. The following proposition shows that the addition of types behaves as one would expect with respect to set-theoretic unions.

Proposition 10.3. *Suppose A and B are bounded subsets of X^*. Then $[A] + [B] \geq [A \cup B]$, with equality if $A \cap B = \varnothing$.*

Proof. Consider the last assertion first. If B' is any upward shift of B, then $B \sim B'$, so $A \cup B \sim A \cup B'$, using two pieces. Hence $A \cup B$ is in $[A] + [B]$, as required. It follows from this that arbitrary bounded sets satisfy $[A] + [B] = [A \cup B] + [A \cap B]$, whence $[A] + [B] \geq [A \cup B]$. □

In the context of the type semigroup, the Banach–Schröder–Bernstein Theorem takes on the simple form, if $\alpha, \beta \in \mathcal{S}$, then $\alpha \leq \beta$ and $\beta \leq \alpha$ imply $\alpha = \beta$. And a type can be called *paradoxical* if $2\alpha = \alpha$; hence $E \subseteq X$ is G-paradoxical if and only if $[E] = 2[E]$ in $\mathcal{S}$. A subtle point arises here because from a measure-theoretic point of view, it might make sense to say $\alpha \in \mathcal{S}$ is paradoxical if, for some k, $(k+1)\alpha \leq k\alpha$ (which is equivalent to $(k+1)\alpha = k\alpha$). This latter condition on α would, like $2\alpha = \alpha$, imply the nonexistence of certain measures normalizing α, but in a general semigroup the two conditions are not equivalent. One would need a certain sort of Cancellation Law in order to deduce $2\alpha = \alpha$ from $(k+1)\alpha = k\alpha$. One of the pleasant aspects of equidecomposability theory is that an appropriate Cancellation Law (Thm. 10.19) is valid for the type semigroup, and these two conditions on α are equivalent. The proof, however, is quite intricate, unlike that of the other major algebraic law, the Banach–Schröder–Bernstein Theorem.

Our first application of this formalism is to give a different proof that a locally commutative action of a free group of rank 2 on a set X yields a paradoxical decomposition. This approach will be much simpler than the proof given in Theorem 5.5, although it does not yield the stronger result that four pieces suffice. First, we generalize Proposition 1.10 slightly, and it is here that the type semigroup is useful.

Definition 10.4. *Suppose the groups $H_1, \ldots, H_n$ all act on X. These actions are called* jointly free *if, for each $x \in X$, there is at least one $i = 1, \ldots, n$ such that x is not a fixed point of any nonidentity element of H_i.*

Proposition 10.5 (AC). *Suppose G acts on X in such a way that the actions of $H, \ldots, H_n$ on X are jointly free, where each H_i is a free subgroup of G of rank 2. Then X is G-paradoxical.*

Proof. Let $D_i = \{x \in X : h(x) = x \text{ for some } h \in H_i \setminus \{e\}\}$, and let $\delta_i = [X \setminus D_i]$ with respect to G^* and X^*. By Proposition 1.10, each $X \setminus D_i$ is H_i-paradoxical, so $\delta_i = 2\delta_i$. Because the actions are jointly free, $X = \bigcup(X \setminus D_i)$, and so by Proposition 10.3, $[X] \leq \sum \delta_i$. Now for each i,

$$[X] = [(X \setminus D_i) \cup D_i] = \delta_i + [D_i] = 2\delta_i + [D_i] = [X] + \delta_i.$$

Therefore

$$[X] = [X] + \delta_1 = [X] + \delta_2 + \delta_1 = \cdots = [X] + \delta_n + \cdots + \delta_1 \geq 2[X] \geq [X],$$

whence $[X] = 2[X]$ and X is G-paradoxical. □

Corollary 10.6 (AC). *If F is a free group of rank 2 and is locally commutative in its action on X, then X is F-paradoxical.*

Proof. We shall show that if G, a free group of rank 4 freely generated by ρ_0, ρ_1, ρ_2, ρ_3, acts on X and is locally commutative, then X is G-paradoxical. Because F contains such a free subgroup ($\rho_i = \sigma^i\tau^i$, where σ, τ generate F), this suffices.

Let H_1 be the subgroup of G generated by ρ_0 and ρ_1, and let H_2 be the subgroup generated by ρ_2 and ρ_3. By Proposition 10.5, it suffices to show that the actions of H_1, H_2 are jointly free. If not, then some $x \in X$ is fixed by w_1 and w_2 in H_1, H_2, respectively, with $w_i \neq e$. But G's action is locally commutative, so w_1 commutes with w_2, contradicting the independence of the ρ_i. □

10.1.2 Continuous Equidecomposability

The Banach–Tarski Paradox in its classic form has two nonphysical aspects: The pieces are, obviously, not physically realizable, and the motions are used abstractly, with no thought to whether the pieces would crash into each other during their motion. Thus a famous question of de Groot [Dek58a] was whether, in the classic Banach–Tarski Paradox, the pieces could be moved continuously so that they remained disjoint throughout the transformation. In 2005, Trevor Wilson, a California Institute of Technology undergraduate, showed that this question has a positive answer. We present his proof, which uses the ideas of the type semigroup, here.

Definition 10.7. *The sets A, B are* continuously G-equidecomposable, *written $A \approx_G B$, if there exist finite partitions $\{A_i\}$ and $\{B_i\}$ of A and B, respectively, and a family of G-paths $\{\gamma_t^i\}$ $(0 \leq t \leq 1)$ such that, for all i, γ_0^i is the identity and $\gamma_1^i(A_i) = B_i$, and for all $0 \leq t \leq 1$ and $i \neq j$, $\gamma_t^i(A_i) \cap \gamma_t^j(A_j) = \varnothing$. We will often suppress the G in $\approx_G$.*

The definition provides motions of the pieces that can be realized physically, because the pieces are disjoint throughout the transformation, that is, at each time value t; for classic equidecomposability, the sets are transported instantaneously to their destinations. It is easy to show that $\approx$ is an equivalence relation.

The key step is defining a semigroup in which we can add sets and so reduce the problem of decomposition of large sets to decomposition of small ones. The construction is a variation on the type semigroup of §10.1.1; the addition will be well defined on the continuous equidecomposability classes.

For $n \geq 2$, let $\mathcal{B}$ be the ideal of bounded subsets of $\mathbb{R}^n$, and let G be any group of isometries that contains all translations in the first two coordinates; for $\vec{v} \in \mathbb{R}^2$, let $T_{\vec{v}}$ denote the corresponding translation in $\mathbb{R}^n$ using 0 in coordinates beyond the second. We will use this group to show that continuous equidecomposability is no stronger than the usual discrete notion of equidecomposability.

When working in $\mathcal{B}$, addition of equidecomposability types is defined as the union of disjoint translates, but there is no guarantee that this operation is well defined when restricted to continuous equidecomposability. Thus we restrict the choice of translates.

We denote by $[A]$ the set $\{A_0 \in \mathcal{B} : A_0 \approx A\}$ and define addition of these classes by $[A] + [B] = [A \cup T_{(v_1,v_2)}(B)]$, where $(v_1, v_2) \in \mathbb{R}^2$ is chosen so that $v_1 > \sup\{(a-b)_1 : a \in A, b \in B\}$ (the subscript denotes projection onto the first coordinate). The definition simply says that A lies strictly to the left of the translated copy of B.

We need some basic properties of this operation.

Proposition 10.8. *Addition of continuous equidecomposability types is well defined, associative, and commutative.*

Proof. Independence of the choice of (v_1, v_2). We have to show $A \cup T_{(v_1,v_2)}(B) \approx A \cup T_{(w_1,w_2)}(B)$, where v_1 and w_1 are such that A is strictly to the left of $T_{(v_1,v_2)}(B)$ and also left of $T_{(w_1,w_2)}(B)$. We can get the desired equidecomposability by translating $T_{(v_1,v_2)}(B)$ to $T_{(w_1,w_2)}(B)$ along horizontal and vertical line segments (those translations lie in G).

Addition is independent of choice of representatives of $[A]$ and $[B]$. Let $A \approx A'$ by subsets $\{A_i\}$ and $\{A'_i\}$ and paths α^i. Because A and B are bounded and $[0, 1]$ is compact, choose $\vec{z} \in \mathbb{R}^2$ so that $\alpha^i_t(A_i)$ is strictly to the left of $T_{\vec{z}}(B_j)$ for all i, j, t. Because classes are independent of the translation vector, what we need to show is $A \cup T_{\vec{z}}(B) \approx A' \cup T_{\vec{z}}(B)$. But this is clear: We use the pieces $\{A_i\} \cup \{T_{\vec{z}}(B)\}$ and $\{A'_i\} \cup \{T_{\vec{z}}(B)\}$ and the paths α^i_t and, for $T_{\vec{z}}(B)$, the identity. Now the case of B being replaced by B' is handled the same way, and transitivity then yields well-definedness.

Commutativity. Choose a vector $(v, 0)$ such that $v > \text{diam}\,(A \cup B)$. Then independence of the choice of translation in the definition of $[A] + [B]$ means that $[A] + [B] = [A \cup T_{(v,0)}(B)]$ and $[B] + [A] = [T_{(v,0)}(A) \cup B]$. But the two representative sets on the right-hand sides are continuously equidecomposable using two pieces: Move $T_{(v,0)}(B)$ left back to B and move A up, around B, and then back down on B's right.

Associativity is left as an exercise. □

Definition 10.9. *A pair of disjoint sets $A, B \in \mathcal{B}$ is* extricable *if* $[A] + [B] = [A \cup B]$. *More generally, a finite family $\{A_i\}$ of pairwise disjoint sets in $\mathcal{B}$ is* extricable *if* $\sum_i [A_i] = [\bigcup_i A_i]$.

Intuitively, two sets are extricable if they can be separated from each other by a physical motion using a finite number of pieces. Note that if $\{A_i\}$ is extricable, then so is $\{B_i\}$, where $B_i \subseteq A_i$; just restrict the functions used in the decomposition to the smaller sets. We call this *extricable by restriction.*

Let the family $\mathcal{E}$ consist of those sets $C \in \mathcal{B}$ such that any two disjoint subsets of C are extricable or, equivalently, such that any finite, pairwise disjoint collection of subsets of C is extricable. We will show that $\mathcal{E}$ is in fact all of $\mathcal{B}$. We need two lemmas.

Lemma 10.10. *The family $\mathcal{E}$ is closed under subsets. Moreover, if $\{C_i\}$ is a finite subset of $\mathcal{E}$ and the family $\{C_i\}$ is an extricable family, then $C = \bigcup C_i$ is in $\mathcal{E}$.*

Proof. The first assertion is clear from the definition of extricable. For the second, one breaks up the given sets according to the partition defining C and extricates the smaller pieces. That is, given subsets $A, B \subseteq C$, let $A_i = A \cap C_i$, and the same for B_i. Let $D_i = A_i \cup B_i \subseteq C_i$; then each family $\{D_i\}$, $\{A_i\}$, $\{B_i\}$ is extricable by restriction. The following sequence now completes the proof, using commutativity and the fact that each $C_i \in \mathcal{E}$ to extricate A_i and B_i:

$$[A \cup B] = \left[\bigcup D_i\right] = \sum [D_i] = \sum ([A_i] + [B_i]) = \sum [A_i] + \Sigma\, [B_i]$$
$$= \left[\bigcup A_i\right] + \left[\bigcup B_i\right] = [A] + [B]. \qquad \square$$

Lemma 10.11 (AC). *There is a partition S_1, S_2 of $\mathbb{R}$ such that, for $i = 1, 2$, the algebraic differences $\Delta S_i = S_i - S_i$ have a dense complement $\mathbb{R}$.*

Proof. Define K to be the dense set $\bigcup_{n\in\mathbb{N}} \frac{1}{3^n}\mathbb{Z}$ and let $H = K + \frac{1}{2}\mathbb{Z}$; H is an additive subgroup of $\mathbb{R}$ and $H \setminus K = K + \frac{1}{2}$. Select elements from the cosets of $\mathbb{R}/H$ to obtain the choice set $\{r_\alpha\}$. Then define $S_1 = \bigcup_\alpha (r_\alpha + K)$ and $S_2 = S_1 + \frac{1}{2}$. These two sets are easily seen to be disjoint, and the coset decomposition implies that every real is in one of the two sets. To finish, we have that for all $a, b \in S_i$ if $a - b \in H$, then $a - b \in K$, and so ΔS_i is disjoint from the set $H \setminus K$, which is the dense set $K + \frac{1}{2}$. $\square$

Theorem 10.12 (AC). *If $n \geq 2$, then every bounded set in $\mathbb{R}^n$ is in $\mathcal{E}$.*

Proof. Let $A \in \mathcal{B}$, and let S_i be the sets of Lemma 10.11. Define S_{ij} to be $S_i \times S_j \times \mathbb{R}^{n-2}$, and let $A_{ij} = A \cap S_{ij}$; then $\bigcup S_{ij} = \mathbb{R}^n$. Choose $r > \operatorname{diam}(A)$. Observe first that the 4-set family $\{A_{ij}\}$ is extricable. This is shown by linearly translating A_{ij} by ir in the second coordinate and then following that by a translation through $(2i + j)r$ in the first coordinate. These motions never lead to a crash, because they are restricted to single coordinates. Consider A_{11} and A_{21}: $(s_1, s_1', \vec{a}\,) \in A_{11}$ translates to $(s_1, s_1' + y, \vec{a}\,)$, which, because of the first coordinate, cannot equal $(s_2, s_1'' + y', \vec{a}\,)$, a similar translate of a point in A_{21}. For the subsequent translation in the first coordinate, a crash is impossible because the choice of r guarantees that the second coordinates differ. Continuing with the example, we would have $(s_1 + x, s_1' + r, \vec{a}\,)$, which cannot equal $(s_2 + x', s_1'' + 2r, \vec{a}\,)$, because $s_1' - s_1'' < r$. The extrication performed by the translations shows that the sets satisfy the condition of Definition 10.9.

Now, by Lemma 10.10, it suffices to show that each $A_{ij} \in \mathcal{E}$. Indeed, for any i and j, there is a single path (a one-dimensional family of translations) $\{\gamma_t : 0 \leq t \leq 1\}$, where each γ_t is a translation in the first two coordinates and the identity in the others, that can extricate any disjoint subsets $B, C \subseteq A_{ij}$. Define γ_t as follows, where γ_t is defined to be a point in $\mathbb{R}^2$, interpreted as a translation vector in $\mathbb{R}^n$. Let $\gamma_0 = (0, 0)$. Let $\{a_k\}$, $\{b_k\}$ be sequences in $\mathbb{R} \setminus \Delta S_i$, $\mathbb{R} \setminus \Delta S_j$, respectively, that converge to 0, and define a sequence $\{v_k\}$ in $\mathbb{R}^2$ by $v_0 = (r, b_0)$, $v_{2k+1} = (a_k, b_k)$, and $v_{2k+2} = (a_k, b_{k+1})$ (Fig. 10.1).

Then let γ_t be the result of linear interpolation between v_{k+1} and v_k during the time interval $2^{-k-1} \leq t \leq 2^{-k}$. For $t > 0$, $\gamma_t \in (\{a_k\} \times \mathbb{R}) \cup (\mathbb{R} \times \{b_k\})$, and so $\gamma_t \notin \Delta S_i \times \Delta S_j$; this means that $\gamma_t(A_{ij}) \cap A_{ij} = \varnothing$ and therefore $\gamma_t(C)$ is always

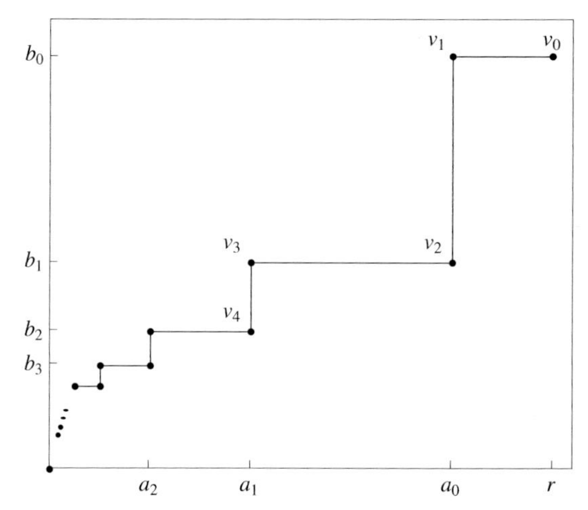

Figure 10.1. The meshing of two 0-converging sequences used for extrication.

disjoint from B. Moreover, $\gamma_1(C) = C + (r, b_0, \vec{0}\,)$ and so lies strictly to the right of B. This means that the path $\{\gamma_t\}$ extricates the pair B, C. □

The next result follows from the preceding proof.

Corollary 10.13. *If $n \geq 2$, then any finite partition of a bounded subset of $\mathbb{R}^n$ is extricable by translations.*

Because the isometry group is a topological group, we can talk about path-connected subgroups.

Theorem 10.14. *If $n \geq 2$ and G is a path-connected group of isometries of $\mathbb{R}^n$ containing all translations in two dimensions, then any two G-equidecomposable bounded subsets of $\mathbb{R}^n$ are continuously G-equidecomposable.*

Proof. Let A and B be equidecomposable using the partitions $\{A_i\}$ and $\{B_i\}$ and isometries $g_i \in G$. Using a translation if necessary, assume A and B are disjoint. Then choosing a path from the identity to g_i in G, we get $A_i \approx_G B_i$ (using only one piece). Then Proposition 10.8 and Theorem 10.12 yield that $[A] = [\bigcup A_i] = \Sigma[A_i] = \Sigma[B_i] = \bigcup[B_i] = [\bigcup_i B_i] = [B]$. □

Corollary 10.15. *If $n \geq 3$, then any two bounded subsets of $\mathbb{R}^n$ with non-empty interior are continuously equidecomposable using orientation-preserving isometries.*

In particular, the solid ball in $\mathbb{R}^3$ has a continuous paradoxical decomposition, and so this corollary answers the question of de Groot posed in [Dek58a, p. 25]. The work of Chapter 9 leads to this slightly stronger squaring of the circle.

Corollary 10.16 (AC). *A disk and a square having the same area are continuously equidecomposable by translations.*

This abstract approach does not easily work in more general situations; here are some unsolved problems. In short, one can ask about continuous equidecomposability for any of the known paradoxical situations. The first is related to an extension of the Banach–Tarski Paradox discussed in §11.2. Wilson [Wil05] points out that the two sets S_i of Lemma 10.11 cannot have the Property of Baire, and that is why his method fails in that context. Because of the Dougherty–Foreman work on Baire sets (§11.2), we have the following question.

Problem 10.17. Does there exist a continuous version of the Banach–Tarski Paradox using pieces having the Property of Baire?

Problem 10.18. Does there exist a continuous version of the Banach–Tarski Paradox in the hyperbolic plane $\mathbb{H}^2$? That is, does Theorem 4.17 hold when restricted to continuous equidecomposability? Is there a continuous version of the Banach–Tarski Paradox of the sphere $\mathbb{S}^2$? Is there a continuous version of any of the linear paradoxes of Chapter 8?

10.2 A Cancellation Law

We now discuss a Cancellation Law for $\mathcal{S}$ that, like the Banach–Schröder–Bernstein Theorem, is a powerful tool for proving sets equidecomposable. Let $\mathcal{S}$ denote the type semigroup of an arbitrary group action on some set.

Theorem 10.19 (Cancellation Law) (AC). *If, for $\alpha, \beta \in \mathcal{S}$ and a positive integer n, $n\alpha = n\beta$, then $\alpha = \beta$.*

By the Banach–Schröder–Bernstein Theorem, the law follows from this weaker version.

Theorem 10.20 (Weak Cancellation Law) (AC). *If, for $\alpha, \beta \in \mathcal{S}$ and a positive integer n, $n\alpha \leq n\beta$, then $\alpha \leq \beta$.*

Before diving into the details of the proof, we illustrate the law's power by deriving some corollaries. First, we show that finitely many copies of a set cannot be packed into fewer copies of the set, unless the set is paradoxical.

Corollary 10.21 (AC). *If $\alpha \in \mathcal{S}$ and $n \in \mathbb{N}$ satisfy $(n+1)\alpha \leq n\alpha$, then $2\alpha = \alpha$*

Proof. Substituting the hypothesized inequality into itself yields $n\alpha \geq (n+1)\alpha = n\alpha + \alpha \geq (n+1)\alpha + \alpha = n\alpha + 2\alpha$. Repeating, we eventually obtain that $n\alpha \geq n\alpha + n\alpha = 2(n\alpha)$. Because $n\alpha \leq 2(n\alpha)$, we have $n\alpha = (2n)\alpha = n(2\alpha)$ and hence cancellation yields $\alpha = 2\alpha$. □

Proof. Substituting the hypothesized inequality into itself yields $n\alpha \geq (n+1)\alpha = n\alpha + \alpha \geq (n+1)\alpha + \alpha = n\alpha + 2\alpha$. Repeating, we eventually obtain that $n\alpha \geq n\alpha + n\alpha = 2(n\alpha)$. So Theorem 10.20 yields $\alpha \geq 2\alpha$, which suffices. □

In Theorem 3.12 we showed how the Banach–Tarski Paradox for balls can be combined with the Banach–Schröder–Bernstein Theorem to get the equidecomposability of arbitrary bounded subsets of $\mathbb{R}^3$ with nonempty interior. But that technique does not suffice to get the $SO_3(\mathbb{R})$-equidecomposability of arbitrary subsets of $\mathbb{S}^2$ with interior. The problem is that an open subset of $\mathbb{S}^2$ does not contain a sphere and so it is not evident that it contains any paradoxical sets. The Cancellation Law allows us to show rather easily that any open subset of $\mathbb{S}^2$ is paradoxical.

Corollary 10.22 (AC). *If G acts on X and $E \subseteq X$ is G-paradoxical, then E is G-equidecomposable with any subset A of E with the property that $g_1A \cup \cdots \cup g_nA \supseteq E$ for some $g_i \in G$. Hence any such A is G-paradoxical and any two such subsets of E are G-equidecomposable.*

Proof. Work in $\mathcal{S}$. The hypothesis on A yields $n[A] \geq [E]$, while $[E] = 2[E] = \cdots = n[E]$ because E is paradoxical. So $n[A \geq n[E] \geq n[A]$ $(A \subseteq E)$, whence $n[A] = n[E]$, and cancellation now implies that $[A] = [E]$, that is, $A \sim E$. Hence A is G-paradoxical because E is, and if B is another subset of E rich enough so that finitely many copies cover E, then $B \sim E$ too, whence $B \sim A$. □

Corollary 10.23 (AC). *If $n \geq 2$, then any two subsets of $\mathbb{S}^n$, each of which has nonempty interior, are $SO_{n+1}(\mathbb{R})$-equidecomposable, and any such subset is $SO_{n+1}(\mathbb{R})$-paradoxical.*

Proof. This follows from Corollary 10.22 and Theorem 6.1, because sufficiently many copies of a nonempty open subset of $\mathbb{S}^n$ cover $\mathbb{S}^n$. □

It is not completely clear which subsets E of $\mathbb{S}^2$ are $SO_3(\mathbb{R})$-equidecomposable with all of $\mathbb{S}^2$. Recall (Thm. 3.10) that $\mathbb{S}^2 \sim \mathbb{S}^2 \setminus D$ for any countable D, and so a set with empty interior can be equidecomposable with $\mathbb{S}^2$. On the other hand, if $E \sim \mathbb{S}^2$, then E cannot be nowhere dense, nor can E have Lebesgue measure zero; this is because these properties are preserved under equidecomposability.

The Cancellation Law, like the Banach–Schröder–Bernstein Theorem, is motivated by the corresponding result for cardinality, that is, arbitrary bijections, namely; $n|X| = n|Y|$ implies $|X| = |Y|$. This more basic result is due to Bernstein [Ber05], who proved it without using the Axiom of Choice (see [Jec73, p. 158] for the beautiful proof in the case $n = 2$). Note that both Bernstein's Theorem and the classical Schröder–Bernstein Theorem ($|X| \leq |Y| \leq |X|$ implies $|X| = |Y|$) are quite easy if the Axiom of Choice is assumed. For, in the case of the former, Choice implies that $n|X| = |X|$ for X infinite, and for the latter, $|X| \leq |Y| \leq |X|$ implies $\aleph_\alpha \leq \aleph_\beta \leq \aleph_\alpha$, where $\aleph_\alpha$, $\aleph_\beta$ are the cardinals corresponding to X, Y respectively, whence $\aleph_\alpha = \aleph_\beta$. What is noteworthy about the two results is that they are theorems of ZF rather than ZFC. In the case of the Schröder–Bernstein Theorem, the Choiceless treatment proved extremely valuable, as it led to Banach's useful generalization to equidecomposability; the proof using $\aleph_\alpha$, $\aleph_\beta$ is of no help with equidecomposability types. With the Bernstein

Cancellation Law, too, the proof using Choice ($n|X| = |X|$) does not help us deal with arbitrary types in $\mathcal{S}$, but the more constructive proof does. There is a catch, though. Bernstein's proof, unlike that of the Schröder–Bernstein Theorem, does not carry over verbatim to the context of $\mathcal{S}$. The necessity of *finite* decompositions when working in $\mathcal{S}$ means that the Axiom of Choice is used at one point in the proof. The interested reader should compare Kuratowski's proof of Theorem 10.19 for $n = 2$ [Kur24] with Bernstein's Theorem as presented by Jech [Jec73, p. 158].

We follow a rather different route, pioneered by D. König. He saw [Kon16] that results on matchings in infinite graphs (which use Choice) could be applied to Cancellation Laws, and after Kuratowski published a proof of Theorem 10.7 for $n = 2$, König [Kon26] showed how results of Valkó and himself could be used to prove Theorem 10.19 for all $n \geq 1$.

The proof of the Cancellation Law that we now give will be based on the Hall–Rado–Hall Infinite Marriage Theorem (App. C), which is a variation on the König method. In fact, just like the Schröder–Bernstein Theorem, the proof of the Cancellation Law works for the situation where equidecomposability is replaced by any relation on sets satisfying properties (a) and (b) of Theorem 3.6's proof.

Proof of 10.20. If $n\alpha \leq n\beta$, then there are two disjoint, bounded, G^*-equidecomposable sets $E, F \subseteq X^*$ with partitions $E = A_1 \cup \cdots \cup A_n$, $F = B_1 \cup \cdots \cup B_n$ such that each $[A_i] = \alpha$ and each $[B_i] = \beta$. Let $\chi : E \to E' \subseteq F$ be the piecewise G^*-congruence witnessing $E \sim E'$, and let $\phi_i : A_1 \to A_i$, $\psi_i : B_1 \to B_i$ likewise witness G^*-equidecomposability (take ϕ_1 and ψ_1 to be the identity). For each $a \in A_1$, let $\overline{a} = \{a, \phi_2(a), \ldots, \phi_n(a)\}$, and for $b \in B_1$, let $\overline{b} = \{b, \psi_2(b), \ldots, \psi_n(b)\}$. Note that $\{\overline{a} : a \in A_1\}$,$\{\overline{b} : b \in B_1\}$ form partitions of E, E', respectively.

Form a bipartite graph by letting the first part of the vertex set be $A = \{\overline{a} : a \in A_1\}$, while $B = \{\overline{b} : b \in B_1\}$ is the other part. There will be n edges emanating from each $\overline{a}$: For each $i = 1, \ldots, n$, form the edge $\overline{a} \multimap \overline{b}$ whenever $\chi\, \phi_i(a) \in \overline{b}$. This graph, which might have multiple edges, is n-regular in the first part and of maximum degree n in the second part. Furthermore, it satisfies the marriage condition on the first part. For suppose Y is a k-sized subset of A; let Z be the set of neighbors of vertices in Y. The number of edges going from Y to Z is nk. But if $|Z| < k$, then the number of edges going from Z to Y is less than kn, contradiction.

The Hall–Rado–Hall Theorem (Thm. C.2) then provides a matching M for the graph that covers A. For each vertex $\overline{a}$, there is a unique edge $\overline{a} \multimap \overline{b}$ in M, and the edge exists by virtue of $\chi\phi_i(a) = \psi_j(b)$ for some i, j. Let $C_{ij} = \{a \in A_1 : \overline{a} \multimap \overline{b} \in M$ and $\chi\phi_i(a) = \psi_j(b)\}$: similarly, let $D_{ij} = \{b \in B_1 : \overline{a} \multimap \overline{b} \in M$ and $\chi\, \phi_i(a) = \psi_j(b)\}$. Then $\psi_j^{-1}\chi\phi_i$ maps C_{ij} into D_{ij} and is a piecewise G^*-congruence because each of ϕ_i, χ, ψ_j^{-1} are. Because $\{C_{ij}\}$ partitions A_1, this shows $A_1 \sim \bigcup D_{ij} \subseteq B$, or $\alpha \leq \beta$, as desired. □

Because the proof of the Cancellation Law uses the Axiom of Choice, it cannot be applied in situations (as in §10.3) where there are restrictions on the sets. Tarski developed a theory of cardinal algebras (such being a commutative semigroup with identity and having an additional infinitary operation that satisfies several axioms) and in [Tar49, §16] he relates that theory to the equidecomposability relation and derives, without AC, several identities that are interesting, but weaker than the full Cancellation Law.

10.3 Restrictions on the Pieces

We have seen that it can be natural when dealing with equidecomposability to restrict the pieces in some way. Indeed, this goes back to the historical method of understanding areas using the cutting apart of shapes described by the method of congruence by dissection (§3.1), but one can also restrict to pieces that are Lebesgue measurable (Theorem 9.29) or have the Property of Baire (the Marczewski Problem, §11.2). In such situations a semigroup of types can still be formed, but its algebraic structure may differ in important ways from the case of arbitrary pieces. Suppose a group G acts on a set X so as to leave a certain subalgebra $\mathcal{A}$ of $\mathcal{P}(X)$ invariant; that is, $\sigma(A) \in \mathcal{A}$ whenever $A \in \mathcal{A}$ and $\sigma \in G$. Examples include the ones just mentioned, as well as letting $\mathcal{A}$ consist of all Borel subsets of a metric space (where G is a group of isometries), or all Borel subsets of a topological group acting on itself by left multiplication. In these examples $\mathcal{A}$ is a σ-algebra; that is, $\mathcal{A}$ is closed under countable unions and intersections. For such $\mathcal{A}$ it is easy to see that the proof of the Banach–Schröder–Bernstein Theorem (Thm. 3.6) applies to G-equidecomposability in $\mathcal{A}$; the proof uses a countable union at one point, and so the sets it constructs remain in $\mathcal{A}$ provided $\mathcal{A}$ is a σ-algebra.

To form a semigroup of equidecomposability types for $\mathcal{A}$, let X^* and G^* be as in §10.1.1, but consider only those bounded subsets A of X^* such that each level of A corresponds to a set in $\mathcal{A}$. Precisely, let

$$\mathcal{A}^* = \{A \subseteq X^* : \text{ for some } n \in \mathbb{N}, A = \bigcup_{m<n} A_m \times \{m\} \text{ where each } A_m \in \mathcal{A}\},$$

and define the type of $A \in \mathcal{A}^*$ to consist of all sets $B \in \mathcal{A}^*$ such that $B \sim_{G^*} A$ using pieces in $\mathcal{A}^*$. Let the collection of such types be denoted by $\mathcal{S}(\mathcal{A})$. If we define $+$ in $\mathcal{S}(\mathcal{A})$ as in §10.1.1, then $\mathcal{S}(\mathcal{A})$ is a commutative semigroup. The proof of the Cancellation Law, however, does not carry over to $\mathcal{S}(\mathcal{A})$, even if $\mathcal{A}$ is a σ-algebra. This is because the proof used the Axiom of Choice, which in general produces sets outside $\mathcal{A}$: The sets C_{ij} and D_{ij}, used for equidecomposability in the proof, were defined from the matching M, which need not lie in $\mathcal{A}$ because it was nonconstructively provided by the marriage theorem. We shall see in §10.3.1 some examples showing that the Cancellation Law can indeed fail, and in §10.3.2 some interesting contexts in which it is not known if the Cancellation Law holds.

10.3.1 Failure of the General Cancellation Law

In this section we show how to modify the Laczkovich matching construction (Thm. 9.31) to show the failure of a general Cancellation Law; this work is due to Gardner and Laczkovich [GL90]. In what follows we use a one-dimensional version of the concept of geometric dissection, introduced in Chapter 3 so the pieces will be simply intervals.

The purpose of the next theorem is twofold. First, it shows that there are two plane squares (extend the intervals into the second dimension) that are equidecomposable with respect to a finite group of plane isometries but are not congruent by dissection using the same isometries. Second, the ideas lead to an algebra where the Cancellation Law (10.19) fails.

Theorem 10.24. *There is a set T of four isometries of the line and two intervals such that the intervals are equidecomposable with respect to isometries in T, but are not congruent by dissection with respect to the group generated by T.*

Proof. Start with an irrational $u \in (0, 1)$ and consider the four simple isometries $\{\sigma_1, \sigma_2, \sigma_3, \sigma_4\}$ defined from u in Theorem 9.31. Let $v \in \mathbb{R}$ be such that 1, u, and v are linearly independent over $\mathbb{Q}$; let T be the isometries given by $\tau_i(x) = \sigma_i(x) + v$. Then the intervals $[0, 1]$ and $[v, v + 1]$ are equidecomposable using T by claim 1 of Theorem 9.31. Observe that every element in the group generated by the τ_i has the form $g(x) = \pm x + ku + jv + 2m$, where $k, j, m \in \mathbb{Z}$ and $k + j$ is even. Now let G be the group consisting of all such functions g. We will prove that these intervals are not congruent by dissection using G (which contains the τ_i).

Suppose the congruence by dissection exists; then there are partitions into closed subintervals of $[0, 1]$ and $[v, v + 1]$ via the numbers $0 = a_0 < a_1 < \ldots < a_n = 1$ and $v = b_0 < b_1 < \cdots < b_n = v + 1$, respectively, such that each $[a_i, a_{i+1}]$ is g_i-congruent to $[b_i, b_{i+1}]$ for some $g_i \in G$. Now define a bipartite graph Γ as follows: The vertices of one part are the numbers a_i, and of the other part are the b_j. And $a_i \leftrightarrow b_j$ is an edge if $g_i(a_i) = b_j$, where $i < n$, or $g_{i-1}(a_i) = b_j$ and $i > 0$. The graph can have multiple edges; we observe that all vertices have degree 2, except the vertices a_0, a_n, b_0, and b_n, which have degree 1. Let Γ_0 be the connected component containing a_0; this component must have a path from a_0 to one of the three leaves. But any vertex in the path has the form $ku + jv + 2m$, with $k + j$ even. Because the end of the path is 1, v, or $v + 1$, this contradicts the assumption of algebraic independence. □

A modification of the preceding result yields an example showing that $2x = 2y$ in the type semigroup for equidecomposability using a set-algebra does not imply that $x = y$. By an *algebra*, we mean a collection of subsets of X containing $\varnothing$, X and closed under finite union and intersection, and complementation in X. Here G is the group of Theorem 10.24.

Theorem 10.25. *There is a subset X of $\mathbb{R}$ and a G-invariant subalgebra $\mathcal{A}$ of $\mathcal{P}(X)$ such that the Cancellation Law fails for G-equidecomposability using pieces in $\mathcal{A}$.*

Proof. Let u, v be the reals from Theorem 10.24; let $H = \{ku + jv + m : k, j, m \in \mathbb{Z}\}$ and $X = \mathbb{R} \setminus H$. Let $(a, b)_X$ denote the intersection of the interval with X, and then let $\mathcal{A}$ be the algebra generated by the *basis intervals*: sets $(a, b)_X$, where $a, b \in H \cup \{\pm\infty\}$. Because of the use of the complement of H, $\mathcal{A}$ is exactly the set of all finite unions of the basis intervals and complements of such sets.

Now we can show that the sets $B = (0, 1)_X \cup (2, 3)_X$ and $C = (v, v + 1)_X \cup (v + 2, v + 3)_X$ are G-equidecomposable with pieces in $\mathcal{A}$. Note that G maps H to H and X to X, and hence $\mathcal{A}$ is G-invariant. We can partition B into $B_1 = (0, u)_X$, $B_2 = (u, 1)_X$, $B_3 = (2, 3 - u)_X$, and $B_4 = (3 - u, 3)_X$ and C into $C_1 = (v, v + u)_X$, $C_2 = (v + 2, v + 3 - u)_X$, $C_3 = (v + u, v + 1)_X$, and $C_4 = (v + 3 - u, v + 3)_X$. Then $g_i(B_i) = C_i$, where $g_1(x) = -x + u + v$, $g_2(x) = x - u + v + 2$, $g_3(x) = x + u + v - 2$, and $g_4(x) = -x - u + v + 6$.

On the other hand, translation by 2 witnesses G-equidecomposability using pieces in $\mathcal{A}$ of the sets $(0, 1)_X$ and $(2, 3)_X$, and also $(v, v + 1)_X$ and $(v + 2, v + 3)_X$. So we have, in the type semigroup, $2[(0, 1)_X] = 2[(v, v + 1)_X]$. But the sets $(0, 1)_X$ and $(v, v + 1)_X$ are not G-equidecomposable using pieces in $\mathcal{A}$, because such a decomposition would use only basis intervals and so, after extending the isometries from X to $\mathbb{R}$, would show that the corresponding real intervals were congruent by dissection using elements of G, in contradiction to the proof of Theorem 10.24. □

The counterexamples to the Cancellation Law found by Gardner and Laczkovich (and also a different example by J. Truss [Tru90]; his example uses a σ-algebra and does not use isometries) used noncommutative groups. But Laczkovich [Lac91a] showed that such failures are possible even with commutative groups. Here and later in this section we use *interval* of $\mathbb{R}^n$ to mean a set of the form $I_1 \times I_2 \times \cdots \times I_n$, where each I_i is a bounded interval on the line. For the next example, all the components will be half-open. Working in $\mathbb{R}^2$, let $\mathcal{R}$ be the set of rectangles $[a, b) \times [c, d)$, where $a < b$ and $c < d$, and let $\mathcal{A}$ be the algebra generated by $\mathcal{R}$. For two fixed positive reals α, β with α/β irrational, define the set of vectors $V = \{(x + n\alpha + k\beta, x) : x \in \mathbb{R}, n, k \in \mathbb{Z}\}$. This set determines a group T_V of planar translations.

Theorem 10.26 (AC). *The Cancellation Law fails for the triple* $(\mathbb{R}^2, T_V, \mathcal{A})$.

Proof. Let $A = [0, \alpha) \times [0, \beta/2)$ and $B = [0, \alpha/2) \times [0, \beta)$; we will show that $2[A] = 2[B]$ and $[A] \neq [B]$ in the type semigroup. Define $A' = [0, \alpha) \times [\beta/2, \beta)$. We have $A \sim_{T_V} A'$ in $\mathcal{A}$. Indeed, choose a positive integer n such that $n\alpha < \beta/2 <$

$(n+1)\alpha$; then $A = A_1 \cup A_2$ and $A' = A'_1 \cup A'_2$, where

$$A_1 = \left[0, (n+1)\alpha - \frac{\beta}{2}\right) \times \left[0, \frac{\beta}{2}\right), A_2 = \left[(n+1)\alpha - \frac{\beta}{2}, \alpha\right) \times \left[0, \frac{\beta}{2}\right)$$

and

$$A'_1 = \left[\frac{\beta}{2} - n\alpha, \alpha\right) \times \left[\frac{\beta}{2}, \beta\right), A'_2 = \left[0, \frac{\beta}{2} - n\alpha, \alpha\right) \times \left[\frac{\beta}{2}, \beta\right).$$

Observe that $A'_1 = A_1 + t_1$, where $t_1 = (\beta/2 - n\alpha, \beta/2) \in T_V$, and $A'_2 = A_2 + t_2$, where $t_2 = (\beta/2 - (n+1)\alpha, \beta/2) \in T_V$. This implies that $A \sim_{T_V} A'$ in $\mathcal{A}$. Similarly, $B \sim_{T_V} B'$ in $\mathcal{A}$, where $B' = [\alpha/2, \alpha) \times [0, \beta)$. Because $A \cup A' = B \cup B'$, we have that $2[A] = 2[B]$ in $\mathcal{A}$.

Next we will show that $[A] \neq [B]$ in $\mathcal{A}$. For this we will construct a function Φ from $\mathcal{R}$ to the circle group $\mathbb{R}/[0, 1)$ that is T_v -invariant modulo 1 and assigns different values to $[A]$ and $[B]$.

Let $F : \mathbb{R}^2 \to \mathbb{R}$ be any function. It is easy to see that Φ, defined by

$$\Phi([a, b) \times [c, d)) = F(a, c) - F(b, c) + F(b, d) - F(a, d),$$

is a finitely additive function on $\mathcal{R}$. Now we have to choose F so that Φ is as desired. For this, we take a Hamel basis H (a basis of $\mathbb{R}$ over $\mathbb{Q}$) containing α and β and define functions $f, g : \mathbb{R} \to \mathbb{R}$ such that $f(x+y) = f(x) + f(y)$ and $g(x+y) = g(x) + g(y)$ for any $x, y \in \mathbb{R}$ as follows: $f(x)$ and $g(x)$ are the coefficients of α and β in the representation $x = \sum r_u \cdot u$, where $u \in H$, $r_u \in \mathbb{Q}$, and $r_u = 0$ for all but finitely many r_u. Note that because $\alpha/\beta \notin \mathbb{Q}$, we can always construct these two functions.

Next define the function $F(x, y) = f(x-y)\lfloor g(x-y)\rfloor$, where x, y are real. It follows that $F(0, 0) = F(\alpha/2, 0) = F(\alpha, 0) = F(0, \beta/2) = F(0, \beta) = 0$, $F(\alpha/2, \beta) = -1/2$, and $F(\alpha, \beta/2) = -1$; therefore $\Phi(A) = -1$ and $\Phi(B) = -1/2$, which gives $[A] \neq [B]$.

For T_v-invariance we need to show that for any $R \in \mathcal{R}$ and $t \in T_V$, $\Phi(R+t) \equiv \Phi(R) \pmod 1$. For this it is enough to check that it is true in the following cases: $t = (z, z)$, where $z \in \mathbb{R}$, $t = (\alpha, 0)$, and $t = (\beta, 0)$. Because $F(x+z, y+z) = F(x, y)$, the first case is clear. If $t = (\alpha, 0)$, then

$$\begin{aligned} F(x + \alpha y) &= f(x - y + \alpha)\lfloor g(x - y + \alpha)\rfloor = (f(x-y) + 1)\lfloor g(x-y)\rfloor \\ &= f(x-y)\lfloor g(x-y)\rfloor \equiv F(x, y) (\text{mod } 1), \end{aligned}$$

which easily implies T_V-invariance. Let $t = (0, \beta)$; then

$$\begin{aligned} F(x+\beta, y) &= f(x - y + \beta)\lfloor g(x - y + \beta)\rfloor = f(x-y)\lfloor g(x - y + 1)\rfloor \\ &= f(x-y)\lfloor g(x-y)\rfloor + f(x-y) = F(x-y) + f(x-y). \end{aligned}$$

If $R = [a, b) \times [c, d)$, then by the preceding we get $\Phi(R + (\beta, 0)) = \Phi(R) + (f(a-c) - f(b-c) + f(b-d) - f(a-d)) = \Phi(R)$, and this concludes the construction. □

Based on the preceding construction, Laczkovich [Lac91a] gave two additional examples of the failure of the Cancellation Law, obtained by using Abelian group actions (see [Lac91a, Exs. 2 and 3]). In the first of these examples the underlying space is a compact Abelian group (the torus $\{(x, y) : 0 \leq x < \alpha, 0 \leq y < \beta\}$ such that α/β is irrational and addition is defined mod α in the first coordinate and mod β in the second) and the acting group is a subgroup of translations. In the second example the space is $\mathbb{R}$ and the acting group is the group of all translations. This work led Laczkovich to the following question.

Question 10.27. Let X be a compact Abelian group, let G be the group of all translations in X, and let $\mathcal{A}$ be any subalgebra of $\mathcal{P}(X)$. Does the Cancellation Law hold for $(X, G, \mathcal{A})$?

We have also the following old question.

Question 10.28. Does the Cancellation Law hold for Borel equidecomposability in every locally compact group?

Laczkovich [Lac91a] found a very interesting relation between Cancellation Law and signed measures. Recall that signed measures, unlike classical measures, are allowed to take any value in $\mathbb{R} \cup \{\infty\}$ as a value (usually such measures are allowed to be $-\infty$, but we restrict to only ∞ here). More precisely, let a group G act on a set X and let $\mathcal{A}$ be a G-invariant subring of $\mathcal{P}(A)$. We will say that θ is a finitely additive, G-invariant signed measure on $(X, G, \mathcal{A})$ if

(a) $\theta : \mathcal{A} \to \mathbb{R} \cup \{\infty\}$
(b) for any $A \in \mathcal{A}$, $\theta(A) = \theta(gA)$
(c) for any disjoint $A, B \in \mathcal{A}$, $\theta(A \cup B) = \theta(A) + \theta(B)$

Now we will need the type semigroup defined for a given ring of sets: It is defined in the same way as the type semigroup for a subalgebra of $\mathcal{P}(X)$ in §10.1. As usual, we use $[A]$ for the type of $A \in \mathcal{A}$. Laczkovich [Lac91a] proved the following theorem.

Theorem 10.29 (AC). *Suppose G acts on X and $A, B \in \mathcal{A}$, where $\mathcal{A}$ is a subalgebra of $\mathcal{P}(X)$. Then $\theta(A) = \theta(B)$ for any G-invariant, finitely additive, signed measure on $\mathcal{A}$ if and only if there is a positive integer n such that $n[A] = n[B]$.*

Consider now the following two conditions for a triple $(X, G, \mathcal{A})$:

(a) $A, B \subseteq X$ are G-equidecomposable with pieces in $\mathcal{A}$.
(b) $\theta(A) = \theta(B)$, whenever θ is any finitely additive G-invariant signed measure defined on $\mathcal{A}$.

Theorem 10.31 leads to the following corollary.

Corollary 10.30 (AC). *For every triple (X, G, A), conditions (a) and (b) are equivalent if and only if the Cancellation Law holds for equidecomposability with respect to G and $\mathcal{A}$.*

Proof. The forward direction follows from Theorem 10.31. For the reverse, observe that if the Cancellation Law fails in $\mathcal{A}$, then (b) does not imply (a). Indeed, let n be a positive integer such that $n[A] = n[B]$ and $[A] \neq [B]$. By the theorem, $n[A] = n[B]$ implies (b), and therefore the implication (b) $\Rightarrow$ (a) does not hold. □

Because the Cancellation Law holds when $A = \mathcal{P}(X)$, we immediately get the following corollary.

Corollary 10.31 (AC). *Let G be a group acting on a set X. Then for any $A, B \subseteq X$, $A \sim_G B$ holds if and only if $\theta(A) = \theta(B)$ is true for any finitely additive, G-invariant signed measure on $\mathcal{P}(X)$.*

Finally, we observe that in Corollary 10.33 we cannot use finitely additive measures or even signed measures with finite values. First we observe that in that corollary, we cannot use a finitely additive, G-invariant measure. Indeed, let $X = \mathbb{Q}$; then $\mu([0, 1) \cap \mathbb{Q}) = \infty$ implies $\mu([0, 1] \cap \mathbb{Q}) = \infty$. And if $\mu([0, 1) \cap \mathbb{Q}) < \infty$, then for any $x \in \mathbb{Q}$, $\mu(\{x\}) = 0$. Therefore $\mu([0, 1] \cap \mathbb{Q}) = \mu([0, 1) \cap \mathbb{Q})$. But, by a result of Sierpiński [Sie54, Thm. 17], $[0, 1) \cap \mathbb{Q}$ is not equidecomposable to $[0, 1] \cap \mathbb{Q}$.

Now observe that the signed measure θ in Corollary 10.3 cannot have only finite values. Let $X = \mathbb{Z}$ and let G be the group of integer translations. If θ is a finitely additive G-invariant signed measure with finite values, then $\theta(\mathbb{N}) = \theta(\mathbb{N} \setminus \{0\})$ (as one is a translate of the other) and so $\theta(\{0\}) = 0 = \theta(\varnothing)$. But $\{0\}$ and $\varnothing$ are not equidecomposable.

10.3.2 Regular-Open Sets and Geometric Bodies

In $\mathbb{R}^n$ the restriction to Borel pieces eliminates paradoxical decompositions of sets such as balls or cubes, because of Lebesgue measure. But there is another way of looking at equidecomposability in $\mathbb{R}^n$ that, when $n \geq 2$, leads to the intriguing Marczewski Problem. This problem was open for more than 60 years before being solved by Dougherty and Foreman in 1992 (§11.2).

An open subset of $\mathbb{R}^n$ is called *regular-open* if it equals the interior of its closure; loosely speaking, the open set can have no cracks. The union of two regular-open sets is not necessarily regular-open, but there is another natural binary operation that is very similar to union. If A and B are regular-open, let $A \vee B$ be $\operatorname{int}(\overline{A \cup B})$, where $\operatorname{int}(E)$ is the interior of E; it is easy to check that $A \vee B$ is regular-open and $\vee$ is associative. Moreover, $A_1 \vee A_2 \vee \cdots \vee A_n = \operatorname{int}(\overline{A_1 \cup \cdots \cup A_n})$. The importance of this notion is that the regular-open sets in any topological space X form a complete Boolean algebra, with $A \wedge B = A \cup B$, $A' = \operatorname{int}(X \setminus A)$, and $\sum\{A_i : i \in I\} = \operatorname{int}\overline{\bigcup\{A_i : i \in \mathbb{N}\}}$ [CN74, p. 53]. We are interested in the collection of bounded regular-open subsets of $\mathbb{R}^n$, which we denote by $\mathcal{R}$. Now, we can define equidecomposability in $\mathcal{R}$ with respect to the isometries of $\mathbb{R}^n$ in the usual way, using $\vee$ instead of $\cup$ and with the restriction that all pieces come from $\mathcal{R}$. Then let $[A]$, for $A \in \mathcal{R}$, be the collection of sets that are equidecomposable in $\mathcal{R}$ with A, and let $\mathcal{S}(\mathcal{R}) = \{[A] : A \in \mathcal{R}\}$. Unlike the general case, where it was necessary to expand X to $X \times \mathbb{N}$, this definition of $\mathcal{S}(\mathcal{R})$ is adequate for adding types. Translations can be used to obtain as many disjoint copies of a bounded set as required, whence addition of types may be defined by $[A] + [B] = [A \cup \tau(B)]$, where τ is a translation taking B to a set disjoint from A.

The Banach–Schröder–Bernstein Theorem holds in $\mathcal{S}(\mathcal{R})$ by the proof given in Theorem 3.6, where one replaces $A \cup B$ by $A \vee B$, $\bigcup_{n=0}^{\infty} C_n$ by $\sum C_n$ and $A \setminus B$ by $A \wedge B'$. But it is unclear whether the Cancellation Law holds in $\mathcal{S}(\mathcal{R})$. Because of the paradox inherent in the solution to the Marczewski Problem, the semigroup becomes trivial (two elements) in dimensions 3 and greater. But the situation in dimensions 1 or 2 is not clear.

Question 10.32. Does $2\alpha = 2\beta$ imply $\alpha = \beta$ in $\mathcal{S}(\mathcal{R})$, where $\mathcal{R}$ is the regular-open algebra in $\mathbb{R}^1$ or $\mathbb{R}^2$?

We know that any two bounded nonempty open sets in $\mathbb{R}^n$ $(n \geq 3)$ are equidecomposable—the strong form of the Banach–Tarski Paradox—but that uses arbitrary sets in the decomposition. Equidecomposability in $\mathcal{R}$ requires that regular-open sets be used as pieces. A first clue that paradoxes might exist comes from the fact that Lebesgue measure is not additive with respect to the join operation that defines equidecomposability in $\mathcal{R}$. We show why such additivity fails in $\mathbb{R}^1$, but the example generalizes easily to $\mathbb{R}^n$.

Proposition 10.33. *For any $\epsilon > 0$, the open unit interval $(0, 1)$ contains disjoint regular-open sets A_1 and A_2 such that $A_1 \vee A_2 = (0, 1)$ but $\lambda(A_1) + \lambda(A_2) < \epsilon$.*

Proof. Let $\{q_n : n \in \mathbb{N}\}$ enumerate the rationals in $(0, 1)$. Choose open intervals I_n such that I_n is centered at q_n, has an irrational radius less than $\epsilon 2^{-(n+1)}$, and is disjoint from the previously defined intervals. But if q_n already lies in some I_m with $m < n$, let $I_n = \varnothing$. Let $A = \bigcup I_n$, an open dense set with $\lambda(A) < \epsilon$. Let I_n' be the open interval forming the left half of I_n, while I_n'' is the (open) right half. Finally, let $A_1 = \bigcup I_n'$, $A_2 = \bigcup I_n''$. Because A_i is open, to show that it is regular-open, it suffices to show that any point x in the interior of $\overline{A_i}$ lies in A_i. Consider A_1. Let I be an open interval containing x and contained in $\overline{A_1}$. Then I is disjoint from A_2. Let $q < x$ be a rational in I; because $I \cap A_2 = \varnothing$, the interval I_n containing q must be centered at a rational to the right of I, which implies that $x \in A_1$. Similarly, $A_2 \in \mathcal{R}$.

Because $A_1 \cup A_2$ contains all rationals in $(0, 1)$, $\overline{A_1 \cup A_2} = [0, 1]$ and $A_1 \vee A_2 = (0, 1)$. Moreover, $\lambda(A_1) = \lambda(A_2) = \frac{1}{2}\lambda(A)$, so $\lambda(A_1) + \lambda(A_2) < \epsilon$. □

And in fact a cube and a cube twice the size are equidecomposable in $\mathcal{R}$. This is a consequence of the Dougherty–Foreman solution to the Marczewski Problem (§11.2). In dimensions 1 and 2, there is a measure on $\mathcal{R}$ that implies a paradox does not exist. The main point is that, because the boundary of an open set is nowhere dense, a finitely additive measure will be additive with respect to $\vee$ if, unlike λ, it vanishes on all nowhere dense sets. This is because the difference between $A \cup B$ and $A \vee B$ is nowhere dense. The techniques of Part II (Cor. 12.9) will show how, in $\mathbb{R}$ and $\mathbb{R}^2$, the solvability of the isometry group allows one to construct a finitely additive, isometry-invariant measure, μ, on all bounded subsets of $\mathbb{R}^n$ ($n \leq 2$) such that $\mu(J) = 1$ and $\mu(E) = 0$ for all meager sets E. Such a measure μ, unlike λ, is additive with respect to $\vee$, and hence yields that neither an open interval in $\mathbb{R}$ nor a square in the plane is paradoxical in the sense of $\mathcal{R}$. This technique does not apply to $\mathbb{R}^3$, and in fact the open unit cube in $\mathbb{R}^3$ is paradoxical in the sense of $\mathcal{R}$.

Paradoxical decompositions involving the join operation on regular-open sets can certainly be eliminated by a further restriction on the pieces. Let $\mathcal{R}_\upsilon$ be the algebra of bounded, regular-open, Jordan measurable sets (sometimes called *geometric bodies*), and let υ, for volume, denote Jordan measure. Recall that Jordan measure is the precursor of Lebesgue measure, using coverings by finite collections of intervals rather than countable ones. A bounded set A is Jordan measurable if and only if its boundary ∂A has Jordan measure zero (see App. B). The fact that $\mathcal{R}_\upsilon$ is an algebra—closed under $\vee$—follows from this: $A \vee B \in \mathcal{R}$ because $\mathcal{R}$ is an algebra; and $\partial(A \vee B) \subseteq \partial(A \cup B)$ (easy proof), and the latter has Jordan measure 0 because $A \cup B$ is Jordan measurable.

Define equidecomposability in $\mathcal{R}_\upsilon$ using the join operation, pieces in $\mathcal{R}_\upsilon$, and the full isometry group G_n. Because the sets in $\mathcal{R}_\upsilon$ are a mathematical analog of physical bodies, we say that A and B are *equidecomposable as geometric bodies* when they are equidecomposable in $\mathcal{R}_\upsilon$ as just defined. The finite additivity and isometry invariance of υ yields that υ is an invariant for equidecomposability: If $A, B \in \mathcal{R}_\upsilon$ are equidecomposable as geometric bodies, then $\upsilon(A) = \upsilon(B)$. What about the converse? A positive answer would be a nice extension of the Bolyai–Gerwien Theorem about geometric dissection, but Laczkovich [Lac03, Thm.3] showed that the answer is negative by proving the following theorem.

Theorem 10.34. *There are two Jordan curves in $\mathbb{R}^2$ such that the defining functions are differentiable with bounded derivative, the corresponding Jordan domains each have Lebesgue measure 1, and the interiors of the domains are not equidecomposable as sets using isometries.*

The connection between Jordan domains and geometric bodies is a little delicate. For example, there is a Jordan domain whose boundary has positive Lebesgue measure [Osg03]; thus continuity is not sufficient to get a Jordan-measure zero boundary.

Lemma 10.35. *If K is a Jordan curve that is differentiable with bounded derivative, then K is Jordan measurable and has Jordan measure 0.*

Proof. Let B bound the norm of the derivative vector; assume $B \geq 1$; let f with domain $[0, 1]$ define K. Given positive ϵ (assumed less than 1), split the unit interval into interior-disjoint subintervals, each of length at most $\epsilon/(2B^2)$. Using the f-value of the midpoints of these as centers of orthogonal squares of side

$\epsilon/(2B)$, we then have a covering of K by squares of area $\epsilon^2/(4B^2)$. The number of these is at most $\lceil 2B^2/\epsilon \rceil$. So the total area of the squares is at most $(\frac{2B^2}{\epsilon}+1)\frac{\epsilon^2}{4B^2}$, which is under ϵ. This shows K has Jordan measure 0. □

Corollary 10.36. *There are two geometric bodies in $\mathbb{R}^2$ that have the same Jordan measure but are not equidecomposable as geometric bodies.*

Similar ideas work in $\mathbb{R}^1$, as Laczkovich, building on the ideas of [Lac93], proved that there are two one-dimensional geometric bodies of the same length but not equidecomposable as geometric bodies. We present a sketch of the proof. In what follows we will use a function δ_f that relates to a necessary condition for translation equidecomposability of two subsets of the unit interval $J = [0, 1)$. Let $f: J \to \mathbb{R}$ be bounded; let $\omega(f)$ be the diameter of the set $f(J)$: $\sup f - \inf f$. Furthermore, T_c denotes the translation operator: $T_c(f)$ takes x to $f(x+c)$. Now, for a given bounded f as just defined, we can define the following function for reals $t \geq 1$:

$$\delta_f(t) = \inf\left\{\omega\left(\frac{1}{n}\sum_{i=1}^{n} T_{c_i}(f)\right) : n \leq t \text{ and } c_i \in J\right\}.$$

For $A \subseteq J$, we use just δ_A for δ_{χ_A}. Note that δ_f is both nonnegative and decreasing on its domain. If f is Riemann integrable and the sequence c_i is uniformly distributed in J, then the function $\frac{1}{n}\sum_{i=1}^{n} T_{c_i}(f)$ converges uniformly to $\int_J f = dx$ (see [KN74]) and therefore $\lim_{t\to\infty} \delta_f(t) = 0$. Also for any $A \subseteq J$, the function $\frac{1}{n}\sum_{i=1}^{n} T_{c_i}(\chi_A)$ takes on only the values i/n, $(i = 1, \ldots, n)$ and so $\delta_A(t)$ is in fact an exact minimum. Finally, if $A \subseteq J$ is Jordan measurable, then χ_A is Riemann integrable and so $\lim_{t\to\infty} \delta_A(t) = 0$.

Laczkovich also proved the following two theorems.

Theorem 10.37. *If $A, B \subseteq J$ and A and B are translation equidecomposable, then there are positive numbers K and ϵ such that, for $t \geq 1$, $\delta_B(t) < \delta_A(\sqrt{t}) + Kt^{-\epsilon}$.*

Theorem 10.38 (AC). *There are one-dimensional geometric bodies A, B of Jordan measure 1 that are not equidecomposable as geometric bodies.*

Sketch of proof of Theorem 10.38. The construction uses two sequences of positive numbers: $1/4 = \epsilon_0 > \epsilon_1 > \ldots$ and an unbounded sequence $1 = t_0 < t_1 < \ldots$, and a sequence of sets $A_n \subset (\epsilon_n, \epsilon_{n+1})$, $n \geq 1$. Each A_n is a union of closed intervals. The sequences satisfy the following condition.

For any positive integer n, if $D_1, D_2 \subseteq \{1, \ldots, n-1\}$, $B_1, B_2 \subset [0, \epsilon_n]$, and J_1, J_2 are subintervals of $[1/2, 1]$, then

1. $\delta_{C_1}(t_n^2) > \frac{\epsilon_{n-1}}{4} > \delta_{C_2}(t_n) + t_n^{-1/n}$, where
2. $C_1 = B_1 \cup (\bigcup_{i\in D_1} A_i) \cup A_n \cup J_1$ and $C_2 = B_2 \cup (\bigcup_{i\in D_2} A_i) \cup J_2$.

Then for any infinite subset H of $\mathbb{N}$ we define $A(H) = (\bigcup_{i \in H} A_i) \cup K$, where the interval K is chosen to fulfill the requirement that $\lambda(A(H)) = 1$. We have the following.

Claim. If $H_1 \setminus H_2$ is infinite, then $A(H_1)$ is not translation equidecomposable to $A(H_2)$.

Proof. Suppose that $A(H_1) \sim_T A(H_2)$; by Theorem 10.39, there are positive K and ϵ such that for any $t \geq 1$, $\delta_{A(H_1)}(t) < \delta_{A(H_2)}(\sqrt{t}) + Kt^{-\epsilon}$. Because $H_1 \setminus H_2$ is infinite, we can choose $n_r \in H_1 \setminus H_2$ such that $\frac{1}{n_r} < \epsilon$ and $t_{n_r}^{\epsilon - 1/n_r} > K$. Then applying (1) and (2) with $n = n_r$, $D_i = H_i \cap \{1, \ldots, n_r - 1\}$, $B_i = A(H_1) \cap [0, \epsilon_r]$, and $C_i = A(H_i)$ (for $i = 1, 2$), we get

$$\delta_{A(H_1)}(t_{n_r}^2) > \delta_{A(H_2)}(t_{n_r}) + t_{n_r}^{-1/n_r} > \delta_{A(H_2)}(t_{n_r}) + Kt_{n_r}^{-\epsilon},$$

which leads to a contradiction.

We now conclude the proof. Define A_1 to be $A(H_1) \cup (-A(H_1))$ and $B_2 = A(H_2) \cup (-A(H_2))$. By the Cancellation Law these two sets are not translation equidecomposable. Moreover the sets $A(H_1) = A_1 \setminus (-1, 0)$ and $A(H_2) = B_2 \setminus (-1, 0)$ are not equidecomposable using G_1. Now define the sets $A = \operatorname{Int} A(H_1)$ and $B = \operatorname{Int} A(H_2)$. Clearly the sets A and B are not equidecomposable as geometric bodies. □

C. A. Rogers asked whether the set A, defined as the regular-open set $(1/3, 2/3) \cup (7/9, 8/9) \cup (25/27, 26/27) \cup \ldots$, is equidecomposable to $(0, 1/2)$. If arbitrary sets are allowed as pieces, then the answer is affirmative (§9.3). But the following remains unsolved.

Question 10.39. (a) Are A and $(0, 1/2)$ equidecomposable as one-dimensional geometric bodies?

(b) Is it true that geometric bodies in the line or plane that are equidecomposable with arbitrary pieces are equidecomposable as geometric bodies?

(c) If geometric bodies A, B in $\mathbb{R}^n$ $(n \geq 3)$ have the same Jordan measure, then are they equidecomposable as geometric bodies?

Regarding (c), Corollary 10.38 and Theorem 10.40 show that it is false in the line and plane. Moreover, Laczkovich's work in that area is valid in all dimensions, provided the group in question is amenable. Because the isometry group in dimension 3 or greater is not amenable, Question 10.39(c) is unresolved.

The following question is also unsolved.

Question 10.40. Does the Cancellation Law hold for equidecomposability in $\mathcal{R}_\upsilon$ in the line or plane?

We close this chapter with a result that shows that equidecomposability does hold for very simple geometric bodies.

Proposition 10.41. *Any two open intervals of $\mathbb{R}^n$ having the same volume are equidecomposable as geometric bodies. Moreover, any two intervals of the same volume are equidecomposable in the usual sense, using pieces that are Borel sets.*

Proof. By transitivity, it suffices to show that any open interval is equidecomposable in $\mathcal{R}_\upsilon$ to a cube (of the same volume). For $\mathbb{R}^1$ this is trivial, and for $\mathbb{R}^2$ it follows by the part of the Bolyai–Gerwien Theorem that deals with rectangles. For higher dimensions, assume inductively that the result is valid for intervals of dimension $n-1$, and let $I = I_1 \times \cdots \times I_n$ be an interval in $\mathbb{R}^n$. Denote the length of an interval by ℓ. Then, by the two-dimensional case, $I \sim K_1 \times K_2 \times I_3 \times \cdots \times I_n$ in $\mathcal{R}_\upsilon$, where $\ell(K_1) = \upsilon(I)^{1/n}$ and $\ell(K_2) = \ell(I_1)\ell(I_2)/\ell(K_1)$. And the induction hypothesis yields that $K_2 \times I_3 \times \cdots \times I_n$ is equidecomposable in $\mathcal{R}_\upsilon$ to a cube of volume $\upsilon(I)^{(n-1)/n}$. It follows that $K_1 \times K_2 \times I_3 \times \cdots \times I_n$, and hence I, is equidecomposable in $\mathcal{R}_\upsilon$ to a cube of volume $\upsilon(I)$. The assertion about Borel sets follows by replacing the phrase "equidecomposable in $\mathcal{R}_\upsilon$" in the proof with "equidecomposable using Borel sets," and invoking Theorem 3.9 instead of the Bolyai–Gerwien Theorem. Note that the proof of Theorem 3.9 uses only Borel sets. The one-dimensional case calls on Lemma 8.16. □

Notes

The idea of expanding the action of G on X to allow addition of equidecomposability types and form a semigroup is due to Tarski [Tar38b, p. 60]. He discovered a fundamental result on measures in semigroups that he applied to this particular semigroup (see Thm. 11.1 and Cor. 11.2).

Jointly free actions stem from work of von Neumann [Neu29], who essentially proved Proposition 10.5 with the conclusion that: X is G-negligible. The idea of decomposing a locally commutative action into two jointly free actions (Cor. 10.6) is due to Wagon [Wag81a].

The Cancellation Law for equidecomposability (Thm. 10.19) was proved for $n = 2$ by Kuratowski [Kur24], who modified a proof of Bernstein's analogous law for cardinality, due to Sierpiński [Sie22]. Banach and Tarski [BT24] deduced the case $n = 2^m$ from the case $n = 2$, and this was sufficient for some applications, for example, Corollaries 10.22 and 10.23. See [Tar49, p. 33] for further references to Bernstein's Theorem and for an abstract treatment of many algebraic laws derived from cardinality considerations. Dénes König [Kon26] realized that Kuratowski's result followed from his matching theorem (see App. C), which he had proved much earlier for all finite graphs and all infinite 2-regular graphs [Kon16]. König also knew that the countable case of his matching theorem for n-regular graphs yielded the general case. Finally, König and Valkó [KV25] proved the countable case, yielding the complete matching theorem and Cancellation Law. The technique used by König and Valkó is based on a result that has come to be known as the König Tree Lemma: Every infinite, finitely branching tree has an infinite branch (see [Nas67, p. 287]). In a later paper, König [Kon27] isolated this lemma and gave several other applications, including a proof that a countable graph is n-colorable ($n < \infty$) if and only if all of its finite subgraphs are.

Corollary 10.23, the analog of the strong form of the Banach–Tarski Paradox for $\mathbb{S}^2$ rather than $\mathbb{R}^3$, is due to Banach and Tarski [BT24]. Their proof was quite different than the one presented here, as it used the analog of the Bolyai–Gerwien Theorem for spherical polygons [Ger83].

Equidecomposability using regular-open sets, $\mathcal{R}$, and regular-open Jordan measurable sets, $\mathcal{R}_\upsilon$, was introduced by Mycielski [Myc77a], who posed Questions 10.32 and 10.39(b) and (c).

Proposition 10.41 is a special case of a result of Hadwiger [Had57, p. 25], who gave necessary and sufficient conditions for the geometric equidecomposability of two n-dimensional parallelotopes using translations. The extension of his result to set-theoretic equidecomposability was provided by Kummer [Kum56].

PART TWO

Finitely Additive Measures, or the Nonexistence of Paradoxical Decompositions

11

Transition

This chapter discusses a remarkable theorem of Tarski that tightly links the two parts of this book. The theme of the first part was the construction of paradoxical decompositions; the second part is devoted to constructing invariant measures, which provide a way of showing that paradoxical decompositions do not exist. Tarski's result is that this reason for the lack of a paradox is essentially the only one: If a paradoxical decomposition does not exist in a certain context, then a finitely additive invariant measure does. Or, recalling Definition 2.4; if G acts on X, then a subset of X is G-paradoxical if and only if it is G-negligible.

In this chapter we shall prove Tarski's beautiful theorem and discuss some applications. Then we summarize the state of two extensions of equidecomposability theory: to the topological context, where pieces must have the Property of Baire, and to the case where a countably infinite number of pieces is allowed.

11.1 Tarski's Theorem

To prove the nontrivial direction of Tarski's Theorem—the construction of a measure in the absence of a paradoxical decomposition—we shall work in $\mathcal{S}$, the type semigroup of G's action on X. Because any finitely additive, G-invariant measure on $\mathcal{P}(X)$ is necessarily invariant under G-equidecomposability, such a measure μ induces a measure ν from $\mathcal{S}$ into $[0, \infty]$: If $[A] \in \mathcal{S}$, let $\upsilon([A]) = \sum \mu(A_n)$, where $A = \bigcup (A_n \times \{n\})$. Note that $\nu(\alpha + \beta) = \nu(\alpha) + \nu(\beta)$ for any $\alpha, \beta \in \mathcal{S}$. Conversely, any function $\nu : \mathcal{S} \to [0, \infty]$ with this additivity property induces a finitely additive, G-invariant measure on $\mathcal{P}(X)$: Let $\mu(A) = \nu([A])$. Thus the type semigroup is a natural context for the construction of a G-invariant measure. Tarski saw how to formulate and prove a very general theorem on measures in semigroups that he applied to the type semigroup to obtain his characterization.

Recall that any commutative semigroup with an identity $(\mathcal{T}, +, 0)$ admits an ordering defined by $\alpha \le \beta$ if $\alpha + \delta = \beta$ for some $\delta \in \mathcal{T}$, and there is a natural multiplication $n\alpha$ of elements of $\mathcal{T}$ by natural numbers. If $\epsilon \in \mathcal{T}$ is fixed, then an element α of $\mathcal{T}$ will be called *bounded* if, for some $n \in \mathbb{N}$, $\alpha \le n\epsilon$. In the type semigroup of a group action, the ordering $\leq$ is a partial ordering (Banach–Schröder–Bernstein Theorem), but the following theorem applies to arbitrary commutative semigroups, even those where $\alpha \leq \beta$ and $\beta \leq \alpha$ might hold for distinct α, β.

Theorem 11.1 (AC). *Let $(\mathcal{T}, +, 0, \epsilon)$ be a commutative semigroup with identity 0 and a specified element ϵ. Then the following are equivalent:*

(a) *For all $n \in \mathbb{N}$, $(n+1)\epsilon \nleq n\epsilon$.*
(b) *There is a measure $\mu : \mathcal{T} \to [0, \infty]$ such that $\mu(\epsilon) = 1$ and $\mu(\alpha + \beta) = \mu(\alpha) + \mu(\beta)$ for all $\alpha, \beta \in \mathcal{T}$. (Note that μ is a homomorphism of semigroups, from $\mathcal{T}$ to $([0, \infty], +)$).*

That (b) implies (a) is clear, because any μ satisfying (b) is such that $\mu(\alpha) \leq \mu(\beta)$ if $\alpha \leq \beta$, and $\mu(n\epsilon) = n$; hence $(n+1)\epsilon \leq n\epsilon$ would imply $n + 1 \leq n$. For the other direction, we shall use a technique that will be one of the main tools for constructing measures in later chapters. The key is to exploit the compactness of the product space $[0, \infty]^{\mathcal{T}}$ (all functions from $\mathcal{T}$ into $[0, \infty]$) together with the observation that if μ fails to be a measure on $\mathcal{T}$ as in (b), then the failure is evident in some finite subset of $\mathcal{T}$. Note that the compactness of $[0, \infty]^{\mathcal{T}}$ is a consequence of Tychonoff's Theorem that products of compact spaces are compact (see [Roy68]), which requires the Axiom of Choice. Without loss of generality, we assume that all elements of $\mathcal{T}$ are bounded (with respect to ϵ). For once we have a measure on the bounded elements, it may be extended by assigning the unbounded elements measure ∞. Most of the work will be in the proof of the following claim.

Claim. If $\mathcal{T}_0$ is a finite subset of $\mathcal{T}$ that contains ϵ, then there is a function $\rho : \mathcal{T}_0 \to [0, \infty]$ such that (i) $\rho(\epsilon) = 1$, and (ii) if $\phi_i, \theta_j \in \mathcal{T}_0$ satisfy $\phi_1 + \cdots \phi_m \leq \theta_1 + \cdots + \theta_n$, then $\sum \rho(\phi_i) \leq \sum \rho(\theta_j)$.

Before proving the claim, we show how it can be combined with compactness to produce the desired measure on all of $\mathcal{T}$. For any $\mathcal{T}_0$ as in the claim, let $\mathcal{M}(\mathcal{T}_0)$ consist of all $f \in [0, \infty]^{\mathcal{T}}$ satisfying (1) $f(\epsilon) = 1$, and (2) $f(\alpha + \beta) = f(\alpha) + f(\beta)$ whenever $\alpha, \beta, \alpha + \beta \in \mathcal{T}_0$. This last additivity property is an easy consequence of property (ii) of the claim, whence the claim implies that each $\mathcal{M}(\mathcal{T}_0)$ is nonempty. Now, the compactness of the product space $[0, \infty]^{\mathcal{T}}$ may be interpreted as follows: If a collection of closed subsets of $[0, \infty]^{\mathcal{T}}$ has the finite intersection property (any intersection of finitely many members of the collection is nonempty), then the intersection of all sets in the collection is nonempty. Each $\mathcal{M}(\mathcal{T}_0)$ is closed. This is because whether $f \in \mathcal{M}(\mathcal{T}_0)$ depends only on the finitely many coordinates mentioned in $f \restriction \mathcal{T}_0$, and if f fails to satisfy (1) or (2), then either $f(\epsilon) \neq 1$ or $f(\alpha + \beta) \neq f(\alpha) + f(\beta)$. In either case, there is an open subset of $[0, \infty]^{\mathcal{T}_0}$ containing $f \restriction \mathcal{T}_0$ on which the condition is violated—just vary the ϵ-coordinate, the α-coordinate, or the β-coordinate a little bit. Therefore $\mathcal{M}(\mathcal{T}_0)$ is closed. Note that $\mathcal{M}(\mathcal{T}_1) \cap \ldots \cap \mathcal{M}(\mathcal{T}_n) \supseteq \mathcal{M}(\mathcal{T}_1 \cup \ldots \cup \mathcal{T}_n)$, where each $\mathcal{T}_i$ is a finite subset of $\mathcal{T}$ containing ϵ. Because $\bigcup \mathcal{T}_i$ is also finite, the claim implies that $\mathcal{M}(\bigcup \mathcal{T}_i) \neq \varnothing$; it follows that $\{\mathcal{M}(\mathcal{T}_0) : \mathcal{T}_0$ is a finite subset of $\mathcal{T}$ containing $\epsilon\}$ has the finite intersection property. By compactness, there must be a μ that lies in each $\mathcal{M}(\mathcal{T}_0)$, and such a μ is a measure as desired: because $\mu \in \mathcal{M}(\epsilon)$, $\mu(\epsilon) = 1$, and to see that $\mu(\alpha + \beta) = \mu(\alpha) + \mu(\beta)$, use the fact that $\mu \in \mathcal{M}(\{\epsilon, \alpha, \beta, \alpha + \beta\})$.

Proof of claim. The proof is by induction on the size of $\mathcal{T}_0$. If $|\mathcal{T}_0| = 1$, then $\mathcal{T}_0 = \{\epsilon\}$, and $\epsilon \mapsto 1$ is the desired function. In this case, property (ii) of the claim reduces to: if $m\epsilon \leq n\epsilon$, then $m \leq n$. But this is a consequence of (a), the theorem's hypothesis on ϵ; for if $m\epsilon \leq n\epsilon$ and $m \geq n+1$, then $(n+1)\epsilon \leq m\epsilon \leq n\epsilon$, a contradiction. This is the only place in the proof where the hypothesis on ϵ is used.

Now, suppose $|\mathcal{T}_0| > 1$, and let α be any element of $\mathcal{T}_0 \setminus \{\epsilon\}$. Use the induction hypothesis to get a function ν on $\mathcal{T}_0 \setminus \{\alpha\}$ satisfying the claim. By our boundedness assumption, all elements are bounded by some $n\epsilon$; hence ν takes on only finite values. Define ρ on $\mathcal{T}_0$ by letting ρ agree with ν on $\mathcal{T}_0 \setminus \{\alpha\}$ and setting $\rho(\alpha) = \inf\{(\sum \nu(\beta_k) - \sum \nu(\gamma_l))/r\}$, where the inf is over all positive integers r and $\beta_1, \ldots, \beta_p, \gamma_1, \ldots, \gamma_q \in \mathcal{T}_0 \setminus \{\alpha\}$ satisfying $\gamma_1 + \cdots + \gamma_q + r\alpha \leq \beta_1 + \cdots + \beta_p$. Because α is bounded, $\rho(\alpha)$ is the greatest lower bound of a nonempty set: If $\alpha \leq n\epsilon$, then $\rho(\alpha) \leq n$. Note the similarity with the classical outer measure construction: $\rho(\alpha)$ is defined to be as small as possible subject to bounds it must have if the claim is to be satisfied.

Because a consequence of property (ii) of the claim is that $\rho(\alpha) \geq 0$ ($\epsilon \leq \epsilon + \alpha$ so $1 \leq 1 + \rho(\alpha)$), it remains only to prove that ρ continues to satisfy property (ii). So suppose that $\phi_1 + \cdots + \phi_m + s\alpha \leq \theta_1 + \cdots + \theta_n + t\alpha$, where ϕ_i, $\theta_j \in \mathcal{T}_0 \setminus \{\alpha\}$ and $s, t \in \mathbb{N}$. If both s and t are 0, then the desired inequality follows from the fact that ν satisfies the claim on $\mathcal{T}_0 \setminus \{\alpha\}$. Consider first the case $s = 0$ and $t > 0$; we must show that $\sum \nu(\phi_i) \leq t\mu(\alpha) + \sum \nu(\theta_j)$, that is, that $\rho(\alpha) \geq w = (\sum \nu(\phi_i) - \sum \nu(\theta_j))/t$. Let $\gamma_1 + \cdots + \gamma_q + r\alpha \leq \beta_1 + \cdots + \beta_p$ be a typical inequality defining $\rho(\alpha)$; it suffices to show that $(\sum \nu(\beta_k) - \sum \nu(\gamma_\ell))/r \geq w$. Multiplying the given inequality $\phi_1 + \cdots + \phi_m \leq \theta_1 + \cdots + \theta_n + t\alpha$ by r and adding the same quantity to both sides yields

$$r\phi_1 + \cdots + r\phi_m + t\gamma_1 + \cdots + t\gamma_q \leq r\theta_1 + \cdots + r\theta_n + tr\alpha + t\gamma_1 + \cdots + t\gamma_q q.$$

Substituting the inequality involving γ, α, β yields

$$r\phi_1 + \cdots + r\phi_m + t\gamma_1 + \cdots + t\gamma_q \leq r\theta_1 + \cdots + r\theta_n + t\beta_1 + \cdots + t\beta_q.$$

The induction assumption about ν now yields

$$r \sum \nu(\phi_i) + t \sum \nu(\gamma_\ell) \leq r \sum \nu(\theta_j) + t \sum \nu(\beta_k),$$

which implies that $(\sum \nu(\beta_k) - \sum \nu(\gamma_\ell))/r \geq w$, as desired.

Finally, suppose $\phi_1 + \cdots + \phi_m + s\alpha \leq \theta_1 + \cdots + \theta_n + t\alpha$ where $s > 0$. It suffices to show that $s\rho(\alpha) + \sum \upsilon(\phi_i) \leq z_1 + \cdots + z_t + \sum \nu(\theta_j)$, where

$z_1, \ldots, z_t$ are any of the numbers whose greatest lower bound defines $\rho(\alpha)$. By considering the smallest of $z_1, \ldots, z_t$ we may assume that these numbers are all the same. In other words, suppose that $\gamma_1 + \cdots + \gamma_q + r\alpha \le \beta_1 + \cdots + \beta_p$, where $\gamma_\ell, \beta_k \in \mathcal{T}_0 \setminus \{\alpha\}$, and let $z = (\sum \nu(\beta_k) - \sum \nu(\gamma_\ell))/r$; then we must prove that $s\rho(\alpha) + \sum \upsilon(\phi_i) \le tz + \sum \nu(\theta_j)$. Multiplying the ϕ, α, θ-inequality by r and adding the same quantity to both sides yields

$$\begin{aligned} &r\phi_1 + \cdots + r\phi_m + rs\alpha + t\gamma_1 + \cdots + t\gamma_q \\ &\quad \le r\theta_1 + \cdots + r\theta_n + rt\alpha + t\gamma_1 + \cdots + t\gamma_q, \end{aligned}$$

and substituting the γ, α, β-inequality yields

$$r\phi_1 + \cdots + r\phi_m + t\gamma_1 + \cdots + t\gamma_q + rs\alpha \le r\theta_1 + \cdots + r\theta_n + t\beta_1 + \cdots + t\beta_p.$$

This last inequality is a typical one used to define $\mu(\alpha)$, and it follows that $s\mu(\alpha) + \sum \nu\,(\phi_i)$ is bounded by

$$\sum \nu(\phi_i) + s\left(\frac{1}{rs}\right)\left(r \sum \nu(\theta_j) + t\sum \nu(\beta_k) - r\sum \nu(\phi_i) - t\sum \nu(\gamma_\ell)\right),$$

which equals $tz + \sum \nu(\theta_j)$, as required to prove the claim, and hence the theorem. □

The reader familiar with transfinite induction will see that the proof could be shortened a little by using that technique rather than Tychonoff's Theorem. The claim allows one to construct μ by treating the elements of $\mathcal{T}$ one at a time, starting with ϵ and proceeding inductively through $|\mathcal{T}|$ steps; this was Tarski's original approach. We chose to illustrate the use of compactness, because this technique will appear several times in Part II.

In his original construction, Tarski used both outer and inner measure defined on the semigroup $\mathcal{T}$ (with distinguished element ϵ as in Thm. 11.1, satisfying (a) of the theorem, and with all elements of $\mathcal{T}$ bounded). His construction was as follows: let $\mathcal{T}_1 \subseteq \mathcal{T}$. Let $g: \mathcal{T}_1 \to [0, \infty)$ satisfy (i) and (ii) of the claim in the preceding proof:

(i) $g(\epsilon) = 1$
(ii) if $\beta_1 + \cdots + \beta_p \le \gamma_1 + \cdots + \gamma_q$, where $\beta_i, \gamma_j \in \mathcal{T}_1$, then $\sum g(\beta_i) \le \sum g(\gamma_j)$

The existence of such a g when $\mathcal{T}_1$ is finite follows from the claim in Theorem 11.1. In the general case, one would use transfinite induction. Then for any $\alpha \in \mathcal{T}$, we can define two numbers as follows, where β and γ lie in $\mathcal{T}_1$:

$$\mu_i(\alpha) = \sup\left\{\sum(g(\beta_k) - \sum g(\gamma_l))/r : \beta_1 + \cdots + \beta_p \le \gamma_1 + \cdots + \gamma_q + r\alpha\right\}$$

$$\mu_e(\alpha) = \inf\left\{\sum(g(\beta_k) - \sum g(\gamma_l))/r : \gamma_1 + \cdots + \gamma_q + r\alpha \le \beta_1 + \cdots + \beta_p\right\}.$$

These values are, respectively, an inner and outer measure on $\mathcal{T}$.

Put $\mathcal{T}_1 = \{\epsilon\}$; then Tarski called the elements $\alpha \in T$ such that $\mu_i(\alpha) = \mu_e(\alpha)$ *absolutely measurable*. In the case of type semigroups (using the full isometry group) for $\mathbb{R}^n$ (with ϵ being the type of the unit cube), these measurable elements have the following property. Let $A \subset \mathbb{R}^n$ be any representative of α; then any two Banach measures (see §12.3) defined on $\mathcal{P}(\mathbb{R}^n)$ agree on A; this explains the term "absolutely measurable." Furthermore, the construction of Tarski's absolute measure has geometric roots. Namely, it is related to the following construction of a geometric absolute measure on the line obtained by Tarski [Tar38a]. For any bounded subset $A \subset \mathbb{R}$, define the numbers

$$\mu_a(A) \text{ to be } \sup\{|I| : I \sim B, B \subseteq A, I \text{ is an interval}\}$$

$$\mu^a(A) \text{ to be } \inf\{|I| : A \sim C, C \subseteq I, I \text{ is an interval}\}.$$

These are the *geometric absolute inner measure* and *geometric absolute outer measure*, respectively, of A. And any set $A \subset \mathbb{R}$ such that $\mu_a(A) = \mu^a(A)$ is called *geometrically absolutely measurable*. As Tarski [Tar38a], observed every Jordan measurable set of reals is geometrically absolutely measurable; the measure is an attempt to extend Jordan measure using equidecomposability in the spirit of the Lebesgue extension of Jordan measure leading to Lebesgue measure. Note, however, that we cannot use countable equidecomposability, because any two intervals of $\mathbb{R}$ are countably equidecomposable (Thm. 11.21). Moreover, this geometric measure is not a measure but a set function. In fact, take a Cantor set $C \subset [0, 1]$ of positive Lebesgue measure and the set $A = ([0, 1] \setminus C) \cup C_1$, where $C_1 = C + 1$. Then the sets $[0, 1]$ and A have Tarski absolute measure 1 but $A \cup [0, 1]$ is not Tarski measurable. Therefore the geometrically measurable sets do not form an algebra.

Let G be a group of isometries of R^n under which the unit cube is not paradoxical. Then it can be shown [MT∞a] that any representative of an absolutely measurable element α of measure 0 has the following property: For any ball $K \subset \mathbb{R}^n$, there is a set $A_1 \subset K$ such that $A \sim_G A_1$. Mycielski and Tomkowicz introduced in [MT∞a] the following notion: A bounded $S \subset \mathbb{R}^n$ is a *small set* if for any ball $K \subset \mathbb{R}^n$ there is a set $S_1 \subset K$ such that $S \sim_G S_1$ and $K \sim_G K \setminus S_1$. Clearly every small set has geometric absolute measure zero (a property also known as being a Tarski null set), but as shown by Mycielski and Tomkowicz [MT∞a], even in $\mathbb{R}$ there are such geometric absolute measure 0 sets that are not small (see [MT∞a]). They also showed that every bounded G-paradoxical subset of $\mathbb{R}^n$ is a small set, provided G contains all the translation of $\mathbb{R}^n$. See [MT∞a] for more information about small sets.

Corollary 11.2 (Tarski's Theorem) (AC). *Suppose G acts on X and $E \subseteq X$. Then there is a finitely additive, G-invariant measure $\mu : \mathcal{P}(X) \to [0, \infty]$ with $\mu(E) = 1$ if and only if E is not G-paradoxical.*

Proof. The easy direction was done in Proposition 2.5, so assume E is not G-paradoxical and let $\mathcal{S}$ be the type semigroup of G's action on X. Then $2\epsilon \neq \epsilon$, where $\epsilon = E \times \{0\}$, and hence Corollary 10.21 to the Cancellation Law for equidecomposability implies that $\mathcal{S}$, with the distinguished element ϵ, satisfies condition (a) of Theorem 11.1. Hence that theorem yields a measure ν on $\mathcal{S}$ normalizing ϵ, whence $\mu(A) = \nu([A \times \{0\}])$ is the desired G-invariant measure on $\mathcal{P}(X)$. □

Theorem 11.1 may also be applied to the case of equidecomposability where the pieces are restricted in some way. Suppose G acts on X and $\mathcal{A}$ is a G-invariant subalgebra of $\mathcal{P}(X)$. Let $\mathcal{S}(\mathcal{A})$ be the type semigroup of the induced action of G on $\mathcal{A}$, where only pieces in $\mathcal{A}$ can be used to witness the equidecomposability of sets in $\mathcal{A}$ (see §10.3). We do not have a general Cancellation Law in this context, and if Corollary 10.21 fails, it may be that, say, $3[E] = 2[E]$ even though $2[E] \neq [E]$ ($[E]$ stands for $[E \times \{0\}]$ in $\mathcal{S}(\mathcal{A})$). In such a case, no G-invariant measure on $\mathcal{A}$ normalizing E could exist, despite the fact that E is not G-paradoxical using pieces in $\mathcal{A}$. Nevertheless, Theorem 11.1 yields the following slightly weaker analog of Corollary 11.2

Corollary 11.3 (AC). *Let $G, X, \mathcal{A}$ be as in the preceding paragraph and suppose $E \in \mathcal{A}$. Then the following are equivalent:*

(a) *For all $n \in \mathbb{N}$, $(n+1)[E] \not\leq n[E]$.*
(b) *There is a finitely additive, G-invariant measure $\mu : \mathcal{A} \to [0, \infty]$ with $\mu(E) = 1$.*

Proof. If μ is as in (b), then μ induces a measure on $\mathcal{S}$ $(\mathcal{A})$ satisfying condition (b) of Theorem 11.1 (see remarks preceding Thm. 11.1) and hence, as in that theorem, $(n+1)[E] \not\leq n[E]$. For the other direction, apply Theorem 11.1 to $\mathcal{S}$ $(\mathcal{A})$, with $\epsilon = [E]$, and use the measure on $\mathcal{S}(\mathcal{A})$ so obtained to induce the desired measure on $\mathcal{A}$. □

Tarski proved a more general version of these results dealing with the case of partial transformations of a set. We refer the reader to [Tar49] for details, mentioning here one important application (see [Tar49, p. 233]). Let $\mathcal{A}$ be the Borel subsets of a compact metric space X and let $\mathcal{F}$ consist of all partial isometries, that is, distance-preserving bijections with domain (and hence image; see remarks preceding Question 3.13) in $\mathcal{A}$. Then Tarski's work yields the equivalence of the existence of a finitely additive, congruence-invariant (i.e., $\mathcal{F}$-invariant) Borel measure on X of total measure 1 with the assertion that for any n, $(n+1)[X] \not\leq n[X]$ in the type semigroup formed from $\mathcal{F}$-equidecomposability. The Banach–Ulam Problem asks if such a measure exists in every compact metric space; see the discussion following Question 3.13 for some partial results.

Because Corollary 11.3 does not appeal to the Cancellation Law for equidecomposability, the only point that requires the Axiom of Choice is the compactness of $[0, \infty]^{\mathcal{S}(\mathcal{A})}$. If it happens that $\mathcal{A}$ is a countable subalgebra of $\mathcal{P}(X)$, then $\mathcal{S}(\mathcal{A})$ is countable too, and the compactness of a countable product of copies of $[0, \infty]$ can be proved without using the Axiom of Choice. Hence, in such a case,

Corollary 11.3 is valid without Choice. This idea will be important in some later applications of the compactness technique of Theorem 11.1, so we give the effective compactness proof here. It should be pointed out that the more general assertion that X^S is compact whenever X is compact and S is countable is equivalent to the Axiom of Choice for countable families of sets.

Proposition 11.4. *If S is a countable set, then the product space $[0, \infty]^S$ is compact.*

Proof. Because $[0, \infty]$ and $[0, 1]$ are homeomorphic, it suffices to show that $[0, 1]^{\mathbb{N}}$ is compact. Suppose that $\{U_i\}$ is an open cover of $[0, 1]^{\mathbb{N}}$ without a finite subcover; assume each U_i is a basic open set in $[0, 1]^{\mathbb{N}}$. Consider the two sets $[0, 1/2] \times [0, 1] \times [0, 1] \times \ldots$ and $[1/2, 1] \times [0, 1] \times [0, 1] \times \ldots$. Choose the leftmost of these two sets that is not coverable by finitely many U_i and get four subsets of it by splitting the first coordinate in two again and splitting the second coordinate in two as well. Thus a typical set at this second stage might be $[1/4, 1/2] \times [1/2, 1] \times [0, 1] \times [0, 1] \times \ldots$. Now choose the (lexicographically) least one of these four sets that cannot be covered by finitely many U_i and get eight subsets of it by halving in three coordinates. Continue for $\aleph_0$ steps, obtaining a decreasing sequence that determines a unique element r of $[0, 1]^{\mathbb{N}}$. One of the U_i contains r. But U_i is determined by open sets in finitely many coordinates, whence U_i contains one of the sets in the sequence defining r, contradiction. □

For an approach to Tarski's Theorem that uses pseudogroups, see §12.4.1.

11.2 The Marczewski Problem: A Paradox Using Baire Sets

One of the most famous problems about paradoxes was formulated in 1930 by E. Marczewski (before 1940, his name was E. Szpilrajn) and concerns $\mathcal{B}$, the sets having the Property of Baire. Although he was thinking in terms of measures, his problem is most easily stated as follows: Is there a paradox of the ball, like the Banach–Tarski Paradox but using pieces in $\mathcal{B}$? Recall that a Baire set differs from an open set (equivalently, a Borel set) by a meager set, whereas a Lebesgue measurable set differs from a Borel set by a measure zero set. So the algebras $\mathcal{B}$ and $\mathcal{L}$ are related to Borel sets in a similar way, but using different ideals of small sets.

The similarities and differences between $\mathcal{B}$ and $\mathcal{L}$ are fascinating. The most famous phenomenon concerns independence results. Let PB (resp., LM) be the assertion that all sets have the Property of Baire (resp., are Lebesgue measurable); DC is the Axiom of Dependent Choice, a weaker version of AC that is sufficient for the basics of analysis and topology. The consistency of ZF is equivalent to the consistency of ZF + DC + PB; but the consistency of ZF + DC + LM is *stronger than* the consistency of ZF: It is equivalent to the consistency of ZF+ the existence of inaccessible cardinal (see §15.1). This is a remarkable and fundamental distinction between sets that are topologically nice and sets that are measure-theoretically nice. Now, it is obvious that there is no paradox using measurable sets. But what about sets in $\mathcal{B}$? That is the Marczewski Problem, and it came as a great surprise when, in 1992, R. Dougherty and M. Foreman [DF92, DF94] proved that such paradoxes do exist.

Let $\mathcal{B}$ denote the family of subsets of $\mathbb{R}^n$ (or $\mathbb{S}^n$) having the Property of Baire, and let a *$\mathcal{B}$-measure* denote a finitely additive, G_n-invariant (or O_{n+1}-invariant) measure on $\mathcal{B}$ that normalizes the unit cube (or $\mathbb{S}^n$). The existence of a $\mathcal{B}$-measure on $\mathbb{R}^n$ implies that the unit cube is not paradoxical using pieces in $\mathcal{B}$. Moreover, a $\mathcal{B}$-measure on $\mathbb{R}^n$ induces a $\mathcal{B}$-measure on $\mathbb{S}^{n-1}$ using the technique of adjoining radii and so implies that $\mathbb{S}^{n-1}$ is not paradoxical using pieces with the Property of Baire. Thus Marczewski's problem is equivalent to asking whether there is a $\mathcal{B}$-measure in either $\mathbb{R}^3$ or $\mathbb{S}^2$. For lower dimensions, such a measure exists (Cor. 13.3). It is also useful to define a *Marczewski measure* as a $\mathcal{B}$-measure that vanishes on all bounded meager sets.

Because we do not know whether the Cancellation Law holds for equidecomposability in $\mathcal{B}$, it is not immediate (via Cor. 11.2) that the existence of a $\mathcal{B}$-measure can be equated with the lack of a paradoxical decomposition. Nevertheless, such an equivalence is true, at least in the case of $\mathbb{R}^n$. For one can, in $\mathcal{B}$, deduce the important Corollary 10.21 of the Cancellation Law. Theorem 11.7 shows how to do this, and shows as well the connection between Marczewski's Problem for $\mathbb{R}^n$ and the existence of paradoxes in $\mathcal{R}$, the algebra of bounded, regular-open subsets of $\mathbb{R}^n$ (see §10.3.2).

The Marczewski Problem concerns a topological version of the Banach–Tarski Paradox, so the following definition is useful.

Definition 11.5. *Two open subsets A, B of $\mathbb{R}^n$ are called* densely equidecomposable *if there are finitely many disjoint open subsets A_i of A and isometries σ_i such that $\bigcup A_i$ is dense in A, $\sigma_i(A_i)$ are pairwise disjoint, and $\bigcup \sigma_i(A_i)$ is dense in B.*

There is a Banach–Schröder–Bernstein Theorem for dense equidecomposability; the proof is similar to the classic case (Thm. 3.6), using some modifications to deal with the dense sets; see [DF94, Prop. 2.9] for details.

Theorem 11.6. *If A and B are open subsets of $\mathbb{R}^n$ such that A is densely equidecomposable to a subset of B, and vice versa, then A and B are densely equidecomposable.*

The next theorem shows that a diverse set of topological properties of Euclidean space are equivalent.

Theorem 11.7 (AC). *Let $\mathcal{S}(\mathcal{B})$, $\mathcal{S}(\mathcal{R})$ be the type semigroups with respect to G_n-equidecomposability in $\mathcal{B}$, $\mathcal{R}$, respectively, as defined in Chapter 10. Let J denote the (open) unit cube in $\mathbb{R}^n$. The following are equivalent:*

(a) *There is $m \in \mathbb{N}$ such that $(m+1)[J] \le m[J]$ in $\mathcal{S}(\mathcal{B})$.*
(b) *J is paradoxical in $\mathcal{B}$.*
(c) *There exists no $\mathcal{B}$-measure on $\mathbb{R}^n$.*
(d) *There is $m \in \mathbb{N}$ such $(m+1)[J] \le m[J]$ in $\mathcal{S}(\mathcal{R})$.*

(e) *J is paradoxical in $\mathcal{R}$.*
(f) $|\mathcal{S}(\mathcal{R})| = 2$*; that is,* $\mathcal{S}(\mathcal{R}) = \{\{\varnothing\}, [J]\}$.
(g) *There is no finitely additive (with respect to $\vee$), G_n-invariant measure $\mu : \mathcal{R} \to [0, \infty)$ with $\mu(J) = 1$.*
(h) *Any bounded, nonempty, open set $U \subseteq \mathbb{R}^n$ has finitely many pairwise disjoint open subsets U_i such that, for isometries ρ_i, $\bigcup \rho_i(U_i)$ is dense in J.*
(i) *There is an open dense subset of J that can be packed into a proper subcube of J using finitely many pieces in $\mathcal{B}$.*

Proof. Corollary 11.3 yields $(a) \Leftrightarrow (c)$, and $(a) \Rightarrow (b)$ can be proved as follows. Suppose $(m+1)[J] \leq m[J]$ in $\mathcal{S}(\mathcal{B})$; using a similarity shows that this is true for any cube. Choose a cube K small enough that m copies of K fit into J. Then choose k large enough that $(m+k)[K] \geq 2[J]$. Because $(m+1)[K] \leq m[K]$ yields $(m+k)[K] \leq m[K]$, we have $m[K] \leq [J] \leq 2[J] \leq (m+k)[K] \leq m[K]$; hence the Banach–Schröder–Bernstein Theorem yields $[J] = 2[J]$, giving (b). Because $(b) \Rightarrow (a)$ is obvious, we have $(a) \Leftrightarrow (b)$, and hence the equivalence of (a), (b), and (c). □

Exactly the same technique, using Theorem 11.1 in the semigroup $\mathcal{S}(\mathcal{R})$, yields the equivalence of (d), (e), and (g). Statement (f) implies that $[J] = 2[J]$ in $\mathcal{S}(\mathcal{R})$, yielding (e). For the converse, we can prove that if J is paradoxical in $\mathcal{R}$, then $[U] = [J]$ for any nonempty $U \in \mathcal{R}$, which yields (f). For this, use similarities as done at the end of the proof of Theorem 8.5. Hence statements (d)–(g) are equivalent.

These two groups of statements will be tied together by considering the assertions about measures, (c) and (g). The proof that $(c) \Rightarrow (g)$ hinges on the fact that any set $A \in \mathcal{B}$ can be uniquely represented as $E \Delta P$ where E is regular-open and P is meager (see [Oxt71, p. 20]). Suppose μ is a measure as in (g); extend μ to all regular-open sets by setting $\mu(A) = \infty$ if A is unbounded. Then define ν on $\mathcal{B}$ by $\nu(A) = \mu(E)$ where $A = E \Delta P$ is the representation just mentioned. The uniqueness of this representation and the fact that $(E_1 \vee E_2) \setminus (E_1 \bigcup E_2)$ is nowhere dense yield that ν is a $\mathcal{B}$-measure; more precisely, one shows that $(A \cup A') \Delta (E \vee E')$ is meager, which implies that the $\mathcal{R}$-representation of $A_1 \cup A_2$ is $E_1 \vee E_2$. Also E_1 and E_2 are disjoint (details in next paragraph). The converse needs some auxiliary results and will be discussed after we treat (h) and (i). We shall show that a $\mathcal{B}$-measure, when restricted to $\mathcal{R}$, satisfies the conditions of (g).

To deal with (h) and (i), we prove that $(f) \Rightarrow (h) \Rightarrow (i) \Rightarrow (e)$. To see that $(i) \Rightarrow (e)$, suppose $A_i \in \mathcal{B}$ and isometries ρ_i witness the packing of an open dense subset of J into a proper subcube, K. As in the proof of $(c) \Rightarrow (g)$, write A_i as $E_i \Delta P_i$, where E_i is regular-open and P_i is meager. Then $(E_i \cap E_j) \setminus (P_i \cup P_j) \subseteq A_i \cap A_j = \varnothing$ if $i \neq j$, so $E_i \cap E_j \subseteq P_i \cup P_j$. But $P_i \cup P_j$ is meager and so contains no nonempty open set (Baire Category Theorem; see [Oxt71, p. 2]); hence $E_i \cap E_j = \varnothing$ if $i \neq j$. Similarly, the disjointness of $\rho_i(E_i)$ and $\rho_i(E_j)$ follows from that of $\rho_i(A_i)$ and $\rho_j(A_j)$. The Baire Category Theorem also implies that $E_i \subseteq \overline{A}_i$; hence $\rho_i(E_i) \subseteq \rho_i(\overline{A}_i) = \overline{\rho_i(A_i)} \subseteq \overline{K}$, so $\rho_i(E_i) \subseteq \overline{K}$. The preceding remarks, together with the easy fact that the join of the E_i in $\mathcal{R}$ must equal J, pack J into K in a way that yields $[J] \leq [K]$ in $\mathcal{S}(\mathcal{R})$. Hence, if L is a cube contained in $J \setminus K$, $[K] + [L] \leq [K]$, and it follows that $[K] + m[L] \leq [K]$ for any m. Choose m so large that m copies of L cover K. Then $2[K] = [K] + [K] \leq [K] + m[L] \leq [K]$, so K is paradoxical in $\mathcal{R}$. Using a similarity yields that J is paradoxical in $\mathcal{R}$, as desired. For $(h) \Rightarrow (i)$, let U in (h) be a proper subcube of J. Then U contains sets as in (h), but the $\rho_i(U_i)$ might not be disjoint. Make them disjoint by taking complements in sequence and then take preimages back in U. This packs a dense subset of J into U using sets in $\mathcal{B}$. Finally, for $(f) \Rightarrow (h)$, let U be a bounded open set and let $E = \text{Int}(\overline{U})$, a regular-open set. By (f), $E \sim J$ in $\mathcal{R}$, so there are disjoint regular-open subsets E_i of E whose join is E and, for some isometries, $\text{Int}(\overline{\bigcup \rho_i(E_i)}) = J$. The sets $E_i \cap U$ are then open subsets of U, and $\bigcup \rho_i(E_i \cap U)$ is dense in J.

We now return to $(g) \Rightarrow (c)$, the proof of which uses an interesting and useful observation of Tarski. To place it in its proper context, we mention the following general problem. If μ is a finitely additive, G_n-invariant measure on the Lebesgue measurable subsets of $\mathbb{R}^n$, and $\mu(J) = 1$ where J is the unit cube, then to what extent must μ agree with Lebesgue measure, λ? This problem, which goes back to Lebesgue, was settled in the early 1980s and will be discussed in more detail later (Thm. 13.13). Note that to have any hope of concluding that μ agrees with λ, one must restrict the discussion to bounded sets; for if μ is defined to agree with λ on the bounded Lebesgue measurable sets, but to give measure ∞ to unbounded measurable sets, then μ satisfies the hypothesis of the problem, but $\mu \neq \lambda$. First we observe that even under the weaker assumption of translation invariance, one can at least be assured that μ agrees with λ on certain elementary sets, namely, those that are Jordan measurable.

Proposition 11.8. *Jordan measure υ is the unique finitely additive, translation-invariant measure on the Jordan measurable sets in $\mathbb{R}^n$ that normalizes J.*

Proof. Let μ be a finitely additive, translation-invariant measure on the Jordan measurable sets with $\mu(J) = 1$. We will show that $\mu(K) = \upsilon(K)$ for any orthogonal closed cube K. This suffices, for suppose E is Jordan measurable; then for any $\epsilon > 0$, there are sets D, F such that $D \subseteq E \subseteq F$, D and F are each a union of finitely many interior disjoint closed cubes with edges parallel to the coordinate axes, and $\upsilon(F) - \upsilon(D) < \epsilon$ (see App. B). It follows from this that $|\upsilon(E) - \mu(E)| < \epsilon$.

Now we will prove the assertion about μ for the plane, but the same idea works in all dimensions. Translation invariance implies that any finite set of bounded line segments gets μ-measure zero, because one could pack arbitrarily many translates of such a set (broken into pieces) into the unit square, J. This means we can ignore the boundaries of squares, and therefore because $\mu(J) = 1$, Figure 11.1 shows that μ is correct on any properly oriented square of side-length $1/2$. Subdividing repeatedly yields the correctness of μ on any properly oriented square of side-length 2^{-n}, and hence those of side-length $m/2^n$, m, $n \in \mathbb{N}$. Now, these

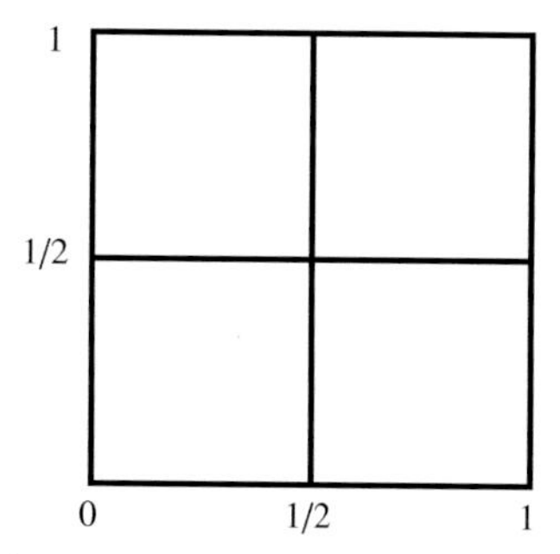

Figure 11.1. A translation-invariant measure that agrees with area on a square must also agree with area on the subsquares.

dyadic rationals are dense in the positive reals; approximating an arbitrary properly oriented square by one with side-length $m/2^n$ yields the correctness of μ on all orthogonal squares, as desired. □

The previous result is valid in all dimensions; Tarski observed that the Banach–Tarski Paradox could be used to provide an improvement in $\mathbb{R}^3$ and beyond. One of the main points of interest of the following lemma is that it turns the negative connotations of the Banach–Tarski Paradox to positive advantage. By shifting the focus from all subsets of $\mathbb{R}^n$ to the bounded Lebesgue measurable ones, the lemma allows us to view the Banach–Tarski Paradox as giving additional information about the uniqueness of Lebesgue measure, rather than just saying that no measures of a certain sort exist on all of $\mathcal{P}(\mathbb{R}^n)$. In fact, the solution of the Ruziewicz Problem (discussed in §13.1) yields the uniqueness of Lebesgue measure as a finitely additive, invariant measure on the bounded Lebesgue measurable sets, if $n \geq 3$. Lemma 11.9, and hence the Banach–Tarski Paradox, is a small but crucial step in the solution (Thm. 13.13; Cor. 13.2 shows that Lemma 11.9, and hence the uniqueness of Lebesgue measure as a finitely additive measure, fails in $\mathbb{R}^1$ and $\mathbb{R}^2$).

Lemma 11.9 (AC). *If $n \geq 3$ and μ is a finitely additive, G_n-invariant measure on the bounded, Lebesgue measurable subsets of $\mathbb{R}^n$ normalizing J (or, a $\mathcal{B}$-measure in $\mathbb{R}^n$), then μ vanishes on the bounded sets of Lebesgue measure zero (or, on the bounded meager sets). The same is true for measures on $\mathbb{S}^n$ having total measure 1 when $n \geq 2$.*

Proof. Suppose E is a bounded subset of $\mathbb{R}^n$ with $\lambda(E) = 0$, and K is a cube large enough to contain E. For any $\epsilon > 0$, K is G_n-equidecomposable with a cube L of volume ϵ, by the strong form of the Banach–Tarski Paradox (Thm. 6.1(c)). Suppose $A_1, \ldots, A_m \subseteq K$ are the pieces witnessing this decomposition. Now, by Proposition 11.8, μ agrees with volume on cubes; so $\mu(L) = \epsilon$. However, this does not imply that $\mu(K) = \epsilon$ because the pieces, A_i, will not be Lebesgue measurable, and hence not μ-measurable. But the sets $A_i \cap E$ must be μ-measurable, as they are subsets of E, which has Lebesgue measure zero, and hence are

Lebesgue measurable. It then follows from the packing of E into L, and the finite additivity and G_n-invariance of μ, that $\mu(E) \leq \mu(L) = \epsilon$. Because ϵ was arbitrary, this means $\mu(E) = 0$. This proof works just as well in the case of meager and Property of Baire sets, because a subset of a meager set is meager, and hence has the Property of Baire and is μ-measurable. The extension to spheres has essentially the same proof, using the strong paradox of Corollary 10.23 and some small open set in place of L. □

The case of the preceding lemma dealing with $\mathcal{B}$-measures and meager sets allows us to prove $(g) \Rightarrow (c)$ of Theorem 11.7 when $n \geq 3$. For if μ is a $\mathcal{B}$-measure, then Lemma 11.9 implies that μ vanishes on all meager sets. Hence all nowhere dense sets get μ-measure zero, and it follows from the remarks following Proposition 10.35 that μ itself, when restricted to $\mathcal{R}$, is additive with respect to $\vee$. In $\mathbb{R}^1$ and $\mathbb{R}^2$, $\mathcal{B}$-measures do exist but do not necessarily vanish on the meager sets (see discussion before Thm. 13.4). But $\mathcal{B}$-measures that do vanish on the meager sets can be constructed (Cor. 13.3), and they yield the negation of (c) (and hence $(c) \Rightarrow (g)$) in those dimensions.

We repeat that the situation regarding cancellation is still open: Does the Cancellation Law hold for $\mathcal{B}$-equidecomposality for subsets of $\mathbb{R}^n$ or $\mathbb{S}^n$?

Now, one of the most noteworthy and surprising results in the area of paradoxes was proved by Randall Dougherty and Matthew Foreman in 1992 [DF92, DF94]. They used an intricate but self-contained argument to prove a general result in Polish spaces that can be applied to $\mathbb{R}^n$ and $\mathbb{S}^{n-1}$ ($n \geq 3$). Inspired by Theorem 11.7, the key step in their solution to the Marczewski Problem is the following result, which does not use the Axiom of Choice. Recall that a *Polish space* is a complete separable metric space; any G_δ subset of $\mathbb{R}^n$ or $\mathbb{S}^n$ is a Polish space [Kur66].

Theorem 11.10. *Let X be a Polish space and suppose σ_i, τ_i ($i = 1, 2, 3$) are independent homeomorphisms of X such that the group G they generate has no nontrivial fixed points. Then there are disjoint open sets U_i, V_i such that $\bigcup \sigma_i(U_i)$ and $\bigcup \tau_i(V_i)$ are each dense in X.*

The idea of the proof is as follows. Let C be a countable, dense, G-invariant subset of X. Give C the induced topology: Open sets are $U \cap C$, where U is an open subset of X. Now it suffices to construct U_i and V_i, six pairwise disjoint open subsets of C, such that $\bigcup \sigma_i(U_i)$ and $\bigcup \tau_i(V_i)$ are each dense in C. This is because taking the interiors of the closures of the six sets will give the open sets we seek. The construction of the six sets uses a complicated induction argument.

The theorem yields a topological paradox on $\mathbb{S}^2$ (i.e., a paradox up to a meager set), which is constructive in that the Axiom of Choice is never used. Such a paradox solves the Marczewski Problem, still without using AC. This is analogous to how the Hausdorff Paradox is used (Thms. 2.3 and 2.6), though that does require AC.

Corollary 11.11. *There are two disjoint subsets of $\mathbb{S}^2$ such that each is in $\mathcal{B}$ and equidecomposable to a dense subset of $\mathbb{S}^2$ using pieces in $\mathcal{B}$.*

Proof. Let G be a free group of rotations having rank 2 (Thm. 2.1) and let D be the set of fixed points of the action. Because D is countable, $\mathbb{S}^2 \setminus D$ is a G_δ set and so is a Polish space with the induced topology. Moreover, G acts by homeomorphisms without fixed points on this space. Choose six independent elements σ_i, τ_i of G (see start of §7.1) and apply the theorem to get U_i, V_i, open (in the induced topology) subsets of $\mathbb{S}^2 \setminus D$; these sets are in $\mathcal{B}$. Let $A = \bigcup U_i$ and $B = \bigcup V_i$. Then A is equidecomposable as claimed to a dense subset of $\mathbb{S}^2 \setminus D$. But a dense subset of $\mathbb{S}^2 \setminus D$ is necessarily a dense subset of $\mathbb{S}^2$, so A is as claimed; the same is true for B. □

Corollary 11.12. (a) *There is no Marczewski measure on $\mathbb{S}^2$.*
(b) *(AC) There is no $\mathcal{B}$-measure on $\mathbb{S}^2$.*

Proof. (a) Suppose μ is such a measure normalizing $\mathbb{S}^2$. Let A and B be as in the preceding corollary; then $\mu(A) + \mu(B) \leq 1$. But A has the same μ-measure as the dense set A^* it is equidecomposable to. And, because μ vanishes on meager sets, A^*, being in $\mathcal{B}$, has the same measure as a dense open set. But a dense open set has μ-measure 1, because its complement is nowhere dense and so of μ-measure 0. So A, and similarly B, have μ-measure 1, a contradiction.
(b) Apply Lemma 11.9. □

Now we can bring in AC and get the pure Baire-set strengthening of the Banach–Tarski Paradox.

Corollary 11.13 (AC). *The sphere $\mathbb{S}^2$ is $SO_3(\mathbb{R})$-paradoxical in $\mathcal{B}$. The same is true, with isometries, for higher-dimensional spheres and for balls and cubes in $\mathbb{R}^n$, $n \geq 3$.*

Proof. Proceed as in Corollary 11.11 to get rotations σ_i and τ_i and pairwise disjoint open sets U_i, V_i. Furthermore, assume that the sets $\sigma_i(U_i)$ are disjoint, and the same for $\tau_i(V_i)$; this can be done without leaving $\mathcal{B}$. Define $L = \bigcup \sigma_i(U_i) \cap \bigcup \tau_i(V_i)$, a comeager set, and let $E = \bigcap_{g \in G} g(L)$. Then G acts on E, which, by the Baire Category Theorem, is a comeager subset of $\mathbb{S}^2$. Put $\hat{U}_i = E \cap U_i$ and $\hat{V}_i = E \cap V_i$, yielding $E = \bigcup \sigma_i(\hat{U}_i) = \bigcup \tau_i(\hat{V}_i)$. This shows that E is paradoxical in $\mathcal{B}$. Now let D be the set of fixed points of G; then $F = \mathbb{S}^2 \setminus (E \cup D)$ is meager and G-invariant. Hence Proposition 1.10 yields that F is G-paradoxical in $\mathcal{B}$; this is because any subset of a meager set is in $\mathcal{B}$. To conclude, we observe that $\mathbb{S}^2$ is G-equidecomposable in $\mathcal{B}$ to $\mathbb{S}^2 \setminus D$ (Thm 3.10), which equals $E \cup (\mathbb{S}^2 \setminus (E \cup D))$, and because E and $\mathbb{S}^2 \setminus (E \cup D)$ are G-paradoxical in $\mathcal{B}$, so is $\mathbb{S}^2$. The extension to the additional cases is easily carried out by using radii and absorption of the origin; the case of unit cubes is implied by the case of unit balls together with the Banach–Schröder Bernstein for $\mathcal{B}$. Scaling yields it for all cubes and balls. □

Now, much more is true; remarkably, everything about the classic paradox holds for Baire sets, except that the classic four-piece result in the Banach–Tarski case becomes a six-piece result when restricted to Baire sets. And a strong paradox holds: Any two bounded subsets of $\mathbb{R}^n$ or $\mathbb{S}^{n-1}$ ($n \geq 3$) having nonempty interior are Baire equidecomposable [DF94, Cors. 2.7, 5.2]. For the first part of the next theorem, see [DF94, Thm. 5.11]; the proof that six is best possible is due to Wehrung [Weh94; see also [DF94, Thms. 5.7, 5.9]. Recently a new approach to Baire paradoxes was discovered by Marks and Unger [MU∞]. Using ideas of matchings in infinite graphs and a generalization of Hall's condition, they proved that whenever a group G acts on a Polish space X so that (a) the functions defined by the action are Borel functions and (b) X is G-paradoxical, then X is G-paradoxical using Baire pieces. This implies Corollary 11.13.

Theorem 11.14 (AC). *The sphere $\mathbb{S}^2$ is $SO_3(\mathbb{R})$-paradoxical in $\mathcal{B}$ using six pieces; a paradox using five or fewer pieces does not exist.*

We next present a very surprising theorem, really a corollary of Theorem 11.10. This theorem, which does not use the Axiom of Choice, says that a pea can be cut into pieces that can be moved by isometries to fill the sun in a way that leaves no hole of positive radius.

Theorem 11.15. *Any two bounded nonempty open subsets of $\mathbb{R}^3$ are densely equidecomposable.*

Proof. Let the sets be A, B, and let K be the open unit ball. Proceed as in Corollary 11.11, using the fact that D, the fixed point set of the free group, is a countable union of lines and so is an F_σ subset of K. Thus we have six independent elements σ_i, τ_i, and six disjoint open sets U_i, V_i of $K \setminus D$ such that $\bigcup \sigma_i U_i$ and $\bigcup \tau_i V_i$ are each dense in $K \setminus D$. Define $\hat{U}_i$, $\hat{V}_i$ to be the interiors of the closures of U_i, V_i, respectively; let $U = \bigcup \hat{U}_i$ and $V = \bigcup \hat{V}_i$. It is easy to see that U and V are disjoint and that each is densely equidecomposable to K. Because the same reasoning applies to any open ball centered at the origin, we may replace K by a small ball contained in A and repeatedly duplicate using the preceding ideas to get that this small ball is densely equidecomposable to a superset of a ball containing B. Therefore A is densely equidecomposable to a superset of B. Because this works the same way in the opposite direction, an application of the Banach–Schröder–Bernstein Theorem (Thm. 11.6) completes the proof. □

By Theorem 11.7(b) $\Rightarrow$ (e) and Corollary 11.13, the Dougherty–Foreman work leads to a paradox using regular-open sets: The cube in $\mathbb{R}^3$ is paradoxical in the regular-open algebra $\mathcal{R}$, using the join of regular-open sets. There is also a paradox using regular-open sets and union (as opposed to join in $\mathcal{R}$), which we now present. These regular-open paradoxes are much more counterintuitive than the original Banach–Tarski Paradox because they involve sets that have a strong geometrical content, unlike the amorphous nature of the nonmeasurable sets in the original 1924 paradox.

Theorem 11.16. *Any two bounded nonempty open subsets A, B of $\mathbb{R}^3$ are densely equidecomposable using regular-open pieces.*

Proof. Assume first that the given sets are regular-open. Start with the sets A_i, B_i that witness the dense equidecomposability of A and B. Recall that for any open set X, there is a (unique) regular-open *shadow* $S(X) \supseteq X$ such that $S(X) \setminus X$ is closed and nowhere dense ([Oxt71, Thm. 4.5]; for open sets, $S(X)$ is $\text{Int}(\overline{X})$). Now, use $S(A_i)$ for the regular-open pieces to get dense equidecomposability. Each $S(A_i)$ is contained in A (because A is regular-open), the nowhere dense differences imply that $\bigcup S(A_i)$ is dense in A, and the $S(A_i)$ are pairwise disjoint (if not, then $S(A_i) \cap S(A_j) \setminus ([S(A_i) \setminus A_i] \cup [S(A_j) \setminus A_j])$ is a nonempty subset of $A_i \cap A_j$). The isometric image of each $S(A_i)$ is $S(B_i)$, which shows that the images are disjoint, regular-open, and dense in B.

Now the general case of open sets follows from the fact that any bounded open set A in $\mathbb{R}^n$ can be densely filled by two disjoint regular-open sets. Let $B = S(A) \supseteq A$ be the regular-open shadow of A; so $B \setminus A = F$ is closed and nowhere dense. Let d be the diameter of A. Let B_m be the collection of points in A that are within distance d/m of a point in F. Let R_m be the annular object $B_m \setminus B_{m+1}$ and let $P = \bigcup_{m\text{odd}} \text{Int}(R_m)$; $Q = \bigcup_{m\text{even}} \text{Int}(R_m)$. It is easy to see that P and Q are disjoint regular-open sets and $P \cup Q$ is dense in A. □

The preceding two results are also valid for spheres: Corollary 2.8 and Theorem 5.1 of [DF94] show that any two open subsets of $\mathbb{S}^n$, $n \geq 2$, are densely equidecomposable. The work in this section indicates how the Marczewski Problem is solved via paradoxes in $\mathbb{R}^3$ and higher dimensions. But in $\mathbb{R}^1$ and $\mathbb{R}^2$, Marczewski measures do exist (Cor. 13.3). Mycielski wondered if the regular-open paradoxes are possible with the movement of the pieces being disjoint (see §10.1.2 for the proof that this is possible for the classic Banach–Tarski Paradox; see also Question 10.17).

Question 11.17. Can the paradox of Theorem 11.16 be realized with continuous equidecomposability; that is, can the pieces be moved to their new positions so that, at each moment, they are disjoint?

11.3 Equidecomposability with Countably Many Pieces

Ever since Lebesgue, countably additive measures have been more prevalent in mathematics than finitely additive ones, and it is natural to ask if the theory of paradoxical decompositions can be applied to such measures. Recall (Thm. 1.5) that the first example of a paradox using isometries arose from Vitali's proof of the nonexistence of a countably additive, translation-invariant measure defined on all sets of reals. That decomposition used infinitely many pieces, and we shall now study some further consequences of allowing countably many pieces in the theory of equidecomposability. First, we note that the question of Lebesgue measure's uniqueness among countably additive measures is much simpler than the case where λ is viewed as a finitely additive measure.

Proposition 11.18. *Lebesgue measure λ is the unique countably additive, translation-invariant measure on the Lebesgue measurable subsets of $\mathbb{R}^n$ that normalizes the unit cube.*

Proof. If μ is another such measure, then by Proposition 11.8, μ agrees with volume on cubes. It follows that μ agrees with λ on all open sets, because an open set is a union of countably many pairwise disjoint half-open cubes (of side-length 2^{-m}). This fact can be proved by subdividing space repeatedly, retaining the cubes that are contained in the open set; see [Coh80, p. 28] for a detailed proof. It follows that μ agrees with λ on bounded closed sets too. Now, if A is measurable and bounded, then for any $\epsilon > 0$, there are F, G with F closed (and bounded), G open, $F \subseteq A \subseteq G$, and $\lambda(G \setminus F) < \epsilon$. It follows easily that $\mu(A) = \lambda(A)$. Because an unbounded measurable set is a union of countably many bounded measurable sets, μ agrees with λ on all measurable sets. □

Uniqueness of countably additive measures holds in more general contexts. Let $(X, \mathcal{A}, m)$ be a measure space with $m(X) = 1$ (i.e., m is a countably additive measure on the σ-algebra $\mathcal{A}$) and suppose m is G-invariant where G, a group acting on X, preserves $\mathcal{A}$. Suppose further that G's action is *ergodic*; that is, if $A \in \mathcal{A}$ and $0 < m(A) < 1$, then there is some $g \in G$ such that $m(A \triangle g(A)) > 0$. Then, using the Radon–Nikodym Theorem, one can show that any other countably additive, G-invariant measure on $\mathcal{A}$ that normalizes X and vanishes on the sets of m-measure 0 (i.e., is *absolutely continuous* with respect to m) must equal m. *Proof sketch:* the Radon–Nikodym derivative f is unique up to a set of measure 0. But for any $g \in G$, $f(gx)$ is easily shown to be a Radon–Nikodym derivative. So f is g-invariant almost everywhere. But then, by ergodicity, f is constant almost everywhere with value $\inf\{q : q \in \mathbb{Q} \text{ and } q \geq 0 \text{ and } m(\{x : f(x) > q\}) = 1\}$. By normalization, the constant is 1.

Next we formalize the notion of countable equidecomposability.

Definition 11.19. *Suppose a group G acts on X and $A, B \subseteq X$. Then A and B are countably G-equidecomposable, $A \sim_\infty B$ (the G in this notation will be clear from the context), if there is a partition of A into countably many sets A_i ($i \in \mathbb{N}$) and elements $\sigma_i \in G$ such that the sets $\sigma_i(A_i)$ form a partition of B. A subset E of X is countably G-paradoxical if E contains disjoint sets A, B such that $A \sim_\infty E$ and $B \sim_\infty E$. If G is the group of isometries of X, then the prefix "G-" will be omitted. If all the pieces are required to lie in $\mathcal{A}$, a G-invariant σ-algebra of subsets of X, the sets will be called countably G-equidecomposable in $\mathcal{A}$, or countably paradoxical in $\mathcal{A}$.*

It is easy to see that countable equidecomposability is an equivalence relation. Moreover, because this relation satisfies properties (a) and (b) of the proof of Theorem 3.6, the proofs of the Banach–Schröder–Bernstein Theorem and the

Cancellation Law carry over without modification. Indeed, as we shall discuss further, the Cancellation Law is valid even for the case where the pieces are restricted to a σ-algebra of sets. But Tarski's Theorem (Cor. 11.2) fails for countable equidecomposability, as the following example shows.

Theorem 11.20. *Let G be the group of all permutations π of ω_1, the first uncountable ordinal, with the property that $\{\alpha < \omega_1 : \pi(\alpha) \neq \alpha\}$ is finite. Then ω_1 is not countably G-paradoxical, but there is no countably additive, G-invariant measure μ on $\mathcal{P}(\omega_1)$ with $\mu(\omega_1) = 1$.*

Proof. Because all singletons are congruent under G's action, a measure μ as in the theorem must assign them all measure zero. But it is a well-known result of Ulam [Oxt71, p. 25] that there exists no countably additive measure on ω_1 having total measure 1 and vanishing on singletons. To see that ω_1 is not countably G-paradoxical, suppose A, B are disjoint subsets of ω_1 with $A \sim_\infty \omega_1 \sim_\infty B$, where the equidecomposabilities are witnessed by the functions f, g. Because only countably many permutations are used, only countably many points are moved by f or g. If $\alpha \in \omega_1$ is not such a moved point, then $f(\alpha) = \alpha$, so $f^{-1}(\alpha) = \alpha$ and $\alpha \in A$; similarly $\alpha \in B$. So A and B are not disjoint. □

Despite the previous example, there are some ways to extend Tarski's Theorem to the countably additive case. Recall that a *probability measure* is a countably additive measure of total measure 1. The ideal extension would be, If X is not countably G-paradoxical, then there is a G-invariant probability measure on X. This fails because of Theorem 11.20. Chuaqui searched for an additional hypothesis and conjectured that the following is sufficient: There is a probability measure on $\mathcal{A}$ (a given σ-algebra) that normalizes X and vanishes on the countably G-paradoxical (in $\mathcal{A}$) sets. P. Zakrzewski [Zak93a, pp. 343–352] proved that this does indeed suffice to give a G-invariant probability measure on $\mathcal{A}$.

We now present some of the details. Suppose G acts on X and $\mathcal{A}$ is a G-invariant σ-algebra of subsets of X. Define the subfamily $\mathcal{I}$ of $\mathcal{A}$ to consist of all sets $A \in \mathcal{A}$ that are *ω-negligible*: sets for which X contains infinitely many "copies" of A (i.e., X has pairwise disjoint subsets $A_i \in \mathcal{A}$, $i \in \mathbb{N}$, with each $A_i \sim_\infty A$ in $\mathcal{A}$). Then $\mathcal{I}$ is closed under taking subsets in $\mathcal{A}$ and under countable unions, and so $\mathcal{I}$ is a σ-ideal in $\mathcal{A}$. Any countably paradoxical set lies in $\mathcal{I}$, and in fact $\mathcal{I}$ is generated by all countably G-paradoxical (with pieces in $\mathcal{A}$) sets in $\mathcal{A}$ [Chu77]. It is clear that any G-invariant probability measure on $\mathcal{A}$ must assign measure zero to all sets in $\mathcal{I}$; also $X \in \mathcal{I}$ iff X is countably G-paradoxical in $\mathcal{A}$.

In the example of Theorem 11.20, $\mathcal{A} = \mathcal{P}(\omega_1)$ and $\mathcal{I}$ consists of all countable subsets of ω_1. The proof of that theorem uses the fact that there is no countably additive measure on $\mathcal{A}$ that vanishes on $\mathcal{I}$ and normalizes ω_1, even without the additional requirement of G-invariance. This phenomenon was the motivation for Chuaqui's conjecture [Chu69] that this is the only barrier to a generalization of Tarski's Theorem.

A set $A \in \mathcal{A}$ is *weakly wandering* if there are elements $g_i \in G$ such that, for i, j distinct, $g_i(A) \cap g_j(A) = \varnothing$. Let $\mathcal{W}$ be the family of such sets; clearly, $\mathcal{W} \subseteq \mathcal{I}$. The following theorem was proved by Zakrzewski [Zak93a].

Theorem 11.21 (AC). *Let G act on X. Suppose that $(X, \mathcal{A}, m)$ is a measure space where $\mathcal{A}$ is a G-invariant σ-algebra and m is a σ-finite countably additive measure. The following are equivalent:*

(a) *There is a G-invariant probability measure μ on $\mathcal{A}$ such that m is absolutely continuous with respect to μ.*
(b) *The measure m vanishes on $\mathcal{W}$.*

Because $\mathcal{W} \subseteq \mathcal{I}$, we have the following affirmative resolution of Chuaqui's conjecture.

Corollary 11.22 (AC). *Suppose G acts on X and let $\mathcal{A}$ be a G-invariant σ-algebra of subsets of X. The following are equivalent:*

(a) *There is a G-invariant probability measure on $\mathcal{A}$.*
(b) *There is a probability measure on $\mathcal{A}$ that vanishes on $\mathcal{I}$.*

It is worth noting that Theorem 11.21 is closely related to the following problem of classic ergodic theory: Let m be a G-invariant, σ-finite measure defined on a σ-algebra $\mathcal{A}$ in $\mathcal{P}(X)$. Find necessary and sufficient conditions for the existence of a G-invariant probability measure μ defined on $\mathcal{A}$ such that m is absolutely continuous with respect to μ (see [Zak93b] for more on this point).

Another positive result is due to Becker and Kechris [BK96], who showed that, for Borel actions of Polish groups on Polish spaces, an exact extension of Tarski's Theorem to the countably additive case does exist.

As was done for finite equidecomposability, one can form a semigroup, $\mathcal{S}_\infty(\mathcal{A})$, of types with respect to the relation of countable equidecomposability in $\mathcal{A}$. Instead of $X \times \mathbb{N}$, one uses $X \times \omega_1$ and considers bounded sets to be those that have only countably many levels. An important aspect of countable equidecomposability is that $\mathcal{S}_\infty(\mathcal{A})$ has a richer algebraic structure than $\mathcal{S}(\mathcal{A})$: $\mathcal{S}_\infty(\mathcal{A})$ is a cardinal algebra, as defined by Tarski [Tar49]. The structure of these algebras is worked out in detail in [Tar49]; in particular, the Cancellation Law is valid. See [Chu69, Chu76, Chu77] for a variety of applications of this point of view to countable equidecomposability and countably additive measures.

Turning now to the specific case of $\mathbb{R}^n$ and isometries, we shall see that countable equidecomposability is much simpler than finite equidecomposability. For instance, the existence of paradoxes is not dependent on dimension. The following result may be viewed as a generalization of Vitali's nonmeasurable set, showing even more emphatically why translation-invariant, countably additive measures defined on all subsets of $\mathbb{R}^n$ do not exist.

Theorem 11.23 (AC). *Any two subsets of $\mathbb{R}^n$ with nonempty interior are countably equidecomposable.*

Proof. By transitivity and the Banach–Schröder–Bernstein Theorem, it suffices to show that for any half-open interval K, $K \sim_\infty \mathbb{R}$. Vitali's construction (Thm. 1.5), restated for K, shows that K may be partitioned into countably many sets A_i such that $A_i \sim_2 A_j$ for each i, j. (In fact, by a result of von Neumann—see comments before Corollary 5.16—it can be guaranteed that the sets A_i are all congruent to each other.) It follows that for any infinite subset S of $\mathbb{N}$, $\bigcup\{A_i : i \in S\} \sim_\infty K$. Now, by splitting the family $\{A_i : i \in \mathbb{N}\}$ into infinitely many infinite subfamilies, one gets that K is countably equidecomposable with infinitely many copies of itself. It follows, upon using some additional translations, that $K \sim_\infty \mathbb{R}$. One can get a Vitali partition of $[0, 1]^n$ by simply using the Vitali partition on the first coordinate; this allows the proof for $\mathbb{R}$ to be repeated for $\mathbb{R}^n$. □

The condition on interiors in the preceding result is necessary because, for example, a meager set cannot be countably equidecomposable with a nonmeager set.

The Vitali partition shows that a countably additive measure on $\mathcal{P}([0, 1])$ that has total measure 1 cannot be invariant with respect to translations (modulo 1). This leads to the question of whether there can be a countably additive measure on $\mathcal{P}([0, 1])$ that has total measure 1, but is not required to satisfy any invariance condition whatsoever. Of course, the principal measure determined by a point is such a measure; hence we add the natural condition (a consequence of translation invariance) that points get measure zero. Because this problem does not refer at all to the geometry of the line, it is really a problem in set theory: If f is a bijection from X to Y, then X bears such a measure if and only if Y does. So let us say that a cardinal κ is *real-valued measurable* if κ bears a κ- additive measure defined on all subsets, having total measure 1, and vanishing on singletons.

Ulam [Ula30] showed that real-valued measurable cardinals must be large (weakly inaccessible, and more), and so ω_1 is not real-valued measurable (this result is used in the proof of Thm. 11.20); thus, if the Continuum Hypothesis is true, then there is no real-valued measure on the real numbers. But unlike CH, which is independent of ZFC, the usual axioms of set theory (i.e., the addition of either CH or its negation yields a noncontradictory system), the statement that the continuum is real-valued measurable has a more complicated status. Its negation is consistent because it follows from CH, but the statement in its positive form is connected with large cardinal axioms for set theory. It follows that the consistency of the statement cannot be derived from the consistency of ZFC alone. In fact, the existence of a real-valued measure on the continuum is equiconsistent with the existence of a (2-valued) measurable cardinal. In short, if it is noncontradictory to add a certain axiom about the existence of very large cardinals to ZFC, then (and only then) is it noncontradictory to add the statement that the continuum is real-valued measurable. See [Jec78] for a complete discussion of the connection between this problem in measure theory and large cardinal axioms in set theory. Other connections between large cardinals and consistency results in set theory and in measure theory are discussed in Chapter 15.

Even if there is a real-valued measure on $\mathcal{P}([0, 1])$, such a measure need not extend Lebesgue measure. For example, one can use a bijection of $[0, 1]$ with $[0, 1/2]$ to transform the measure into one that assigns measure 0 to $[1/2, 1]$. Nevertheless, if the continuum is real-valued measurable, then there is a countably additive measure on $\mathcal{P}([0, 1])$ that extends Lebesgue measure; see [Jec78, p. 302] for a proof.

Of course, Theorem 11.23 fails if the pieces in the decomposition are required to be Lebesgue measurable. On the other hand, the result remains valid if the pieces are restricted to lie in $\mathcal{B}$, the algebra of subsets of $\mathbb{R}^n$ having the Property of Baire. While the following result for $n \geq 3$ and bounded sets follows from the much stronger Dougherty–Foreman work, the following proof is simple, applies to the unbounded case, and works in all dimensions.

Theorem 11.24 (AC). *Any two subsets of $\mathbb{R}^n$, each of which has the Property of Baire and nonempty interior, are countably equidecomposable in $\mathcal{B}$.*

Proof. We shall show that $K \sim_\infty \mathbb{R}^n$ in $\mathcal{B}$, where K is any cube in $\mathbb{R}^n$. By transitivity and the Banach–Schröder–Bernstein Theorem, this suffices. First we show that K can be packed into any other cube L, using countably many pieces in $\mathcal{B}$. Let E be an open dense subset of K with $\lambda(E) \leq \lambda(L)/2$; simply get E as a union of sufficiently small pairwise disjoint intervals I_m, so that the first k points of a countable dense subset of K are contained in the union of the first k intervals. Let $M = K \setminus E$, a nowhere dense set. Now, split L into two halves, L_1 and L_2, and pack E and M into L_1 and L_2, respectively, as follows (see Fig. 11.2). By Proposition 10.41, each I_m is finitely equidecomposable, using Borel pieces, with an open strip of L having the same volume (see Fig. 11.2), and it follows that $E = \bigcup I_m$ can be packed into L_1 using countably many Borel pieces.

To pack M into L_2, use Theorem 11.23 to pack all of $\mathbb{R}^n$ into L_2 with no restriction on the pieces. But a subset of a nowhere dense set is nowhere dense, and therefore the packing of M into L_2 induced by the packing of $\mathbb{R}^n$ has its pieces in $\mathcal{B}$, as desired.

Now, we may cover $\mathbb{R}^n$ with infinitely many closed cubes K_i, each having the same volume as a given cube K, and K itself contains infinitely many pairwise disjoint open cubes L_i. The result in the previous paragraph shows how to pack each K_i into L_i, and this yields that $\mathbb{R}^n \preceq_\infty K$. Obviously $K \preceq_\infty \mathbb{R}^n$, so $K \sim_\infty \mathbb{R}^n$ as desired. □

A consequence of Theorem 11.24 is that no invariant measure on $\mathcal{B}$ that normalizes J can be countably additive. Recall that the Marczewski Problem was whether a finitely additive, invariant measure on $\mathcal{B}$ that normalizes J exists if $n \geq 3$. To put it another way, while the unit cube in $\mathbb{R}^n$, $n \geq 3$, is paradoxical in $\mathcal{B}$ using finitely many pieces (Cor. 11.13), the use of countably many pieces yields that it is $\mathcal{B}$-paradoxical even in $\mathbb{R}^1$ and $\mathbb{R}^2$. Note that the result about the nonexistence of a countably additive measure on $\mathcal{B}$ could not have been obtained by the classical Vitali approach (Thm. 1.5). Using the representation of sets in $\mathcal{B}$ in

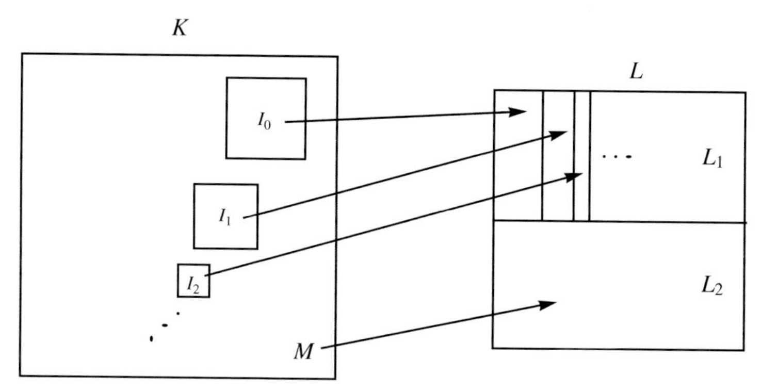

Figure 11.2. A packing of a square into a smaller square using pieces with the Property of Baire.

terms of regular-open sets (see proof of Thm. 11.7, $(c) \Rightarrow (g)$), it is easy to see that the unit cube cannot be partitioned into countably many sets with the Property of Baire that are pairwise congruent using translations modulo 1. Note that the condition on interiors in Theorem 11.24 cannot be eliminated. M. Penconek [Pen91, Cor. 1.5] showed that there is a comeager set A in $\mathbb{R}^n$ that is not countably paradoxical. Because $\mathbb{R}^n$ is countably paradoxical, Proposition 3.5, extended to the countable case, shows that A and $\mathbb{R}^n$ are not countably equidecomposable.

In $\mathbb{R}^n$, $n \geq 3$, one can simplify the part of Theorem 11.24's proof that packs M into L_2 by invoking the Banach–Tarski Paradox instead of Theorem 11.23. Because M is bounded, the strong form of the Banach–Tarski Paradox yields that a cube containing M, and hence M itself, may be packed into L_2 using finitely many pieces. But all subsets of M are nowhere dense and hence in $\mathcal{B}$. If E could also be handled with finitely many pieces in $\mathcal{B}$, we would essentially have a paradoxical decomposition in $\mathcal{B}$ as in Theorem 11.7(b). But the existence of an E that could be so handled is just Theorem 11.7(i) and is equivalent to the existence of a Marczewski measure.

Of course, Theorem 11.24 fails if $\mathcal{L}$, the algebra of Lebesgue measurable sets, is used instead of $\mathcal{B}$, because of the existence of the countably additive invariant measure λ. Even if two sets have the same Lebesgue measure, they need not be countably equidecomposable in $\mathcal{L}$: Simply consider the empty set and any nonempty set of measure zero. But this example is essentially the only one. The following definition allows us to show that Lebesgue measure is as complete an invariant for countable equidecomposability as is possible.

Definition 11.25. *Two measurable sets $A, B \in \mathcal{L}$ are* almost countably equidecomposable *in $\mathcal{L}$ if there are measure zero sets E_1, E_2 such that $A \setminus E_1 \sim_\infty B \setminus E_2$ in $\mathcal{L}$.*

The reader may prove, as an exercise, that the relation of almost countable equidecomposability in $\mathcal{L}$ is transitive. (This also follows from the discussion of quotient algebras in §13.2, because the relation in question is equivalent to countable equidecomposability in the quotient algebra $\mathcal{L}/N$, where N is the ideal of sets of measure zero.) This new relation clearly preserves λ and eliminates the problem of sets of measure zero. As we now show, λ is a complete invariant for this modified notion of equidecomposability.

Theorem 11.26. *If A and B are Lebesgue measurable subsets of $\mathbb{R}^n$ (or of $\mathbb{S}^n$), then $\lambda(A) = \lambda(B)$ if and only if A and B are almost countably equidecomposable in $\mathcal{L}$.*

Proof. Only the forward direction requires proof. Assume $\lambda(A) < \infty$. For if $\lambda(A) = \lambda(B) = \infty$, we may partition each of A, B into countably many bounded sets such that corresponding sets have the same measure. Then the fact that corresponding summands are almost equidecomposable yields that the same is true for A and B. Now, the Lebesgue density theorem asserts that if $\lambda(E) > 0$, then for almost all $P \in E$, $\lim_{h\to 0} \lambda(E \cap C(h))/h^n = 1$, where $C(h)$ is a cube of side-length h centered at P. Choose P_1, P_2 in A, B, respectively, so that for some $h > 0$, the quotients of the preceding limit are each greater than $1/2$. Then $B \cap (A + (P_2 - P_1))$ has positive measure. This shows that whenever two sets have positive measure, some translate of one intersects the other in a set of positive measure.

Now, let ℓ be the supremum of $\lambda(\tau(A) \cap B)$ over all translations τ, and choose τ_0 so that $\lambda(\tau_0(A) \cap B) = \ell_0 \geq \ell/2$. Let $B_0 = \tau_0(A) \cap B$ and $A_0 = \tau_0^{-1} B_0$. Then replace A, B by $A \setminus A_0$, $B \setminus B_0$, respectively, and (stopping if $\lambda(A \setminus A_0) = \lambda(B \setminus B_0) = 0$) repeat to obtain A_1, B_1, and ℓ_1. Continue as long as possible, always choosing τ_m so that $\ell_m = \lambda(\tau_m(A \setminus \bigcup_{k\leq m-1} A_k) \cap (B \setminus \bigcup_{k\leq m-1} B_k)$ is at least half as large as is possible. If this procedure stops in finitely many steps—that is, if one of $A \setminus \bigcup_{k\leq m} A_k$, $B \setminus \bigcup_{k\leq m} B_k$ has measure zero—then both of these sets have measure zero (because $\lambda(A) = \lambda(B)$, $\lambda(A_k) = \lambda(B_k)$) and A and B are almost finitely equidecomposable in $\mathcal{L}$. Otherwise, $\bigcup A_m \sim_\infty \bigcup B_m$ in $\mathcal{L}$, and because the ℓ_m correspond to measures of pairwise disjoint subsets of B, $\sum \ell_m \leq \lambda(B) = \lambda(A)$. Therefore the sequence $\{\ell_m\}$ converges to 0. But this means that $\sum \ell_m = \lambda(\bigcup A_m) = \lambda(A)$, for if $\lambda(A \setminus \bigcup A_m) > 0$, then there would be a translation τ such that $\lambda(\tau(A \setminus \bigcup A_m) \cap (B \setminus \bigcup B_m)) = r > 0$. Choosing m such that $r > 2\ell_m$ yields that τ contradicts the choice of τ_m, because ℓ_m is not greater than $r/2$.

The same proof works for subsets of $\mathbb{S}^n$, because the Lebesgue density theorem is valid on spheres and any point can be taken to any other by a rotation. □

It is interesting to note that in the preceding proof, it is sufficient to choose τ_m so that $\ell_m > 0$. The procedure might then continue on beyond ω steps, but one could still define τ_α, ℓ_α, A_α, and B_α for infinite ordinals α. Because the ℓ_α correspond to disjoint sets, the sets A and B must be exhausted, in measure, by the A_α and B_α corresponding to $\alpha < \beta$ for some countable ordinal β. This is a consequence of the fact that a set of finite measure cannot contain uncountably many pairwise disjoint subsets of positive measure (otherwise, for some n, infinitely many of the sets would have measure greater than $1/n$).

Theorem 11.26 may be generalized to Haar measure on the Borel subsets of a locally compact topological group: If A and B are Borel sets with the same finite Haar measure, then A and B are almost countably equidecomposable using Borel pieces. The proof is the same as the one just given, once it is established that whenever $\mu(A)$ and $\mu(B)$ are nonzero (μ denotes Haar measure), for some $g \in G$, $\mu(B \cap gA) > 0$. This, in turn, may be derived from Fubini's theorem, together with the following two facts about Haar measure [Coh80, Chap. 9]:

(a) $\mu(A^{-1}) > 0$ if $\mu(A) > 0$.
(b) The function $F(s, t) = (s, t^{-1}s^{-1})$ is a measure-preserving (with respect to $\mu \times \mu$) homeomorphism of $G \times G$ to $G \times G$.

Because $\mu(B)\mu(A^{-1}) = (\mu \times \mu)(F^{-1}(B \times A^{-1})) = (\mu \times \mu)(\{(h, g) : h \in B \cap gA\})$, the set of g such that $\mu(B \cap gA) > 0$ must itself have positive measure.

It might even be true that two bounded subsets of $\mathbb{R}$ with the same Lebesgue measure are almost finitely equidecomposable in $\mathcal{L}$.

One can also consider almost countable equidecomposability with no restrictions on the pieces. This yields the following variation of Theorem 11.26: Any two subsets of $\mathbb{R}^n$ with positive Lebesgue inner measure are almost countably equidecomposable. For the proof of this, see [BT24, Thm. 41].

Another way to avoid the problem caused by sets of measure zero is to consider just the open sets or, more generally, sets with nonempty interior, rather than all measurable sets. We shall show that such sets are countably equidecomposable in $\mathcal{L}$ provided they have the same measure. The proof requires the following lemma, part (a) of which can also be deduced from the "sack of potatoes" theorem (see [Mau81, p. 74]); that theorem states that given a countable sequence of convex bodies in $\mathbb{R}^n$, with bounded diameters and with bounded total volume, there is a cube such that the potatoes fit disjointly into the cube.

Lemma 11.27. (a) *If $E \subseteq \mathbb{R}^n$ has Lebesgue measure 0 and U is a given nonempty open subset of $\mathbb{R}^n$, then E is countably equidecomposable with a subset of U.*

(b) *The same result is valid for subsets E of $\mathbb{S}^n$ having Lebesgue measure zero.*

Proof. (a) Choose a cube $K \subseteq U$. Then cover E with rectangles R_i whose volumes sum to less than the volume of K. Choose pairwise disjoint rectangles $R_i' \subseteq K$ such that $\lambda(R_i') = \lambda(R_i)$ and apply Proposition 10.41 to obtain the (Borel) equidecomposability of R_i with R_i'. It follows that E is countably equidecomposable to a subset of $\bigcup R_i'$.

(b) The result is clear for $\mathbb{S}^1$ and may be proved for $\mathbb{S}^2$ in the same way as part (a), using the Bolyai–Gerwien Theorem for spherical polygons [Ger83]. To get it for all dimensions, use a stereographic projection, from the north pole, of the southern hemisphere of $\mathbb{S}^n$ onto a ball in $\mathbb{R}^n$. Then one can apply the "sack of potatoes" theorem in $\mathbb{R}^n$ mentioned prior to the lemma. The key point is that this projection sends spherical caps to balls, and vice versa. One advantage of this constructive approach that pertains to both cases is that if the set E is Borel, then the packing can be accomplished with Borel pieces. □

The preceding proof for spheres does not use the Axiom of Choice. If one is willing to use AC, then one can do it, for $\mathbb{S}^n$ with $n \geq 2$, by appealing to the strong form of the Banach–Tarski Paradox; such an approach will use only finitely many pieces. But if E is Borel, then this approach might introduce non-Borel sets.

Lemma 11.27 raises the question whether the packings can be effected using only finitely many pieces. As pointed out in the proof (see also Lemma 11.9), the Banach–Tarski Paradox yields that any bounded null set in $\mathbb{R}^n$ or $\mathbb{S}^{n-1}$ ($n \geq 3$) can be packed into any nonempty open set using finitely many (necessarily measurable) pieces. In $\mathbb{R}^1$, $\mathbb{R}^2$, and $\mathbb{S}^1$, however, there exist null sets that cannot be finitely packed into any open set of measure less than 1; see remarks following Corollary 13.3. And the solution to the Marczewski Problem tells us that we can do the same in the topological context, without the necessity of the restriction to small (i.e., meager) sets. Any bounded Baire set can be packed into any cube using Baire pieces! The given cube is, by Corollary 11.13, scaling, and repeated duplication, $\mathcal{B}$-equidecomposable to a cube large enough to contain the given set.

Lemma 11.27 can be combined with Theorem 11.26 to yield the following theorem.

Theorem 11.28. *Any two measurable subsets of $\mathbb{R}^n$ (or of $\mathbb{S}^n$) with nonempty interior and with the same Lebesgue measure are countably equidecomposable in $\mathcal{L}$.*

Proof. First note that if in Theorem 11.26, A and B have nonempty interior, then we may construct the decomposition so that the equidecomposable sets $A \setminus E_1$ and $B \setminus E_2$ also have nonempty interior. Simply choose the first translation τ_0 so that $\tau_0(A) \cap B$ contains an open set; the condition $\ell_0 \geq 1/2$ may fail for this one case, but this will not affect the proof, because the sequence $\{\ell_m\}$ still converges to 0. Hence if A and B are as hypothesized, there are measure zero subsets E_1, E_2, of A, B, respectively, such that $A \setminus E_1 \sim_\infty B \setminus E_2$ in $\mathcal{L}$ and $A \setminus E_1$, $B \setminus E_2$ each contain a nonempty open set.

Now, it is sufficient to show that $A \sim_\infty A \setminus E_1$, in $\mathcal{L}$ (and the same for B and $B \setminus E_2$). But this is a consequence of the previous lemma. Choose countably many pairwise disjoint open subsets U_i of $A \setminus E_1$, and choose $F_i \subseteq U_i$ so that $F_i \sim_\infty E_1$. Then partition A into $A \setminus (E_1 \cup \bigcup F_i)$, $E_1, F_0, F_1, \ldots$. Because $E_1 \sim_\infty F_0$, $F_0 \sim_\infty F_1, \ldots$, this shows that $A \sim_\infty A \setminus E_1$ in $\mathcal{L}$ as required. Similarly, $B \sim_\infty B \setminus E_2$ in $\mathcal{L}$, completing the proof. □

If A and B are Borel subsets of $\mathbb{R}^n$ and have nonempty interior and the same measure, then the proof of the previous theorem, together with the remark at the end of Lemma 11.27's proof, yields a decomposition using Borel sets. It follows that some of the decomposition problems posed earlier have affirmative answers if countably many pieces are allowed: The circle can be squared and a regular tetrahedron cubed using Borel sets.

As with Theorem 11.26, it seems reasonable to expect that the previous result is valid for sets of the same finite Haar measure, using Borel sets as pieces. Even for Lie groups, this problem is unsolved (although Freyhoffer [Fre78] has obtained a solution in commutative Lie groups). Because the generalization of Theorem 11.26 to Haar measure is valid, the method of proof of Theorem 11.28 reduces the problem to generalizing Lemma 11.27.

Question 11.29. Is it true that whenever E is a Borel set of Haar measure zero in a locally compact topological group G, and U is a nonempty open subset of G, then E is countably equidecomposable with a subset of U using left translations and Borel pieces?

Notes

Tarski's Theorem is remarkably elegant, and it provided the original motivation for the first edition of this book in 1985. The theorem is presented in [Tar38b], although he obtained the result much earlier [Tar29]. Tarski used transfinite induction. The use of the Tychonoff Theorem stems from work of Łoś and Ryll-Nardzewski [LR51], who used it to prove the Hahn–Banach Theorem. Tarski's Theorem is not as well known as it deserves to be, and the result has been rediscovered by J. Sherman [She79] (in a weaker form than Tarski's) and by Emerson [Eme79]. For a proof based on the Hahn–Banach Theorem, see [HS86].

Marczewski was the first to investigate measures vanishing on the meager sets. The equivalences of Theorem 11.7 are due to Mycielski [Myc77a, Myc80]. The 1980 paper, however, deals with $\mathbb{S}^2$ rather than $\mathbb{R}^3$ and assumes that Corollary 11.2 is valid for $\mathcal{B}$. This is not known, and so Theorem 11.7 is not known to be true as stated for $\mathbb{S}^2$. The trick of using similarities to sidestep the problem in $\mathbb{R}^n$ (Thm. 11.7(a) $\Rightarrow$ (b)) is due to Mycielski.

Lemma 11.9 is due to Tarski [Tar38b, p. 65] and has turned out to be important in recent work on characterizing Lebesgue measure (see remarks preceding Defn. 13.9). Proposition 11.18 was essentially known to Lebesgue [Leb04].

The fact that the Dougherty–Foreman work yields a general equidecomposability result about open sets using regular-open pieces (Thm. 11.16) is due to Mycielski.

Chuaqui [Chu69, p. 74] conjectured that Tarski's Theorem is valid for countably additive measures and $\sim_\infty$, but then he discovered the counterexample of Theorem 11.20 [Chu73]. (A counterexample was first discovered by Lang [Lan70], but he assumed the Continuum Hypothesis.)

The notions of countable equidecomposability and almost countable equidecomposability were first investigated by Banach and Tarski [BT24], who proved Theorems 11.22 and 11.26. Their proof of Theorem 11.26 was more elementary and more complicated than the proof presented here, which is due to Mycielski.

The extension of Theorem 11.26 to Haar measure is due to von Neumann (unpublished; see [Chu76] and [Mah42]).

Lemma 11.27 is due to Banach and Tarski [BanTar24, p. 277] for $\mathbb{R}^n$ (see also [Chu69]), and the fact that the result is valid on spheres without having to use the Axiom of Choice was pointed out to the authors by B. Grünbaum. The fact that Lemma 11.27 could be used to obtain the characterization of Theorem 11.28 is due to Chuaqui [Chu69], although for open sets the result was known to Banach and Tarski. The possibility that Theorem 11.28 is valid for Haar measure, that is, Question 11.29, was first investigated by Chuaqui [Chu76].

12

Measures in Groups

In Part One we saw that the main idea in the construction of a paradoxical decomposition of a set was to first get such a decomposition in a group acting on the set and then transfer it to the set. A similar theme pervades the construction of invariant measures on a set X acted upon by a group G. If there is a finitely additive, left-invariant measure defined on all subsets of G, then it can be used to produce a finitely additive, G-invariant measure defined on all subsets of X. Such measures on X yield that X (also certain subsets of X) is not G-paradoxical.

It was von Neumann [Neu29] who realized that such a transference of measures was possible, and he began the job of classifying the groups that bear measures of this sort. In this chapter we first study some properties of the class of groups having measures and show that it is fairly extensive, containing all solvable groups. We then give the important application to the case of isometries acting on the line or plane, obtaining the nonexistence of Banach–Tarski-type paradoxes in these two dimensions.

12.1 Amenable Groups

An amenable group is one that bears a certain measure; the name, put forward by Mahlon Day, is a pun when the "*e*" is long, as in "ameanable."

Definition 12.1. *If, for a group G, μ is a finitely additive measure on $\mathcal{P}(G)$ such that $\mu(G) = 1$ and μ is left-invariant ($\mu(gA) = \mu(A)$ for $g \in G$, $A \subseteq G$), then μ will be called simply a* measure *on G. An* amenable *group is one that bears such a measure; AG denotes the class of all amenable groups.*

If a group G is paradoxical with respect to left translations, then G cannot be amenable, because a measure would have to satisfy $\mu(G) = 2\mu(G)$. Thus, for example, any group that has a free subgroup of rank 2 is not amenable (by Thm. 1.2 and Prop. 1.10). In fact, an application of Tarski's Theorem (Cor. 11.2) to a group's action on itself yields that groups that are not paradoxical are necessarily amenable, and hence AG coincides with the class of nonparadoxical groups.

While this characterization of amenability can be used to establish the amenability of some families of groups (e.g., the amenability of Abelian groups follows from Thm. 14.21(a)), we shall establish the existence of a measure more directly whenever possible, avoiding the rather difficult theorem of Tarski.

Note that if G is an infinite group, there cannot be a left-invariant measure on $\mathcal{P}(G)$ that has total measure 1 and is countably additive. This is easily proved by the Vitali technique (Thm. 1.5), using instead the equivalence relation based on any countable subgroup of G. In fact, there cannot even be a σ-finite, left-invariant, countably additive measure on $\mathcal{P}(G)$ (see [EM76, AP80]).

The basic result on amenability is that all *elementary groups* are amenable, where the elementary groups are the smallest family containing all finite and Abelian groups and closed under subgroup, quotient group, group extension, and direct limits. Before proving this, we discuss a few simple consequences of the definition of amenability.

For a group G, let $B(G)$ denote the collection of bounded real-valued functions on G; $B(G)$ is a vector space under pointwise addition and scalar multiplication of functions. If μ is a measure on G, then the standard construction of an integral from a measure (see, e.g., [Roy68, Chap. 11]) can be applied to the system $(G, \mathcal{P}(G), \mu)$. This construction, which proceeds by defining the integral first on simple functions, then, via suprema, on all measurable functions, is somewhat simplified in this context because all sets, and hence all functions, are measurable. Thus, $\int f\, d\mu$ defines a linear functional on all of $B(G)$. On the other hand, because the measure is not necessarily countably additive, certain standard theorems, such as the monotone and dominated convergence theorems, cannot be proved. In summary, then, a real number, $\int f\, d\mu$, is assigned to each $f \in B(G)$, and this integral satisfies the following properties:

(a) $\int af + bg\, d\mu = a \int f\, d\mu + b \int g\, d\mu$ if $a, b \in \mathbb{R}$
(b) $\int f\, d\mu \geq 0$ if, for all $g \in G$, $f(g) \geq 0$
(c) $\int \chi_G\, d\mu = 1$
(d) for each $g \in G$, $f \in B(G)$, $\int {}_g f\, d\mu = \int f\, d\mu$, where $({}_g f)(h) = f(g^{-1}h)$; that is, the integral is left-invariant

It is easy to see that conditions (b) and (c) may be replaced by the single condition $\inf\{f(g) : g \in G\} \leq \int f\, d\mu \leq \sup\{f(g) : g \in G\}$. Because of this, a linear functional on $B(G)$ that satisfies (a)–(d) is called a *left-invariant mean* on G; therefore an amenable group always bears a left-invariant mean. Conversely, if $F : B(G) \to \mathbb{R}$ is a left-invariant mean, then defining $\mu(A) = F(\chi_A)$ yields a measure on G. Hence a group is amenable if and only if it bears a left-invariant mean.

While the existence of a measure is sufficient to prove some properties of amenable groups, in some proofs the corresponding mean is the more useful object. The following proposition is an example. Defining a right-invariant measure in the obvious way, one sees that a left-invariant measure, μ, on G yields a right-invariant measure, μ_0, defined by $\mu_0(A) = \mu(A^{-1})$. Note that the integral with respect to μ_0 is invariant under the right action of G on $B(G)$ given by $f_g(h) = f(hg^{-1})$. A (left-invariant) measure on an amenable group is not necessarily right-invariant (see [HR63, p. 230]), but at least some measures on an amenable group are two-sided invariant.

Proposition 12.2. *If G is amenable, then there is a measure ν on $\mathcal{P}(G)$ that is left-invariant and right-invariant.*

Proof. Let μ be a measure on G and μ_0 the corresponding right-invariant measure. If $A \subseteq G$, define f_A in $B(G)$ by $f_A(g) = \mu(Ag^{-1})$; f_A is bounded by 1. Then define ν on $\mathcal{P}(G)$ by $\nu(A) = \int f_A \, d\mu_0$. It is easy to check that $\nu(G) = 1$ and ν is finitely additive ($f_{A\cup B} = f_A + f_B$ if $A \cap B = \varnothing$). Moreover, $f_{gA} = f_A$ and $f_{Ag} = (f_A)_g$, so ν is both left- and right-invariant. □

A much more important application of means is the Invariant Extension Theorem (Thm. 12.8). The next result is a very simple special case. It shows why a set cannot be paradoxical if the group acting on it is amenable.

Theorem 12.3. *Suppose the amenable group G acts on X. Then there is a finitely additive, G-invariant measure on $\mathcal{P}(X)$ of total measure 1; hence X is not G-paradoxical.*

Proof. Choose any $x \in X$ and, if μ is a measure on G, define $\upsilon : \mathcal{P}(X) \to [0, 1]$ by $\nu(A) = \mu(\{g \in G : g(x) \in A\})$. It is easy to check that ν, which gives measure 1 to the orbit of x, is as required. □

Because of Tarski's Theorem, this result may be stated as; if G acts on X and G is not paradoxical, then X is not G-paradoxical. In fact, it is easy (see remarks after Prop. 1.10) to give a direct proof of this latter result, which is a strong converse to Proposition 1.10.

We now turn our attention to proving that all elementary groups—in particular, all Abelian and solvable groups—are amenable. Note, however, that Theorem 12.3, together with the amenability of the solvable groups G_1 and G_2, yields only that $\mathbb{R}^n$ is not G_n-paradoxical if $n = 1$ or 2. We are more interested in the fact that no interval or square is paradoxical, and this requires a measure that does not vanish on intervals or squares; the measures provided by Theorem 12.3 necessarily vanish on all bounded sets. The proof that a measure that agrees with Lebesgue measure exists (if $n \leq 2$) requires a stronger form of Theorem 12.3, the Invariant Extension Theorem (Thm. 12.8).

Theorem 12.4. (a) *Finite groups are amenable.*

(b) *Abelian groups are amenable.*

(c) *(AC) A subgroup of an amenable group is amenable.*

(d) *If N is a normal subgroup of the amenable group G, then G/N is amenable.*

(e) *If N is a normal subgroup of G, and each of N, G/N are amenable, then G is amenable.*

(f) *(AC) If G is the direct union of a directed system of amenable groups, $\{G_\alpha : \alpha \in I\}$, then G is amenable.*

Before proving this theorem, we note some consequences. By parts (b) and (e), every solvable group is amenable; in particular, O_1, G_1, and G_2 are amenable. Also, because any group is the direct union of its finitely generated subgroups, it follows from (c) and (f) that a group is amenable if and only if all of its finitely generated subgroups are. Thus amenability may be regarded as being primarily a property of finitely generated groups.

Proof of 12.4. (a) If $|G| = n < \infty$, then defining $\mu(A) = |A|/n$ yields the desired measure on G.

(b) This part is the most difficult and will be proved after the others.

(c) Let μ be a measure on G, and suppose H is a subgroup of G. Let M be a set of representatives (choice set) for the collection of right cosets of H in G. Then define ν on $\mathcal{P}(H)$ by $\nu(A) = \mu(\bigcup\{Ag : g \in M\})$. It is easy to check that ν is a measure on H.

(d) If μ is a measure on G, then define $\nu : \mathcal{P}(G/N) \to [0, 1]$ by setting $\nu(A) = \mu(\bigcup A)$. Again, it is routine to check that ν is as desired.

(e) Let ν_1, ν_2, be measures on N, G/N, respectively. For any $A \subset G$ let $f_A : G \to \mathbb{R}$ be defined by $f_A(g) = \nu_1(N \cap g^{-1}A)$. Then if g_1 and g_2 define the same coset of N in G, $f_A(g_1) = f_A(g_2)$. For if $g_2^{-1}g_1 = h \in N$, then

$$\begin{aligned} f_A(g_2) &= \nu_1(N \cap g_2^{-1}A) = \nu_1(N \cap hg_1^{-1}A) = \nu_1(h(N \cap g_1^{-1}A)) \\ &= \nu_1(N \cap g_1^{-1}A) = f_A(g_1). \end{aligned}$$

This means that f_A induces $\hat{f}_A$, a (bounded) real-valued function with domain G/N. Define $\mu(A)$ to be $\int \hat{f}_A \, d\nu_2$. Because $f_G = \chi_G$, $\mu(G) = 1$; and if $A, B \subseteq G$, $A \cap B = \varnothing$, then for any $g \in G$, $g^{-1}A \cap g^{-1}B = \varnothing$ whence $f_{A \cup B}(g) = f_A(g) + f_B(g)$, and so $\hat{f}_{A \cup B}(gN) = \hat{f}_A(gN) + \hat{f}_B(gN)$. This yields the finite additivity of μ. Finally, $f_{gA}(g_0) = \nu_1(N \cap g_0^{-1}gA) = f_A(g^{-1}g_0) = {}_g(f_A)(g_0)$, and so the left-invariance of the integral defined by ν_2 yields that $\mu(gA) = \mu(A)$.

(f) We are given that $G = \bigcup\{G_\alpha : \alpha \in I\}$ where each G_α is amenable (with measure μ_α), and for each $\alpha, \beta \in I$ there is some $\gamma \in I$ such that G_α and G_β are each subgroups of G_γ. Consider the compact topological space $[0, 1]^{\mathcal{P}(G)}$; this is where the Axiom of Choice, via Tychonoff's Theorem, is used. For each $\alpha \in I$, let $\mathcal{M}_\alpha$ consist of those finitely additive $\mu : \mathcal{P}(G) \to [0, 1]$ such that $\mu(G) = 1$ and $\mu(gA) = \mu(A)$ whenever $g \in G_\alpha$. Then each $\mathcal{M}_\alpha$ is nonempty, as can be seen by defining $\mu(A) = \mu_\alpha(A \cap G_\alpha)$. And, as in the proof of Theorem 11.1, one may check that each $\mathcal{M}_\alpha$ is a closed subset of $[0, 1]^{\mathcal{P}(G)}$. Because $\mathcal{M}_\alpha \cap \mathcal{M}_\beta \supseteq \mathcal{M}_\gamma$ if $G_\alpha, G_\beta \subseteq G_\gamma$, the collection $\{\mathcal{M}_\alpha : \alpha \in I\}$ has the finite intersection property. By compactness, then, there is some $\mu \in \bigcap\{\mathcal{M}_\alpha : \alpha \in I\}$, and such a μ witnesses the amenability of G.

(b) As pointed out before, any group is the direct union of its finitely generated subgroups; hence, by (f) it suffices to consider finitely generated Abelian groups. So suppose G is Abelian, with generating set $\{g_1, \ldots, g_m\}$. We claim that it suffices to show that for all $\epsilon > 0$, there is a function $\mu_\epsilon : \mathcal{P}(G) \to [0, 1]$ such that

1. $\mu_\epsilon(G) = 1$
2. μ_ϵ is finitely additive
3. μ_ϵ is almost invariant with respect to the generators in the sense that for each $A \subseteq G$ and generator g_k, $|\mu_\epsilon(A) - \mu_\epsilon(g_k A)| \leq \epsilon$

Once the existence of such a μ_ϵ is established, we may let $\mathcal{M}_\epsilon$ denote the set of functions from $\mathcal{P}(G)$ to $[0, 1]$ satisfying (1)–(3). Then each $\mathcal{M}_\epsilon$ is nonempty and closed, because if a function fails to lie in $\mathcal{M}_\epsilon$, then that failure is evident from finitely many values of the function (see Thm. 11.1's proof); more precisely, if $\mu(G) < 1$, then the inequality holds for an open set containing μ; and if $|\mu(A) - \mu(g_k A)| > \epsilon$ for some A, k, ϵ, then again the inequality is valid on an open set containing μ; and the same for failure of finite additivity. Furthermore, the collection of the sets $\mathcal{M}_\epsilon$ has the finite intersection property: $\bigcap \mathcal{M}_{\epsilon_i} = \mathcal{M}_{\min \epsilon_i}$, which is nonempty. Therefore, by compactness of $[0, 1]^{\mathcal{P}(G)}$, there is some μ lying in each $\mathcal{M}_\epsilon$. Such a μ is left-invariant with respect to each g_k and hence is left-invariant with respect to any member of G, as desired.

The idea behind the construction of a single μ_ϵ is really very simple. For example, consider the case where G has the single generator g_1. Choose N so large that $2/N \leq \epsilon$, and let $\mu_\epsilon(A) = |\{i : 1 \leq i \leq N \text{ and } g_1^i \in A\}|/N$. Then μ_ϵ differs from $\mu_\epsilon(gA)$ by no more than $2/N \leq \epsilon$. For the general case, choose N as before and let $\mu_\epsilon(A)$ be $|\{(i_1, \ldots, i_m). : 1 \leq i_1, \ldots, i_m \leq N \text{ and } g_1^{i_1} g_2^{i_2} \cdots g_m^{i_m} \in A\}.|/N^m$. Then $\mu_\epsilon(G) = 1$, μ is finitely additive, and because g_k commutes with the other generators, $\mu_\epsilon(g_k A)$ differs from $\mu_\epsilon(A)$ by no more than $|\{(i_1, \ldots, i_m) : 1 \leq i_1, \ldots, i_{k-1}, i_{k+1}, \ldots, i_m \leq N \text{ and } i_k = 1 \text{ or } N+1\}|/N^m = 2N^{m-1}/N^m = 2/N \leq \epsilon$, as desired. □

12.2 Classes of Groups

The concept of amenability in groups is central to the theory of paradoxical decompositions, because the key idea from the beginning has been to take a paradox in a group and lift it to a set on which the group acts. The class of amenable groups, AG, includes EG, the class of elementary groups, which we now define formally. The elementary groups are the smallest class of groups containing all finite groups and all Abelian groups and satisfying (a)–(d). Recall that $\{G_i : i \in I\}$ is a *directed system of subgroups* if for all $i, j \in I$, there is some $k \in I$ such that G_i and G_j are subgroups of G_k.

(a) If H is a subgroup of EG, then $H \in EG$.
(b) If H is a normal subgroup of $G \in EG$, then $G/H \in EG$.
(c) If H is a normal subgroup of G, and both H and G/H are in EG, then $G \in EG$.

(d) If $\{G_i : i \in I\}$ is a directed system and each $G_i \in EG$, then the union of the G_i is a group in EG. Because of (a), this is equivalent to the following: If all finitely generated subgroups of G are in EG, then $G \in EG$.

This class can be defined more explicitly using transfinite induction (see [Cho80]). Consider also the class NF, which consists of all groups without a free subgroup of rank 2. Now, Theorem 1.2 tells us that F_2 is a paradoxical group under left multiplication; hence F_2 has no invariant measure and is not amenable. By Theorem 12.4(c), this means that no group with F_2 as a subgroup can be amenable; hence $AG \subseteq NF$. Theorem 12.4 implies that every elementary group is amenable, so we have $EG \subseteq AG \subseteq NF$.

The question as to whether these inclusions are proper is difficult but has been resolved; they are both proper. The assertion that $AG = NF$ is sometimes known as the von Neumann Conjecture or the von Neumann–Day Problem. To rephrase, this is asking; Is it true that every paradoxical group contains F_2? or; Is it true that every nonelementary group is paradoxical? The first result in this area stems from the 1968 solution to the Burnside Problem: Is each Burnside group $B(m, n)$ finite? (The group $B(m, n)$ is $\langle x_1, x_2, \ldots, x_m : w^n = 1\rangle$, where w ranges over all words in the $x_i^{\pm 1}$.) Any Burnside group is a *periodic* group: Each element has finite order; clearly any periodic group is in NF.

Theorem 12.5. *There is a finitely presented amenable group that is not elementary. There is a nonamenable periodic group; such a group is in $NF \setminus AG$.*

The existence of a nonamenable group in NF was established in 1968 by Adian and Novikov (see [Adi79]). Also Adian resolved the Burnside Conjecture by showing that $B(2, 665)$ is infinite. Now, $B(2, 665)$ obviously has no free subgroup, and it is not hard to see that $B(2, 665) \notin EG$. This is a consequence of the fact [Cho80, Thm. 2.3] that there are no infinite groups that are elementary, finitely generated, and periodic. And more is true, as Adian [Adi83] has proved that $B(2, 665)$ is not amenable; indeed, this applies to $B(m, n)$ whenever $m \geq 2$ and $n \geq 665$. So this is a concrete example of a group in $NF \setminus AG$. And there is even an example, due to Lodha and Moore, that is finitely presented: It has three generators and nine relations (Thm. 14.38).

That AG is strictly larger than EG follows from Grigorchuk's counterexample to the Milnor–Wolf Conjecture (Thm. 14.28). That group is finitely generated but not finitely presented (a finitely presented group is one defined by a finite generating set together with finitely many relations). Grigorchuk later found [Gri96] a finitely presented example in $AG \setminus EG$; it is given as follows:

$$H = \langle a, b, c, d, t : a^2 = b^2 = c^2 = d^2 = bcd = (ad)^4 = (adacac)^4 = e,$$
$$t^{-1}at = aca, t^{-1}bt = d, t^{-1}ct = b, t^{-1}dt = c\rangle.$$

It should be noted that the three classes EG, AG, and NF share many properties; for instance, they all satisfy the closure properties (a)–(d) that define elementary groups.

Note that if a group is paradoxical in the same way that a non-Abelian free group is paradoxical—that is, there is a 4-piece paradox—then in fact the group contains such a free subgroup (Cor. 5.9).

Theorem 12.5 notwithstanding, there is a wide class of groups in which the nonamenable groups do coincide with the groups having a free subgroup of rank 2. A theorem of Tits [Tit72] yields that, when restricted to a certain class of groups, the classes *EG*, *AG*, and *NF* do coincide. Moreover, this class includes all groups of isometries of $\mathbb{R}^n$, and hence the result is applicable to the central examples of this book. If a group has a normal subgroup of finite index with a certain property, we shall say that the group *almost* has the property.

Theorem 12.6. *Let G be any group of $n \times n$ nonsingular matrices (under multiplication) with entries in a field K; that is, G is a subgroup of $GL_n(K)$.*

(a) *If K has characteristic 0, then either G has a free subgroup of rank 2 or G is almost solvable.*

(b) *If K has nonzero characteristic, then either G has a free subgroup of rank 2 or G has a normal solvable subgroup H such that G/H is locally finite (meaning that every finite subset generates a finite subgroup).*

The interested reader is referred to [Tit72] for a proof of this powerful result and some further applications. Further expositions can be found in [Dix73] and [Weh73], and an illuminating discussion appears in [Har83]. A refinement of part (a) of the theorem appears in [Wan81]. Because a locally finite group is the direct limit of its finite subgroups and hence is elementary, this result yields that a matrix group without a free subgroup of rank 2 is elementary and hence amenable. Note that the extra complication of G/H in part (b) is necessary. If K is infinite and $n \geq 2$, then $GL_n(K)$ is not almost solvable, and if, in addition, K is an algebraic extension of a finite field, then $GL_n(K)$ is locally finite and hence has no free subgroup of rank 2. Thus, for example, $GL_n(K)$, where K is the algebraic closure of a finite field, shows that the result of part (a) is not valid in the case of nonzero characteristic. Now, the Euclidean affine group A_n is isomorphic to a subgroup of $GL_{n+1}(\mathbb{R})$ (see App. A); it follows that any group of isometries of $\mathbb{R}^n$ either has a free subgroup of rank 2 or is almost solvable.

Tits's Theorem is an algebraic one. A topological approach was used by Balcerzyk and Mycielski [BM57] to show that any locally compact, connected topological group either is solvable or contains a free subgroup of rank the continuum. It follows that the three classes *EG*, *AG*, and *NF* coincide when restricted to this collection of topological groups. The work of Balcerzyk and Mycielski builds upon the earlier investigation into free subgroups of Lie groups by Kuranishi [Kur51].

12.3 Invariant Measures

For applications to equidecomposability theory, the most important question is whether a group is amenable. But there are many interesting questions concerning the types of measures on an amenable group. For instance, it is clear that the normalized counting measure is the only possible measure on a finite group, and it is now known ([Gra63]; see also [Gre69, App. 1]) that any infinite amenable group carries more than one measure. Indeed, if G is infinite and amenable, then there are as many measures on G as there are real-valued functions on $\mathcal{P}(G)$ ([Cho76]). The question of measure uniqueness for an action of a group on a set is discussed in §13.1. The unique measure on a finite group is also inverse-invariant ($\mu(A) = \mu(A^{-1})$), and it is easy to see that an amenable group always bears an inverse-invariant measure: Just use $(\mu(A) + \mu(A^{-1}))/2$, where μ is both left- and right-invariant (use Prop. 12.2). It may happen that all measures are necessarily inverse-invariant, however. For instance, if G is Abelian, then all measures are inverse-invariant if and only if $2G$ (which is $\{g + g : g \in G\}$) is finite [RW68].

To prove several results about the existence of invariant measures in as uniform a way as possible, it is most convenient to work in the context of Boolean algebras. There are other approaches, however, and we shall indicate after the proof of Theorem 12.11 how the important Corollaries 12.9 and 12.10 can be proved using linear functionals and the Hahn–Banach Theorem. But Boolean algebras are more natural for some of the other applications of amenability. We now summarize the basic facts about Boolean algebras that will be needed. The reader who is familiar with the Hahn–Banach Theorem and desires a short proof that analogs of the Banach–Tarski Paradox do not exist in $\mathbb{R}^1$ or $\mathbb{R}^2$ might skip ahead to Theorem 12.11, $(a) \Rightarrow (e)$. For a more detailed treatment of Boolean algebras, see [Sik69].

A *Boolean algebra* $\mathcal{A}$ is a nonempty set (also denoted by $\mathcal{A}$) together with three operations defined on elements of $\mathcal{A}$: $a \vee b$ (join), $a \wedge b$ (meet), and a' (complement). These operations, which generalize the set-theoretic operations of union, intersection, and complement, must satisfy the following axioms:

(a) $\vee$ and $\wedge$ are commutative and associative.
(b) $(a \wedge b) \vee b = (a \vee b) \wedge b = b$.
(c) $a \wedge (b \vee c) = (a \wedge b) \vee (a \wedge c)$, $a \vee (b \wedge c) = (a \vee b) \wedge (a \vee c)$.
(d) $(a \wedge a') \vee b = (a \vee a') \wedge b = b$.

We assume that Boolean algebras are nondegenerate: They have at least two elements. It follows that there are two distinguished elements, $\mathbf{0}$ and $\mathbf{1}$, defined by $\mathbf{0} = a \wedge a'$ and $\mathbf{1} = a \vee a'$ (the choice of a is immaterial). The *complement* of b in a, $a \wedge b'$, is denoted simply by $a - b$. If A is a finite subset of $\mathcal{A}$, $A = \{a_0, \ldots, a_n\}$, then $\sum A$ denotes $a_0 \vee a_1 \vee \ldots \vee a_n$. Two elements $a, b \in \mathcal{A}$ are called *disjoint* if $a \wedge b = \mathbf{0}$. There is a natural partial ordering in any Boolean algebra, defined by $a \leq b$ if $a \wedge b = a$; $\mathbf{1}$ is the greatest element and $\mathbf{0}$ the least with respect to this ordering.

A subset $\mathcal{A}_0$ of $\mathcal{A}$ is a *subalgebra* if $\mathcal{A}$ is closed under $\vee$, $\wedge$, and $'$. For any subset X of $\mathcal{A}$, there is a smallest subalgebra of $\mathcal{A}$ containing X. It is called the subalgebra of $\mathcal{A}$ generated by X. See [Sik69] for an explicit description of the subalgebra generated by a set; an important fact is that the subalgebra generated by a finite set is finite. An equally important concept is that of relativization—the restriction of a Boolean algebra to elements smaller than a fixed $b \in \mathcal{A}$. More precisely, let $\mathcal{A}_b = \{a \in \mathcal{A} : a \leq b\}$, with the same join and meet as in $\mathcal{A}$ and with a' in $\mathcal{A}_b$ defined to be $a' \wedge b$ (in $\mathcal{A}$). Then $\mathcal{A}_b$ is a Boolean algebra whose zero is the same as the zero of $\mathcal{A}$ but whose unit is b. Note that $\mathcal{A}_b$ is not a subalgebra of $\mathcal{A}$. An element $b \in \mathcal{A}$ is called an *atom* if there are no elements below b except for $\mathbf{0}$ and b, that is, if $\mathcal{A}_b = \{\mathbf{0}, b\}$, a two-element Boolean algebra. Any finite Boolean algebra is isomorphic to the algebra of all subsets of a finite set. It follows that any nonzero element b of a finite Boolean algebra satisfies $b = \sum\{a : a \leq b$ and a is an atom$\}$; in particular, there is at least one atom a such that $a \leq b$.

An *automorphism* of a Boolean algebra $\mathcal{A}$ is a bijection $g: \mathcal{A} \to \mathcal{A}$ that preserves the three Boolean operations. For instance, if a group G acts on a set X, then each $g \in G$ induces an automorphism of the Boolean algebra $\mathcal{P}(X)$ by $g(A) = \{g(x) : x \in A\}$. A *measure* on a Boolean algebra $\mathcal{A}$ is a function $\mu: \mathcal{A} \to [0, \infty]$ such that μ is finitely additive (meaning $\mu(a \vee b) = \mu(a) + \mu(b)$ if $a \wedge b = \mathbf{0}$) and $\mu(\mathbf{0}) = \mathbf{0}$. If G is a group of automorphisms of $\mathcal{A}$, then a measure μ on $\mathcal{A}$ is *G-invariant* if $\mu(g(b)) = \mu(b)$ for all $b \in \mathcal{A}$, $g \in G$.

Finally, we shall need the notion of a subring of a Boolean algebra. A nonempty subset $\mathcal{A}_0$ of a Boolean algebra $\mathcal{A}$ is a *subring* if $a_1 \vee a_2$ and $a_1 - a_2$ are in $\mathcal{A}_0$ whenever $a_1, a_2 \in \mathcal{A}_0$. For instance, the collection of bounded subsets of $\mathbb{R}^n$ is a subring of $\mathcal{P}(\mathbb{R}^n)$. Note that a subalgebra is always a subring. A measure on a subring $\mathcal{A}_0$ means a finitely additive function $\mu: \mathcal{A}_0 \to [0, \infty]$ with $\mu(\mathbf{0}) = 0$.

The next theorem is fundamental to the study of measures in Boolean algebras. For instance, letting $\mathcal{A}_0 = \{\mathbf{0}, \mathbf{1}\}$ and $\mu(\mathbf{0}) = 0$, $\mu(\mathbf{1}) = 1$, the theorem yields that every Boolean algebra admits a measure of total measure 1. Or, letting $\mathcal{A}_0 = \mathcal{L}$, the subalgebra of $\mathcal{P}(\mathbb{R})$ consisting of all Lebesgue measurable sets, and $\mu = \lambda$, one gets a finitely additive extension of Lebesgue measure to all sets. Such a measure will not necessarily have any invariance properties, but Theorem 12.8 will show how invariance can be guaranteed, provided the group in question is amenable.

Theorem 12.7 (Measure Extension Theorem) (AC). *Suppose $\mathcal{A}_0$ is a subring of the Boolean algebra $\mathcal{A}$, and μ is a measure on $\mathcal{A}_0$. Then there is a measure $\overline{\mu}$ on $\mathcal{A}$ that extends μ.*

Proof. We first prove the theorem under the additional assumption that $\mathcal{A}$ is finite, proceeding by induction on the number of atoms in $\mathcal{A}$. If $\mathcal{A}$ has one atom, then $\mathcal{A} = \{\mathbf{0}, \mathbf{1}\}$, and the assertion is trivial. In general, choose s to be a minimal (with respect to $\leq$) element of $\mathcal{A}_0 \setminus \{\mathbf{0}\}$ (if $\mathcal{A}_0$ consists only of $\mathbf{0}$, simply let $\overline{\mu}$ be identically 0). Let $c = s'$ and consider the relativized algebra $\mathcal{A}_c$ with associated subring $\mathcal{A}_0 \cap \mathcal{A}_c$ and measure $\mu \restriction (\mathcal{A}_0 \cap \mathcal{A}_c)$. Let a_0 be an atom of $\mathcal{A}$ such that $a_0 \leq s$. Then $a_0 \notin \mathcal{A}_c$, whence $\mathcal{A}_c$ has fewer atoms than $\mathcal{A}$ and the induction hypothesis can be used to obtain a measure ν on $\mathcal{A}_c$ that extends $\mu \restriction \mathcal{A}_0 \cap \mathcal{A}_c$. Now, define $\overline{\mu}$ on $\mathcal{A}$ by first defining $\overline{\mu}$ on all atoms of $\mathcal{A}$, and then extending to all of $\mathcal{A}$ by

$\overline{\mu}(b) = \sum\{\overline{\mu}(a) : a \leq b, a \text{ an atom}\}$; $\overline{\mu}$ will then automatically be finitely additive on $\mathcal{A}$. Let a be any atom of $\mathcal{A}$, and define $\overline{\mu}(a)$ by $\overline{\mu}(a) = \nu(a)$ if $a \leq c$, $\overline{\mu}(a) = \mu(s)$ if $a = a_0$, and $\overline{\mu}(a) = 0$ if $a \leq s$ but $a \neq a_0$. Note that $\overline{\mu}$ must agree with ν on all of $\mathcal{A}_c$; it remains to show that $\overline{\mu}$ agrees with μ on $\mathcal{A}_0$. Observe that by the minimality of s, if $d \in \mathcal{A}_0$, then either $d \wedge s = \mathbf{0}$ or $d \geq s$ (otherwise, $s - d \in \mathcal{A}_0 \setminus \{\mathbf{0}\}$). If the former, then $d \leq c$, so $\overline{\mu}(d) = \nu(d) = \mu(d)$. If, instead, $d \geq s$, then $d = s \vee (d - s)$ and $\overline{\mu}(d) = \mu(s) + \overline{\mu}(d - s) = \mu(s) + \nu(d - s) = \mu(s) + \mu(d - s) = \mu(d)$.

The result will be extended to infinite Boolean algebras by the usual compactness technique applied to the product space $[0, \infty]^{\mathcal{A}}$, viewed as consisting of functions from $\mathcal{A}$ to $[0, \infty]$. For each finite subalgebra, $\mathcal{C}$, of $\mathcal{A}$, let $\mathcal{M}(\mathcal{C}) = \{\nu \in [0, \infty]^{\mathcal{A}} : \mu \restriction \mathcal{C}$ is *a* measure extending $\mu \restriction (\mathcal{A}_0 \cap \mathcal{A}_c)\}$. Each $\mathcal{M}(\mathcal{C})$ is easily seen to be closed and, as shown, nonempty. Moreover, finitely many finite subalgebras of $\mathcal{A}$ generate a finite subalgebra; it follows that the family of all $\mathcal{M}(\mathcal{C})$ has the finite intersection property. By compactness, then, the intersection of all the $\mathcal{M}(\mathcal{C})$ over all finite subalgebras $\mathcal{C}$ is nonempty, and any element of this intersection satisfies the conclusion of the theorem. □

As shown in Proposition 11.4, the compactness of $[0, \infty]^{\mathcal{A}}$ does not require the Axiom of Choice if $\mathcal{A}$ is countable; hence neither does the Measure Extension Theorem in this case. Some form of choice is definitely required for the general Measure Extension Theorem just proved, however. This is because if the theorem is applied to $\mathcal{A} = \mathcal{P}(\mathbb{N})$, $\mathcal{A}_0 = \{A \in \mathcal{A} : A \text{ or } \mathbb{N} \setminus A \text{ is finite}\}$, and μ, the measure assigning measure 0 to finite sets and measure 1 to cofinite sets, one obtains a finitely additive measure on $\mathcal{P}(\mathbb{N})$ that has total measure 1. In fact, by using $\{0, 1\}^{\mathcal{A}}$ instead of $[0, \infty]^{\mathcal{A}}$ in the latter part of the proof, one obtains a $\{0, 1\}$-valued measure on $\mathcal{P}(\mathbb{N})$ of total measure 1 that vanishes on singletons. Such a measure (more precisely, the sets having measure 1 with respect to such a measure) is called a *nonprincipal ultrafilter* on $\mathbb{N}$. In Chapter 15 (Thms. 15.4 and 15.5) it will be indicated why the existence of such an ultrafilter, or even of a finitely additive measure on $\mathcal{P}(\mathbb{N})$ that vanishes on singletons, cannot be proved in ZF alone. It will also be shown in that chapter why parts (c), (f), and, with an additional hypothesis about inaccessible cardinals, (b) of Theorem 12.4 cannot be proved in ZF.

However, the Measure Extension Theorem is not strong enough to prove the full Axiom of Choice. This is because the proof of the Measure Extension Theorem used the fact that a product of compact Hausdorff spaces is compact, rather than the full Tychonoff Theorem. This Hausdorff version of the Tychonoff Theorem is equivalent to the Boolean Prime Ideal Theorem (every Boolean algebra has a $\{0, 1\}$-valued measure of total measure 1; see [Jec73]), and it is known that these assertions do not imply the Axiom of Choice. In fact, the Measure Extension Theorem is equivalent (in ZF) to the Hahn–Banach Theorem on extending linear functionals [Lux69], and it is known that these statements are strictly weaker than the Boolean Prime Ideal Theorem; see Figure 15.1.

Now, it is a slightly more complex situation that interests us. Namely, if G is a group of automorphisms of $\mathcal{A}$, $\mathcal{A}_0$ is a G-invariant subring ($a \in \mathcal{A}_0$ and $g \in G$ imply $g(a) \in \mathcal{A}_0$), and μ is a G-invariant measure on $\mathcal{A}_0$, can we be sure that there is a G-invariant extension of μ to all of $\mathcal{A}$? This can be done if G is amenable: First use the preceding theorem to obtain some extension ν, and then use a left-invariant mean on G to average out the noninvariance of ν.

Theorem 12.8 (Invariant Extension Theorem) (AC). *If, in the Measure Extension Theorem, G is an amenable group of automorphisms of $\mathcal{A}$, and $\mathcal{A}_0$ and μ are G-invariant, then $\overline{\mu}$ can be chosen to be G-invariant as well.*

Proof. Use the Measure Extension Theorem to get a measure ν on $\mathcal{A}$ extending μ. Let θ be a measure on the amenable group G. Now, if $b \in \mathcal{A}$, define $f_b : G \to \mathbb{R}$ by $f_b(g) = \nu(g^{-1}(b))$. Then define $\overline{\mu}$ by $\overline{\mu}(b) = \int f_b \, d\theta$, if $f_b \in B(G)$, that is, if f_b is bounded, and $\overline{\mu}(b) = \infty$ if f_b is unbounded (where a function that takes on the value ∞ is considered to be unbounded). It is easy to see that $\overline{\mu}$ is a G-invariant extension of μ, using the fact that for any $g \in G$, $f_{g(b)} = {}_g(f_b)$ and property (d) at the start of §12.1 □

In fact, it is easy to see that the amenable groups are the only ones for which the Invariant Extension Theorem holds. If the theorem holds for G, apply it to $\mathcal{A} = \mathcal{P}(G)$, $\mathcal{A}_0 = \{\varnothing, G\}$, $\mu(\varnothing) = 0$, and $\mu(G) = 1$, with G acting on $\mathcal{A}$ by left translation to obtain a measure on G.

The Invariant Extension Theorem can be applied to extend Lebesgue measure to an invariant, finitely additive measure defined on all subsets of $\mathbb{R}^1$ or $\mathbb{R}^2$. Hence the Banach–Tarski Paradox has no analog using isometries of the line or plane. Because of the Banach–Tarski Paradox (more precisely, by Thm. 2.6), such invariant extensions of Lebesgue measure do not exist in $\mathbb{R}^3$ and beyond; but if invariance with respect to an amenable group of isometries, rather than the full group, is desired, such extensions exist in all dimensions. Finitely additive, isometry-invariant extensions of Lebesgue measure to all sets are called *Banach measures*.

Corollary 12.9 (AC). *If G is an amenable group of isometries of $\mathbb{R}^n$ (resp., $\mathbb{S}^n$), then there is a finitely additive, G-invariant extension of Lebesgue measure λ to all subsets of $\mathbb{R}^n$ (resp., $\mathbb{S}^n$). In particular, Lebesgue measure on $\mathbb{S}^1$, $\mathbb{R}^1$, or $\mathbb{R}^2$ has an isometry-invariant, finitely additive extension to all sets.*

Proof. Apply the Invariant Extension Theorem to $\mathcal{A} = \mathcal{P}(\mathbb{R}^n)$, $\mathcal{A}_0 = \mathcal{L}$, the subalgebra of Lebesgue measurable sets, and $\mu = \lambda$. The second assertion follows from the fact that the corresponding isometry groups are solvable (see App. A). □

Recall (Cor. 1.6) that there is no countably additive, translation-invariant measure on all subsets of $\mathbb{R}^n$ that normalizes a cube. But the translation group is Abelian, so Corollary 12.9 shows that if only finite rather than countable additivity is desired, then such measures do exist.

Corollary 12.10 (AC). *If G is an amenable group of isometries of $\mathbb{R}^n$, then no bounded subset of $\mathbb{R}^n$ with nonempty interior is G-paradoxical. In particular, no bounded subset of $\mathbb{R}^1$ or $\mathbb{R}^2$ with nonempty interior is paradoxical.*

Proof. Use the preceding corollary to obtain μ, a finitely additive, G-invariant extension of λ to all subsets of $\mathbb{R}^n$. Then $0 < \mu(A) < \infty$ for any bounded set A with nonempty interior, so A is not G-paradoxical. □

Corollary 12.10 actually applies to a wider class of sets. For if A has positive inner Lebesgue measure and bounded outer Lebesgue measure, then if μ is as in the proof of Corollary 12.10, it is easy to see that $0 < \mu(A) < \infty$, whence A is not paradoxical. But some conditions on A are necessary because of the Sierpiński–Mazurkiewicz Paradox (Thm. 1.7), which provides an example of a paradoxical subset of $\mathbb{R}^2$. Recall that the set of that paradox is countable and unbounded. However, there are bounded, uncountable paradoxical subsets of the plane (§14.2).

The situation in $\mathbb{R}^1$ is quite different, because the isometry group is much simpler. In fact, in $\mathbb{R}^1$, Corollary 12.10 is valid without any conditions on the set, that is, no nonempty subset of the line is paradoxical. This will follow from Theorem 14.21 and Proposition 14.24, which show that this is the case whenever the group acting on the set is almost Abelian.

Although paradoxes using isometries are missing in the plane, recall that the Von Neumann Paradox (Thm. 8.5) shows that a square is paradoxical using area-preserving affine transformation. Of course, this enlargement of the isometry group is nonsolvable and nonamenable.

In the next chapter, we investigate the necessity of amenability in these two corollaries. It turns out that in the case of spheres, amenability is a necessary condition for the existence of a measure as in Corollary 12.10. But amenability is not necessary for the absence of paradoxes, that is, the existence of a finitely additive, G-invariant measure on $\mathcal{P}(\mathbb{R}^n)$ that normalizes the unit cube but does not necessarily agree with Lebesgue measure.

We have used the Axiom of Choice to prove that the analog of the Banach–Tarski Paradox fails in $\mathbb{R}^1$ and $\mathbb{R}^2$. Thus Choice sits on both sides of the fence—it is used to construct the paradoxes in the higher dimensions and to get rid of them in the lower dimensions. If one desires only the nonexistence of paradoxes rather than the stronger (in the absence of Choice) existence of total measures, one can get by without using the Axiom of Choice (see Cors. 14.25 and 15.9). Thus a more accurate portrayal of the axiom's role, at least in $\mathbb{R}^1$ and $\mathbb{R}^2$, is that it is used to destroy countably additive measures but to construct finitely additive measures. In the next chapter, we give some further applications of amenability to the construction of measures.

12.4 Characterizations of Amenability

We now present some more properties of amenable groups, including an alternate proof of Corollary 12.10 based on the use of linear functionals rather than Boolean algebras. The following result gives several characterizations of amenability, showing how natural this notion is.

Theorem 12.11 (AC). *For a group G, the following are equivalent:*

(a) *G is amenable.*
(b) *There is a left-invariant mean on G.*
(c) *G is not paradoxical.*
(d) *G satisfies the Invariant Extension Theorem: A G-invariant measure on a subring of a Boolean algebra may be extended to a G-invariant measure on the entire algebra.*
(e) *G satisfies the Hahn–Banach Extension Property: Suppose*
  i) *G is a group of linear operators on a real vector space V;*
  ii) *F is a G-invariant linear functional on V_0, a G-invariant subspace of V; and*
  iii) *$F(v) \leq p(v)$ for all $v \in V_0$, where p is some real-valued function on V such that $p(v_1 + v_2) \leq p(v_1) + p(v_2)$ for $v_1, v_2 \in V$, $p(\alpha v) = \alpha p(v)$ for $\alpha > 0$, $v \in V$, and $p(g(v)) \leq p(v)$ for $g \in G$, $v \in V$.*

  Then there is a G-invariant linear functional $\overline{F}$ on V that extends F and is dominated by p.
(f) *G satisfies Følner's Condition: For any finite subset W of G and every $\epsilon > 0$, there is a nonempty finite subset W^* of G such that for any $g \in W$, $|gW^* \triangle W^*|/|W^*| \leq \epsilon$.*
(g) *G satisfies Dixmier's Condition: If $f_1, \ldots, f_n \in B(G)$ and $g_1, \ldots, g_n \in G$, then for some $h \in G$, $\sum f_i(h) - f_i(g^{-1}(h)) \leq 0$.*
(h) *G satisfies the Markov–Kakutani Fixed Point Theorem: Let K be a compact convex subset of a locally convex linear topological space X, and suppose G acts on K in such a way that each transformation $g: K \to K$ is continuous and affine ($g(\alpha x + (1-\alpha)y) = \alpha g(x) + (1-\alpha)g(y)$ whenever $x, y \in K$ and $0 \leq \alpha \leq 1$). Then there is some x in K that is fixed by each $g \in G$.*

Proof. The equivalence of (a)–(d) follows from previous work. The equivalence of (a) and (b) was discussed at the beginning of this chapter; that of (a) and (c) follows from Tarski's Theorem (Cor. 9.2); and that of (a) and (d) follows from Theorem 12.8 and the remarks following its proof. We shall prove that *(a)* ⇔ *(e)* and *(a)* ⇒ *(h)* ⇒ *(b)* ⇒ (g) ⇒ *(b)*, yielding the equivalence of all statements but (f). That (f) implies (a) will be proved here, and the converse will be given a complete proof in §12.4.1.

(a) ⇒ *(e)*. Use the standard Hahn–Banach Theorem [Roy68, p. 187] to obtain a linear functional F_0 on V that extends F and is dominated by p. Then, for any $v \in V$, define $f_v : G \to \mathbb{R}$ by $f_v(h) = F_0(h^{-1}(v))$. Because $F_0(h^{-1}(v)) \leq p(h^{-1}(v)) \leq p(v)$, f_v is bounded by $p(v)$. Hence, choosing a measure μ on G, we may define $\overline{F}(v)$ to be $\int f_v \, d\mu$. Then $\overline{F}(v) \leq p(v)$ and $\overline{F}$ is a linear functional

on V. Moreover, $\overline{F}$ extends F, and because $f_{g(v)} = {}_g(f_v)$, the G-invariance of $\overline{F}$ follows from that of μ.

(e) ⇒ *(a)*. Let $V = B(G)$, with V_0 taken to be the subspace of constant functions. The action of G on $B(G)$ by $f \mapsto {}_gf$ is linear, and V_0 is G-invariant. Letting $F(\alpha\chi_G) = \alpha$ and $p(f) = \sup\{f(g) : g \in G\}$, we see that the hypotheses of (e) are satisfied. Hence there is a left-invariant linear functional $\overline{F}$ on $B(G)$ with $\overline{F}(\chi_G) = 1$. To prove that $\overline{F}$ is a left-invariant mean (see (a)–(d) at the beginning of §12.1), it remains to show that $\overline{F}(f) \geq 0$ if $f(g) \geq 0$ for each $g \in G$. But $p(-f) \leq 0$ for such f, whence $\overline{F}(-f) \leq p(-f) \leq 0$ and $\overline{F}(f) = -\overline{F}(-f) \geq 0$.

(b) ⇒ *(g)*. If F is a left-invariant mean on $B(G)$, then $F(\sum(f_i - {}_{g_i}f_i)) = 0$; by the inf-sup condition after the definition of invariant mean, this means that for some $h \in G$, $\sum(f_i(h) - ({}_{g_i}f_i)(h)) \leq 0$.

(g) ⇒ *(b)*. Let V_0 be the subspace of $B(G)$ generated by the constant functions and all functions of the form $f - {}_gf$, where $f \in B(G)$ and $g \in G$. An element of V_0 has the form $f - {}_gf + \alpha\chi_g$ for some real α that, by the hypothesis that each $f - {}_gf$ takes on a nonpositive value, is unique; hence we can define a linear functional on V_0 by $F(f - {}_gf + \alpha\chi_g) = \alpha$. We claim that $F(v) \leq \sup v$ for each $v \in V_0$. For if $v = f - {}_gf + \alpha\,\chi_g$, then because $-(f - {}_gf) = -f - {}_g(-f)$ takes on a nonpositive value, $v = \alpha\chi_g - (-(f - {}_gf))$ takes on a value no smaller than $\alpha = F(v)$. So, letting $p(v) = \sup v$ for $v \in B(G)$, we may apply the Hahn–Banach Theorem to obtain a linear functional $\overline{F}$ on $B(G)$ that extends F and is dominated by p. The definition of F, and the fact that and $f - {}_gf$ lies in V_0, guarantees that $\overline{F}$ is left-invariant and normalizes χ_G. And if $f(h) \geq 0$ for all $h \in G$, then $\overline{F}(f) = -\overline{F}(-f) \geq -p(-f) \geq 0$. Hence F is a left-invariant mean on $B(G)$.

(h) ⇒ *(b)*. Turn $B(G)$ into a normed linear space by using the sup norm, $\|f\| = \sup\{|f(g)| : g \in G\}$ and let X be the dual space of $B(G)$ (all bounded linear functionals on $B(G)$), equipped with the weak* topology. With this topology, X is a locally convex linear topological space. Let K be the subset of X consisting of all functionals F satisfying $\inf f \leq F(f) \leq \sup f$. Note that each $F \in K$ satisfies $|F(f)|/\|f\| \leq 1$; hence K is contained in the unit ball of X. Because this ball is compact (Banach–Alaoglu Theorem) and K is closed (for this, one shows that every net of elements of K, defined by reverse inclusion of neighborhoods, converges to some element in K), K is a compact subset of X. Moreover, K is convex. Now, each $g \in G$ acts on X by $({}_gF)(f) = F({}_{g^{-1}}f)$, and the transformation of X induced by G is linear (hence affine) and continuous. Moreover, because $\inf({}_gf) = \inf(f)$ and $\sup({}_gf) = \sup(f)$, each G maps K into K. So, by (h), there is some $F \in K$ that is fixed by each $g \in G$; such an F is a left-invariant mean on $B(G)$.

(a) ⇒ *(h)*. Let μ be a measure on G and choose any point $y \in K$. We will show how μ can be used to get an average value of the function $f : G \to K$ defined by $f(g) = g(y)$. This average will be the desired fixed point. Let D be the directed set consisting of all finite open covers $\pi = \{U_i\}$ of K, ordered by refinement (i.e., $\pi' > \pi$ iff every set in the net π' is contained in a set in π), where it is assumed that each $U_i \cap K$ is nonempty. If V is a neighborhood of the origin, then π will be called V*-fine* if each set in π can be translated to fit into V.

Define a net $\phi : D \to K$ as follows: For $\pi \in D$, choose points $s_i \in U_i \cap K$ and let $\phi(\pi) = \sum \mu(E_i)s_i$, where $E_i = f^{-1}(U_i) \setminus \bigcup\{E_j : j < i\}$. Then $\phi(\pi)$ is a convex combination of the s_i, and therefore $\phi(\pi) \in K$.

Claim 1. If V is a convex and symmetric ($V = -V$) neighborhood of the origin, π is $V/2$-fine, and π' refines π, then $\phi(\pi') - \phi(\pi) \in V$.

Proof. For simplicity assume $\pi = \{U_1, U_2\}$ with $s_i \in U_i$ and $\pi' = \{W_1, W_2, Z_1, Z_2\}$ with designated points s'_i and $W_i \subseteq U_1$ and $Z_i \subseteq U_2$. Then $\phi(\pi) = \alpha_1 s_1 + \alpha_2 s_2$ and $\phi(\pi') = \beta_1 s'_1 + \beta_2 s'_2 + \beta_3 s'_3 + \beta_4 s'_4$ where $\alpha_1 + \alpha_2 = \beta_1 + \beta_2 + \beta_3 + \beta_4 = 1$. But the coefficents are measures and the containments imply that $\beta_1 + \beta_2 \le \alpha_1$ and $\beta_3 + \beta_4 \le \alpha_2$, which yields $\beta_1 + \beta_2 = \alpha_1$ and $\beta_3 + \beta_4 = \alpha_2$. We use "$v$" for a generic element of $V/2$; it can stand for different elements of $V/2$. Fineness gives $s_i = t_i + v$; because of the containments, it gives also $s'_1 = t_1 + v$, $s'_2 = t_1 + v$, $s'_3 = t_2 + v$, and $s'_4 = t_2 + v$. Therefore $\beta_1 s'_1 + \beta_2 s'_2 = \alpha_1 t_1 + v$ and $\beta_3 s'_3 + \beta_4 s'_4 = \alpha_2 t_2 + v$. So $\phi(\pi') - \phi(\pi) = (\alpha_1 t_1 + \alpha_2 t_2 + v) - (\alpha_1 t_1 + \alpha_2 t_2 + v) \in V$.

The next claim gets the "average value" we seek.

Claim 2. There is a unique $x \in K$ such that the net ϕ converges to x.

Proof. The compactness of K yields that ϕ has at least an accumulation point x in K. To prove that $\phi \to x$ it suffices to show that if U is any convex symmetric neighborhood of the origin, then there is some $\pi \in D$ such that $\phi(\pi') \in x + U$ whenever π' refines π. To this end, let π be any $U/4$-fine cover in D (such exists by compactness) and, using the fact that ϕ accumulates at x to refine π if necessary, assume $\phi(\pi) \in x + U/2$. Now, if π' refines π, then $\phi(\pi') - \phi(\pi) \in U/2$ by claim 1. Because $\phi(\pi) - x \in U/2$, this yields that $\phi(\pi') - x \in U$, as required. Because X is a Hausdorff space, limits of nets are unique.

The proof that x from claim 2 is the desired fixed point requires the following assertion.

Claim 3. If ρ is a net on D defined in the same way as ϕ, but with different designated points, then ρ converges to x.

Proof. Because $\phi \to x$, for any convex symmetric neighborhood of the origin, U, there is some $\pi \in D$ such that, for all refinements π' of π, $\phi(\pi') \in x + U$. Now, given such a U, refine the corresponding π to so that it is $U/4$-fine; do this by choosing a finite subcover of K from the cover $\{(t + U/4) \cap U_i\}_{t \in K, U_i \in \pi}$. Let π' be any refinement of π. By claim 1, $\phi(\pi') \in x + U/2$. Suppose $\phi(\pi')$ is defined using s_i, while $\rho(\pi')$ uses r_i. Then there are points t_i such that $r_i, s_i \in t_i + U/4$. Because $U = -U$, it follows that $r_i - s_i \in U/2$, and because $U/2$ is convex, this implies that $\rho(\pi') - \phi(\pi') = \sum \alpha_i r_i - \sum \alpha_i s_i \in U/2$. Now, $\phi(\pi') - x \in U/2$ so $\rho(\pi') - x = (\rho(\pi') - \phi(\pi')) - (\phi(\pi') - x) \in U/2 - U/2 \subseteq U$. Hence $\rho(\pi') \in x + U$, proving $\rho \to x$.

Now, g is continuous, so the net $g\phi$ defined by $(g\phi)(\pi) = g(\phi(\pi))$ converges to $g(x)$. To complete the proof, we shall show that $g\phi \to x$; uniqueness of limits then yields $x = g(x)$. Note that each $g \in G$ induces an order-preserving map from D to D by $g(\pi) = \{\{z : g(z) \in U_i\} : U_i \in \pi\}$. Use this mapping to define a net ψ on D by $\psi(\pi) = \phi(g(\pi))$; it is easy to prove that if $\psi \to z$, then $\phi \to z$ too. We claim that the net $g\psi$ converges to x, the limit of ϕ. This is because $(g\psi)(\pi) = g(\psi(\pi)) = g(\phi(g(\pi))) = g(\sum \alpha_i r_i)$, where r_i is some point on the ith set of the cover $g(\pi)$ and α_i is the μ-measure of the appropriate subset of G. Because $g{:}X \to X$ is affine, $g(\sum \alpha_i r_i) = \sum \alpha_i g(r_i)$ and the left-invariance of μ yields that α_i is the measure of the subset of G arising from π. Because $g(r_i) \in U_i$, this means that the net $g\psi$ satisfies the condition of claim 3. Hence $g\psi \to x$. This yields that $\psi \to g^{-1}(x)$, which implies by the preceding remark that $\phi \to g^{-1}(x)$. Therefore $g\phi \to x$, as desired. (For a rather different proof of the Markov–Kakutani Theorem in the Abelian case, but without the assumption that X is locally convex, see [DS67, p. 456].)

Følner's Condition is a very interesting property that abstracts the essential fact about Abelian groups that makes them amenable. Loosely speaking, it states that for any finite subset of the group, there is another finite subset that is almost invariant with respect to translation on the left by elements of the first set. It is not hard to see that Følner's Condition yields amenability. The surprising fact is that all amenable groups satisfy Følner's Condition. This was discovered by Følner [Fol55]; we will present a complete proof in §12.4.1. See also [Nam64], or [Gre69], where a topological version of Følner's Condition (involving Haar measure) is proved equivalent to topological amenability for locally compact topological groups (see the remarks at the end of §12.5). Because Haar measure on a discrete group is just the counting measure, the equivalence for abstract groups is a consequence of the topological generalization. Another aspect of Følner's Condition is its connection to expander graphs and Property (T) (Thm. 13.11).

$(f) \Rightarrow (a)$. The proof of amenability from Følner's Condition is similar to the proof of Theorem 12.4(b). For each $\epsilon > 0$ and finite $W \subseteq G$ that is closed under inversion, let $\mathcal{M}_{W,\epsilon}$ consist of those finitely additive functions $\mu : \mathcal{P}(G) \to [0, 1]$ such that $\mu(G) = 1$, and for each $g \in W$ and $A \subseteq G$, $|\mu(A) - \mu(gA)| \leq \epsilon$. Then $\mathcal{M}_{W,\epsilon}$ is a closed subset of $[0, 1]^{\mathcal{P}(G)}$; for any $A \subseteq G$, $g \in G$, and $\epsilon > 0$, the condition $|\mu(A) - \mu(gA)| > \epsilon$ defines an open subset of the product space, and so the union over all A, g, and ϵ yields an open set. The collection of $\mathcal{M}_{W,\epsilon}$ has the finite intersection property, and so Tychonoff's Theorem gives an element in all the sets, which will complete the proof once it is shown that each $\mathcal{M}_{W,\epsilon}$ is nonempty. For this, simply define $\mu(A)$ to be $|A \cap W^*|/|W^*|$, where W^* is as provided by Følner's Condition. Then μ is finitely additive and $\mu(G) = 1$. For the ϵ-invariance, argue as follows:

Given A, we want $||A \cap W^*| - |(gA) \cap W^*|| \leq \epsilon |W^*|$. Consider first the case that $|A \cap W^*| \geq |(gA) \cap W^*|$. We have $|A \cap W^*| = |g(A \cap W^*)|$. Now,

$$\begin{aligned} g(A \cap W^*) &= (g(A \cap W^*) \cap W^*) \cup (g(A \cap W^*) \setminus W^*) \\ &\subseteq ((gA) \cap W^*) \cup (g(W^*) \setminus W^*) \\ &\subseteq ((gA) \cap W^*) \cup (g(W^*) \triangle W^*). \end{aligned}$$

This means $|A \cap W^*| \le |((gA) \cap W^*)| + |(g(W^*) \triangle W^*)| \le |((gA) \cap W^*)| + \epsilon |W^*|$, as desired.

Finally, suppose $|(gA) \cap W^*| \ge |A \cap W^*|$. Then $|(gA) \cap W^*| = |g^{-1}((gA) \cap W^*)|$ and

$$\begin{aligned} g^{-1}((gA) \cap W^*) &= (g^{-1}((gA) \cap W^*) \cap W^*) \cup (g^{-1}((gA) \cap W^*) \setminus W^*) \\ &\subseteq (g^{-1}(gA) \cap W^*) \cup ((g^{-1}W^*) \setminus W^*) \\ &\subseteq (A \cap W^*) \cup (g^{-1}W^* \triangle W^*), \end{aligned}$$

which means $|(gA) \cap W| \le |A \cap W^*| + |(g^{-1}W^* \triangle W^*)| \le |A \cap W^*| + \epsilon |W^*|$. □

Define D to consist of all finite sums $\Sigma f_i - ({}_{g_i}f_i)$, where $f_i \in B(G)$. Then 12.11 $(a) \Leftrightarrow (g)$ says that G is nonamenable iff there is $f \in D$ such that $\inf f > 0$. G. A. Willis proved in [Wil88] that more is true: G is amenable if and only if $D \ne B(G)$.

Several other properties of a group are equivalent to amenability. One such is due to Kesten [Kes59a,Kes59b]. He proved, using Følner's Condition, that amenability is equivalent to an assertion about recurrence for random walks in a group with respect to symmetric probability distributions on the group. See [Day64] for a different, shorter proof of Kesten's characterization. More recently, Cohen [Coh82] has obtained a very useful characterization of amenability in terms of a growth condition on the group; this characterization is discussed in §14.4.

Now, by considering the problem of extending Lebesgue measure to all sets as one of extending the Lebesgue integral to all functions, we can use the Hahn–Banach Extension Property for amenable groups to give an alternate proof of Corollaries 12.9 and 12.10. The only part of the previous theorem that is required is $(a) \Rightarrow (e)$.

Another proof of Corollary 12.9. Suppose G is an amenable group of isometries of $\mathbb{R}^n$. Let V_0 be the space of all Lebesgue integrable real-valued functions on $\mathbb{R}^n$, and V the space of all functions $f: \mathbb{R}^n \to \mathbb{R}$ such that for some $g \in V_0$ and all x, $-g(x) \le f(x) \le g(x)$. Then any group of isometries of $\mathbb{R}^n$ acts on V and V_0 in the obvious way $(f \mapsto {}_gf)$, and we may let F be the G-invariant linear functional on V_0 defined by the Lebesgue integral, that is, $F(g) = \int g\, d\lambda$. Finally, define a G-invariant sublinear function p on V that dominates F by $p(f) = \inf\{F(g) : g \in V_0$ and for all x, $g(x) \ge f(x)\}$; recall that each $f \in V$ is dominated by some $g \in V_0$.

Now, use the Hahn–Banach Extension Property to obtain a G-invariant linear functional $\overline{F}$ on V, and use $\overline{F}$ to define the desired measure, μ, on $\mathcal{P}(\mathbb{R}^n)$ by $\mu(A) = \overline{F}(\chi_A)$ if $\chi_A \in V$; $\mu(A) = \infty$ otherwise. It is easy to see that μ is finitely additive and G-invariant. And because $\overline{F}(\chi_A) = \int \chi_A\, d\lambda = \lambda(A)$ if A has finite Lebesgue measure, μ extends λ. Finally, note that $\mu(A) \ge 0$; for if $f(x) \ge 0$ for all $x \in \mathbb{R}^n$, then $-f(x) \le 0$, whence $\overline{F}(-f) \le p(-f) \le 0$ and $\overline{F}(f) = -\overline{F}(-f) \ge 0$. □

Much of the preceding discussion of amenability is valid in semigroups. The definition is the same, and many of the applications remain valid. For instance, the Hahn–Banach Extension Property and Følner's Condition are valid for precisely the amenable semigroups. See [Day57] for more on amenable semigroups.

12.4.1 Pseudogroups and Følner's Condition

In this section we present a result about equidecomposability that plays a role in an approach to Følner's Condition based on the theory of pseudogroups. In particular, we will present a proof of Ceccherini-Silberstein, Grigorchuk, and de la Harpe [CGH99] that G satisfies Følner's Condition iff G is amenable, thus completing the proof of Theorem 12.11.

We start with a result of Laczkovich [Lac01] that generalizes the work of Deuber, Simonovits, and Sós [DSS95]. Recall that §9.2.1 introduced the concept of a bounded bijection (its spread is finite) transforming a subset of $\mathbb{R}^2$ onto $\mathbb{Z}^2$. The idea easily generalizes to arbitrary metric spaces. A *wobbling bijection* is a bounded bijection from one subset of a metric space to another. We will say that two subsets A, B of a metric space (X, d) are *wobbling equivalent* if there is wobbling bijection $\Phi: A \to B$. And a set $A \subseteq X$ is *wobbling paradoxical* if A can be partitioned into A_1 and A_2 such that A_1 and A_2 are wobbling equivalent to A.

For $r > 0$, the *closed r-neighborhood* of $H \subseteq X$ is $U(H, r) = \{x \in X : \text{dist}(x, H) \leq r\}$, where $\text{dist}(x, H)$ is the distance of x from H (the infimum of all $d(x, h)$ for $h \in H$). The following characterization of Laczkovich [Lac01] is central.

Theorem 12.12 (AC). *A metric space X is* wobbling paradoxical *iff there is $r > 0$ such that for any finite $H \subseteq X$, $|U(H, r)| \geq 2|H|$.*

Deuber, Simonovits, and Sós proved this theorem for discrete, countable metric spaces. Their result was applied by Ceccherini-Silberstein, Grigorchuk, and de la Harpe to prove the Følner/amenable equivalence. Here we give the details of an extension of the result due to Laczkovich.

We look at wobbling equivalence in the context of group actions. Let G_w be the group of all wobbling bijections from X to itself; G_w acts on X. We can as usual consider G_w-equidecomposability and G_w-paradoxical sets. In fact, wobbling equivalence and G_w-equivalence are equivalent. Before proving that, we introduce the notion of a *bounded space* introduced by Laczkovich [Lac01]. For a nonempty set X, define the diagonal $\Delta = \{(x, x) : x \in X\}$, and if $A, B \subseteq X \times X$, define A^{-1} to be $\{(y, x) : (x, y) \in A\}$ (inversion) and $A \circ B$ to be $\{(x, y) : \exists z(x, z) \in A \text{ and } (z, y) \in B\}$ (composition). An *ideal* in a Boolean algebra is a nonempty subset of the algebra such that $a \vee b \in \mathcal{I}$ whenever $a, b \in \mathcal{I}$, and $a \leq b \in \mathcal{I}$ implies $a \in \mathcal{I}$; the main example here is when the algebra is the family of all subsets of a set. (Any Boolean algebra can be turned into a ring using $a \triangle b$ as addition and $a \wedge b$ as multiplication [Sik69, §17]; the definition of ideal just given corresponds to the usual notion in rings.)

Definition 12.13. *For any set X, the pair $(X, \mathcal{B})$ is a* bounded space *if $\mathcal{B}$ is an ideal of subsets of $X \times X$ containing Δ and closed under inversion and composition.*

The motivation is that $\mathcal{B}$ generalizes the family of subsets of $X \times X$ that are boundedly close to the diagonal, by which is meant sets so that each point (x, y) in the set has $d(x, y) < r$. For any metric space, there is an induced bounded space as follows

$$\mathcal{B} = \{A \subseteq X \times X : \exists r > 0 \text{ such that } \forall (x, y) \in A, d(x, y) < r\}.$$

Note that Δ always lies in $\mathcal{B}$. But there are bounded spaces that do not arise from a metric space in this way [Lac01].

We can interpret the ideas related to wobbling bijections in the language of bounded spaces. Let $(X, \mathcal{B})$ be a bounded space and let $A, B \subseteq X$. We say that f is an *$(X, \mathcal{B})$-wobbling bijection from A to B* if $f : A \to B$ is a bijection and the graph of f is in $\mathcal{B}$. And A and B are *$(X, \mathcal{B})$-wobbling equivalent* if there is an $(X, \mathcal{B})$-wobbling bijection from A to B. It is easy to see that the set $G_{\overline{w}}$ of all $(X, \mathcal{B})$-wobbling bijections from X to itself is a group. Therefore we can define equidecomposability and paradoxical sets with respect to the group $G_{\overline{w}}$. Moreover, if $H \subseteq X$ and $V \subseteq X \times X$, then we can define the *V-neighborhood of H* as $U(H, V) = \{x \in X : \exists y \in H \text{ such that } (x, y) \in V\}$; this is $\pi_1(\pi_2^{-1}(H) \cap V)$, where π_i is the standard projection.

Proposition 12.14. *The sets $A, B \subseteq X$ are wobbling equivalent if and only if $A \sim_{G_{\overline{w}}} B$.*

Proof. For the reverse direction, suppose that $A \sim_{G_{\overline{w}}} B$; so A can be partitioned into finitely many sets $\{A_i\}$ such that $\{f_i(A_i)\}$ partitions B, where the f_i are wobbling bijections. The union of the graphs of the f_i lies in $\mathcal{B}$ because $\mathcal{B}$ is an ideal, so the piecewise definition of $f : A \to B$ using f_i and A_i leads to the desired wobbling bijection.

For the forward direction, let $f : A \to B$ be a wobbling bijection and let V be the graph of f. Define $A_0 = A \setminus B$ and $A_n = f^n(A_0)$, where f^n is the nth iterate of f and $f^n(A_0) = \{f^n(x) : x \in A_0 \cap \text{dom}(f^n)\}$. One can easily check that the A_n are pairwise disjoint. Then define g as follows:

$$g(x) = \begin{cases} f^{-1}(x) & \text{if } x \in A_1, \\ f^{-2}(x) & \text{if } x \in A_n, \text{ where } n \geq 3 \text{ is odd}, \\ f^{-1}(x) & \text{if } x \in A_n, \text{ where } n \text{ is even and } x \in B \setminus A, \\ f^2(x) & \text{if } x \in A_n, \text{ where } n \text{ is even}, x \in A, \text{ and } f(x) \in A, \\ f(x) & \text{if } x \in A_n, \text{ where } n \text{ is even}, x \in A, \text{ and } f(x) \in B \setminus A. \end{cases}$$

Then g is a bijection from $A' = \bigcup_{n \geq 0} A_n$ to itself. Extend g to X by $g(x) = x$ for $x \in X \setminus A'$. Then $g \in G_{\overline{w}}$ because the graph of g is contained in $\Delta \cup V \cup V^{-1} \cup V \circ V \cup V^{-1} \circ V^{-1}$.

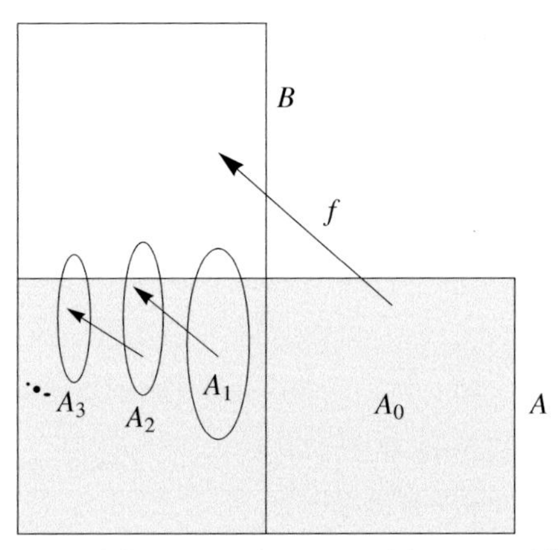

Figure 12.1. A schematic of the sets used to get equidecomposability from a wobbling bijection $f : A \to B$.

Now define $C = \bigcup_{n \text{ even}} A \cap A_n$ and $D = \bigcup_{n \text{ odd}} A \cap A_n$. Then $g(C) \cap D = \varnothing$. In fact, if $x \in C, x \in A_n$, and $f(x) \in A$, then $g(x) = f^2(x) \in A_{n+2}$, where n is even, and therefore $g(x) \notin D$. If $x \in C$ and $f(x) \notin A$, then $g(x) = f(x) \notin D$, because $D \subseteq A$. Observe now that $C \cup D \cup (A \setminus A')$ partitions A and $g(C), D \cup (A \setminus A')$ are disjoint subsets of B. Thus A is $G_{\overline{w}}$-equidecomposable to a subset of B. Similarly, B is $G_{\overline{w}}$-equidecomposable to a subset of A, and an application of the Banach–Schröder–Bernstein Theorem finishes the proof. □

For $V \subseteq X \times X$, define the cross section $V[x] = U(\{x\}, V) = \{y \in X : (y, x) \in V\}$. This is just $\pi_1 \pi_2^{-1}(\{x\})$. If $V \in \mathcal{B}$ in a bounded space, let $V^\infty = \{x \in X : V[x] \text{ is infinite}\}$. We need the following lemma; only the case involving 4 is needed, but it holds for any positive integer in place of 4.

Lemma 12.15 (AC). *Given a bounded space $(X, \mathcal{B})$, if $V \in \mathcal{B}$, then, in the type semigroup for $G_{\overline{w}}$, $4[V^\infty] \le [X]$.*

Proof. Because $V^\infty \subseteq (V \cup V^{-1} \cup \Delta)^\infty$, we may assume that $V = V^{-1}$ and $\Delta \subseteq V$. Define D to be a maximal subset of V^∞ such that the sets $V[y]$, for $y \in D$, are pairwise disjoint; $D \ne \varnothing$. Thus for any $x \in V^\infty$, there is a $y \in D$ such that $V[x] \cap V[y] \ne \varnothing$. This implies that $(x, y) \in V \circ V$ and $x \in (V \circ V)[y]$. Therefore $V^\infty \subseteq \bigcup_{y \in D} (V \circ V)[y]$. Put $C = \bigcup_{y \in D} (V \circ V)[y]$; then $[V^\infty] \le [C]$ (in the type semigroup). To finish, it is enough to show that $4[C] \le [X]$.

Define, for any $y \in D$, the set $H_y = (V \circ V)[y] \setminus \left(\bigcup_{z \in D} V[z]\right)$. Using a well-ordering of D, one can get pairwise disjoint K_y such that $K_y \subseteq H_y$ for any $y \in D$, and $\bigcup_{y \in D} K_y = \bigcup_{y \in D} H_y$. Put $L_y = K_y \cup V[y]$ for $y \in D$; then the sets L_y are pairwise disjoint and, because any $V[z]$ omitted from H_y is picked up by L_z, we have $C = \bigcup_{y \in D} L_y$. Because $V[y] \subseteq L_y$ and $y \in D \subseteq V^\infty$, every L_y is infinite. Therefore, for any $y \in D$, there is a decomposition of L_y into $L_y^1, L_y^2, L_y^3, L_y^4$ such that

$|L_y| = |L_y^i|$. Then define, for each $i \leq 4$, a bijection f_i from C onto $\bigcup_{y \in D} L_y^i$ such that $f_i(L_y) = L_y^i$ for any $y \in D$.

Because $L_y \subseteq (V \circ V)[y]$ (because $\Delta \subseteq V$), we get that $(x, f_i(x)) \in (V \circ V) \circ (V \circ V) \in \mathcal{B}$ for any $x \in C$; that is, the graph of f is a set in the ideal $\mathcal{B}$. This implies that C and $\bigcup_{y \in D} L_y^i$ are wobbling equivalent, proving $4\,[C] = [C] \leq [X]$. □

Lemma 12.15 yields the following very general theorem of Laczkovich, which has Theorem 12.12 as a corollary.

Theorem 12.16 (AC). *Let $(X, \mathcal{B})$ be a bounded space. Then X is $G_{\overline{w}}$-paradoxical iff there is $V \in \mathcal{B}$ such that for any finite $H \subset X$, $|U(H, V)| \geq 2|H|$.*

Proof. For the forward direction, if X is $G_{\overline{w}}$-paradoxical, then X can be partitioned into X_1, X_2 with $X_1 \sim_{G_{\overline{w}}} X \sim_{G_{\overline{w}}} X_2$. By Proposition 12.14, X_1, X_2, and X are pairwise $(X, \mathcal{B})$-wobbling equivalent. Let f_i be an $(X, \mathcal{B})$-wobbling bijection from X to X_i, and let V_i be the graph of f_i. Because $\mathcal{B}$ is an ideal, $V = V_1 \cup V_2 \in \mathcal{B}$. For any finite $H \subset X$, $f_1(H) \cup f_2(H) \subseteq U(H, V)$, which, because $f_1(X) \cap f_2(X) = \varnothing$, implies $2|H| = |f_1(H_1)| + |f_2(H_2)| \leq |U(H, V)|$.

Suppose now that $V \in \mathcal{B}$ and $|U(H, V)| \geq 2|H|$ for any finite $H \subseteq X$. Let $A = (V \circ V)^\infty$ and $B = X \setminus A = \{x \in X : (V \circ V)[x] \text{ is finite}\}$. Define the types $[A] = a, b = [B]$, and $c = [X]$; then $a + b = c$. Lemma 12.15 gives $4a \leq c = a + b$; the key is showing $4b \leq c$.

Claim. For any finite $H \subseteq X$, $U(H, V \circ V)$ has at least $4|H|$ elements.

Proof of claim. It is easy to see that $U(H, V \circ V) = U(U(H, V), V)$. If $U(H, V)$ is finite, then, by hypothesis, $|U(U(H, V), V)| \geq 2|U(H, V)| \geq 4|H|$. Otherwise, choose $H_0 \subset U(H, V)$ so that $|H_0| = 2|H|$ and apply the hypothesis to H_0 to get $|U(H_0, V)| \geq 2|H_0| = 4|H|$. This means $|U(U(H, V), V)| \geq |U(U(H_0, V), V)| \geq 4|H|$.

Let Γ be the bipartite graph with edges defined by $V \circ V$: The parts are B and X (where B here is a copy of B, to avoid the complication that $B \subseteq X$) and each $(b, x) \in V \circ V$ (where $b \in B$) determines an edge $b \leftrightarrow x$. Every vertex in B has finite degree. The claim means that H, an arbitrary finite subset of B, is adjacent in the graph to at least $4|H|$ vertices in X. An extension of the (infinite) marriage theorem (Thm. C.4) then yields four disjoint matchings, showing that B is $(X, \mathcal{B})$-wobbling equivalent to four disjoint subsets of X; that is, $4b \leq c$. Because $4a \leq c$, we have $2a + 2b \leq 4a + 4b \leq 2c = 2a + 2b$. Therefore $4c = 4a + 4b = 2a + 2b = 2c$, and by the Cancellation Law, $2c = c$; that is, X is $G_{\overline{w}}$-paradoxical. □

Proof of Theorem 12.12. Let $(X, \mathcal{B})$ be the bounded space induced by X. It is easy to see that the wobbling bijections of (X, d) and of $(X, \mathcal{B})$ coincide. Moreover, if $V \in \mathcal{B}$, then there is an $r > 0$ such that $U(H, V) \subseteq U(H, r)$, and also for any $r > 0$, there is a $V \in \mathcal{B}$ such that $U(H, r) \subseteq U(H, V)$. Now it is enough to apply Theorem 12.16. □

The preceding work will lead to a proof that a group satisfying Følner's Condition is amenable, completing the proof of Theorem 12.11. We will do this using pseudogroups, a useful tool that allows the study of partial isometries, thus leading to the generalization of many important results about group actions.

Definition 12.17. *We say that $\mathcal{G}$ is a* pseudogroup of transformations *of X (denoted $(\mathcal{G}, X)$) if $\mathcal{G}$ is a set of bijections $\gamma : S \to T$ between subsets of X that satisfies*

(a) *the identity $e : X \to X$ is in $\mathcal{G}$*
(b) *if $\gamma : S \to T$ is in $\mathcal{G}$, then $\gamma^{-1} : T \to S$ is in $\mathcal{G}$*
(c) *if $\gamma : S \to T$ and $\delta : T \to U$ are in $\mathcal{G}$, then so is $\delta\,\gamma : S \to U$*
(d) *if $\gamma : S \to T$ is in $\mathcal{G}$ and $S_1 \subseteq S$, then the restriction $\gamma \restriction S_1 : S_1 \to \gamma(S_1)$ is in $\mathcal{G}$*
(e) *if $\gamma : S \to T$ is a bijection of subsets of X and if there is a finite partition $\{S_i\}$ of S such that each $\gamma \restriction S_i$ is in $\mathcal{G}$, then $\gamma \in \mathcal{G}$*

Conditions (d) and (e) tell us that two operations that are common in the context of equidecomposability will work in any pseudogroup.

Any action of a group G on a nonempty set X generates a pseudogroup $\mathcal{G}_{G,X}$ by placing a bijection $\gamma : S \to T$ in $\mathcal{G}_{G,X}$ iff there is a finite partition $\{S_i\}$ of S and $g_i \in G$ such that γ is the piecewise bijection defined by S_i and g_i. In this case, S is G-equidecomposable to T. A special case of this construction is the left action of a group G on itself; this induces a pseudogroup denoted just $\mathcal{G}_G$.

Suppose X is a metric space. Then $\mathcal{G}_{P,X}$ denotes the pseudogroup of piecewise isometries of X. This pseudogroup can be larger than $\mathcal{G}_{G,X}$, where G is the isometry group, because there can be a piecewise isometry that does not extend to all of X [CGH99, ex (ii)].

The set of all partial wobbling bijections defined on a subset of a metric space X is a pseudogroup, denoted by $W(X)$. De la Harpe and Skandalis [HS86] investigated pseudogroups of transformations between the subsets in an algebra (or σ-algebra) of subsets of given space X. This led them to an alternative proof of Tarski's theorems (Cor. 11.2 and 11.3; see Thm. 12.23).

Let $\mathcal{G}$ be a pseudogroup of transformations of X. For $\gamma \in \mathcal{G}$, $\mathrm{dom}(\gamma)$ and $\mathrm{rng}(\gamma)$ refer to its domain and range, respectively. For $A \subseteq X$ and $R \subseteq \mathcal{G}$, we define the R-boundary of A as

$$\partial_R A = \{x \in X \setminus A : \exists \gamma \in R \cup R^{-1} \text{ and } x \in \text{ dom } (\gamma) \text{ such that } \gamma(x) \in A\}.$$

One can define Følner's Condition for a pseudogroup in a slightly different way than for groups; amenability is defined in the most natural way.

Definition 12.18. *A pseudogroup $(\mathcal{G}, X)$ satisfies* Følner's Condition *if for any finite $R \subseteq \mathcal{G}$ and $\epsilon > 0$, there is a nonempty finite $F \subseteq X$ such that $|\partial_R F| < \epsilon |F|$. And $(\mathcal{G}, X)$ is an* amenable pseudogroup *if there is a finitely additive $\mathcal{G}$-invariant measure on $\mathcal{P}(X)$ having total measure 1, where invariance means that $\mu(A) = B$ whenever $\gamma : A \to B$ is in $\mathcal{G}$.*

Lemma 12.19. *If G is an amenable group, then $\mathcal{G}_G$ is an amenable pseudogroup. And if $\mathcal{G}_G$ satisfies Følner's Condition, then so does G.*

Proof. The first is easy because the same measure works: For $X \subseteq G$, just let the new measure equal the given measure from the amenable group.

Now assume Følner's Condition for $\mathcal{G}_G$ and consider a finite $R \subseteq G$ and positive ϵ. Apply the pseudogroup condition to R and $\epsilon/2$ to get F so that $|\partial_R F| < (\epsilon/2)|F|$. The following claim suffices because $|\partial_R F| = |g(\partial_R F)|$.

Claim. For any $g \in R$, $gF \triangle F \subseteq \partial_R F \cup g(\partial_R F)$.

Proof of claim. Suppose $\sigma \in gF \setminus F$. Then $\sigma \in \partial_R F$ because $g^{-1}\sigma \in F$ and $g^{-1} \in R^{-1}$. Suppose $\sigma \in F \setminus gF$. Then $g^{-1}\sigma \in g^{-1}F \setminus F$, and the preceding argument yields $g^{-1}\sigma \in \partial_R F$, so $\sigma \in g(\partial_R F)$. This proves the claim, and the lemma. □

Pseudogroups are a convenient way to study equidecomposability. Ceccherini-Silberstein, Grigorchuk, and de la Harpe proved the following theorem, which, by the preceding lemma, yields the equivalence in groups (Thm. 12.22).

Theorem 12.20 (AC). *A pseudogroup is amenable if and only if it satisfies Følner's Condition.*

Proof of reverse direction. The reverse direction mimics Theorem 12.11 $(f) \Rightarrow (a)$. Given $(\mathcal{G}, X)$, for each finite $R \subseteq \mathcal{G}$ and $\epsilon > 0$, let $\mathcal{M}_{R,\epsilon}$ consist of those finitely additive $\mu : \mathcal{P}(X) \to [0, 1]$ such that $\mu(X) = 1$ and, for each $g \in R$ and $A \subseteq \text{dom}(g) \subseteq X$, $|\mu(A) - \mu(gA)| \leq \epsilon$. Then $\mathcal{M}_{R,\epsilon}$ is a closed (see proof of Thm. 12.4(b)) subset of $[0, 1]^{\mathcal{P}(G)}$ and compactness completes the proof once we know that each $\mathcal{M}_{R,\epsilon}$ is nonempty. For this, let $F \subseteq X$ be from Følner's Condition for R and define $\mu(A)$ to be $|A \cap F|/|F|$; then $\mu(X) = 1$ and μ is finitely additive. We need that $\mu \in \mathcal{M}_{R,\epsilon}$, which requires $|\mu(A) - \mu(gA)| \leq \epsilon$ whenever $A \subseteq \text{dom}(g)$. The proof of this is essentially the same as the proof of the same point in Theorem 12.11$(f) \Rightarrow (a)$ and is left as an exercise. □

For the proof of the difficult forward direction of this theorem, we need several ideas. Let $(\mathcal{G}, X)$ be a pseudogroup; then X is *$\mathcal{G}$-paradoxical* if there is a partition of X into A_1 and A_2 such that $\mathcal{G}$ contains bijections $\gamma_i : A_i \to X$. Skandalis and de La Harpe [HS86] proved that Tarski's Theorem (Cor. 11.2) extends to pseudogroups; we will prove that here using the methods of [CGH99].

Let (X, d) be a discrete metric space. By $\mathcal{C}(X)$ we mean the family of all maps of bounded spread from a subset of X to another subset (these maps are not necessarily bijective). Recall that $W(X)$ is the set of wobbling bijections between subsets of X and $U(A, r)$ is the closed r-neighborhood of A. We will say that $\Phi \in \mathcal{C}(X)$ satisfies the *Gromov condition* if, for any $x \in X$, $|\Phi^{-1}(x)| \geq 2$. And the condition appearing in the statement of Theorem 12.12 will be called the *doubling condition*.

Let Γ be a bipartite graph on parts A and B; a *perfect double matching* is an edge subgraph Γ_0 of Γ such that every vertex in A (resp., B) has degree 2 (resp., 1) in Γ_0. The *boundary* $\partial_\Gamma F$ of a set of vertices of Γ is the set of vertices not in F but connected by a Γ-edge to a vertex in F. For a metric space (X, d) and a real $r > 0$, we can construct a bipartite graph Γ_r where each part is a copy of X. The edge $x \bullet\!\!\rightarrow\!\bullet\, y$ is in Γ_r iff $d(x, y) < r$. Note that (X, d) is discrete iff for any positive r, Γ_r is locally finite.

Theorem 12.21 (AC). *If X is a discrete metric space, then the following are equivalent:*

(a) *X is $W(X)$-paradoxical.*
(b) *There is a mapping $\Phi: X \to X$ in $C(X)$ such that, for $x \in X$, $|\Phi^{-1}(x)| = 2$.*
(c) *There is a mapping $\Phi: X \to X$ in $C(X)$ that satisfies the Gromov Condition.*
(d) *X satisfies the doubling condition.*
(e) *$W(X)$ does not satisfy Følner's Condition.*

Proof. We first prove $(a) \Leftrightarrow (b)$. For the forward direction, use the two wobbling bijections that witness the paradox to define, in a piecewise manner, a function satisfying (b). For the opposite, use AC to select, for each x, a point $a_x \in \Phi^{-1}(x)$, and let b_x be the other point so that $\Phi(b_x) = x$. Then $\{a_x\}$ and $\{b_x\}$ together with the function Φ in each case give the paradox.

The implications $(b) \Rightarrow (c) \Rightarrow (d)$ are immediate. To finish the proof, we show $(a) \Rightarrow (e) \Rightarrow (d) \Rightarrow (a)$. The last implication is just Theorem 12.12. For $(a) \Rightarrow (e)$, if $W(X)$ satisfies Følner's Condition, then (Thm. 12.20) it is an amenable pseudogroup. But amenability gives a measure μ on $\mathcal{P}(X)$, and a paradox would yield $2 = 2\mu(X) = \mu(X) = 1$.

Finally, $(e) \Rightarrow (d)$. For Følner's Condition to fail for $W(X)$, there is $\epsilon > 0$ and a finite $R \subseteq W(X)$ so that, for any nonempty finite $F \subseteq X$, $|\partial_R F| \geq \epsilon |F|$, whence $|F \cup \partial_R F| \geq (1 + \epsilon)|F|$. Let K be large enough to bound the spread of any $\sigma \in R \cup R^{-1}$. Then, for any finite F, $|U(F, K)| \geq (1 + \epsilon)|F|$. Choose m so that $(1 + \epsilon)^m \geq 2$. Now $|U(F, 2K)| \geq |U(U(F, K), K)| \geq (1 + \epsilon)^2 |F|$, where the discreteness of the space is used to get finiteness of $U(F, K)$, which is used in this inequality. Repeating gives $|U(F, mK)| \geq (1 + \epsilon)^m |F| \geq 2|F|$, which is the doubling condition. □

Proof of forward direction of Theorem 12.20. Consider first the case of the pseudogroup $W(X)$, where X is a discrete metric space. The result then follows by Theorem 12.21 $(e) \Rightarrow (a)$. For if Følner's Condition fails, then (a) implies that there is no $W(X)$-invariant measure on $\mathcal{P}(X)$, so $W(X)$ is nonamenable. Now suppose $(\mathcal{G}, X)$ is a pseudogroup for which Følner's Condition fails, witnessed by R and ϵ. Define a discrete metric d_R on X by letting $d_R(x, y)$ be the smallest positive integer n such that there are n functions $\rho_i \in R \cup R^{-1}$ so that $y = \rho_n \rho_{n-1} \cdots \rho_1(x)$; if there are no such functions, set $d_R(x, y)$ to ∞. By concatenation of the ρ-strings, d_R satisfies the triangle inequality and so is a metric. Then, for any finite $F \subseteq X$ and using the neighborhood for this metric, we have

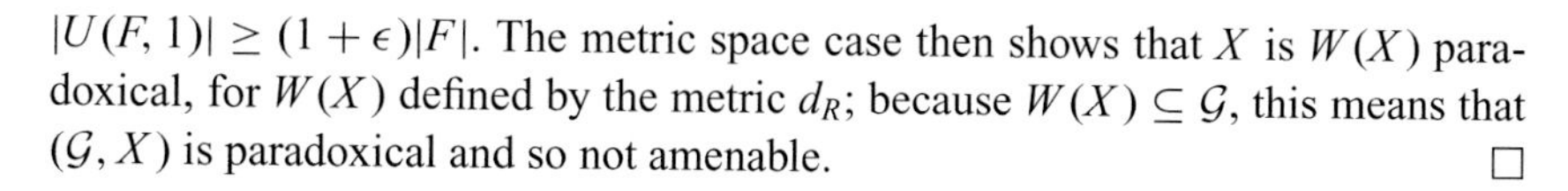

$|U(F, 1)| \geq (1 + \epsilon)|F|$. The metric space case then shows that X is $W(X)$ paradoxical, for $W(X)$ defined by the metric d_R; because $W(X) \subseteq \mathcal{G}$, this means that $(\mathcal{G}, X)$ is paradoxical and so not amenable. □

Theorem 12.22 (AC). *An amenable group satisfies Følner's Condition.*

Proof. By Theorem 12.20 and Lemma 12.19. □

Theorem 12.23 (AC). *For any pseudogroup, $(\mathcal{G}, X)$ is amenable iff $(\mathcal{G}, X)$ is not paradoxical.*

Proof. The forward direction is clear: A measure destroys any paradox. For the reverse, suppose $(\mathcal{G}, X)$ is not paradoxical. We will prove that $(\mathcal{G}, X)$ satisfies Følner's Condition, which, by Theorem 12.20, suffices. Suppose not, as witnessed by ϵ and R. Define the metric d_R as in the proof of Theorem 12.20, thus turning X into a discrete metric space with corresponding pseudogroup $W(X)$. Now, using the finiteness of R, the fact that $\mathcal{G}$ is not paradoxical implies that X is not $W(X)$-paradoxical. And this means that $W(X)$ satisfies Følner's Condition by Theorem 12.21.

Now, suppose $R = \{\rho_1, \rho_2, \ldots, \rho_k\}$. Each ρ_i can be viewed as an element of $W(X)$ with wobbling bound 1. So Følner's Condition gives $F \subseteq X$ so that $|\partial_R F| < \epsilon |F|$, where the boundary is formed in $W(X)$. To show that the same F witnesses Følner's Condition in $\mathcal{G}$, consider the inequality $|\partial_R F| < \epsilon |F|$ in $\mathcal{G}$. Because $\partial_R F$ in $\mathcal{G}$ is contained in $\partial_R F$ in $W(X)$, the inequality holds in $\mathcal{G}$. □

Using Lemma 12.19, the preceding theorem yields the same result for groups, and so the work in this section gives an independent proof of Tarski's Theorem: Corollary 11.2 in the case that $E = X$. In fact, these results hold in the case of measures normalizing a subset of X as well [CGH99, §7].

Pseudogroups are a flexible tool in studying invariant measures and paradoxical decompositions. Besides group actions, they allow the investigation of measures that are invariant with respect to partial bijections (such as partial isometries in metric spaces). It is possible that the proof of Theorem 12.20, with a necessary modification of Følner's Condition that replaces finite sets by Borel sets of X, works also in the case of pseudogroups of transformations between sets in a given algebra (or σ-algebra). If so, the concept might apply to Borel sets, and so pseudogroups might well be relevant to an investigation of the Banach-Ulam Problem (Question 3.13).

12.5 Topological Amenability

An important generalization of amenability is a topological one. In the summary that follows, all topological groups will be assumed to be locally compact and Hausdorff; some of the results are valid without the assumption, but the theory is much more coherent in this class of groups. A topological group is called *topologically amenable* if there is a finitely additive, left-invariant measure on the Borel subsets of the group that has total measure 1. In what follows, we use the word *amenable* in this topological sense. To say that G is amenable as a discrete group (i.e., with the discrete topology, which is locally compact) is the same as saying that G is amenable in the usual, nontopological sense: In the discrete case, all sets are Borel. Note that a topological group that is amenable as an abstract group is topologically amenable: Simply restrict the measure on $\mathcal{P}(G)$ to the Borel sets. If G is compact, then there is a countably additive, left-invariant Borel measure with total measure 1, namely, Haar measure; hence compact groups are topologically amenable.

The book by Greenleaf [Gre69] contains a detailed account of the theory of topological amenability, which is much richer than the theory of amenability in abstract groups and has many applications to diverse areas of analysis. Theorem 12.4 is valid after making appropriate topological modifications; for example, a closed subgroup of an amenable locally compact group is amenable. It is not surprising that G is topologically amenable if and only if there is a left-invariant mean on the bounded, Borel measurable functions from G to $\mathbb{R}$, but in fact it is sufficient to have a mean on the bounded, continuous real-valued functions on G. The following version of Følner's Condition is equivalent to amenability. If K is a compact subset of G and $\epsilon > 0$, then there is a Borel set $K^* \subseteq G$ with $0 < \theta(K^*) < \infty$ such that $\theta(g(K^*) \triangle K^*)/\theta(K^*) \leq \epsilon$ for all $g \in K$ (here θ denotes left-invariant Haar measure). Also, there is a topological version of the Markov–Kakutani Fixed Point Theorem that is equivalent to amenability in locally compact groups. One simply adds the condition that the function from $G \times K$ to K induced by the action is continuous, that is, the map $(g, x) \rightarrow g(x)$ is jointly continuous. Many more characterizations of topological amenability are given in [Gre69].

If a locally compact group G has a closed subgroup H that is a free group on two generators, then G is not amenable. This is because H is countable and Hausdorff, so all of its subsets are Borel; hence the standard paradox of a free group of rank 2 yields that H is not topologically amenable, and therefore, because H is closed, G is not amenable. Just as with abstract groups, this leads to the question whether a locally compact group is amenable if and only if it has no closed subgroup that is free of rank 2. This optimistic characterization is false, because by Theorem 12.5, it is false already in the discrete case. But yet, just as in the discrete case, it may be that this characterization is valid for large classes of interesting locally compact groups (just as Tits's Theorem yields the discrete version for a large class of groups). In fact, Rickert [Ric67] has shown that if the locally compact group G is almost connected (i.e., G/N is compact where N is the component of the identity), then G is amenable if and only if G does not have a closed subgroup that is free of rank 2.

Clearly a group cannot be amenable if it is paradoxical using Borel pieces. But the converse is not clear. Tarski's Theorem (Cor. 11.3) yields that a group is amenable if and only if, for each n, it is not possible to pack $n + 1$ copies of the group into n copies using Borel pieces (and left multiplication). But, because no cancellation law for Borel equidecomposability is known, it is not clear that the assertion referring to all n is equivalent to the one for $n = 1$, that is, the assertion that G is paradoxical. Nevertheless, A. Paterson [Pat86; see also Pat88, p. 123] has shown that a locally compact group that is not Borel paradoxical is amenable.

Theorem 12.24 (AC). *A locally compact group is amenable iff it is not paradoxical using Borel sets.*

For some recent work that uses a weakening of Følner's Condition to relate amenability to a condition connected to Ramsey theory, see [Moo13].

Notes

The material of this chapter has its origin in the seminal paper of Banach [Ban23]. In that paper Banach developed the main ideas of the Hahn–Banach Theorem and showed that the isometry groups of $\mathbb{R}^1$ and $\mathbb{R}^2$ satisfy the Hahn–Banach Extension Property of Theorem 12.11. Thus he proved Corollary 12.9 for the line and plane, using the technique of the proof given after the proof of Theorem 12.11. Von Neumann [Neu29] realized that Banach's results could be formulated more abstractly. He introduced the definition of an amenable group and proved Theorem 12.4(a), (b), and (e). Von Neumann was the first to state explicitly that a free group of rank 2 is not amenable, and the problem whether $AG = NF$ (§12.2) has often been attributed to him. Agnew and Morse [AM38] showed how von Neumann's ideas applied to the Hahn–Banach Theorem in its modern form. They showed that all solvable groups satisfied the Hahn–Banach Extension Property and obtained corresponding results about extensions of Lebesgue measure.

The class of amenable groups was studied extensively by Day [Day49, Day57]. He proved Proposition 12.2; parts (c), (d), and (f) of Theorem 12.4 (parts (c) and (d) were obtained independently by Følner [Fol55]); and many other results about amenable groups and semigroups.

The Measure Extension Theorem (Thm. 12.7) is due to Horn and Tarski [HT48]. See [Lux69, Pin72, Pin74, PS77] for various results related to the axiomatic strength of the Measure Extension Theorem and comparisons with the Hahn–Banach Theorem; foundational considerations are also discussed in Chapter 15. The idea of the Invariant Extension Theorem (Thm. 12.8) goes back to Banach and von Neumann; the formulation given here, using subrings, appears in Mycielski [Myc79].

Probability theory is mostly concerned with countably additive measures. But the restriction to finitely additive measures can be of value; see [BH15, Chap. 9], where such finitely additive measures are used in the context of Benford's Law.

Theorem 12.11 contains only a sample of the many properties of a group now known to be equivalent to amenability. The Hahn–Banach Extension Property was considered by Agnew and Morse [AM38], Klee [Kle54], and Silverman [Sil56a, Sil56b]. Silverman proved that the property implies amenability and investigated generalizations of the property to vector-valued linear functionals. Følner's Condition was introduced by Følner [Fol55], whose proof of the sufficiency of amenability was substantially simplified by Namioka [Nam64]. Dixmier's Condition was introduced in [Dix50]; see [Eme78] and [Gre69]. The fact that the Markov–Kakutani Fixed Point Theorem is valid for amenable groups was proved by Day [Day61], and this paper also contains the converse, which is due to Granirer. Chen [Che78] presents a weaker version of the Markov–Kakutani Fixed Point Property that holds for groups without a free non-Abelian subgroup.

There has been a large amount of work on amenability done in the Soviet Union. For a summary, see [Ver82].

Pseudogroups were first studied in the context of topology and differential geometry. Their use in the area of equidecomposability is by de la Harpe and Skandalis [HS86]. The reverse direction of the proof of Theorem 12.20 is due to Tomkowicz and Wagon.

A variety of results, applications, and references on the subject of topological amenability is contained in Greenleaf [Gre69] and Pier [Pie84]. The equivalence of topological amenability with a topological version of the Markov–Kakutani Fixed Point Theorem can be found in [Ric67] (see also [Day64]). Emerson [Eme79], unaware of Tarski's work, obtained the characterization of topological amenability discussed just before Theorem 12.24 and raised the question that that theorem answers.

13

Applications of Amenability

This chapter contains some refinements and further applications of the basic technique of constructing finitely additive measures that was introduced in Chapter 12. After constructing certain "exotic" measures in $\mathbb{R}^1$ and $\mathbb{R}^2$, we survey several results about an old problem about the uniqueness of Lebesgue measure that shows that such strange measures do not exist in $\mathbb{R}^3$ or beyond. Much of the emphasis here is on measures on the algebra of measurable sets rather than measures on all sets. Finally, we discuss the problem of characterizing groups of Euclidean isometries (and more general group actions) for which paradoxes exist.

13.1 Exotic Measures

Corollary 12.9 showed how to construct finitely additive, invariant extensions of Lebesgue measure. Such measures might seem unnatural because they mix the two types of additivity; λ is countably additive, but the extension is finitely additive. To prove that paradoxical decompositions do not exist, all that is required is a finitely additive, isometry-invariant measure that normalizes the unit cube. Of course, if we are considering all isometries, then we know (by Cor. 12.9 and the Banach–Tarski Paradox) that such measures exist in $\mathbb{R}^n$ if and only if $n \leq 2$.

As we shall see, it is worthwhile to study the properties of a finitely additive, isometry-invariant measure that normalizes the cube. Recall from Proposition 11.8 that such a measure must agree with Jordan measure, υ, on the Jordan measurable sets, $\mathcal{J}$. A famous problem, settled for all $\mathbb{R}^n$ in 1982 (Thm. 13.11), is whether such a measure must agree with Lebesgue measure. If $\mathcal{B}$ denotes the sets with the Property of Baire, then $\mathcal{J} \subseteq \mathcal{L} \cap \mathcal{B}$ (see App. B). If $A \in \mathcal{J}$, then $\upsilon(A) = 0$ if and only if A is nowhere dense (or meager) if and only if $\lambda(A) = 0$. So, Jordan measure is unbiased regarding measure and category: The Jordan null sets are precisely those sets in $\mathcal{J}$ that are small in the sense of both measure and category. Lebesgue measure is an extension of υ from $\mathcal{J}$ to $\mathcal{L}$ that is biased toward the measure side: λ gives positive measure to some nowhere dense sets (consider the complement in $[0, 1]$ of a small open cover of the rationals). We may also view

λ as a finitely additive, invariant Borel measure that vanishes on the (Borel) sets of Lebesgue measure zero.

The Marczewski Problem (§11.2) is about the existence of finitely additive, invariant Borel measures that are biased in the opposite direction, that is, they vanish on all meager sets (and hence necessarily disagree with λ on some closed, nowhere dense sets). Such measures may be viewed as measures on $\mathcal{B}$ (because if $A \in \mathcal{B}$, then $A = C \triangle E$, where C is Borel, E is meager, and $X \triangle Y = (X \setminus Y) \cup (Y \setminus X)$). The techniques of Chapter 12 can be used to produce Marczewski measures in $\mathbb{R}^1$ and $\mathbb{R}^2$, while the Dougherty–Foreman work shows that they do not exist in higher dimensions. In Corollary 13.3, we construct these measures in low dimensions and then examine the consequences of their existence for questions about Lebesgue measure's uniqueness.

The Marczewski construction and some of the other results to be presented in this chapter are most easily understood in the context of quotient Boolean algebras with respect to an ideal. Two prominent examples of ideals are the collection of meager sets, to be denoted by $\mathcal{M}$, and the collection of sets of Lebesgue measure zero (also called *null sets*), to be denoted by $\mathcal{N}$. In $\mathcal{P}(\mathbb{R}^n)$, both $\mathcal{M}$ and $\mathcal{N}$ are ideals, while $\mathcal{M}$ is an ideal in $\mathcal{B}$ (sets with the Property of Baire) and $\mathcal{N}$ is an ideal in $\mathcal{L}$. We shall also use the notion of ideal in a subring $\mathcal{C}$ of a Boolean algebra $\mathcal{A}$, defined by $a \vee b \in \mathcal{I}$ whenever $a, b \in \mathcal{I}$, and $a \leq b \in \mathcal{I}$ implies $a \in \mathcal{I}$ whenever $a, b \in \mathcal{C}$ (see §12.3 for a definition of *subring*). Note that if $\mathcal{I}$ is an ideal in $\mathcal{A}$ and $\mathcal{C}$ is a subring of $\mathcal{A}$, then $\mathcal{I} \cap \mathcal{C}$ is an ideal in $\mathcal{C}$. Whenever μ is a measure on a Boolean algebra, the collection of μ-measure-zero sets is an ideal.

If $\mathcal{I}$ is an ideal in $\mathcal{A}$, a quotient algebra can be defined using the equivalence relation $a_1 \sim a_2$ if $a_1 \triangle a_2 \in \mathcal{I}$. Let $\mathcal{A}/\mathcal{I}$ denote the collection of equivalence classes and define the Boolean operations in the obvious way using representatives, that is, $[a_1] \vee [a_2] = [a_1 \vee a_2]$, and so on. These operations are well-defined [Sik69, §10] and turn $\mathcal{A}/\mathcal{I}$ into a Boolean algebra. The same idea works for ideals in rings. The next theorem shows that in the Measure Extension Theorem and the Invariant Extension Theorem (§12.3), the property of "vanishing on an ideal" can be preserved.

Theorem 13.1 (AC). *Suppose $\mathcal{A}$ is a Boolean algebra, G is a group of automorphisms of $\mathcal{A}$, $\mathcal{I}$ is a G-invariant ideal in $\mathcal{A}$, $\mathcal{C}$ is a G-invariant subring of $\mathcal{A}$, and μ is a G-invariant measure on $\mathcal{C}$ that vanishes on $\mathcal{C} \cap \mathcal{I}$. Suppose further that G is amenable. Then there is a G-invariant extension of μ to a measure on $\mathcal{A}$ that vanishes on $\mathcal{I}$.*

Proof. Form the quotients $\mathcal{A}_{\mathcal{I}} = \mathcal{A}/\mathcal{I}$ and $\mathcal{C}_{\mathcal{I}} = \mathcal{C}/(\mathcal{C} \cap \mathcal{I})$. Because $\mathcal{I}$ is G-invariant (if $a \in \mathcal{I}$, then $g(a) \in \mathcal{I}$), G may be regarded as acting on these two quotients; moreover, $\mathcal{C}_{\mathcal{I}}$ induces a G-invariant subring $\hat{\mathcal{C}}_{\mathcal{I}}$ of $\mathcal{A}_{\mathcal{I}}$, where $\hat{\mathcal{C}}_{\mathcal{I}}$ consists of the classes containing an element of $\mathcal{C}$. The hypothesis on μ implies that μ induces a G-invariant measure on $\hat{\mathcal{C}}_{\mathcal{I}}$ by $\mu([c]) = \mu(c)$ (where $c \in \mathcal{C}$). The Measure Extension Theorem extends μ to ν on $\mathcal{A}_{\mathcal{I}}$, which gives the desired measure $\overline{\mu}$ on $\mathcal{A}$ by $\overline{\mu}(a) = \nu([a])$. □

Corollary 13.2 (AC). *If G is an amenable group of isometries of $\mathbb{R}^n$ (or $\mathbb{S}^n$), then there is a finitely additive, G-invariant measure on $\mathcal{P}(\mathbb{R}^n)$ (or $\mathcal{P}(\mathbb{S}^n)$) that normalizes the unit cube (or $\mathbb{S}^n$) and vanishes on all meager sets.*

Proof. Apply Theorem 13.1 with $\mathcal{A} = \mathcal{P}(\mathbb{R}^n)$, $\mathcal{C} = \mathcal{J}$, $\mu = \upsilon$, and $\mathcal{I} = \mathcal{M}$, the ideal of meager sets. It is shown in Appendix B that Jordan measure υ vanishes on all Jordan measurable, meager sets. □

Corollary 13.3 (AC). *Marczewski measures exist in $\mathbb{R}^1$, $\mathbb{R}^2$, and $\mathbb{S}^1$.*

Proof. The isometry groups in these cases are solvable, hence amenable. If μ is the total measure that exists by the previous corollary, then $\mu \restriction \mathcal{B}$ is a $\mathcal{B}$-measure that vanishes on $\mathcal{M}$, that is, a Marczewski measure. □

Because of the Banach–Tarski Paradox (really, the Hausdorff Paradox), measures as in Corollary 13.2 cannot exist if G is the group of all isometries of $\mathbb{R}^n$ or $\mathbb{S}^{n-1}$ ($n \geq 3$). Corollary 11.13 shows much more: In these dimensions, measures as in Corollary 13.2 cannot exist even on $\mathcal{B}$. Note that a Marczewski measure such as does exist by Corollary 13.3 cannot be countably additive (Thm. 11.24).

Corollary 13.3 was required to prove $(c) \Rightarrow (g)$ of Theorem 11.7, and so that proof is now complete. The existence of Marczewski measures has an interesting geometric consequence. Let A be an open dense subset of $[0, 1]$ with $\lambda(A) < 1$; just choose a small open covering of $\mathbb{Q} \cap [0, 1]$. Then $[0, 1] \setminus A$ is nowhere dense; therefore, if μ is an isometry-invariant measure on all sets, as in Corollary 13.2, $\mu([0, 1] \setminus A) = 0$ and $\mu(A) = 1$. It follows that A is not equidecomposable with any subset of a proper closed subinterval of $[0, 1]$, even using arbitrary pieces. (However, if countably many pieces are allowed, then A and $(0, \lambda(A))$ are Borel equidecomposable; see remark after Thm. 11.28.) For the analogous question in $\mathbb{R}^3$, the result is the opposite: Given any small cube J_0, the unit cube has an open subset A such that the complement of A in the cube is nowhere dense, but A is equidecomposable with a subset of J_0 (Thm. 11.15).

If, in the preceding two corollaries, one uses the ideal of nowhere dense sets rather than $\mathcal{M}$, one gets measures vanishing on the nowhere dense sets. A natural question is whether such a measure necessarily vanishes on the meager sets. The answer is NO. Here is a sketch of the proof for $\mathbb{S}^1$. Let A be a meager F_σ subset of $\mathbb{S}^1$ with $\lambda(A) = 1$ (see [Oxt71, p. 5]; A is obtained as the complement of the intersection of a sequence of successively smaller open covers of a countable dense set). Let $\mathcal{C}$ be the subalgebra of $\mathcal{P}(\mathbb{S}^1)$ generated by $\{\rho(A) : \rho \in O_2(\mathbb{R})\}$; then any element of $\mathcal{C}$ is a finite union of terms, each of which is a finite intersection of terms of the form $\rho(A)$ or $\mathbb{S}^1 \setminus \rho(A)$ (see [Sik69, p. 14]). It follows that $\mathcal{C}$ is contained in the algebra of Borel sets, and so λ may be viewed as a measure on $\mathcal{C}$. The representation of elements of $\mathcal{C}$ implies that λ is $\{0, 1\}$-valued on $\mathcal{C}$, and it follows that λ vanishes on any nowhere dense set in $\mathcal{C}$ (because otherwise there would be a nowhere dense set $A \subset S$ such that $\lambda(A) = 1$). Now, Theorem 13.1 may be applied to get an invariant extension of $\lambda \restriction \mathcal{C}$ to $\mathcal{P}(\mathbb{S}^1)$ that vanishes on all nowhere dense sets. This extension, however, gives A measure 1.

The standard measures (Jordan and Lebesgue) have another property that we can try to preserve in our extensions; namely, these measures react to similarities by multiplying the measure by the nth power of the magnifying factor, where n is the dimension. The following generalization of the Invariant Extension Theorem can be used to show how this property can be preserved.

Theorem 13.4 (AC). *Suppose G is an amenable group of automorphisms of a Boolean algebra $\mathcal{A}$, $\mathcal{A}_0$ is a G-invariant subring of $\mathcal{A}$, and μ is a measure on $\mathcal{A}_0$. Suppose further that $\pi : G \to (0, \infty)$ is a homomorphism into the multiplicative group of positive reals such that whenever $c \in \mathcal{A}_0$ and $g \in G$, $\mu(g(c)) = \pi(g)\mu(c)$. Then there is an extension $\overline{\mu}$ of μ to all of $\mathcal{A}$ that satisfies $\mu(g(a)) = \pi(g)\mu(a)$.*

Proof. The proof is identical to that of the Invariant Extension Theorem (Thm. 12.8), except that for $a \in \mathcal{A}$, $f_a : G \to \mathbb{R}$ is defined by $f_a(g) = \pi(g)\nu(g^{-1}(a))$. Then $\overline{\mu}$, which is defined by integrating f_a over the amenable group G, has the desired property, because $(f)_{g(a)} = \pi(g)\,{}_g(f_a)$. □

Now, let H_n be the group of similarities of $\mathbb{R}^n$; any element of H_n has a unique representation as $h\,\sigma$, where $\sigma \in G_n$ and h is a magnification through the origin: $h(\vec{v}) = \alpha\vec{v}$ for some positive real α. Let the homomorphism $\pi : H_n \to (0, \infty)$ be defined by $\pi(h\,\sigma) = \alpha^n$, where α is h's magnification factor.

Corollary 13.5 (AC). *If $n \leq 2$, then there is a finitely additive extension μ of λ to all of $\mathcal{P}(\mathbb{R}^n)$ such that $\mu(g(A)) = \alpha^n\,\mu(A)$ whenever g is a similarity with magnification factor α.*

Proof. The mapping $h\,\sigma \mapsto \sigma$ is a homomorphism from H_n onto G_n whose kernel is the subgroup consisting of pure magnifications. Because this subgroup is Abelian, H_n is solvable if G_n is. But G_1 and G_2 are solvable, whence H_1 and H_2 are amenable, and the corollary is now an immediate consequence of Theorem 13.4, using the homomorphism $\pi : H_n \to (0, \infty)$. □

Of course, this corollary is valid for larger n, provided H_n is replaced by some amenable group of similarities. Also, Theorems 13.4 and 13.1 can be combined so that one can get a Marczewski measure in $\mathbb{R}^1$ or $\mathbb{R}^2$ that behaves properly with respect to similarities. R. Mabry [Mab10] has investigated Banach measures that are isometry invariant but do not behave properly with respect to similarities. The Axiom of Choice can be eliminated from the next corollary in the same way that, in Corollary 15.11, it is eliminated from Corollary 10.10 when $n \leq 2$.

Corollary 13.6. *Suppose $A \subseteq \mathbb{R}^2$ is Lebesgue measurable, with $0 < \lambda(A) < \infty$. Suppose K is a square such that for any $\epsilon > 0$, A is H_2-equidecomposable with K using similarities (g) that are ϵ-magnifying (meaning: $1 - \epsilon \leq \pi(g) \leq 1 + \epsilon$). Then K has area $\lambda(A)$.*

Proof. Let μ be a measure on all subsets of the plane as in Corollary 13.5. Suppose $\lambda(K) \neq \lambda(A)$ and let $\epsilon = |1 - \sqrt{\lambda(K)/\lambda(A)}|/2$. Then, because A is H_2-equidecomposable with K using ϵ-magnifying similarities, $(1-\epsilon)^2\mu(A) \leq \mu(K) \leq (1+\epsilon)^2\mu(A)$, whence $1 - \epsilon \leq \sqrt{\lambda(K)/\lambda(A)} \leq 1 + \epsilon$, contradicting the choice of ϵ. □

This corollary means that Theorem 9.3, which showed how for any $\epsilon > 0$ the circle could be squared using ϵ-magnifying similarities, is not valid if the square's area is different than that of the circle.

Marczewski measures were a refinement of some measures produced by Banach to answer a question of Ruziewicz and Lebesgue regarding the uniqueness of Lebesgue measure as a finitely additive measure. We have already seen (Prop. 11.18) that λ is the only countably additive, translation-invariant measure on $\mathcal{L}$ normalizing the unit cube. As stated earlier, this is false if λ is viewed as a finitely additive measure: Simply let $\mu(A) = \lambda(A)$ if $A \in \mathcal{L}$ is bounded, and $\mu(A) = \infty$ if $A \in \mathcal{L}$ is unbounded. This led Ruziewicz to pose the problem of λ's uniqueness as a finitely additive, invariant measure on $\mathcal{L}_b$, the bounded Lebesgue measurable subsets of $\mathbb{R}^n$ (on $\mathbb{S}^n$, $\mathcal{L}_b$ is $\mathcal{L}$). Call a finitely additive measure μ on $\mathcal{L}_b$ *exotic* if μ is isometry-invariant and normalizes the unit cube (or, $\mathbb{S}^n$), but $\mu \neq \lambda$. Ruziewicz's Problem asks whether exotic measures exist in $\mathbb{R}^n$ or $\mathbb{S}^n$.

This problem is closely related to questions of amenability and paradoxical decompositions. First of all, note that Banach's solution is an easy consequence of the construction given in Corollary 13.2. If μ is a finitely additive, isometry-invariant measure on $\mathcal{P}(\mathbb{R}^1)$, $\mathcal{P}(\mathbb{R}^2)$, or $\mathcal{P}(\mathbb{S}^1)$ that normalizes the unit interval, square, or sphere, respectively, but vanishes on all meager sets (such measures exist by Cor. 13.2), then μ vanishes on a closed nowhere dense set of positive Lebesgue measure. Hence $\mu \restriction \mathcal{L}_b$ is an exotic measure. The solution of the Ruziewicz Problem in higher dimensions, which makes use of the Banach–Tarski Paradox, was completed only in 1980–1981 and is discussed later. First we examine Banach's solution for $\mathbb{S}^1$ more closely.

The fact that an exotic measure exists for $\mathbb{S}^1$ relies heavily on the amenability of $O_2(\mathbb{R})$. How general is this phenomenon? Suppose $(X, \mathcal{A}, m)$ is an arbitrary nonatomic measure space with $m(X) = 1$ (i.e., $\mathcal{A}$ is a σ-algebra, m is a countably additive probability measure on $\mathcal{A}$, and any set of positive measure splits into two sets, each having positive measure). Suppose further that a group G acts on X so that $\mathcal{A}$ and m are G-invariant (in short, G is measure-preserving). Then a finitely additive, G-invariant measure ν on $\mathcal{A}$ with $\nu(X) = 1$ and $\nu \neq m$ is called *exotic*. If, in addition, ν vanishes on all sets in $\mathcal{A}$ of m-measure zero, then ν is called an *absolutely continuous exotic measure*.

Now, Corollary 13.2 yields an exotic measure for $O_2(\mathbb{R})$'s action on $(\mathbb{S}^1, \mathcal{L}, \lambda)$. Indeed, one can get an absolutely continuous exotic measure by letting $\mathcal{I}$ be the $O_2(\mathbb{R})$-invariant ideal in $\mathcal{L}$ generated by $\mathcal{N}$ together with a single nowhere dense set of positive measure; because a finite union of nowhere dense sets is nowhere dense and hence has Lebesgue measure less than 1, $\mathcal{I}$ is a proper ideal. The desired measure arises by applying Theorem 13.1 with $\mathcal{A} = \mathcal{L}$, $m = \lambda$, $\mathcal{C} = \mathcal{I} \cup \{\mathbb{S}^1 \setminus A : A \in \mathcal{I}\}$, and μ the $\{0, 1\}$-valued measure on $\mathcal{C}$ that vanishes on $\mathcal{I}$. In its use of a nowhere dense set, these arguments use a bit of the topology of $\mathbb{S}^1$.

J. Rosenblatt wondered whether amenability of G was enough to always yield an exotic measure on any nonatomic measure space on which G acts. His problem can be stated both in the context of the Ruziewicz problem (i.e., in terms of invariant means on equivalence classes of functions) and in terms of pure measures. For the latter, there are connections to the axioms of set theory. Let $\mathbb{N}!$ be the group of permutations of $\mathbb{N}$. A group is *locally finite* if any finitely generated subgroup is finite; such a group is amenable. Then the situation for such groups is complicated; part (a) of the next result is due to S. Krasa [Kra88]; (b) is due to Z. Yang [Yan91]; and (c) and the Martin's Axiom assertion of (b), is due to M. Foreman [For89].

Theorem 13.7. (a) *If a solvable group G acts on the infinite set X, then there is an exotic measure on $\mathcal{P}(X)$.*

(b) *Assuming the Continuum Hypothesis (or the weaker Martin's Axiom), there is a locally finite subgroup G of $\mathbb{N}!$ such that there is only one finitely additive G-invariant measure on $\mathcal{P}(\mathbb{N})$ having total measure 1.*

(c) *It is consistent with ZFC that, for any locally finite subgroup G of $\mathbb{N}!$, there are at least two finitely additive G-invariant measures on $\mathcal{P}(\mathbb{N})$ having total measure 1.*

Part (b) shows that (a) cannot be extended from solvable to amenable groups, as it shows (under CH) that uniqueness can occur. The question remains: Can one prove in ZFC that there is an amenable group acting on a set for which there are no exotic measures? The nonuniqueness result is also known to hold for any countable amenable group, but that is better stated in the context of absolutely continuous measures (Thm. 13.8).

Now we turn to the exotic measure problem in Ruziewicz's original context, which was related to Lebesgue measure. Absolute continuity plays a very important role. This is best understood in the context of functional analysis rather than measure theory. A G-invariant measure m on an algebra $\mathcal{A} \subseteq \mathcal{P}(X)$ induces a G-invariant linear functional on $L^\infty(X)$ (the m-equivalence classes of bounded, measurable, real-valued functions on X) that assigns nonnegative values to nonnegative (almost everywhere) functions and normalizes the constant function with value 1, namely, the integral with respect to m. Such a functional is called an *invariant mean* on $L^\infty(X)$. The integral with respect to an exotic measure on $\mathcal{A}$ will be a new invariant mean on $L^\infty(X)$ only if the exotic measure is absolutely continuous with respect to m, for otherwise the new integral will not be well defined on the m-equivalence classes of functions. Thus the existence of an absolutely continuous exotic measure on $\mathcal{A}$ is equivalent to the nonuniqueness of the m-integral as an invariant mean. Most of the results in this area have been motivated by the uniqueness question for invariant means, and the proofs use techniques of functional analysis applied to $L^\infty(X)$.

In the case of countable groups, Rosenblatt's Problem has been solved. Moreover, for arbitrary groups, it is known that the amenability condition in Rosenblatt's Problem is necessary. The proof of the countable case uses the notion of an *asymptotically invariant sequence*, by which is meant a sequence of sets $(A_n)_{n=0}^\infty$ such that $m(A_n) > 0$, and for every $g \in G$, $\lim_{n\to\infty} m(A_n \triangle gA_n)/m(A_n) = 0$. For an example of such a sequence in a slightly different context, see the proof of Theorem 13.16. This notion, with *net* replacing *sequence*, also plays a role in the omitted details of the proof of (a) in the next theorem. The notion of an asymptotically invariant sequence, which plays a central role in the next theorem and also in the solution to Ruziewicz's Problem, to be discussed, is intimately related to Følner's Condition (§12.4). Indeed, another way of phrasing Følner's Condition for a countable group G is; There is a sequence (A_n) of nonempty finite subsets of G that is asymptotically invariant with respect to the counting measure on $\mathcal{P}(G)$.

Theorem 13.8 (AC). *Let G be a group.*

(a) *If every nonatomic G-invariant measure space $(X, \mathcal{A}, m)$ admits an absolutely continuous exotic measure on $\mathcal{A}$, then G is amenable.*
(b) *If G is amenable and countably infinite and $(X, \mathcal{A}, m)$ is nonatomic and G-invariant, then there is an absolutely continuous exotic measure on $\mathcal{A}$; that is, the m-integral is not the unique G-invariant mean on $L^\infty(X)$.*

Proof. (a) (Sketch; see [LR81, Ros81] for more details.) Suppose G is not amenable. Let $X = \mathbb{Z}_2^G$, a compact topological group with Haar measure m on its Borel sets. Consider the natural action of G on X given by $g((s_h)_{h\in G}) = (s_{gh})_{h\in G}$; m is invariant under this action. But it can be shown {LR81, Ex. (d)] that the m-integral is the unique G-invariant mean on $L^\infty(X)$, contradicting the hypothesis for the measure space (X, Borel sets, m).

(b) Del Junco and Rosenblatt [JR79, Thm. 2.4] proved that if T is a finite subset of an amenable group H and $\epsilon > 0$, then there is a set B such that $0 < m(B) < \epsilon$ and, for $g \in T$, $m(B \triangle gB)/m(B) < \epsilon$. Now, enumerate the given G as $\{g_i : i = 0, 1, \ldots\}$ and use the preceding result to get a sequence $\{A_k\}$ so that $m(A_k) < 2^{-(k+1)}$ and A_k is $(1/k)$-invariant for g_i with $i \le k$. Then (A_k) is an asymptotically invariant sequence for G, and it yields a new invariant mean as follows. Define m_n by $m_n(A) = m(A \cap A_n)/m(A_n)$. Let U be a nonprincipal ultrafilter on $\mathbb{N}$ (see remarks after Thm. 12.7) and define $\nu(A) = \int f_A \, dU$, where $f_A(n) = m_n(A)$. Because $\lim_{n\to\infty} f_A(n) - f_{gA}(n) = 0$ (to see this, use $m(A \cap A_n) = m(g(A \cap A_n)) = m(gA \cap gA_n)$ to relate the given limit to the asymptotic invariance of (A_n)) and finite sets have U-measure zero, ν is G-invariant. Clearly ν is absolutely continuous with respect to m, and ν is exotic because $m(\bigcup A_n) < 1/2 + 1/4 + \cdots = 1 = \nu(\bigcup A_n)$. □

For details of the preceding proofs, and various related results on asymptotically invariant sequences and the problem of uniqueness for invariant means, see [CFW81, CW80, Dan85, JR79, LR81, Mar80, Mar82, Ros81, Sch80, Sch81, Sul81, Bek98, Oh05].

The existence of absolutely continuous exotic measures can be given another interpretation. Let us, for the moment, call a measure-preserving action of G on $(X, \mathcal{A}, m)$ *superergodic* if, for any set A of positive measure, there are $g_1, \ldots, g_n \in G$ such that $m(\bigcup g_i A) = 1$; a superergodic action on a nonatomic space is necessarily ergodic (easy exercise). Now, asking for an absolutely continuous exotic measure is equivalent to asking whether any measure-preserving action of an amenable group on a finite, nonatomic measure space fails to be superergodic. For if ν is exotic and absolutely continuous, then as shown by Rosenblatt [Ros81, Prop. 1.1], there is another absolutely continuous exotic measure ν' such that ν' vanishes on a set A of positive m-measure; this set witnesses the nonsuperergodicity of the action because $m(\bigcup g_i A) = 1$ implies $\nu'(\bigcup g_i A) = 1$, contradicting $\nu'(A) = 0$. Conversely, if A is such that $m(A) > 0$ and $m(g_1 A \cup \cdots \cup g_n A) < 1$ for all finite subsets $\{g_i\}$ of G, then $\mathcal{I}$, the collection of sets contained in a set of the form $E \cup (\bigcup g_i A)$, $m(E) = 0$, and $\{g_i\}$ is a finite subset of G, is a proper G-invariant ideal in $\mathcal{A}$. Then Theorem 13.1, with $\mathcal{C} = \mathcal{I} \cup \{X \setminus B : B \in \mathcal{I}\}$ and μ equal to the $\{0, 1\}$-valued measure on $\mathcal{C}$ that vanishes on $\mathcal{I}$, yields an absolutely continuous finitely additive measure on $\mathcal{A}$ that vanishes on $\mathcal{I}$ and hence is exotic. In short, a measure-preserving action of an amenable group G on $(X, \mathcal{A}, m)$ is superergodic if and only if the m-integral is the unique invariant mean on $L^\infty(X)$.

For approximately fifty years, no progress was made on Ruziewicz's Problem in higher dimensions. All that was known was Tarski's observation (Lemma 11.9) that, because of the Banach–Tarski Paradox, any exotic measure necessarily vanishes on the bounded sets of Lebesgue measure zero, that is, is absolutely continuous with respect to Lebesgue measure on $\mathcal{L}_b$. This means that to prove the nonexistence of an exotic measure, it suffices to prove that the Lebesgue integral is the unique invariant mean on $L^\infty(\mathbb{R}^n)$ or $L^\infty(\mathbb{S}^{n-1})$, $n \geq 3$. Then, in 1980–1981, a solution for $\mathbb{S}^n$, $n \geq 4$, was obtained simultaneously in Russia (by G. Margulis) and in the United States (by D. Sullivan and J. Rosenblatt). The solution leans heavily on a property of groups diametrically opposed to amenability called Property (T), first introduced by Kazhdan [Kaz67]. We state the definition for countable groups, although the property can be defined more generally for locally compact topological groups.

Definition 13.9. *A countable group G has* Property (T) *if, whenever a unitary representation π of G on a complex Hilbert space H admits an asymptotically invariant sequence of nonzero vectors (i.e., a sequence of nonzero $v_n \in H$ such that for all $g \in G$, $\lim_{n\to\infty} \|v_n - \pi(g)v_n\| / \|v_n\| = 0$), then there is a nonzero G-invariant vector (i.e., a nonzero $v \in H$ such that for all $g \in G$, $\pi(g)v = v$).*

There is a connection between groups having Property (T) and expander graphs, which play a role in a variety of fields.

Definition 13.10. *Let $\Gamma = (V, E)$ be a finite k-regular graph (undirected, no multiple edges, loops allowed) with n vertices. For $Y \subseteq V$, we define the boundary ∂Y as the set of vertices outside Y and connected by an edge to a vertex in Y. Then Γ is an ϵ-expander if, for any $Y \subseteq V$ with $|Y| \leq |V|/2$, we have $|\partial Y| \geq \epsilon |\partial Y|$.*

For any ϵ-expander Γ, we can define the isoperimetric constant $h(\Gamma)$ as the minimum of $|\partial Y|/|Y|$ over all vertex sets Y with $|Y| \leq |V|/2$. A sequence of graphs (Γ_m) where $\Gamma_m = (V_m, E_m)$ is k-regular and $|V_m|$ approaches infinity is called an *expander family* if there is an $\eta > 0$ such that $h(\Gamma_m) > \eta$ for all Γ_m in the family. The existence of an expander family is closely related to finitely generated groups with Property (T). This is the content of a famous theorem of Margulis [BHV08, Ex. 6.1.12]. Recall that the (right) Cayley graph $C(G, S)$ of a finitely generated group G with generating set S is the graph whose vertices are the elements of G and with an edge (possibly a loop) $v_1 \bullet\!\!-\!\!\bullet\ v_2$ whenever $v_2 = v_1 s$ for some $s \in S^{\pm 1}$.

Theorem 13.11. *Let G be a finitely generated group with Property (T), and let $\mathcal{F}$ be an infinite family of normal subgroups of G having finite index. Then the Cayley graphs $C(G/N, S)$, where $N \in \mathcal{F}$, form an expander family.*

There is also a link between expander families and nonamenable pseudogroups. Negating Følner's Condition for pseudogroups (see Defn. 12.18) and applying Theorem 12.20, we obtain that a pseudogroup $\mathcal{G}$ of transformations of X is not amenable if and only if there is an $\epsilon > 0$ and a finite $R \subseteq \mathcal{G}$ such that for any finite $F \subseteq X$, $|\partial_R F| \geq \epsilon |F|$. Therefore a pseudogroup $\mathcal{G}$ is not amenable if and only if every infinite sequence of finite subsets of X yields a family of expanders by forming the graphs using elements of $\mathcal{G}$. For more information on expanders, see [BHV08, Chap. 6].

The key step in the solution of the Ruziewicz Problem in $\mathbb{S}^4$ and beyond was the identification, using techniques of algebraic group theory, of a countable dense subgroup of $SO_n(\mathbb{R})$ $(n \geq 5)$ that has Property (T). The relevance of Property (T) to the uniqueness of invariant means is given by the following theorem.

Theorem 13.12. *Suppose a countable group G with Property (T) acts on X in a way that is measure-preserving and ergodic, where $(X, \mathcal{A}, m)$ is a nonatomic measure space and $m(X) = 1$. Then the m-integral is the unique G-invariant mean on $L^\infty(X)$.*

Proof. Suppose the m-integral is not the unique mean. Then, by a result of del Junco and Rosenblatt ([JR79]; see [Ros81, Thm. 1.4] for a proof), there is an asymptotically invariant sequence (A_n) of sets of positive measure such that $\lim_{n\to\infty} m(A_n) = 0$. Let $L^2(X)$ denote the Hilbert space of all square-integrable complex functions on X, and let H be the subspace consisting of those f such that $\int f\, dm = 0$. Let π be the natural representation of G as isometries of H: $(\pi(g) f)(x) = f(g^{-1}(x))$. Let $f_n(x) = \chi_{A_n}(x) - m(A_n)$; then $f_n \in H$, $\|f_n\|^2 = m(A_n)(1 - m(A_n))$, and $\|f_n - \pi(g) f_n\|^2 = m(A_n \triangle gA_n)$. It follows that $\{f_n\}$ is an asymptotically invariant sequence in H, so because G has Property (T), some nonzero $f \in H$ satisfies $\pi(g)f = f$ for all $g \in G$. But then set $E_r = \{x \in X : |f(x)| \geq 1/r\}$; E_r has positive measure for some r and $m(E_r \triangle gE_r) = 0$ for all $g \in G$, which contradicts the ergodicity of G's action. □

The converse of the previous theorem is valid too. If invariant means are unique for all actions of a countable group G as in the theorem, then G has Property (T) [CW80, Sch81]. Property (T) can also be characterized in terms of cohomological properties of G [Sch81, Wan74]. For more on the relation between Property (T) and amenability, see the beginning of §11.3.

Now, the action of a countable dense subgroup of $SO_n(\mathbb{R})$ on $\mathbb{S}^{n-1}$ is ergodic, like the action of all of $SO_n(\mathbb{R})$. Hence Theorem 13.12, together with the existence, mentioned earlier, of a countable dense subgroup of $SO_n(\mathbb{R})$, $n \geq 5$, having Property (T), implies that the Lebesgue integral is the unique invariant mean on $L^\infty(\mathbb{S}^n)$, $n \geq 4$, and hence by Tarski's observation, there is no exotic measure on $\mathbb{S}^n$, $n \geq 4$.

The preceding technique can be adapted to $\mathbb{R}^5$ and beyond, but the lower-dimensional cases $\mathbb{R}^3$, $\mathbb{R}^4$, $\mathbb{S}^2$, and $\mathbb{S}^3$ proved more troublesome. Nevertheless, Margulis [Mar82] obtained a solution for all $\mathbb{R}^n$, $n \geq 3$, and Drinfeld [Dri85] solved the remaining two cases, $\mathbb{S}^2$ and $\mathbb{S}^3$, using some deep mathematics, including Deligne's proof of the Petersson Conjecture. For a detailed discussion of the complete solution to the Ruziewicz Problem, see [Lub1994]. To summarize:

Theorem 13.13 (AC). *An exotic measure in $\mathbb{S}^n$ or $\mathbb{R}^n$ exists only in the cases $\mathbb{R}^1$, $\mathbb{R}^2$, and $\mathbb{S}^1$.*

The role of absolute continuity in the proof of Theorem 13.13, via Tarski's observation, shows how the Banach–Tarski Paradox, which is usually interpreted negatively (invariant extensions of Lebesgue measure to all sets do not exist), can be looked upon more positively if the focus is shifted from all of $\mathcal{P}(\mathbb{R}^n)$ to the measurable sets. The paradox yields that exotic measures must be absolutely continuous, which, by the work just outlined, yields a characterization of Lebesgue measure as a finitely additive measure.

Theorem 13.13 points to a close connection between the existence of exotic measures and the amenability of the group in question, because exotic measures exist only in the cases that the isometry group is amenable. How closely are these two properties related? We shall see that for a somewhat trivial reason, exotic measures on $\mathbb{S}^3$, for example, can exist even when the group in question is a non-amenable group of rotations (see the use of a principal measure after Thm. 13.24). Thus it is appropriate to formulate a question in terms of absolutely continuous measures; they lead to invariant means on $L^\infty(\mathbb{S}^n)$.

Question 13.14. Does the nonamenability of a subgroup G of $O_{n+1}(\mathbb{R})$ guarantee the uniqueness of the Lebesgue integral as a G-invariant mean on $L^\infty(\mathbb{S}^n)$? Can the mean's uniqueness be proved under the stronger assumption that $\mathbb{S}^n$ is G-paradoxical?

An affirmative answer to the first part of this question would yield (in fact, is equivalent to) the assertion that if G is a nonamenable subgroup of $O_{n+1}(\mathbb{R})$ and F is a G-invariant mean on $L^\infty(\mathbb{S}^n)$, then F is, in fact, $O_{n+1}(\mathbb{R})$-invariant. Note that if G is an amenable subgroup of $O_{n+1}(\mathbb{R})$ and n is at least 2, then there is a G-invariant mean on $L^\infty(\mathbb{S}^n)$ that is not $O_{n+1}(\mathbb{R})$-invariant: Any G-invariant mean that differs from the Lebesgue integral (such exist by the method preceding Theorem 13.7) is not $O_{n+1}(\mathbb{R})$-invariant because of Theorem 13.13.

All the known results about exotic measures have been obtained by studying the related question about the uniqueness of invariant means on L^∞. But these techniques do not apply if one is considering the family of Borel sets rather than the collection of measurable sets. Lemma 11.9 cannot be used because the proof of that lemma uses subsets of sets of measure zero, a technique that can introduce non-Borel sets. Thus we have the following question, which is completely unresolved.

Question 13.15. Is it true that in $\mathbb{S}^n$, $n \geq 2$ (or $\mathbb{R}^n$, $n \geq 3$), Lebesgue measure is the only finitely additive, isometry-invariant measure on the Borel sets that has total measure 1 (or, that normalizes the unit cube)?

13.2 Paradoxes modulo an Ideal

The nonexistence of exotic measures leads to a new type of paradoxical decomposition, one that uses measurable pieces. Of course, J, the unit cube in $\mathbb{R}^3$, cannot be paradoxical using pieces in $\mathcal{L}$. But it is possible that J is paradoxical in $\mathcal{L}$ provided sets in some ideal $\mathcal{I}$ of $\mathcal{L}$ are ignored. We have already seen the usefulness of considering equidecomposability modulo an ideal; see Definitions 11.5 and 11.25. To fit this generalization into the machinery already developed, we digress briefly to discuss equidecomposability in arbitrary Boolean algebras.

If G is a group of automorphisms of a Boolean algebra $\mathcal{A}$, and $a, b \in \mathcal{A}$, then a and b are *G-equidecomposable* ($a \sim b$) if there are two pairwise disjoint collections of n elements, $\{a_i\}$, $\{b_i\}$, and $g_i \in G$, such that $a = a_1 \vee \cdots \vee a_n$, $b = b_1 \vee \cdots \vee b_n$, and $g_i(a_i) = b_i$. An element a in $\mathcal{A}$ is *G-paradoxical* if there are disjoint $b, c \leq a$ such that $b \sim a$ and $c \sim a$. Most of the algebras considered so far have been fields of sets, but Ruziewicz's Problem leads to the consideration of paradoxical decompositions in quotient algebras of the form $\mathcal{L}/\mathcal{I}$. One consequence of this increased generality is that the proof of the Banach–Schröder–Bernstein Theorem (Thm 3.6) breaks down. The two axioms for equidecomposability used in that proof, (a) and (b), remain valid if suitably reformulated for Boolean algebras (the relativization of $\mathcal{A}$ to an element is defined in §12.3): (a) If $a \sim b$, then there is a Boolean isomorphism $g\colon \mathcal{A}_a \to \mathcal{A}_b$ such that $c \sim g(c)$ for each $c \leq a$, and (b) if $a_1 \wedge a_2 = \mathbf{0} = b_1 \wedge b_2$, $a_1 \sim b_1$, and $a_2 \sim b_2$, then $a_1 \vee a_2 \sim b_1 \vee b_2$. For (a), let $g(c) = g_1(c \wedge a_1) \vee \cdots \vee g_n(c \wedge a_n)$, where a_i, g_i witness $a \sim b$. But the proof of Theorem 3.6 uses an infinite union at one point, and the Boolean counterpart of that operation does not always exist. A

Boolean algebra $\mathcal{A}$ is called *countably complete* if sups and infs (with respect to $\leq$) of countable subsets of $\mathcal{A}$ exist in $\mathcal{A}$. It is easy to see that in countably complete Boolean algebras, the proof of Theorem 3.6 yields the Banach–Schröder–Bernstein Theorem for G-equidecomposability.

An example showing that the Banach–Schröder–Bernstein Theorem is not valid in all Boolean algebras can be constructed as follows. Kinoshita [Kin53] found countable Boolean algebras $\mathcal{A}$, $\mathcal{C}_1$, and $\mathcal{C}_2$ such that $\mathcal{A} \cong \mathcal{A} \times \mathcal{C}_1 \times \mathcal{C}_2$, but $\mathcal{A} \ncong \mathcal{A} \times \mathcal{C}_1$ ($\cong$ is isomorphism of Boolean algebras). This result was improved by Hanf [Han57], who showed that $\mathcal{C}_1$ could be taken isomorphic to $\mathcal{C}_2$. Now, let $\mathcal{B} = \mathcal{A} \times \mathcal{C}_1 \times \mathcal{C}_2 \times \mathcal{A}$, and let G be the group of all automorphisms of $\mathcal{B}$. Let g be the automorphism that uses the isomorphism to take the first factor of $\mathcal{B}$ to the last three, and vice versa; let h be the automorphism that switches the first and last coordinates. Finally, let $a = (\mathbf{1}, \mathbf{0}, \mathbf{0}, \mathbf{0})$ and $b = (\mathbf{1}, \mathbf{1}, \mathbf{0}, \mathbf{0})$. Then $a \leq b$ and $b \sim_G gh(b) = g(\mathbf{0}, \mathbf{1}, \mathbf{0}, \mathbf{1}) \leq g(\mathbf{0}, \mathbf{1}, \mathbf{1}, \mathbf{1}) = a$. If a and b are G-equidecomposable, let f be the piecewise automorphism from $\mathcal{B}_a$ to $\mathcal{B}_b$. For simplicity, assume two pieces, so that $a = a_1 \vee a_2$, $b = b_1 \vee b_2$ (with $a_1 \wedge a_2 = b_1 \wedge b_2 = \mathbf{0}$), and with $g_i \in G$ so that $g_i(a_i) = b_i$. Now we can show that f is actually an isomorphism of $\mathcal{B}_a = \mathcal{A}$ and $\mathcal{B}_b = \mathcal{A} \times \mathcal{C}_1$. Consider the join operation: Suppose $c_1 \vee c_2 \in \mathcal{A}$. In the following derivations, we make use of the fact that g_i is a Boolean automorphishm.

Claim. If $c \leq a$, then $f(c) = (g_1(c) \wedge b_1) \vee (g_2(c) \wedge b_2)$.

Proof. $f(c) = f(c \wedge a_1) \vee f(c \wedge a_2) = g_1(c \wedge a_1) \vee g_2(c \wedge a_2)$
$= (g_1(c) \wedge g_1(a_1)) \vee (g_2(c) \wedge g_2(a_2)) = (g_1(c) \wedge b_1) \vee (g_2(c) \wedge b_2)$. □

Now, to finish,

$$\begin{aligned} f(c_1 \vee c_2) &= (b_1 \wedge g_1(c_1 \vee c_2)) \vee (b_2 \wedge g_2(c_1 \vee c_2)) \text{ [by the claim]} \\ &= [b_1 \wedge (g_1(c_1) \vee g_1(c_2))] \vee [b_2 \wedge (g_2(c_1) \vee g_2(c_2))] \\ &= [(b_1 \wedge g_1(c_1)) \vee (b_1 \wedge g_1(c_2))] \vee [(b_2 \wedge g_2(c_1)) \vee (b_2 \wedge g_2(c_2))] \\ &= [(b_1 \wedge g_1(c_1) \vee (b_2 \wedge g_2(c_1))] \vee [(b_1 \wedge g_1(c_2)) \vee (b_2 \wedge g_2(c_2))] \\ &= f(c_1) \vee f(c_2) \text{ [by the claim]}. \end{aligned}$$

So f preserves join, and similar reasoning applies to the other operations. This shows that $f : \mathcal{A} \to \mathcal{A} \times \mathcal{C}_1$ is an isomorphism, contradiction.

The generalization of equidecomposability to arbitrary Boolean algebras provides a convenient context for studying equidecomposability in quotient algebras, but strictly speaking, this can all be carried out within the framework used earlier, that of G-equidecomposability in an algebra of subsets of some set, where G acts on the set. This is because by the Stone Representation Theorem [Sik69, p. 24], any Boolean algebra $\mathcal{A}$ is isomorphic to an algebra of subsets of X for some set X (X is the set of ultrafilters on $\mathcal{A}$), and a group of automorphisms of $\mathcal{A}$ may be viewed as a group acting on X. Because of these remarks, we may talk of equidecomposability of objects na where $n \in \mathbb{N}$ and $a \in \mathcal{A}$; simply apply the semigroup construction at the start of §10.3.2 to the algebra of sets corresponding to $\mathcal{A}$. Moreover, Tarski's Theorem in the form of Corollary 11.3 applies to equidecomposability in Boolean algebras. Because of the lack of a general cancellation law, we cannot be sure that the stronger Corollary 11.2 is valid.

With all these preliminaries, we can now derive a new type of paradoxical decomposition of a cube in $\mathbb{R}^n$ $(n \geq 3)$ or of $\mathbb{S}^n$ $(n \geq 2)$ from the nonexistence of exotic measures. Let $\mathcal{I}$ be any isometry-invariant ideal in $\mathcal{L}$, the measurable subsets of $\mathbb{R}^n$, $n \geq 3$, or of $\mathbb{S}^n$, $n \geq 2$, and let $\mathcal{A}$ be the Boolean algebra $\mathcal{L}/\mathcal{I}$; let $[A]$ denote the $\mathcal{I}$-equivalence class of a measurable set A. If $\mathcal{I}$ consists only of measure zero sets, then Lebesgue measure induces an isometry-invariant measure on $\mathcal{A}$. But in all other cases, that is, whenever $\mathcal{I}$ is an isometry-invariant ideal in $\mathcal{L}$ that contains a set of positive measure, there is no isometry-invariant measure on $\mathcal{A}$ normalizing $[J]$ (or normalizing $[\mathbb{S}]^n$). This is a consequence of Theorem 13.13, because such a measure would induce an exotic measure on $\mathcal{L}_b$. Note that such ideals are easy to obtain; for example, choose any closed nowhere dense set A of positive measure and let $\mathcal{I} = \{B \in \mathcal{L} : \text{for some } \sigma_1, \ldots, \sigma_m \in G_n, B \subseteq \sigma_1 A \cup \cdots \cup \sigma_m A\}$. Now, an application of Corollary 11.3 yields the following result.

Corollary 13.16 (AC). *Let $\mathcal{L}$ denote the class of measurable subsets of $\mathbb{R}^n$, $n \geq 3$, or $\mathbb{S}^n$, $n \geq 2$; let $\mathcal{I}$ be any proper isometry-invariant ideal in $\mathcal{L}$ that contains a set of positive measure; let $\mathcal{A} = \mathcal{L}/\mathcal{I}$; and let α be the equivalence class of the unit cube in $\mathbb{R}^n$ or of the sphere $\mathbb{S}^n$. Then there is some positive integer m such that $(m+1)\alpha \leq m\alpha$ in $\mathcal{A}$; that is, $m+1$ copies of the cube or sphere can be packed into m copies using measurable pieces, but ignoring sets in $\mathcal{I}$.*

This corollary is noteworthy because it applies to every ideal satisfying the hypothesis; but the conclusion is weaker than one might hope because of the usual problems with the Cancellation Law requiring Corollary 11.3 to be used rather than Corollary 11.2. Nevertheless, even before Ruziewicz's Problem for $\mathbb{S}^2$ had been solved, Rosenblatt [Ros79] had constructed a specific isometry-invariant ideal $\mathcal{I}$ in the algebra of measurable subsets of $\mathbb{S}^2$ such that $\mathbb{S}^2$ is paradoxical modulo $\mathcal{I}$ using measurable pieces. Note also that the case of $\mathbb{R}^n$ in Corollary 13.16 can be improved if $\mathcal{I}$ is assumed, in addition, to be closed under countable unions and invariant under similarities. Then $\mathcal{L}/\mathcal{I}$ is countably complete and satisfies the Banach–Schröder–Bernstein Theorem, whence the proof of $(a) \Rightarrow (b)$ of Theorem 11.7 can be applied to obtain that the unit cube is paradoxical in $\mathcal{L}/\mathcal{I}$. In other words, the extra hypotheses on $\mathcal{I}$ in the case of $\mathbb{R}^n$ allow m to be taken to be 1.

13.3 How to Eliminate Exotic Measures in $\mathbb{R}^2$

Recall from §8.1.2 that even though the square in $\mathbb{R}^2$ is not paradoxical using isometries, the square is paradoxical using a larger group, one that contains some area-preserving linear transformations. Let G be the group generated by G_2 and σ, where $\sigma(x,y) = (x+y,y)$. Then $\left[\begin{smallmatrix} 1 & 2 \\ 0 & 1 \end{smallmatrix}\right] = \sigma^2$ and $\left[\begin{smallmatrix} 1 & 0 \\ 2 & 1 \end{smallmatrix}\right] = \left[\begin{smallmatrix} 0 & -1 \\ 1 & 0 \end{smallmatrix}\right] \sigma^{-1} \left[\begin{smallmatrix} 0 & 1 \\ -1 & 0 \end{smallmatrix}\right]$, so A and B of Proposition 4.4 lie in G, and by the von Neumann Paradox (Thm. 8.5), the square is G-paradoxical. It follows as in Lemma 11.9 that any G-invariant finitely additive measure on $\mathcal{L}_b$, the Lebesgue measurable subsets of $\mathbb{R}^2$, that normalizes the unit square must be absolutely continuous with respect to λ.

Now, Rosenblatt [Ros81] has shown that despite the existence of absolutely continuous exotic measures in $\mathbb{R}^2$, no such measure can be σ-invariant. But, by the preceding remarks, the absolute continuity condition is redundant in the presence of σ-invariance. Thus any finitely additive measure on $\mathcal{L}_b$ that normalizes the unit square and is G-invariant must coincide with Lebesgue measure. In other words, if one modifies the original Ruziewicz Problem by considering area-preserving affine transformations instead of just isometries, then one obtains a characterization of Lebesgue measure as a finitely additive measure that is valid in all $\mathbb{R}^n$ except $\mathbb{R}^1$.

Rosenblatt's result was obtained as a corollary to results on the n-dimensional torus, $\mathbb{T}^n$. He proved that the Lebesgue integral on $L^\infty(\mathbb{T}^n)$, where $n \geq 2$, is the unique mean that is invariant under the natural action of $SL_n(\mathbb{Z})$; hence there is a unique mean on $L^\infty(\mathbb{T}^n)$ that is invariant under all topological automorphisms of $\mathbb{T}^n$. Therefore absolutely continuous $SL_n(\mathbb{Z})$-exotic measures on the measurable subsets of $\mathbb{T}^n$ do not exist. As pointed out in [Ros81], however, finitely additive $SL_n(\mathbb{Z})$-invariant measures differing from Lebesgue measure do exist. Thus the toroidal case shows that the uniqueness of invariant means can be a different question than the nonexistence of exotic measures. For a further analysis of invariant, finitely additive measures both on the family of measurable subsets of $\mathbb{T}^n$ and on $\mathcal{P}(\mathbb{T}^n)$, see [Dan85].

For countably infinite groups, amenability and Property (T) are mutually exclusive. To see this, let $G = (g_i)_{i \in \mathbb{N}}$ be an enumeration of an amenable group. By Følner's Condition (Thm. 12.11), we have, for each n, a finite subset $W_n \subseteq G$ such that $|g_i(W_n) \triangle W_n| \leq |W_n|/n$ for each $i \leq n$. Let $L^2(G)$ be the space induced by Haar measure m on G; then $\{\chi_{W_n}\}$ is an asymptotically invariant sequence of nonzero vectors for the representation of G on $L^2(G)$ induced by G's action on itself (as in proof of Thm. 13.12); so Property (T) yields a G-invariant vector in $L^2(G)$. But the only such G-invariant vectors for this action are constant, and if a constant function is in $L^2(G)$, then $m(G)$ is finite. Because Haar measure on a countable group is a scalar multiple of the counting measure, G must be finite.

These two properties are not exact opposites, however. For example, $SL_1(\mathbb{Z})$ is amenable and $SL_n(\mathbb{Z})$ has Property (T) if $n \geq 3$ (see [Mar80]), but $SL_2(\mathbb{Z})$, which is nonamenable (Prop. 4.4, Thm. 8.1), fails to have Property (T) (see [Sch81, Example 3.7]). Also, a free group of rank n, $n \geq 2$, fails to have Property (T) (see [Ake81, Kaz67]). Of course, such groups are nonamenable.

One can also investigate exotic measures in hyperbolic space. Let $\lambda_{\mathbb{H}}$ be Lebesgue measure in $\mathbb{H}^2$ with a unit disk normalized (see end of §4.6.1). The following is open.

Question 13.17. Is $\lambda_{\mathbb{H}}$ unique as a finitely additive, isometry-invariant measure on the bounded Lebesgue measurable subsets of $\mathbb{H}^2$ and normalizing the unit disk?

13.4 Paradoxes Using Measurable Pieces

Recall from §4.2 that the hyperbolic plane (also $\mathbb{H}^n$, $n \geq 2$) is paradoxical using measurable pieces. The measure-theoretic consequence of this is that there is no finitely additive measure on the measurable subsets of $\mathbb{H}^2$ that has total measure 1 and is invariant under hyperbolic isometries. Analogous paradoxical decompositions of $\mathbb{R}^n$ using measurable pieces do not exist. This is not too difficult to prove geometrically [Myc74, §5], but it can also be derived from the construction of a measure of the type that cannot exist in hyperbolic space. Part of the interest in the following theorem is that it is valid even in the cases where the isometry group is nonamenable. Moreover, it yields the somewhat surprising result that paradoxes using measurable pieces do not exist even if one allows similarities, which can change areas greatly. Measures as in the next theorem are sometimes called *Mycielski measures*.

Theorem 13.18 (AC). *For any n, there is a finitely additive, isometry-invariant measure μ on $\mathcal{L}$ such that $\mu(\mathbb{R}^n) = 1$. Moreover, μ can be chosen so that $\mu(A) = \mu(g(A))$ for any similarity g.*

Proof. Let U be a nonprincipal ultrafilter on $\mathbb{N}$ (see discussion following Theorem 12.7). Let B_m denote the ball of radius m centered at the origin and, for $A \in \mathcal{L}$, define $f_A : \mathbb{N} \to [0, \infty]$ by $f_A(m) = \lambda(A \cap B_m)/\lambda(B_m)$. Then let $\mu(A) = \int f_A \, dU$. It is clear that μ is finitely additive, $\mu(\mathbb{R}^n) = 1$, and μ is $O_n(\mathbb{R})$-invariant. Because $O_n \cup T_n$ generates G_n, it must be shown that for any $v \in \mathbb{R}^n$ and $A \in \mathcal{L}$, $\mu(A) = \mu(A + v)$. But $\lim_{m\to\infty} \lambda(B_m \,\triangle\, (B_m + v)/\lambda(B_m) = 0$, which implies (as in Thm. 13.8's proof) that $\lim_{m\to\infty} f_A(m) - f_{A+v}(m) = 0$. Using the fact that finite subsets of $\mathbb{N}$ have U-measure zero, it follows that $\int f_A \, dU = \int f_{A+v} \, dU$, as required.

For the part of the theorem dealing with H_n-invariance, where H_n is the group of similarities of $\mathbb{R}^n$, modify the measure of the preceding paragraph as follows. Let M be the group of magnifications from the origin; then (see remarks preceding Cor. 13.5) $H_n/G_n \cong M$ and M is Abelian, hence amenable. Let θ be a left-invariant measure on M, and define μ^* by $\mu^*(A) = \int f_A \, d\theta$, where $f_A : M \to \mathbb{R}$ is given by $f_A(d) = \mu(d^{-1}(A))$. Then, as usual, μ^* is finitely additive and M-invariant, and $\mu^*(\mathbb{R}^n) = 1$. But because G_n is a normal subgroup of H_n, μ^* is G-invariant too: $f_{\sigma(A)}(d) = \mu(d^{-1}\,\sigma(A)) = \mu(\sigma_0 d^{-1}(A)) = \mu\,(d^{-1}(A)) = f_A(d)$, where $\sigma_0 = d^{-1}\,\sigma\, d \in G_n$. Because H_n is generated by $G_n \cup M$, μ^* is H_n-invariant. □

Corollary 13.19. *$\mathbb{R}^n$ is not paradoxical using measurable pieces and isometries (or even similarities).*

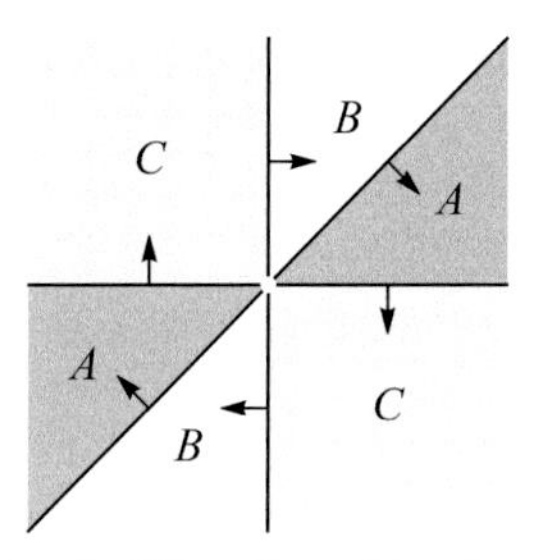

Figure 13.1. A constructive Hausdorff Paradox of the plane using affine transformations.

Note that the key point of the preceding proof is that any unbounded increasing sequence of concentric balls in $\mathbb{R}^n$ is asymptotically G_n-invariant.

Now consider the action of $SA_n(\mathbb{R})$, the group of affine transformations with determinant 1, on $\mathbb{R}^n$. For $\mathbb{R}^1$ this adds nothing new, but in $\mathbb{R}^2$ and beyond $SA_n(\mathbb{R})$ is substantially larger than the isometry group. The techniques of §8.1 (Cor. 8.2(b)) easily yield that if $n \geq 2$, $\mathbb{R}^n$ is $SA_n(\mathbb{R})$-paradoxical. Indeed, if $n \geq 3$, this follows already from Corollary 6.8, while for $\mathbb{R}^2$ this follows from the local commutativity of $SL_2(\mathbb{Z})$ in its action on $\mathbb{R}^2 \setminus \{\mathbf{0}\}$. But can it happen that $\mathbb{R}^n$ is $SA_n(\mathbb{R})$-paradoxical using Lebesgue measurable pieces? In other words, can there be a Mycielski measure when the group is $SA_n(\mathbb{R})$? The answer is NO.

Theorem 13.20. *If $n \geq 2$, then there does not exist a finitely additive, $SA_n(\mathbb{R})$-invariant measure of total measure one on $\mathcal{L}$, the Lebesgue measurable subsets of $\mathbb{R}^n$.*

Proof. Consider the plane first, and let A, B, C be the sets of Figure 13.1. Let σ, τ, ρ be, respectively, the following elements of $SL_2(\mathbb{Z})$: $\left[\begin{smallmatrix} 1 & 1 \\ 0 & 1 \end{smallmatrix}\right]$, $\left[\begin{smallmatrix} 1 & 0 \\ 1 & 1 \end{smallmatrix}\right]$, $\left[\begin{smallmatrix} 0 & -1 \\ 1 & 0 \end{smallmatrix}\right]$. Then $\sigma(A \cup B) = A$, $\tau(A \cup B) = B$, and $\rho(A \cup B) = C$, whence $A \equiv B \equiv C \equiv A \cup B$. This is a Hausdorff decomposition, and it implies that there is no measure as in the theorem. The sets extend to $\mathbb{R}^n$ in the obvious way, and the usual absorption technique to deal with origin then yields that $\mathbb{R}^n$ is $SA_n(\mathbb{R})$-paradoxical using measurable pieces, if $n \geq 2$. □

Although no finitely additive measure on the measurable subsets of $\mathbb{R}^2$ having total measure 1 is $SA_2(\mathbb{R})$-invariant, Belley and Prasad [BP82] have shown that there are interesting subalgebras of the algebra of Borel subsets of $\mathbb{R}^n$ that (a) are invariant under all affine transformations and (b) bear a finitely additive measure of total measure 1 that is invariant under all affine transformations.

13.5 Characterizing Isometry Groups That Yield Paradoxes

We now turn our attention to a finer analysis of paradoxical decompositions and attempt to characterize the groups of isometries of $\mathbb{R}^n$ (resp., $\mathbb{S}^n$) with respect to which the cube (resp., $\mathbb{S}^n$) is paradoxical. In other words, for which subgroups of G_n (resp., $O_n(\mathbb{R})$) are Corollaries 12.9 and 12.10 valid? This leads to two distinct problems. For the first, we are asking to characterize the groups, G, of isometries for which λ has a G-invariant, finitely additive extension to all sets. For the second, we are asking about groups with respect to which no paradox exists, which by Tarski's Theorem is equivalent to asking about groups G for which there exists a finitely additive, G-invariant measure on all sets that normalizes the unit cube, or the sphere.

The following theorem treats the first question for spheres; the case of $\mathbb{R}^n$ follows.

Theorem 13.21 (AC). *Let G be a subgroup of $O_{n+1}(\mathbb{R})$. Then the following are equivalent to each other and to the statements of Theorem 12.11:*

(a) *G is amenable.*
(b) *G has no free subgroup of rank 2.*
(c) *There is a finitely additive, G-invariant extension of λ to all subsets of $\mathbb{S}^n$.*
(d) *There is a finitely additive, G-invariant measure μ on $\mathcal{P}(\mathbb{S}^n)$ with $\mu(\mathbb{S}^n) = 1$ and $\mu(E) = 0$ if $\lambda(E) = 0$.*

Proof. As indicated just before Tits's Theorem (Thm. 12.6), that theorem implies the equivalence of (a) and (b) for groups of Euclidean (and hence spherical) isometries. Because $(a) \Rightarrow (c)$ is just Corollary 12.9 and $(c) \Rightarrow (d)$ is trivial, it remains only to show $(d) \Rightarrow (a)$. Assume μ is as in (d) and G is not amenable. Then, by Theorem 12.4(f) and the fact that any group is the union of its finitely generated subgroups, there is some finitely generated, and hence countable, subgroup H of G that fails to be amenable. Let $D = \{x \in \mathbb{S}^n : \sigma(x) = x$ for some nonidentity $\sigma \in H\}$. Now, the fixed point set (in $\mathbb{R}^{n+1}$) of a nonidentity isometry of $\mathbb{S}^n$ is a linear subspace of dimension at most n. It is easy to see (use radial extension inward) that such a subspace intersects $\mathbb{S}^n$ in a set of surface measure 0. Because H is countable, this implies that $\lambda(D) = 0$, and hence $\mu(D) = 0$ and $\mu(\mathbb{S}^n \setminus D) = 1$. But H acts on $\mathbb{S}^n \setminus D$ without nontrivial fixed points, and H, being nonamenable, is paradoxical by Tarski's Theorem. Therefore, by Proposition 1.10, $\mathbb{S}^n \setminus D$ is H-paradoxical, contradicting $\mu(\mathbb{S}^n \setminus D) = 1$. (Note: The appeal to Tarski's Theorem here can be avoided by using an easy variation of Prop. 1.10 to transfer the measure, μ, on $\mathbb{S}^n \setminus D$ directly to H, contradicting H's nonamenability.) □

We now use the previous theorem to obtain a characterization of subgroups of G_n for which invariant extensions of Lebesgue measure on $\mathbb{R}^n$ exist.

Theorem 13.22 (AC). *Let G be a subgroup of G_n. Then the following are equivalent:*

(a) *G is amenable.*
(b) *G has no free subgroup of rank 2.*
(c) *There is a finitely additive, G-invariant extension μ of λ to all subsets of $\mathbb{R}^n$.*

Proof. As in Theorem 13.21, only $(c) \Rightarrow (b)$ requires proof. Suppose F is a rank-2 free subgroup of G. Consider first the case that F is locally commutative on $\mathbb{R}^n$. Then, by Theorem 5.5, $\mathbb{R}^n$ is F-paradoxical. But this contradicts the existence of a finitely additive G-invariant measure ν on $\mathcal{P}(\mathbb{R}^n)$ having total measure 1, which can be obtained from μ as in Theorem 13.18. More precisely, let U and B_m be as in Theorem 13.18 and define $\nu(A)$ to be $\int \mu(A \cap B_m)/\mu(B_m)\, dU$.

If F is not locally commutative, choose noncommuting $\sigma_1, \sigma_2 \in F$, and $P \in \mathbb{R}^n$ such that $\sigma_1(P) = P = \sigma_2(P)$. Let τ be the translation by $-P$ and define ν on $\mathcal{P}(\mathbb{R}^n)$ by $\nu(A) = \mu(\tau^{-1}(A))$; let $\sigma_i' = \tau\, \sigma_i\, \tau^{-1}$; this fixes the origin and so is in $O_n(\mathbb{R})$. Then ν is a finitely additive extension of λ that is $\langle \sigma_1', \sigma_2' \rangle$-invariant. But $\langle \sigma_1', \sigma_2' \rangle$ is isomorphic to $\langle \sigma_1, \sigma_2 \rangle$, which, because σ_1 and σ_2 do not commute, is freely generated by σ_1, σ_2. Because ν induces a $\langle \sigma_1', \sigma_2' \rangle$-invariant extension of λ on $\mathcal{P}(\mathbb{S}^{n-1})$ by adjunction of radii, this contradicts $(c) \Rightarrow (b)$ of Theorem 13.21. □

A question related to the previous characterization problem asks, Which groups of isometries can arise as the group with respect to which a finitely additive extension of λ to $\mathcal{P}(\mathbb{R}^n)$ is invariant? If μ is such a measure, let Invt(μ) denote the subgroup of G_n containing those isometries σ such that μ is σ-invariant. The following result shows that precisely the amenable groups arise as Invt(μ).

Theorem 13.23 (AC). *Suppose G is a subgroup of G_n. Then there is a finitely additive extension μ of λ to $\mathcal{P}(\mathbb{R}^n)$ such that Invt(μ) $= G$ iff G is amenable.*

Thus, for example, there are total extensions of λ in $\mathbb{R}^1$ that are invariant under all translations, but not under any reflections, or invariant under all rational translations, but no others. In fact, in $\mathbb{R}^1$ and $\mathbb{R}^2$, any group of isometries is realizable as Invt(μ). Note that in other contexts, things may turn out quite differently in that invariance with respect to one group is sufficient to imply invariance with respect to a larger group. For instance, any countably additive, translation-invariant measure on $\mathcal{L}$ in $\mathbb{R}^1$ that normalizes [0, 1] must coincide with Lebesgue measure and hence is necessarily invariant under reflections. (But see [HW83] for a variation on this problem.) Or, letting G be the countable dense subgroup of $SO_5(\mathbb{R})$ having Property (T) (see remarks preceding Thm. 13.12), any G-invariant mean on $L^\infty(\mathbb{S}^4)$ is necessarily $O_5(\mathbb{R})$-invariant (because it must equal the Lebesgue integral).

The necessity of amenability in Theorem 13.23 follows from Theorem 13.22. The proof of sufficiency is due to Wagon [Wag81a]. The main idea is to find a set A such that $B = \{\chi_{\sigma(A)} : \sigma \in G_n\}$ is linearly independent over $L^1(\mathbb{R}^n)$, the space of Lebesgue integrable, real-valued functions on $\mathbb{R}^n$ (this means that no function in B lies in the subspace spanned by $L^1(\mathbb{R}^n)$ and the rest of B). Then the Hahn–Banach Theorem is used to extend the Lebesgue integral to a linear functional F on the space generated by $L^1(\mathbb{R}^n) \cup B$ in such a way that $F(\chi_{\sigma(A)}) = 1$ or 0 according as σ is or is not in G. Then F, which is precisely G-invariant, is extended by Theorem 12.11(e) to a G-invariant linear functional on the space of all real-valued functions on $\mathbb{R}^n$ that are bounded by a function in $L^1(\mathbb{R}^n)$. This extension induces a measure μ on $\mathcal{P}(\mathbb{R}^n)$ such that μ extends λ and $\text{Invt}(\mu) = G$.

We now consider the rather different problem of characterizing the groups of isometries of $\mathbb{S}^n$ for which an invariant, finitely additive measure on $\mathcal{P}(\mathbb{S}^n)$ having total measure 1 exists. By Corollary 11.2, this is equivalent to asking for the subgroups of $O_n(\mathbb{R})$ with respect to which $\mathbb{S}^n$ is paradoxical. For the case of $\mathbb{S}^2$, the local commutativity of $SO_3(\mathbb{R})$ yields an easy solution, as follows.

Theorem 13.24 (AC). *For a subgroup G of $O_3(\mathbb{R})$, the following are equivalent:*

(a) *G is amenable.*
(b) *G has no free subgroup of rank 2.*
(c) *$\mathbb{S}^2$ is not G-paradoxical.*
(d) *There is a finitely additive, G-invariant measure on $\mathcal{P}(\mathbb{S}^2)$ having total measure 1.*

Proof. $(a) \Rightarrow (b)$ follows from Theorem 13.21 and $(c) \Rightarrow (d)$ follows from Corollary 11.2. Moreover, $(a) \Rightarrow (d)$ by Theorem 12.3. Finally, Theorem 5.5 (see also Cor. 10.6) and the local commutativity of $SO_3(\mathbb{R})$ yield $(c) \Rightarrow (b)$ for subgroups of $SO_3(\mathbb{R})$. The result for $O_3(\mathbb{R})$ follows, because if σ, τ are independent in $O_3(\mathbb{R})$, then σ^2 and τ^2 are independent in $SO_3(\mathbb{R})$. □

Property (d) is much more sensitive to an action's fixed points than the corresponding property (c) of Theorem 13.19. Because of this, Theorem 13.24 does not extend to $\mathbb{R}^3$ or to $\mathbb{S}^3$. If $\sigma, \rho \in SO_3(\mathbb{R})$ are independent, then G, the group they generate, satisfies (d) with respect to $\mathbb{R}^3$: Just let $\mu(A)$ be 1 or 0 according as the origin is or is not in A. But G does not satisfy (b). For a similar example in $\mathbb{S}^3$, use the same σ, ρ, but extend them to σ', $\rho' \in SO_4(\mathbb{R})$ by fixing the new coordinate. Then σ' and ρ' are still free generators of G, the group they generate, so G fails to satisfy (b). But there is a G-invariant measure on $\mathbb{S}^3$; just let μ be the principal measure determined by the point $(0, 0, 0, 1)$, which is fixed by the action of G (a *principal measure* assigns 1 to sets containing a given point, 0 to the others).

The point of the examples of the preceding paragraph is that if a group G acting on X fixes some $x \in X$, then the principal measure determined by x is G-invariant. In fact, it is not necessary that G fix a point but only that a certain subgroup of G fix a point. For suppose H is a normal subgroup of G such that G/H is amenable and H has a common fixed point x. Then let E be the G-orbit of x and consider the natural action of G/H on E given by $(gH)(g_0x) = gg_0\,x$. This is well defined because, using the fact that H is normal, $g\,h\,g_0\,x = gg_0h'\,x = gg_0x$. The amenability of G/H yields a G/H-invariant measure on $\mathcal{P}(E)$, and it is easy to see that this yields a G-invariant measure. Giving $X \setminus E$ measure zero then yields a G-invariant measure on $\mathcal{P}(X)$. A result of Dani (who used the technique introduced in [Dan85]) confirmed a conjecture of Wagon [Wag81a] that this condition is necessary as well as sufficient for G-invariant measures on $\mathcal{P}(\mathbb{S}^n)$ to exist, where G is a subgroup of $O_{n+1}(\mathbb{R})$.

Theorem 13.25 (AC). *For a subgroup G of $SO_{n+1}(\mathbb{R})$, the following are equivalent:*

(a) *$\mathbb{S}^n$ is not G-paradoxical.*
(b) *There is a finitely additive, G-invariant measure on $\mathcal{P}(\mathbb{S}^n)$ having total measure 1.*
(c) *There is a normal subgroup H of G such that G/H is amenable and some point in $\mathbb{S}^n$ is fixed by all isometries in H.*

Of course, (a) and (b) are equivalent by Tarski's Theorem, and as proved (c) implies (b). Dani's proof that (b) implies (c) combines Theorem 13.27 with the fact that G's action on $\mathbb{S}^n$ has the following property: If $x \in \mathbb{S}^n$, then G_x denotes the subgroup of G that fixes x; if H is a subgroup of G, then F_H denotes the set of points in $\mathbb{S}^n$ left fixed by each member of H. Then Dani's property is that every intersection of members of $\{G_x : x \in \mathbb{S}^n\}$ (and of $\{F_H : H \text{ a subgroup of } G\}$) is in fact an intersection of finitely many members of the family.

If G, a subgroup of $O_{n+1}(\mathbb{R})$, contains a locally commutative free subgroup of rank 2, then by Theorem 5.5 or Corollary 10.6, $\mathbb{S}^n$ is G-paradoxical. Thus each of (a)–(c) of Theorem 13.25 implies that G has no such subgroup. This leads to the following question.

Question 13.26. Does $O_4(\mathbb{R})$ have a subgroup G such that for any normal subgroup H of G with $G/H \in AG$, H has no common fixed point on $\mathbb{S}^3$, but yet G does not have a free locally commutative subgroup of rank 2? Equivalently, is there a subgroup G of $O_4(\mathbb{R})$ such that $\mathbb{S}^3$ is G-paradoxical, but $\mathbb{S}^3$ is not G-paradoxical using four pieces?

The preceding discussion for spheres brings us to the general question of what conditions on a group action are necessary and sufficient for the existence of an invariant measure. Condition (c) of Theorem 13.25 is sufficient but not necessary. Van Douwen [Dou90] has shown that F, a free group of rank 2, can act on an infinite set X in such a way that the action is transitive and each element of F fixes only finitely many points in X (it follows that condition (c) of Thm. 13.25 fails), but yet a finitely additive, F-invariant measure on $\mathcal{P}(X)$ exists. Another counterexample was presented by Promislow [Pro83]: let G be the group with generators $\{g_1, g_2, \ldots, h_1, h_2, \ldots\}$ and subject to the relations that h_i and h_j commute for all i, j and that g_i and h_j commute if $i \leq j$. Let G act on $G \setminus \{e\}$ by conjugation. Then there is an invariant measure on all subsets of $G \setminus \{e\}$: For any finite subset S of G, let μ_S be the principal measure determined by h_n, where n is so large that $i \leq n$ for all $g_i \in S$. Then μ_S is S-invariant, and applying the usual compactness technique to subsets M_S of $[0, 1]^{\mathcal{P}(G)}$ (see the proof of Thm. 12.4), one gets a G-invariant measure on $\mathcal{P}(G)$. But it is easy to see that Theorem 13.25(c) fails for this action.

For some positive results, recall that a locally commutative action of a free group of rank 2 is paradoxical. In fact, a somewhat stronger result is true; the following result shows that the hypothesis can be weakened to nonamenability.

Theorem 13.27 (AC). *If a nonamenable group G acts on X in such a way that for every $x \in X$, $G_x = \{g \in G : gx = x\}$ is amenable, then there is no finitely additive, G-invariant measure on X having total measure 1 (and hence X is G-paradoxical).*

Proof. The proof is a generalization of that of Theorem 12.4(e). Suppose μ is a finitely additive, G-invariant measure on X. Let $M \subseteq X$ be a set of representatives for the orbits of G's action, and for each $x \in M$, let ν_x be a finitely additive, left-invariant measure on the amenable group G_x. Now, for any $A \subseteq G$, define $f_A : X \to \mathbb{R}$ by setting $f_A(y) = \nu_x(G_x \cap h^{-1}A)$, where x is the unique element of M such that y is in x's orbit and h is any element of G such that $hx = y$. To see that f_A is well defined, note that if $h_1 x = h_2 x$, then $f_A(h_1 x) = \nu_x(G_x \cap h_1^{-1}A) = \nu_x(h_2^{-1} h_1 (G_x \cap h_1^{-1} A)) = f_A(h_2 x)$. Now define ν on $\mathcal{P}(G)$ by $\nu(A) = \int f_A \, d\mu$. Because $f_{gA}(y) = f_A(g^{-1} y)$, ν is a left-invariant finitely additive measure, contradicting the nonamenability of G. □

In the case that G is a free group, another version of the preceding result has been strengthened by Promislow [Pro83]: If m is a positive integer and G, a free group of rank $m + 1$ or greater, acts on X in such a way that each G_x, $x \in X$, is a free group of rank at most m, then there is no finitely additive, G-invariant measure on X having total measure 1. Still, these results leave us quite short of a general solution to the question posed earlier, which we restate in the possibly simpler case of a transitive action. The transitivity condition eliminates trivial cases such as groups whose action is the identity action.

Question 13.28. Suppose a nonamenable group G acts transitively on X. Is there a condition on the action that is both necessary and sufficient that no finitely additive, G-invariant measure on $\mathcal{P}(X)$ exists?

Notes

The existence of Marczewski measures in $\mathbb{R}^1$, $\mathbb{R}^2$, and $\mathbb{S}^1$ was proved by E. Marczewski in the 1930s (see [Myc79, Myc80]); [Myc79] contains the explicit formulation of Theorem 13.1. The modification of the Invariant Extension Theorem to obtain a measure that scales according to a fixed homomorphism from G to $(0, \infty)$ (Thm. 13.4) is due to Klee [Kle54] for solvable groups; as a consequence, he proved Corollary 13.5. For amenable groups in general, the result is given in [Myc79].

The Ruziewicz Problem is investigated in [Ban23], where Banach solved the problem in $\mathbb{R}^1$, $\mathbb{R}^2$, and $\mathbb{S}^1$. In fact, the problem had been posed earlier by Lebesgue, in the case of $\mathbb{R}^1$. Lebesgue had realized [Leb04, p. 106] that the Lebesgue integral was the unique invariant mean on $L^\infty[0, 1]$ that satisfied the Monotone Convergence Theorem, and he asked whether this last, rather complicated condition was necessary for the characterization. Because the Monotone Convergence Theorem is equivalent (via Fatou's Lemma) to countable additivity of the underlying measure, and because countable additivity does characterize λ as an invariant measure, Lebesgue was really posing the same problem as Ruziewicz.

Rosenblatt's Problem (stated before Thm. 13.7), which was an attempt to generalize Banach's negative solution of the Ruziewicz Problem, was posed in [Ros81]. Theorem 13.8(b), which solves this problem for countable groups, is due to del Junco and Rosenblatt [JR79]. The other implication of the theorem is due to Losert and Rindler [LR81, Ex. (d)]; a related result was proved independently by Rosenblatt [Ros81, Thm. 3.8]. Theorem 13.8(b) was first shown to be true for nilpotent groups by Rosenblatt and Talagrand [RT81].

The paper by del Junco and Rosenblatt turned out to be the impetus for the solution of the higher-dimensional cases of Ruziewicz's Problem, a problem that had been dormant for fifty years. Using their result that the nonuniqueness of the invariant mean yields an asymptotically invariant sequence, Margulis [Mar80] discovered the argument of Theorem 13.12 and constructed a countable dense subgroup of $SO_n(\mathbb{R})$, $n \geq 5$, that has Property (T). Simultaneously and independently, Rosenblatt [Ros81] used the work of [JR79] to prove that if $n \geq 2$, then the Lebesgue integral is the unique mean on $L^\infty(\mathbb{R}^n)$ that normalizes χ_J, is invariant under isometries, and is invariant under the shear: $\sigma(x_1, \ldots, x_n) = (x_1 + x_2, x_2, \ldots, x_n)$. After seeing Rosenblatt's paper, Sullivan [Sul81] saw how to solve Ruziewicz's Problem in $\mathbb{S}^n$, $n \geq 4$, using essentially the same technique as Margulis. The solution in $\mathbb{R}^n$, $n \geq 3$, is due to Margulis [Mar82], and V. Drinfeld [Dri85] settled the two remaining cases, $\mathbb{S}^2$ and $\mathbb{S}^3$. It should be mentioned that the fundamental result of [JR79] on uniqueness of means and asymptotically invariant sequences was inspired by Namioka's proof [Nam64] that amenable groups satisfy Følner's Condition.

Many others have studied Property (T), and some of their results have a bearing on the topics considered here. See [Ake81, Bek98, CW80, Kaz67, Oh05, Sch81, Sul81, Wan69, Wan74, Wan75].

The fact that Theorem 13.13 could be used to obtain a paradoxical decomposition modulo an ideal was pointed out by Rosenblatt [Ros81], although that paper asserts that m can be taken equal to 1 in Corollary 13.16, which is not known unless $\mathcal{I}$ is closed under countable unions and invariant under similarities. However, Corollary 13.16 for a particular ideal of measurable subsets of $\mathbb{S}^2$, and with $m = 1$, was proved by Rosenblatt earlier [Ros79].

Theorem 13.18 is due to Mycielski [Myc74, Myc79], although a geometric proof of Corollary 13.19 for isometries that avoids the Axiom of Choice was found by Davies [Myc74]. Some results related to Theorem 13.18 may be found in [Ban83]. Theorem 13.20 is due to Rosenblatt for $n \geq 3$ and to Kallman (unpublished) for $n = 2$.

Theorem 13.21 $(d) \Rightarrow (a)$ is due to Wagon [Wag81]; a version of this result for the action of $SL_n(\mathbb{Z})$ on $\mathbb{T}^n$ is given by Dani [Dan85]. The proof of Theorem 13.22 is due to Mycielski. Theorem 13.23 is due to Wagon [Wag81], and that paper also contains Theorem 13.24, although, as indicated in Theorem 13.24's proof, this characterization follows from known results. Theorem 13.25 is due to

Dani [Dan85b] exploiting the ideas of [Dan85] and answers a question of Wagon [Wag81]. Question 13.26 was suggested by A. Borel.

The general problem of characterizing those actions of a free group of rank 2 that yield a paradox was posed by Greenleaf [Gre69, p. 18]. The paper of van Douwen [Dou90] contains several relevant results. Theorem 13.27, which, because AG is strictly smaller than NF (Thm. 12.5), is a substantial strengthening of Corollary 10.6, is due to Rosenblatt [Ros81, Thm. 3.5]. It is interesting to note that the proof of Theorem 13.27 is this book's third proof that a locally commutative action of a free group of rank 2 is paradoxical (see also Thm. 5.5 and Cor. 10.6). The result was discovered again, with yet a different proof, by Akemann [Ake81]. Promislow's result mentioned at the end of the chapter (see [Pro83]) is noteworthy because its proof combines methods of graph theory and combinatorial group theory.

14

Growth Conditions in Groups and Supramenability

We present some interesting connections between a group's amenability and its rate of growth: the speed at which new elements appear when one considers longer and longer words using letters from a fixed finite subset of the group. This approach sheds light on a fundamental difference between Abelian and solvable groups. Both families are amenable, but their growth properties can be quite different. This will explain why there is a paradoxical subset of the plane (the Sierpiński–Mazurkiewicz Paradox, Thm. 1.7) but no such subset of the real line.

The study of growth conditions also elucidates the amenability of Abelian groups. The proof that Abelian groups are amenable (Thm. 12.4(b)) is somewhat complicated. But one can use growth rates to prove quite simply (without the Axiom of Choice) that an Abelian group is not paradoxical (Thm. 14.21). Then one can deduce amenability simply by calling on Tarski's Theorem (Cor. 9.2).

We shall also discuss the cogrowth of a group, a notion that refines the idea of growth. This leads to a striking and important characterization of amenable groups (Cor. 14.26), a characterization that is central to Ol'shanksii's construction of a group that is periodic (meaning each group element has finite order; hence it is NF) but not amenable.

14.1 Supramenable Groups

The notion of amenability of a group is based on the existence of a measure of total measure 1. But we are often interested in invariant measures that assign specific subsets measure 1. The following definition is the appropriate strengthening of amenability that guarantees the existence of such measures for any nonempty subset of a set on which the group acts.

Definition 14.1. *A group G is* supramenable *if, for any nonempty $A \subseteq G$, there is a finitely additive, left-invariant measure $\mu : \mathcal{P}(G) \to [0, \infty)$ with $\mu(A) = 1$. The class of supramenable groups is denoted AG^+.*

Of course, if G is supramenable, then no nonempty subset of G is paradoxical. By Tarski's Theorem, this latter condition is equivalent to supramenability. Before discussing some simple properties and examples of supramenable groups, we give the main result on supramenable actions.

Theorem 14.2. *Suppose that a supramenable group G acts on X and that A is a nonempty subset of X. Then there is a finitely additive, G-invariant measure $\mu : \mathcal{P}(X) \to [0, \infty]$ such that $\mu(A) = 1$. Hence no nonempty subset of X is G-paradoxical.*

Proof. Fix any point $x \in A$ and then assign a subset B^* of G to every $B \subseteq X$ as follows: $B^* = \{g \in G : g(x) \in B\}$. Note that the identity of G is in A^*. Therefore, by G's supramenability, there is a finitely additive, left-invariant $\nu : \mathcal{P}(G) \to [0, \infty]$ with $\nu(A^*) = 1$. Now, define the desired measure μ on $\mathcal{P}(X)$ by setting $\mu(B) = \nu(B^*)$. Then $\mu(A) = \nu(A^*) = 1$, and the finite additivity of ν easily yields the same for μ. Finally, if $h \in G$, then $h(B)^* = h(B^*)$, whence the G-invariance of μ follows from the left-invariance of ν. □

Of course, any nonamenable group fails to be supramenable, but the Sierpiński–Mazurkiewicz Paradox provides an example of an amenable group that is not supramenable. The planar isometry group, G_2, is solvable and hence amenable, but the existence of a paradoxical subset of the plane, together with the previous theorem, implies that G is not supramenable. This can be shown more directly by using Theorem 1.8, which constructed a free subsemigroup of rank 2 in G_2. In fact, free semigroups of rank 2 play much the same role for supramenability that free groups of rank 2 do for amenability. As we now show, a group that contains a free subsemigroup of rank 2 cannot be supramenable. Recall that a semigroup S is free with free generating set A if S is the semigroup generated by A and distinct words using elements of A as letters yield distinct elements of S.

Proposition 14.3. *If G contains σ and ρ, free generators of a free subsemigroup of G, then G is not supramenable.*

Proof. Let S be the subsemigroup of G generated by σ and ρ. It was shown in Proposition 1.3 that S is G-paradoxical: $S \supseteq \sigma S$, ρS and $S = \sigma^{-1}(\sigma S) = \rho^{-1}(\rho S)$. □

Let NS be the class of groups with no free subsemigroup of rank 2; then Proposition 14.3 states that $AG^+ \subseteq NS$. Although no simple example of a group in $NS \setminus AG^+$ is known, the work on the analogous problem for amenable groups shows that such examples exist. Recall from Theorem 12.5 that there is a nonamenable periodic group G. Then G is not supramenable, but because all elements of G have finite order, $G \in NS$. Another natural question is whether $AG^+ = NS \cap AG$. An example due to Grigorchuk shows that this is false, as it lies in $NS \cap AG \setminus AG^+$. However, AG^+ and NS do coincide when restricted to the subclass EG of AG; see remarks following Theorem 14.18.

The next theorem details the known closure properties of AG^+, and it is easy to see that NS satisfies all of them as well. (For (f), use the fact that if σ, τ freely generate a subsemigroup of G, then for each $n \geq 1$, so does the pair σ^n, τ^n. But for some n, σ^n and τ^n both lie in H.) Furthermore, neither class is closed under general group extension because the solvable group G_2 is not amenable. But, unlike the case of the classes AG, NF, and EG, it is not clear that AG^+ and NS share the same closure properties (see Question 14.5).

Theorem 14.4. (a) *Finite groups are supramenable.*

(b) *(AC) Abelian groups are supramenable.*

(c) *A subgroup of a supramenable group is supramenable.*

(d) *If N is a normal subgroup of a supramenable group G, then G/N is supramenable.*

(e) *(AC) If G is the direct union of a directed system of supramenable groups, $\{G_\alpha : \alpha \in I\}$, then G is supramenable.*

(f) *If H, a subgroup of G, is supramenable and H has finite index in G, then G is supramenable.*

Proof. (a) If $A \subseteq G$ is nonempty, let μ be defined by $\mu(B) = |B|/|A|$.

(b) This will be deduced from the more general result, to be proved later (Thm. 14.21), that groups that do not grow too fast are necessarily supramenable.

(c) If A is a nonempty subset of H which, in turn, is a subgroup of a supramenable group G, then simply restrict a measure on $\mathcal{P}(G)$ that normalizes A to $\mathcal{P}(H)$.

(d) If A is a nonempty set of N-cosets, let μ be a left-invariant measure on $\mathcal{P}(G)$ that normalizes $\bigcup A$. Then define ν on $\mathcal{P}(G/N)$ by $\nu(B) = \mu(\bigcup B)$.

(e) Let a nonempty $A \subseteq G$ be given. Because each G_α is contained in a G_β that intersects A, we may assume that each G_α intersects A; simply delete from the system any subgroups that miss A. Consider the topological space $[0, \infty]^{\mathcal{P}(G)}$, which is compact. For each $\alpha \in I$, let $\mathcal{M}_\alpha$ consist of those finitely additive $\mu : \mathcal{P}(G) \to [0, \infty]$ such that $\mu(A) = 1$ and μ is G_α-invariant. Each G_α is supramenable, so if μ_α is a G_α-invariant measure on $\mathcal{P}(G_\alpha)$ with $\mu_\alpha(A \cap G_\alpha) = 1$, then the measure defined by $\mu(B) = \mu_\alpha(B \cap G_\alpha)$ lies in $\mathcal{M}_\alpha$. As in Theorem 12.4, each $\mathcal{M}_\alpha$ is closed, and because $\mathcal{M}_\alpha \cap \mathcal{M}_\beta \supseteq \mathcal{M}_\gamma$ if $G_\alpha, G_\beta \subseteq G_\gamma$, the collection $\{\mathcal{M}_\alpha : \alpha \in I\}$ has the finite intersection property. Compactness yields $\mu \in \bigcap \mathcal{M}_\alpha$, and such a μ is a G-invariant measure normalizing A.

(f) Suppose A is a nonempty subset of G, and let $\{g_i, \ldots, g_m\}$ represent the right cosets of H in G. Because H is supramenable, we may apply Theorem 14.2 to the action of H on G by left multiplication to obtain an H-invariant measure ν on $\mathcal{P}(G)$ with $\nu(\bigcup g_i A) = 1$. Note that $0 < \sum \nu(g_i A) < \infty$, for otherwise some $\nu(g_i A) = \infty$, contradicting $\nu(\bigcup g_i A) = 1$.

Now, let $a = \sum \nu(g_i A)$ and define μ on $\mathcal{P}(G)$ by $\mu(B) = (1/a) \sum \nu(g_i B)$. Then μ is finitely additive and $\mu(A) = 1$. Moreover, for any $g \in G$, the set $\{g_i g\}$ represents the right cosets of H. Hence $\mu(gB) = (1/a) \sum_i \nu(g_i g B) =$

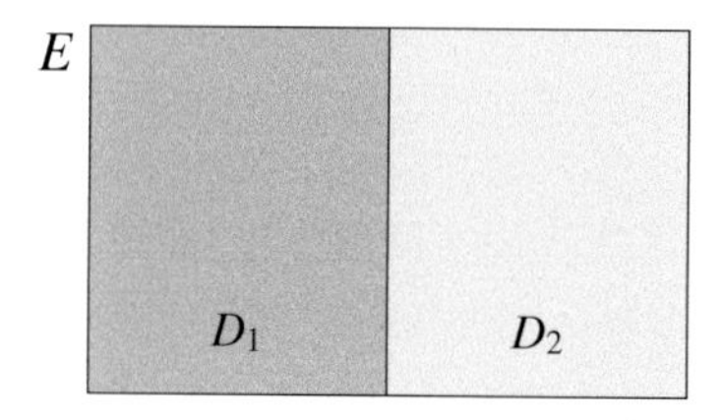

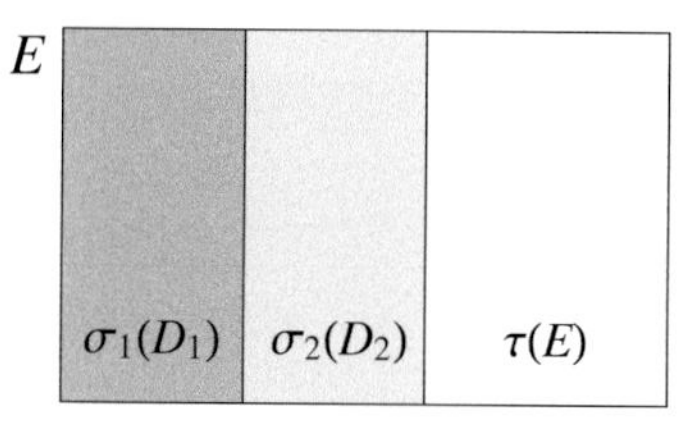

Figure 14.1. In a $(1, 2)$-paradoxical set E, there is a partition of E (on the right) into three pieces that yield two copies of E via three isometries.

$(1/a)\sum_k \nu(h_k g_k B)$, where $h_k \in H$, and the H-invariance of ν then yields that $\mu(gB) = \mu(B)$. □

It can be shown that NS is closed under (finite) direct products (see [Ros74]); therefore the same is true of $NS \cap AG$. But it is not known whether the corresponding closure property for AG^+ holds.

Question 14.5. If G and H are supramenable, is $G \times H$ supramenable?

14.2 Bounded Paradoxical Sets

The Sierpiński–Mazurkiewicz Paradox gives a simple paradoxical subset of the plane, but it is unbounded. A question that was open for many years was whether a bounded example exists. That was answered affirmatively by W. Just [Jus87]. We present here G. A. Sherman's later construction [She90]. We wish to minimize the number of pieces in a paradox, and so we use the following definition of an (m, n)-paradoxical set; this definition requires that the pieces cover the set.

Definition 14.6. *A set E is* (m, n)-paradoxical *if there are isometries ρ_i and σ_j and three partitions of E: $\{A_i : i = 1, \ldots, m\}$, $\{B_j : i = 1, \ldots, n\}$, and $\{\rho_i(A_i) : i = 1, \ldots, m\} \cup \{\sigma_j(B_j) : i = 1, \ldots, n\}$.*

In particular, a $(1, 2)$-paradoxical set E (Fig. 14.1) has a partition $\{D_1, D_2\}$ such that, for isometries σ_1, σ_2, and τ, E is partitioned by $\{\sigma_1(D_1), \sigma_2(D_2), \tau(E)\}$. Note that an (m, n)-paradoxical set is also paradoxical for any pair of larger numbers.

The Sierpiński–Mazurkiewicz set is $(1, 1)$-paradoxical. The bounded set found by Just is $(1, 3)$-paradoxical; Sherman's example, which we give next, is $(2, 2)$-paradoxical. The common feature of Just's and Sherman's constructions is that they both use a rotation of infinite order and translations by algebraic complex numbers.

Theorem 14.7. *There is a bounded* $(2, 2)$*-paradoxical subset of the plane.*

Proof. Working in $\mathbb{C}$, let ρ be counterclockwise rotation by 1 radian ($\rho(z) = e^i z$, though any transcendental number on the unit circle can replace e^i), let τ be translation by 1, and let E be the orbit of $\mathbf{0}$ under the semigroup action of ρ and τ, all

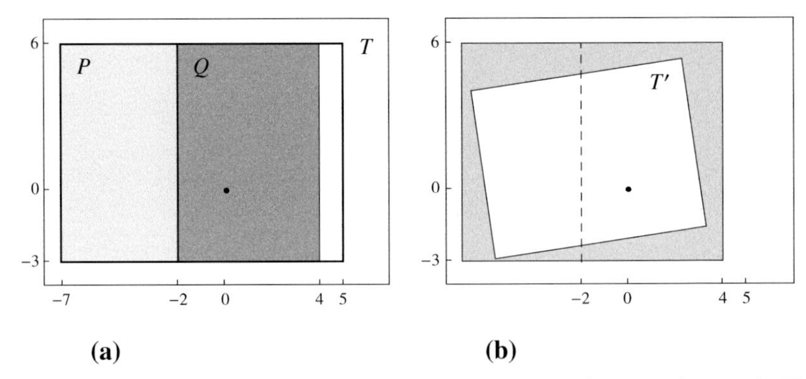

Figure 14.2. The rectangles used to get a bounded paradoxical subset of rectangle T'. Rectangle T contains Q and is interior-disjoint from P; it rotates to T', which is contained in $P \cup Q$.

as in the Sierpiński–Mazurkiewicz Paradox. Define three closed rectangles as follows: $T = [-2, 5] + i[-3, 6]$, $P = [-7, -2] + i[-3, 6]$; $Q = (-2, 4] + i[-3, 6]$ (Fig. 14.2(a)).

Let $\sigma = \rho^8$ and $T' = \sigma(T)$. Eight radians is about 98°, close to a quarter-turn, so we have that $T' \subset P \cup Q$ (Fig. 14.2(b)). Because Q, $Q + 1$, $P + 6$, and $P + 7$ are all contained in T, we have that $\sigma(Q)$, $\sigma\tau(Q)$, $\sigma\tau^6(P)$, and $\sigma\tau^7(P)$ are all subsets of T'. Now define $X = \bigcup X_n \subset T'$ inductively as follows: $X_0 = \{\mathbf{0}\}$; $X_1 = \{\sigma\tau(\mathbf{0})\}$; and X_{n+1} is defined by taking each point $z \in X_n$ and

- if $z \in P$, put $\sigma\tau^6(z)$ and $\sigma\tau^7(z)$ in X_{n+1}
- if $z \in Q$, put $\sigma(z)$ and $\sigma\tau(z)$ in X_{n+1}

Any point placed in X is either $\mathbf{0}$ or has the form $\sigma\,\square\square\ldots\square\,\tau(\mathbf{0})$, where each $\square$ is one of $\{\rho, \tau\}$. Note that $X \subseteq E$, the Sierpiński–Mazurkiewicz set. Let $X_P = X \cap P$, and same for Q. We have that X is the disjoint union of X_P and X_Q, and the following four sets also partition X:

$$\mathcal{P} = \{\pi_1, \pi_2, \pi_3, \pi_4\} = \{\sigma\tau^6(X_P), \sigma(X_Q), \sigma\tau^7(X_P), \sigma\tau(X_Q)\}.$$

The sets in $\mathcal{P}$ clearly cover all points of X, by construction. Note that $\mathbf{0} \in \pi_2$. Because e^i is transcendental, we have that every point in E other than $\mathbf{0}$ (which is both identity($\mathbf{0}$) and $\rho(\mathbf{0})$) has a unique representation as $w(\mathbf{0})$, where w is a word in the free semigroup generated by ρ and τ. This uniqueness can be used to show disjointness of the sets in $\mathcal{P}$ as follows:

- if $z \in \pi_1$, then $z = \sigma\tau^6\sigma\,\square\square\ldots\square\,\tau(0)$
- if $z \in \pi_2$, then $z = 0$ or $z = \sigma^2\square\square\ldots\square\tau(0)$
- if $z \in \pi_3$, then $z = \sigma\tau^7\sigma\,\square\square\ldots\square\,\tau(0)$
- if $z \in \pi_4$, then $z = \sigma\tau\,\sigma\,\square\square\ldots\square\,\tau(0)$

By uniqueness of the words, these representations give disjointness of the four sets in $\mathcal{P}$. Because $\tau^{-6}\sigma^{-1}(\pi_1) = X_P$, $\sigma^{-1}(\pi_2) = X_Q$, $\tau^{-7}\sigma^{-1}(\pi_3) = X_P$, and $\tau^{-1}\sigma^{-1}(\pi_4) = X_Q$, the four sets define a (2, 2)-paradox for the bounded set X. □

Every word w for which $w(\mathbf{0})$ is in Sherman's set X is a word in σ, τ, where $\sigma = \rho^8$, but not all words appear. Note that instead of e^i, we may choose a transcendental $e^{i\theta}$ so that, with ρ being the corresponding rotation, $\rho(T) = T' \subset P \cup Q$; that is, we choose θ to be near 8. Then, instead of ρ^8 in the proof, we can use just ρ. Making such a choice requires knowing that the set of θ for which $e^{i\theta}$ is transcendental is dense in $[0, 2\pi)$. To see this, choose any $\phi \in [0, 1)$ so that $e^{2\pi i\phi}$ is transcendental. Then ϕ is irrational and Kronecker's Theorem [HWHSW, §23.2] says that $\{n\phi \pmod 1) : n \in \mathbb{Z}\}$ is dense in $[0, 1]$. It follows that a power $e^{i2\pi n\phi}$ can be found arbitrarily close to 8.

The set X of Theorem 14.7 is countable and so has measure zero. To understand more about paradoxical sets in the plane, let μ be a Banach measure. Then if X is paradoxical, $\mu(X)$ must be 0 or ∞, and so the same is true of $\lambda(X)$ for measurable X; for bounded paradoxical sets, $\mu(X) = 0$, and so X has empty interior. Sherman strengthened these results by showing that if X is a paradoxical set in the plane, then

- X cannot have nonempty interior
- X cannot be measurable with infinite Lebesgue measure

On the other hand, Burke [Bur04] showed that a bounded paradoxical set can have positive outer measure. We prove these results next, starting with Sherman's work from [She91].

Theorem 14.8 (AC). *If X is a paradoxical subset of $\mathbb{R}^2$, then $\mu(X_0) = 0$ for every bounded set $X_0 \subseteq X$.*

This result has several consequences of interest.

Corollary 14.9 (AC). (a) *Every paradoxical subset of $\mathbb{R}^2$ has empty interior.*
(b) *If X is measurable and paradoxical, then $\lambda(X) = 0$.*
(c) *Let X be a subset of $\mathbb{R}^2$ with nonempty interior or having positive (possibly infinite) Lebesgue measure. Then there is a finitely additive, isometry-invariant measure ν defined on all subsets of $\mathbb{R}^2$ and with $\nu(X) = 1$.*

Note that part (c) is not true in $\mathbb{R}^n$ ($n \geq 3$) as stated because Theorem 6.8 yields a paradoxical decomposition of $\mathbb{R}^n$. And, as shown by Penconek [Pen91, Cor. 1.4], there are sets of infinite measure in every $\mathbb{R}^n$ that are not paradoxical, and so have measures as in (c). In $\mathbb{R}^1$, (c) holds for *any* nonempty set X (see Cor. 14.25).

Proof. (a) If a paradoxical subset were to contain a nonempty open set, then it would contain a disk of positive radius, which has positive μ-measure.

(b) Let X be measurable and paradoxical. Then Theorem 14.8 implies that $\lambda(X \cap D_n) = 0$ for every $n \in \mathbb{N}$, where D_n is an origin-centered disk of radius n. Countable additivity then gives $\lambda(X) = 0$.

(c) By (a) and (b), using Tarski's Theorem (Cor. 11.2). □

Proof of Theorem 14.8. Let X be a paradoxical subset of $\mathbb{R}^2$ with the witnessing sets $A_1, \ldots, A_m, B_1, \ldots, B_n$ and isometries σ_i, τ_j as in the definition of an (m, n)-paradoxical set. So each of $\{A_i\}$, $\{B_j\}$, and $\{\sigma_i(A_i)\} \cup \{\tau_j(B_i)\}$ is a partition of X. Let P be any point in the plane, let D_r denote the closed disk of radius r centered at P, and let ρ stand for any of the isometries σ_i, τ_j. Define s to be $\max_\rho \|P - \rho(P)\|$. Then $\rho(D_r) \subseteq D_{r+s}$ for any ρ, and so $\sigma_i(A_i \cap D_r) \subseteq \sigma_i(A_i) \cap D_{r+s}$ and $\tau_j(B_i \cap D_r) \subseteq \tau_j(B_j) \cap D_{r+s}$.

Using the aforementioned three partitions and the isometry invariance of μ, we have, for each positive r,

$$\begin{aligned} 2\mu(X \cap D_r) &= \sum_{i=1}^{m} \mu(A_i \cap D_r) + \sum_{j=1}^{n} \mu(B_j \cap D_r) \\ &= \sum_{i=1}^{m} \mu(\sigma_i(A_i \cap D_r)) + \sum_{j=1}^{n} \mu(\tau_j(B_j \cap D_r)) \\ &\leq \sum_{i=1}^{m} \mu(\sigma_i(A_i) \cap D_{r+s}) + \sum_{j=1}^{n} \mu(\tau_j(B_j) \cap D_{r+s}) = \mu(D_{r+s}). \end{aligned}$$

Then an easy induction on n shows that $2^n\mu(X \cap D_r) \leq \mu(X \cap D_{r+ns})$ and, finally,

$$\mu(X \cap D_r) \leq \lim_{n\to\infty} \frac{1}{2^n}\mu(D_{r+ns}) = \lim_{n\to\infty} \frac{1}{2^n}\pi(r + ns)^2 = 0.$$

This is true for all r, so, for large enough r and bounded $X_0 \subseteq X$, $\mu(X_0) = \mu(X_0 \cap D_r) \leq \mu(X \cap D_r) = 0$. □

Theorem 14.8 can be strengthened as follows, where μ is still a Banach measure.

Corollary 14.10. *If X is a paradoxical subset of $\mathbb{R}^2$, then $\mu(Y) = 0$ for every subset Y of X having finite outer measure.*

Proof. Let V be a set containing Y and having finite measure. Let $\{V_i\}$ partition V into countably many bounded measurable sets. For each $\epsilon > 0$, there is an integer n such that $\mu(\bigcup_{i=n}^{\infty} V_i) < \epsilon$. Hence, by Theorem 14.8, we have

$$\mu(Y) = \sum_{i=1}^{n-1} \mu(Y \cap V_i) + \mu\left(Y \cap \left(\bigcup_{i=n}^{\infty} V_i\right)\right) < 0 + \epsilon.$$

□

Sherman [She91] showed more: A paradoxical subset of $\mathbb{R}^2$ must have Lebesgue inner measure zero; so there cannot be a paradoxical Lebesgue measurable subset of the plane having infinite measure. So that raises the question whether there is a bounded paradoxical set that does not have Lebesgue measure zero, that is, that has positive outer measure. Such a set was constructed by Burke [Bur04].

Recall that Lebesgue inner measure λ_* is given by $\lambda_*(E) = \sup\{\lambda(K) : K \subseteq E$ and K compact$\}$. As Sherman observed, it follows from Theorem 14.8 that every paradoxical set in the plane has inner measure zero. For suppose E was a counterexample. Then there is a compact set $K \subseteq E$ such that $0 < \lambda(K) = \mu(K)$, which, because compact sets are bounded, contradicts the theorem.

Mycielski and Wagon [MW84] showed that $\mathbb{H}^2$ is paradoxical (Thm. 4.5). The proof of Theorem 14.8 gives us some information about such paradoxes. Recall that there is no Banach measure in $\mathbb{H}^2$, because of Mycielski's hyperbolic analog of the classic Banach–Tarski Paradox (Thm. 4.17). Nevertheless, letting X be a paradoxical subset of $\mathbb{H}^2$ with pieces that are measurable with respect to hyperbolic Lebesgue measure $\lambda_{\mathbb{H}}$, the inequality of Theorem 14.8's proof is valid: $\lambda_{\mathbb{H}}(X \cap D_r) \leq \lim_{n\to\infty} 2^{-n}\lambda_{\mathbb{H}}(D_{r+ns})$. But in the hyperbolic plane, the preceding limit is not always 0; it depends on s. Using the fact that the area of a hyperbolic disk of radius r is $4\pi \sinh^2(r/2)$,

$$\lim_{n\to\infty} \frac{1}{2^n}\lambda_{\mathbb{H}}(D_{r+ns}) = \begin{cases} 0 & \text{if } s < \ln 2, \\ \pi e^r & \text{if } s = \ln 2, \\ \infty & \text{if } s > \ln 2. \end{cases}$$

This gives us some information on paradoxes such as the Mycielski–Wagon Paradox.

Theorem 14.11. *Let X be a measurable subset of $\mathbb{H}^2$ having a paradoxical decomposition with measurable pieces and hyperbolic isometries. Then either $\lambda_{\mathbb{H}}(X) = 0$ or, for each point $P \in \mathbb{H}^2$, at least one of the isometries moves P to a point at hyperbolic distance $\ln 2$ or more from P.*

Proof. Suppose some point P_0 was not moved distance $\ln 2$ or greater for any isometry used in the paradox. Then, let this P_0 be the point P of the proof of Theorem 14.8, and define s as in that proof. Then s, being a maximum over a finite set, would be under $\ln 2$, and the limit displayed earlier would be 0, which means that $\lambda_{\mathbb{H}}(X) = 0$. □

G. A. Sherman [She91] then raised the following question.

Question 14.12. Is there a constant larger than $\ln 2$ for which Theorem 14.11 holds?

Now we turn to M. R. Burke's construction of paradoxical sets with positive outer measure [Bur04]. The following theorem is the key.

Theorem 14.13 (AC). *Let E_0 be a countable (m, n)-paradoxical planar set and let $\{K_\xi : \xi < 2^{\aleph_0}\}$ be a family of uncountable compact subsets of $\mathbb{C}$. Then there exists a planar (m, n)-paradoxical set E such that, for each ξ, $E \cap K_\xi \neq \emptyset$.*

Proof. Work in $\mathbb{C}$. We may assume by translation that $\mathbf{0} \in E_0$. Let the witnessing sets of the (m, n)-paradox be the partitions $\{A_i : 1 \leq i \leq m\}$ and $\{A_i : m+1 \leq i \leq m+n\}$, with isometries σ_j of the form $z \mapsto a_j z + b_j$ with $|a_j| = 1$. Let $\mathcal{H}$ be the group generated by $\{a_j\}$ using multiplication; $\mathcal{H}$ is a subgroup of the multiplicative group of the unit circle in $\mathbb{C}$. Recall that a closed set obeys the Continuum Hypothesis, and so each $|K_\xi| = 2^{\aleph_0}$. Inductively choose, for $\xi < 2^{\aleph_0}$, $z_\xi \in K_\xi$ so that $z_\xi \notin \mathbb{Q}(\mathcal{H} \cup E_0 \cup \{z_\eta : \eta < \xi\})$. This is possible because the cardinality of the field $\mathbb{Q}(\mathcal{H} \cup E_0 \cup \{z_\eta : \eta < \xi\})$ is less than $2^{\aleph_0}$. Define $E = \{az_\xi + y : \xi < 2^{\aleph_0}, a \in \mathcal{H}, y \in E_0\}$.

The representation of an element of E in the form $az_\xi + y$ is unique. For suppose $a_1 z_{\xi_1} + y_1 = a_2 z_{\xi_2} + y_2$, where $\xi_1 \leq \xi_2$; then we have $z_{\xi_2} \in \mathbb{Q}(\mathcal{H} \cup E_0 \cup \{z_{\xi_1}\})$, contradiction. So $\xi_1 = \xi_2$, and we have $(a_1 - a_2) z_{\xi_1} = y_2 - y_1$. But $z_{\xi_1} \notin \mathbb{Q}(\mathcal{H} \cup E_0)$, so $a_1 = a_2$ and then $y_1 = y_2$.

For $1 \leq j \leq m + n$, define $B_j = \{az_\xi + y : \xi < 2^{\aleph_0}, a \in \mathcal{H}, y \in A_j\}$. Then, because of the A-partitions and the uniqueness fact just proved, $\{B_i : 1 \leq i \leq m\}$ and $\{B_i : m + 1 \leq i \leq m + n\}$ are partitions of E. And for $j \leq m + n$,

$$\begin{aligned}\sigma_j(B_j) &= \{a_j a z_\xi + a_j y + b_j : \xi < 2^{\aleph_0}, a \in \mathcal{H}, y \in E_0\} \\ &= \{az_\xi + \sigma_j(y) : \xi < 2^{\aleph_0}, a \in \mathcal{H}, y \in E_0\}.\end{aligned}$$

The uniqueness of representations and the fact that $\{\sigma_j(A_j) : j \leq m + n\}$ is a partition of E_0 mean that $\{\sigma_j(B_j) : j \leq m + n\}$ partitions E. Thus E is (m, n)-paradoxical. Because $\mathbf{0} \in E_0$, each $z_\xi \in E$, and that gives the desired nonempty intersections. □

Corollary 14.14 (AC). (a) *There exists an unbounded planar* $(1, 1)$-*paradoxical set having positive outer Lebesgue measure.*

(b) *There exist bounded planar* $(2, 2)$-*paradoxical and* $(1, 3)$-*paradoxical sets having positive outer measure.*

Proof. (a) Apply Theorem 14.13 to the Sierpiński–Mazurkiewicz set and the family of all uncountable compact subsets of $\mathbb{C}$ to get a set E. There are only continuum many bounded open sets (they arise from unions of intervals with rational ends), so there are that many compact sets. Then E intersects every uncountable compact set so that $\lambda_*(\mathbb{C} \setminus E) = 0$. And then E cannot have Lebesgue outer measure 0, for if it did, it would have Lebesgue measure 0, and the complement would have positive or infinite Lebesgue measure and hence positive or infinite λ_*.

(b) Use Theorem 14.13 on the corresponding set E_0 of Just or Sherman, and let K be the family all uncountable compact subsets of the unit disk D. Then the set E constructed by the theorem is bounded, because, in E's definition, a, z_ξ, and y are from bounded sets. Suppose $\lambda(E \cap D) = 0$. Then $\lambda(D \setminus E) = \pi$, and so $D \setminus E$, having positive inner measure, contains a compact set K_0 of positive measure. But $E \cap K_0$ is nonempty, a contradiction. This means $E \cap D$ has positive outer measure. □

The Axiom of Choice, which is used in the preceding results, is not essential to the construction of large bounded paradoxical plane sets. One can use the isometries and partitions in Sherman's construction (Thm. 14.7) to get a continuum-sized bounded paradoxical set as follows.

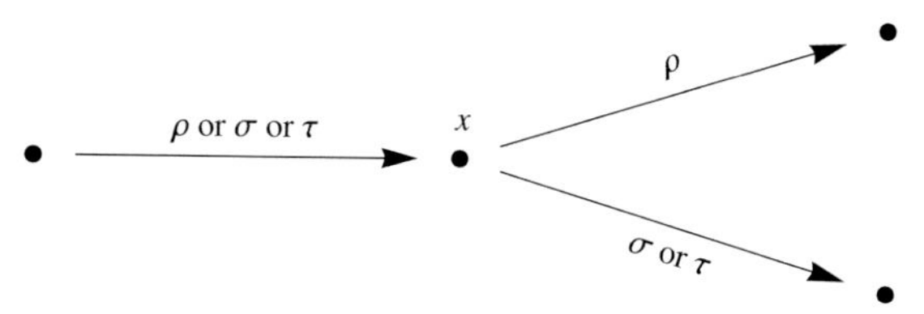

Figure 14.3. All vertices of $\mathcal{G}$ have one edge in and two out; if the point is in D_1, there is a σ-edge going out, and if it is in D_2, a τ-edge leaves. And the incoming edge depends on which of $\rho(X)$, $\sigma(D_1)$, and $\tau(D_2)$ contains the point.

Theorem 14.15. *There is a bounded paradoxical subsets of the plane having cardinality* $2^{\aleph_0}$.

Proof. The coefficients appearing in the proof of Theorem 14.7 are algebraic numbers, and we have to modify the approach to get the desired continuum-sized set. Use Theorem 7.5 to get a nonempty perfect set P of complex numbers in the unit disk that are algebraically independent over the field generated by the algebraic numbers together with the number ρ^8 from Theorem 14.7's proof; this is possible because there are only countably many polynomials with coefficients in the field. Let S be the semigroup defined in Theorem 14.7. Then, as in that theorem, define $E = \{w(z) : w \in S, p \in P, \text{ and } z = \sigma^{-k}(p), k = 0, 1, 2, \ldots\}$, where $\sigma = \rho^8$ is a rotation as in Theorem 14.7 (or as in Just's construction). Then E is a bounded $(2, 2)$-paradoxical (or $(1, 3)$-paradoxical) set; verification is as in the proof of Theorem 7.14. Note that the isometries witnessing the paradox are the same as those in Theorem 14.7. □

Lindenbaum [Lin26] proved that there is no $(1, 1)$-paradoxical bounded subset of $\mathbb{R}^2$, a proof that was improved by Hadwiger, Debrunner, and Klee [HDK64, p. 80]. Using some of the reasoning of the improved proof, Sherman [She90] obtained the same result for $(1, 2)$-paradoxical sets; this combines with Theorem 14.7 and Just's result to resolve the whole story: The only forbidden pairs for bounded paradoxical sets are $(1, 1)$ and $(1, 2)$.

Theorem 14.16 (G. A. Sherman). *There is no* $(1, 2)$-*paradoxical bounded subset of* $\mathbb{R}^2$.

Proof. Suppose there is a $(1, 2)$-paradoxical decomposition of a bounded set X. So we have a partition of X into D_1 and D_2 and isometries ρ, σ, τ, so that $\rho(X)$, $\sigma(D_1)$, $\tau(D_2)$ form a partition of X. Make a directed infinite graph $\mathcal{G}$ with vertex set X and possibly with loops, as follows: edges are $x \bullet\!\!\rightarrow \rho(x)$ for all x, and $x \bullet\!\!\rightarrow \sigma(x)$ where $x \in D_1$, and $x \bullet\!\!\rightarrow \tau(x)$ where $x \in D_2$. Think of the edges as being labeled with the functions ρ, σ, τ. Every vertex has indegree 1 and outdegree 2 (Fig. 14.3). We need an easy lemma. □

Lemma 14.17. *Suppose $\mathcal{H}$ is a directed graph that is connected in the undirected sense and has indegree 1 at each vertex. Then $\mathcal{H}$ has at most one directed cycle.*

Proof. If C and D are distinct directed cycles, then they must be disjoint by the indegree hypothesis. Let $\{v_i : 1 \leq i \leq n\}$ be an undirected path from a vertex in

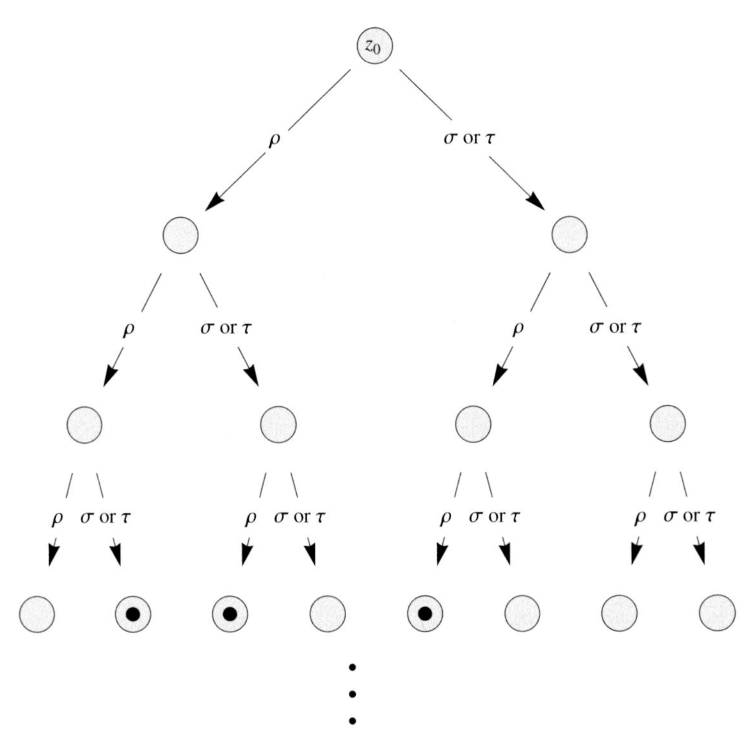

Figure 14.4. The tree forming the graph $\mathcal{H}$. The marked nodes at the bottom are relevant to the proof of case 1.

C to one in D. The last edge in this path must be directed backward: $v_n \bullet\!\!\rightarrow v_{n-1}$. Then let v_j be the first vertex in the path such that the edge to the next vertex is directed backward; v_j has indegree 2, a contradiction. □

Now work in the graph $\mathcal{G}$; the term *component* means component in the undirected sense. If every component of $\mathcal{G}$ has a directed cycle, choose c to be a vertex in such a cycle and edge $c \bullet\!\!\rightarrow z_0$ that is not in the cycle. Otherwise, let z_0 be any vertex in a component having no cycle. Let $\mathcal{H}$ be the subgraph of $\mathcal{G}$ generated by z_0 and all directed paths starting at z_0: the *strong component* of z_0. Then $\mathcal{H}$ has no cycle by Lemma 1 and so is a directed tree (Fig. 14.4). Let U be the vertices of $\mathcal{H}$; note that $\rho(U) \subseteq U$.

Because z_0 was defined to be part of no directed cycle, $\rho(z_0) \neq z_0$ and, furthermore, $\{\rho^n(z_0)\}$ consists of distinct points. So ρ is not a reflection or a rotation of finite order. And because X is bounded, so is $\{\rho^n(z_0)\}$, which means that ρ is not a translation or glide reflection. Therefore ρ is a rotation of infinite order; let O be the fixed point of this rotation and D the smallest closed disk, of radius r, say, centered at O and containing all vertices in $\mathcal{H}$.

Claim. The circle forming ∂D, the boundary of D, is contained in $\overline{U}$, the closure of U.

Proof of claim. The ρ-orbit of each point in U lies in and, by Kronecker's Theorem, is dense in some circle around O. If some $h \in U$ lies on ∂D, then Kronecker's Theorem yields that the orbit of h is dense in ∂D, proving the claim. If not, then U decomposes into countably many ρ-orbits; each such orbit lies on (and is dense in) a circle C_r of radius r. Let $\{r_i\}$ be the radii that arise in this way, with $r_0 = 0$. Let $\bar{r}$ be the supremum of these radii. Then D is the disk of radius $\bar{r}$, because D must contain all the circle C_{r_i} and the disk of radius $\bar{r}$ is the smallest that does so. Find points $h_i \in U$ that have a limit h_∞ lying on ∂D. Now, given any positive ϵ and any point Q on the circle ∂D, there is some $\rho^n(h_\infty)$ within distance $\epsilon/2$ of Q. But there is some h_i within distance $\epsilon/2$ of h_∞. So this point h_i is in U and within ϵ of Q. Hence U is dense in ∂D.

The proof now concludes with geometric arguments for six cases depending on the type of σ and τ and their action on the point O. We present only two cases here; see [She90] for the complete proof.

Case 1. σ and τ are both rotations fixing O.

Case 2. $\sigma(O) \neq O$ and $\tau(O) \neq O$.

The additional four cases are $\sigma(O) \neq O$ and τ is a rotation fixing O; $\sigma(O) \neq O$ and τ is a reflection fixing O; σ and τ are both reflections fixing O; σ is a rotation fixing O and τ is a reflection fixing O.

Case 1 proof. The points in the tree of the form $\gamma\, \rho^2(z_0)$, $\rho\, \gamma\, \rho(z_0)$, and $\rho^2\, \gamma(z_0)$, where in each case γ is one of $\{\sigma, \tau\}$ are all different (these are the black dots on the lowest level in Fig. 14.4). But two of the γs must be the same, and all three rotations fix O and therefore commute, so they are not different.

Case 2 proof. The circle ∂D of radius r cannot be covered by two disks of radius r unless it is covered by one of them. Therefore $\sigma^{-1}(D) \cup \tau^{-1}(D)$ does not cover ∂D. So there is $\upsilon \in U \setminus (\sigma^{-1}(D) \cup \tau^{-1}(D))$. But because $\upsilon \in U$, one of $\sigma(\upsilon)$ or $\tau(\upsilon)$ is in U, and so $\upsilon \in \sigma^{-1}(D) \cup \tau^{-1}(D)$, contradiction. □

Lindenbaum's original proof is quite simple, so we include it here.

Theorem 14.18. *No nonempty bounded subset E of $\mathbb{R}^2$ contains two disjoint subsets, each of which is congruent to E.*

Proof. Suppose $A, B \subseteq E$, $A \cap B = \varnothing$, and σ_1, σ_2 are isometries such that $\sigma_1(E) = A$ and $\sigma_2(E) = B$. Each σ_i^n maps E into a proper subset and therefore is not the identity; hence σ_1 and σ_2 have infinite order. This means that σ_1 and σ_2 are not reflections, but in fact they cannot be translations or glide reflections either. For if σ_i is a translation by $\vec{\upsilon}$, choose P in $\overline{E}$ such that any other point of $\overline{E}$ is to the left of P in the direction determined by $\vec{\upsilon}$. Then $P + \vec{\upsilon} \notin \overline{E}$, contradicting $\sigma_i(\overline{E}) \subseteq \overline{E}$, which follows from $\sigma_i(E) \subseteq E$. An identical argument works if σ_i is a glide reflection; use the fact that the translation vector may be assumed to be parallel to the line of reflection.

So σ_1 and σ_2 must both be rotations; denote their fixed points by P_1, P_2, respectively. We first prove that $P_1 = P_2$. Let C be the smallest circle such that E is contained in C and its interior. Because $A \subseteq E$ and $\sigma^{-1}A = E$, it follows that E is also contained in $\sigma_1^{-1}(C)$ and its interior. But by the choice of C, this means that $\sigma_1^{-1}C = C$, and therefore σ_1 leaves the center of C invariant; that is, P_1 is the center of C. Similarly, the same is true of P_2. Hence $P_1 = P_2$. This means that σ_1 and σ_2 commute. Now choose any $Q \in E$ and obtain a contradiction by observing that $\sigma_1\sigma_2 Q$, which is in A, equals $\sigma_2\sigma_1 Q$, which is in B. □

A general question in this area is: Can one characterize those groups of isometries G of $\mathbb{R}^n$ that lead to the existence of bounded G-paradoxical sets? The question is related to the question at the end of §8.1, but there is another motivation. Here we ask about the properties of a group that suffice for bounded paradoxes. For example, what properties of a subgroup of G_2 give bounded paradoxes like those of Just and Sherman? One possible approach is to investigate the structure of the graph associated with a paradoxical decomposition as in the proof of Theorem 14.16.

14.3 Group Growth

We now discuss growth rates in groups and their connection to supramenability and the proof of Theorem 14.4(b). The *length* of a reduced word $g_1^{m_1} \cdots g_r^{m_r}$ (g_i not necessarily distinct, $g_i \neq g_{i+1}$, $m_i \in \mathbb{Z} \setminus \{0\}$) is $m_1 + \cdots + m_r$; the length of the identity e is 0.

Definition 14.19. *If S is a finite subset of a group G, then the* growth function *(with respect to S) $\gamma_S : \mathbb{N} \to \mathbb{N}$ is defined by setting $\gamma_S(n)$ equal to the number of distinct elements of G obtainable as a reduced word of length at most n using elements of $S \cup S^{-1}$ as letters.*

Of course, γ_S is nondecreasing. Also, $\gamma_S(0) = 1$ and $\gamma_S(1) = |\{e\} \cup S \cup S^{-1}|$. Because $\gamma_S(n+m) \leq \gamma_S(n)\gamma_S(m)$, it follows that $\gamma_S(n) \leq \gamma_S(1)^n$, so γ_S is always bounded by an exponential function. We are interested in whether γ_S really exhibits exponential growth. If G contains a free subsemigroup of rank 2, and S contains two free generators of such a semigroup, then $\gamma_S(n) \geq 2^n$, the number of words in S with positive exponents and length exactly n. Hence if G contains a free subsemigroup (or free subgroup) of rank 2, then the growth function exhibits exponential growth with respect to some choice of S. At the other extreme lie the Abelian groups. For suppose $S = \{g_1, \ldots, g_r\} \subseteq G$, which is Abelian. Any word in S is equal, in G, to a word of the form $g_1^{m_1} g_2^{m_2} \cdots g_r^{m_r}$ where $m_i \in \mathbb{Z}$. Hence the number of group elements that arise from words of length n is at most $(2n+1)^r$ (because each $m_i \in [-n, n]$). Hence $\gamma_S(n)$ is dominated by $(n+1)(2n+1)^r$, a polynomial of degree $r+1$, which shows that Abelian groups have slow, that is, nonexponential, growth for any choice of S. (A more careful word count shows that in fact $\gamma_S(n)$ is dominated by a polynomial in n of degree r [Wol68].)

Definition 14.20. *A group G is* exponentially bounded *if, for any finite $S \subseteq G$ and any $b > 1$, there is some $n_0 \in \mathbb{N}$ such that $\gamma_S(n) < b^n$ whenever $n > n_0$; equivalently, $\lim_{n\to\infty} \gamma_S(n)^{1/n} = 1$. If G fails to be exponentially bounded, G is said to have* exponential growth. *The class of exponentially bounded groups is denoted by EB.*

We shall give more examples of exponentially bounded groups, but first we point out the important connection with supramenability.

Theorem 14.21. (a) *(AC) If G is exponentially bounded, then G is supramenable.*

(b) *If G is exponentially bounded, G acts on X, and A is a nonempty subset of X, then A is not G-paradoxical.*

Proof. We first prove (b) (without using AC). Tarski's Theorem then yields a finitely additive, G-invariant measure on $\mathcal{P}(X)$ that normalizes A; this in turn yields (a), if one considers the left action of G on itself.

Suppose, to get a contradiction, that A is G-paradoxical. Then there are two one-one piecewise G-transformations, $F_1 : A \to A$ and $F_2 : A \to A$, such that $F_1(A) \cap F_2(A) = \varnothing$. Let $S = \{g_1, \ldots, g_r\}$ be all the elements of G occurring as multipliers in F_1 and F_2. Because G is exponentially bounded, there is an integer n such that $\gamma_S(n) < 2^n$. Consider the 2^n functions H_i, obtainable as compositions of a string of n F_1s and F_2s. Each $H_i : A \to A$, and if $i \neq j$, then $H_i(A) \cap H_j(A) = \varnothing$. To see this, let p be the first (i.e., leftmost) of the n positions where H_i and H_j differ. Because F_1 and F_2 map A into disjoint sets, and because the function obtained by restricting H_i and H_j to the first $p-1$ positions is one-one, H_i and H_j must map A into disjoint sets too. If we now choose any $x \in A$, it must be that the set $\{H_i(x) : 1 \leq i \leq 2^n\}$ has 2^n elements. But each $H_i(x)$ has the form wx, where w is a word of length n composed of elements of S, and this contradicts the fact that $\gamma_S(n) < 2^n$. □

We have seen why Abelian groups are exponentially bounded and therefore, by the theorem just proved, such groups are supramenable. In particular, this yields a proof that Abelian groups are amenable that is completely different from the proof of Theorem 12.4(b). This new proof is somewhat more informative, because in addition to proving the stronger conclusion regarding supramenability, it shows quite clearly and effectively why the existence of a G-paradoxical set in any action of G means that G has exponential growth.

By Proposition 14.3 and Theorem 14.21, we have the containments $EB \subseteq AG^+ \subseteq AG \cap NS$. When restricted to elementary groups, these three classes coincide: By Theorem 14.27, they all coincide with the class of groups with the property that all finitely generated subgroups have a nilpotent subgroup of finite index. But it is not known whether the first equality holds in general. For the second, R. Grigorchuk [Gri87, CGH99, Ex. 69] constructed p-groups $G_{G,p}$ showing that AG^+ is properly contained in $AG \cap NS$. Figure 14.5 shows several relationships

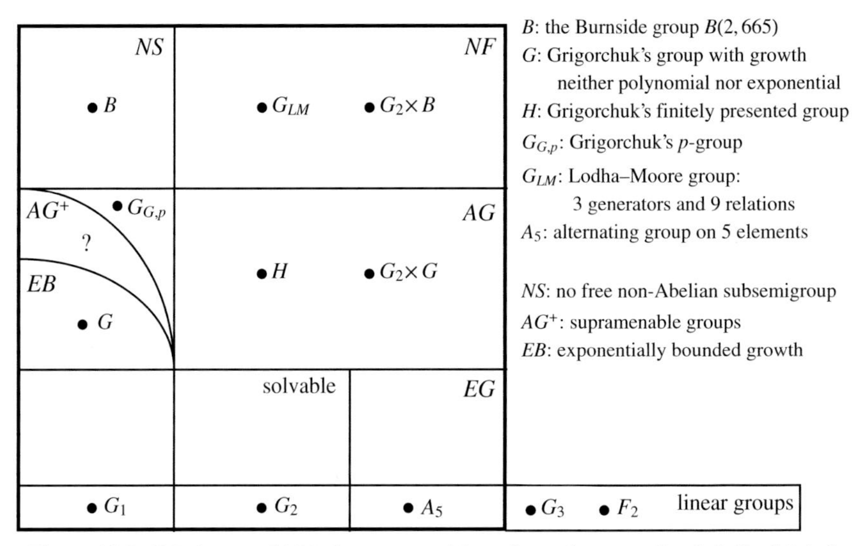

Figure 14.5. Subclasses of NF, the groups with no free subgroup of rank 2. Each label describes the area below and left. G_n is the isometry group of $\mathbb{R}^n$, and H, G, $G_{G,p}$ refer to groups found by Grigorchuk ($G_{G,p}$ is a p-group). The group G_{LM} is presented in §14.4.

and examples. A finite presentation for H is given after Theorem 12.5; H does contain a free semigroup of rank 2 [Gri15].

Question 14.22. Is every supramenable group exponentially bound? That is, does $EB = AG^+$?

The proof of Theorem 14.21 used only the fact that for all finite $S \subseteq G$, $\gamma_S(n) < 2^n$ for n sufficiently large, but there was no loss of generality in assuming the ostensibly stronger hypothesis that for all finite S and all $b > 1$, $\gamma_S(n) < b^n$ for n sufficiently large. The equivalence of these two conditions can be proved quite simply as follows.

Proposition 14.23. (a) *For any finite subset S of a group G, the sequence $\{\gamma_S(n)^{1/n}\}$ converges.*

(b) *If G is not exponentially bounded, then for any $c > 1$, there is some finite $S \subseteq G$ such that $\gamma_S(n) \geq c^n$ for arbitrarily large n.*

Proof. (a) Let γ denote γ_S and fix a positive integer m. Then, for any n, let $k = \left\lfloor \frac{n}{m} \right\rfloor + 1$. It follows that $\gamma(n) \leq \gamma(km) \leq \gamma(m)^k \leq \gamma(m) \cdot \gamma(m)^{\lfloor n/m \rfloor}$, and because $\gamma(m)^{1/n} \to 1$ as $n \to \infty$, this means that $\limsup \gamma(n)^{1/n} \leq \gamma(m)^{1/m}$. Now, letting m vary, $\limsup \gamma(n)^{1/n} \leq \inf\{\gamma(m)^{1/m}\} \leq \liminf \gamma(n)^{1/n}$, yielding the existence of a limit of $\gamma(n)^{1/n}$.

(b) If G is not exponentially bounded, then for some finite $S \subseteq G$, $\gamma_S(n)^{1/n} \to \ell > 1$. Choose r such that $\ell^r > c$ and let S' consist of the identity of G together with all words of length exactly r from S. Any word of length at most rn from

S may be viewed (using $e \in S'$) as a word of length at most n from S'. Hence $\gamma_S(n) \geq \gamma_S(rn)$, and because $\gamma_S(rn)^{1/n} \to \ell^r$ as $n \to \infty$, this means that $\gamma_S(n) > c^n$ for arbitrarily large n. □

The next proposition gives some simple closure properties of the class of exponentially bounded groups, which are sufficient for some geometric applications on the line and in the plane. But there are more exponentially bounded groups than the ones provided by Proposition 14.24; we return to this class later in the chapter and mention a possible characterization of exponentially bounded groups.

Proposition 14.24. (a) *Finite and Abelian groups are exponentially bounded.*

(b) *A subgroup or a homomorphic image of an exponentially bounded group is exponentially bounded.*

(c) *If H is an exponentially bounded subgroup of G and H has finite index in G, then G is exponentially bounded.*

(d) *A direct union of exponentially bounded groups is exponentially bounded; in particular, a group is exponentially bounded if (and only if) all of its finitely generated subgroups are.*

Proof. The case of finite groups is trivial, and the Abelian case has already been discussed. The subgroup case is also trivial, and if S is a finite subset of H, a homomorphic image of G, let S' be a finite subset of G whose image is S. Then $\gamma_S(n)$ in H is bounded by $\gamma_{S'}(n)$ in G. For (c), choose a set of representatives $g_1, \ldots, g_m$ of the right cosets of H in G, with $g_1 = e$. Then, given a finite $S \subseteq G$, let S' consist of the finitely many $h \in H$ that arise when each $g_i s$, where $s \in S$ and $1 \leq i \leq m$, is written in the form hg_k. We claim that each $w \in G$ that arises as a word of length at most n from S may be represented as $w'g_i$, where $1 \leq i \leq m$ and w' is a word of length at most n from S'. Suppose $w = s_1 s_2 \cdots$ with $s_i \in S$. Because $s_1 = es_1 = g_1 s_1$, there is some $h_1 \in H$ and $k_1 \leq m$ such that $s_1 = h_1 g_{k_1}$. Similarly, there are h_2 and k_2 such that $g_{k_1} s_2 = h_2 g_{k_2}$. Continuing to the end of w and substituting into w transforms w to $h_1 \cdots h_r g_{k_r}$, where r is the length of w, as required. This claim yields that $\gamma_S(n)$ in G is at most $m\, \gamma_{S'}(n)$ in H. Therefore $\lim \gamma_S(n)^{1/n} \leq \lim m^{1/n} \gamma_{S'}(n)^{1/n} = 1$, so γ_S, and hence G, is exponentially bounded. Finally, suppose G is the union of the directed system of subgroups $\{G_i : i \in I\}$ and G fails to be exponentially bounded. If this failure is witnessed by γ_S, choose $i \in I$ such that $S \subseteq G_i$. Then, because γ_S only refers to elements in the group generated by S, G_i must have exponential growth too. □

Corollary 14.25. *The isometry group of the line, G_1, is exponentially bounded. Therefore no nonempty subset of $\mathbb{R}$ is paradoxical.*

Proof. If T is the group of translations of $\mathbb{R}$, then T is a normal subgroup of G_1, $G_1/T \cong \mathbb{Z}_2$, and T is Abelian. It follows from Proposition 14.24(a) and (c) that G_1

is exponentially bounded. The rest of the corollary follows from the supramenability of G_1 (Thm. 14.21(a)), but this would use the Axiom of Choice. Instead, use Theorem 14.21(b). □

Note that this corollary, which shows why there is no Sierpiński–Mazurkiewcz Paradox in $\mathbb{R}^1$, also yields Corollary 12.10 for $\mathbb{R}^1$ without using the Axiom of Choice. The elimination of Choice from the $\mathbb{R}^2$ case is more complicated (see Cor. 15.11).

We now return to the problem of classifying the exponentially bounded groups. Groups with a free subsemigroup of rank 2 are the only examples of groups having exponential growth presented so far, but there are more. Adian [Adi79] showed not only that the Burnside group $B(r, e)$ with $r \geq 2$ and $e \geq 665$ is infinite but also that it has exponential growth. (An interesting open question [Sha06] is whether every Burnside group $B(m, n)$ has Property (T); see Definition 13.9.) Indeed, the number of group elements in $B(2, 665)$ corresponding to words in the two given generators of length exactly n is at least $4\,(2.9)^{n-1}$, whereas F_2 has $4 \cdot 3^{n-1}$ such words. Now, because all elements of $B(r, e)$ have finite order, the group has no free subsemigroup. Hence the class of exponentially bounded groups is a proper subclass of the class of groups without a free subsemigroup of rank 2. To study the class of exponentially bounded groups in more detail, it is useful to delineate those groups that, like Abelian groups, have growth functions bounded by a polynomial. Note that the next definition applies only to finitely generated groups.

Definition 14.26. *A finitely generated group G has* polynomial growth *if, for any finite $S \subseteq G$, there are positive constants c, d such that for all n, $\gamma_S(n) \leq cn^d$.*

It is easy to see that for a finitely generated group to have polynomial growth, it is sufficient that γ_S have polynomial growth for a single finite generating set S. For if S' is another finite subset of G, then there is an integer k such that $\gamma_{S'}(n) \leq \gamma_S(kn)$. (This observation also shows that if a finitely generated group has exponential growth, then $\lim \gamma_S(n)^{1/n} > 1$ for *any* finite generating set S.) The proof given earlier that Abelian groups are exponentially bounded shows that a finitely generated Abelian group has polynomial growth. This was extended by Wolf [Wol68] to nilpotent groups. Recall that G is *nilpotent* if there is a sequence of normal subgroups $\{e\} = N_0 \triangleleft N_1 \triangleleft \cdots \triangleleft N_n = G$ such that for each $i \leq n$, N_{i+1}/N_i is contained in the center of G/N_i; G is solvable if each N_{i+1}/N_i is Abelian; therefore nilpotent groups are solvable. Wolf showed that a finitely generated nilpotent group has polynomial growth; hence by Proposition 14.24(c) (modified for polynomial growth), any finitely generated almost nilpotent group has polynomial growth. The converse was proved first for solvable groups ([Mil68c, Wol68]), then for elementary groups ([Cho80]; see Thm. 14.29), and, finally, Gromov [Gro81] showed that it is true in general, thus yielding the following beautiful characterization of groups with polynomial growth.

Theorem 14.27. *A finitely generated group has polynomial growth if and only if it is almost nilpotent.*

Note that the definition of polynomial growth asserts only that $\gamma_S(n)$ is bounded above by a polynomial. It is not clear that there is a single d such that for any S and appropriate positive constants c_i, $c_1 n^d \leq \gamma_S(n) \leq c_2 n^d$. However, it does follow from the remark after Definition 14.26 that the lowest possible degree of a polynomial bound on $\gamma_S(n)$ (as S varies over finite generating sets) is a well-defined integer that is independent of the choice of generators. Bass [Bas72] refined Wolf's work to show that if G is finitely generated and almost nilpotent, then there is an integer d such that for any finite generating set S, there are constants c_1 and c_2 so that $c_1 n^d \leq \gamma_S(n) \leq c_2 n^d$ for all n. Thus the degree of polynomial growth is a well-defined invariant of all groups having polynomial growth.

The characterization of Theorem 14.27 was conjectured by Milnor [Mil68a], but Wolf [Wol68] conjectured even more. A function of n that dominates all polynomials is not necessarily greater than some nontrivial exponential function c^n ($c > 1$); for example, consider $2^{\sqrt{n}}$ or $n^{\log n}$. Wolf conjectured that if a group's growth function dominates all polynomials, then it does dominate a nontrivial exponential function (this became known as the *Milnor–Wolf Conjecture*). This conjecture is true for many interesting groups, but Grigorchuk [Gri83] showed that it is not true in general. It is an open question whether there is a finitely presented example that is exponentially bounded but does not have polynomial growth.

Theorem 14.28. *There exist finitely generated periodic groups that are exponentially bounded but do not have polynomial growth.*

Milnor and Wolf [Mil68c, Wol68] had proved their conjecture for solvable groups, and their work was strengthened by Rosenblatt [Ros74] and Chou [Cho80], who proved the following theorem.

Theorem 14.29. *If G is a finitely generated elementary group, then either G is almost nilpotent (and hence has polynomial growth) or G has a free subsemigroup of rank 2 (and hence has exponential growth).*

This theorem shows that the Milnor–Wolf Conjecture is valid in EG. Therefore the example of Theorem 14.28 is a nonelementary amenable (in fact, supramenable) group.

By Theorem 14.21 and Proposition 14.24(d), the preceding theorem shows that the strongest possible characterization of supramenability is valid in EG, the class of elementary groups: G is supramenable if and only if G has no free subsemigroup of rank 2. It is also known that $AG^+ = NS$ is valid when restricted to connected, locally compact topological groups [Ros74, Cor. 4.20]. Theorem 14.29 also settles the Milnor–Wolf Conjecture in EG in the strongest possible way, because it yields that a finitely generated group in EG fails to have polynomial growth if and only if it has a free subsemigroup of rank 2. Even before Theorem 14.28 was proved, it was known that this strong form of the Milnor–Wolf Conjecture is not valid, because $B(2, 665)$ has exponential growth and is periodic.

To summarize, we have that each of the following statements implies the next (and (a) and (a') are equivalent):

(a) All finitely generated subgroups of G have polynomial growth.
(a') All finitely generated subgroups of G are almost nilpotent.
(b) G is exponentially bounded.
(c) G is supramenable.
(d) G has no free subsemigroup of rank 2.

Moreover, (d) implies (a) in EG, and so these statements are all equivalent for elementary groups. In general, (d) does not imply (c) (Thm. 12.5) and (b) does not imply (a) (by Thm. 14.28). Whether (c) implies (b) is Question 14.22.

Another large class of groups for which the four statements are equivalent is the class of all linear groups, that is, subgroups of $GL_n(K)$ for some positive integer n and field K. For any group G that satisfies (d) can have no free subgroup of rank 2, and hence by Tits's Theorem (Thm. 12.6), a linear group that satisfies (d) is elementary. Hence for linear groups—in particular, all groups of isometries of $\mathbb{R}^n$—both the Milnor–Wolf Conjecture and the equality $AG^+ = AG \cap NS$ are valid. Note that, by the preceding remarks about elementary groups, the lower left square in Figure 14.5 consists of precisely the groups all of whose finitely generated subgroups have polynomial growth (equivalently, all such subgroups are almost nilpotent).

In light of Theorem 14.28, the following result pertaining to the Milnor–Wolf Conjecture is quite interesting. As a consequence of their investigation into Gromov's proof of Theorem 14.27, van den Dries and Wilkie [WD84] constructed a computable nondecreasing unbounded $g : \mathbb{N} \to \mathbb{N}$ such that $n^{g(n)}$ is exponentially bounded, and for any constant k, no group has a growth function lying in the gap between all polynomials and $kn^{g(n)}$.

Just as with actions of free groups, the mere presence of a free subsemigroup in a group acting on a set does not guarantee that the set has a paradoxical subset. Proposition 1.9 gives an extra condition on the action that is sufficient; in general, some condition is necessary, because of the trivial actions where $g(x)$ always equals x. Nevertheless, in the special case of groups of Euclidean isometries acting on $\mathbb{R}^n$, no extra condition is necessary. The following result, which is analogous to Theorem 13.21, gives a complete characterization of the subgroups of G_n with respect to which $\mathbb{R}^n$ has a paradoxical subset.

Theorem 14.30 (AC). *Suppose G is a subgroup of G_n. Then the following statements are equivalent to each other and to statements (a)–(d) following Theorem 14.29:*

(a) *G has no free subsemigroup of rank 2.*
(b) *G is supramenable.*
(c) *For any nonempty $E \subseteq \mathbb{R}^n$, there is a finitely additive, G-invariant measure on $\mathcal{P}(\mathbb{R}^n)$ that normalizes E.*
(d) *No nonempty subset of $\mathbb{R}^n$ is G-paradoxical.*

(e) *No nonempty subset E of $\mathbb{R}^n$ contains two disjoint subsets, each of which is congruent, via G, to E.*

Proof. As pointed out before, Tits's Theorem and Theorem 14.29 yield $(a) \Rightarrow (b)$. That $(b) \Rightarrow (c) \Rightarrow (d)$ is just Theorem 14.2, and $(d) \Rightarrow (e)$ is obvious. For $(e) \Rightarrow (a)$, suppose that G does have a subsemigroup S freely generated by σ and τ. Let H be the subgroup of G generated by σ and τ. Because H is countable and because the set of fixed points of a single $w \in H \setminus \{e\}$ has Lebesgue measure zero, there must be some $P \in \mathbb{R}^n$ that is not fixed by any element of $H \setminus \{e\}$ (see the proof of Thm. 13.21). Now, let $E = \{sP : s \in S\}$ and let $A = \sigma(E), B = \tau(E)$. Then A and B are subsets of E, and each is congruent to E. Moreover, $A \cap B = \varnothing$, for suppose $\sigma s_1(P) - \tau s_2(P)$ for some $s_i \subset S$. By freeness of S, σs_1 and τs_2 must be distinct elements of G, because they have distinct leftmost terms. But then $(\tau s_2)^{-1} \sigma\ s_1$ is a nonidentity element of H fixing P, in contradiction to the choice of P. Because the conditions (a)–(d) after Theorem 14.29 are equivalent for groups of isometries, the theorem is proved. □

This result clarifies the Sierpiński–Mazurkiewicz Paradox because it shows that any free subsemigroup of G_2 of rank 2 yields a paradoxical subset of the plane. The proof of Theorem 1.7 given in Chapter 1, however, gives an especially succinct description of the paradoxical set. Moreover, the set E of that proof splits into two sets, each congruent to E. Theorem 14.30, except for the exceptional case where P is fixed by some nonidentity element of S, yields two subsets whose union is $E \setminus \{P\}$, a proper subset of E.

Recall from Corollary 5.9 that if a group is paradoxical using four pieces, as it is if it has a free subgroup of rank 2, then, indeed, the group contains such a subgroup. For free semigroups an analogous result is valid, which we derive as a corollary to the more general theorem that follows.

Theorem 14.31. *Suppose a group G acts on X and $\sigma, \tau \in G$. Then the following are equivalent:*

(a) *Some nonempty $E \subseteq X$ is such that $\sigma\, E$ and $\tau\, E$ are disjoint subsets of E.*
(b) *There is some $x \in X$ such that whenever w_1 and w_2 are nonidentity (semigroup) words in σ and τ beginning with σ, τ, respectively, then $w_1(x) \neq w_2(x)$. (It follows that σ, τ are free generators of a free subsemigroup.)*

Proof. The fact that (b) implies (a) is just a restatement of Proposition 1.9 and its proof; E is simply the orbit of X using the generated semigroup. For the converse, let X be any point in E. Because of the hypothesis on words, $w_1\, E \subseteq \sigma\, E$ and $w_2 E \subseteq \tau\, E$, or vice versa. In any event, $w_1(x)$ and $w_2(x)$ lie in disjoint subsets of E and are therefore unequal. The fact that the condition on σ, τ expressed in (b) is sufficient to guarantee that they are free generators of a subsemigroup of G was, in essence, proved at the beginning of Theorem 1.8's proof. □

Corollary 14.32. *A group G has a nonempty subset E and two elements, σ, τ, such that σE and τE are disjoint subsets of E if and only if σ and τ are free generators of a free subsemigroup of rank 2.*

In fact, a slightly stronger result is true. If $E \subseteq G$, $E \neq \varnothing$, $\sigma E \cup \tau E \subseteq E$, and σ, τ fail to be free generators of a subsemigroup, then for some $w \in G$, $wE \subseteq \sigma E \cap \tau E$. To prove this, choose w_1 and w_2, two distinct words in σ, τ, so that $w_1 = w_2$ in G. Repeated left cancellation allows us to assume that either (a) $w_1 = \sigma \cdots$ and $w_2 = \tau \cdots$, (b) $w_1 = \sigma \cdots$ and $w_2 = e$, or (c) $w_1 = \tau \cdots$ and $w_2 = e$. In case (a), let g be the common value of w_1 and w_2 in G; in case (b), let $g = \tau$; and in case (c), let $g = \sigma$. It is easy to see, using $\sigma E \cup \tau E \subseteq E$, that these choices work. This result strengthens the corollary by showing that $\sigma E \cap \tau E$ is not only nonempty but contains a set congruent to E.

We have seen that the major difference between an action of a free group on a set X and an action of a group with a free subsemigroup is that the former (under appropriate additional hypotheses) causes a paradoxical decomposition of X while the latter causes such a decomposition of a subset of X. But this analogy cannot be pushed too far. For instance, it is not the case that whenever a group with a free subsemigroup of rank n ($n \geq 2$) acts on X without nontrivial fixed points, given a proper system of n congruences, some nonempty subset of X, can be partitioned to satisfy the congruences. To see this, consider the congruences $A_2 \cong A_2 \cup A_3 \cup A_4$ and $A_1 \cong A_1 \cup A_2 \cup A_4$. If some nonempty subset E of X splits into A_i such that $\sigma(A_2) = A_2 \cup A_3 \cup A_4$ and $\tau(A_4) = A_1 \cup A_2 \cup A_4$, where σ and τ come from a group acting on X, then it can be shown that σ, τ are free generators of a free group. The two equations just mentioned yield $\sigma(A_2) \subseteq X \setminus A_1$, $\sigma^{-1}(A_1) \subseteq X \setminus A_2$, $\tau(A_4) \subseteq X \setminus A_3$, and $\tau^{-1}(A_3) \subseteq X \setminus A_4$; these four relations were proved in Theorem 5.8 to guarantee the independence of σ, τ as group elements. So, for example, in the case of the action of the solvable group G_2 on itself, no subset of G_2 can be partitioned to solve the system despite the existence in G_2 of a free subsemigroup of rank 2 and the lack of fixed points in the action.

Nevertheless, these systems can be solved provided we allow the congruences to be witnessed by a piecewise G-transformation; more precisely, we interpret $\cong$ as $\sim_2$.

Theorem 14.33. *Suppose a group G, acting on X, contains a free subsemigroup S of rank κ generated by $\{\rho_\alpha : \alpha < \kappa\}$, where κ is an infinite cardinal. Suppose further that there is some $x \in X$ satisfying that whenever w_1 and w_2 are semigroup words in $\{\rho_\alpha\}$ beginning on the left with ρ_α, ρ_β, respectively, where $\alpha \neq \beta$, then $w_1(x) \neq w_2(x)$. Then E, the S-orbit of X, may be partitioned into κ sets, $\{A_\beta : \beta < \kappa\}$, such that for any two nonempty subsets of κ, L and R, $\bigcup\{A_\beta : \beta \in L\}$ and $\bigcup\{A_\beta : \beta \in R\}$ are G-equidecomposable using two pieces.*

Proof. By Proposition 7.12, the hypotheses imply that the sets $\rho_\beta(E)$ are pairwise disjoint subsets of E. Let $A_\beta = \rho_\beta(E)$ except when $\beta = 0$; in order that the sets A_β partition E, A_0 is defined to be $E \setminus \bigcup\{A_\beta : 0 < \beta < \kappa\}$. To show that this partition of E works, let L and R be two nonempty subsets of κ and choose γ, δ to lie in L, R, respectively. Because each $\rho_\beta(E)$ is a subset of A_β, it must be that $\rho_\gamma(\bigcup\{A_\beta : \beta \in R\}) \subseteq A_\gamma \subseteq \bigcup\{A_\beta : \beta \in L\}$ and $\rho_\delta(\bigcup\{A_\beta : \beta \in L\}) \subseteq A_\delta \subseteq \bigcup\{A_\beta : \beta \in R\}$. Hence, by the Banach–Schröder–Bernstein Theorem, $\bigcup\{A_\beta : \beta \in L\} \sim_G \bigcup\{A_\beta : \beta \in R\}$ using two pieces. □

Because G_2 satisfies the hypothesis of Theorem 14.33 (Thm. 7.13), it follows that systems of congruences can be solved using subsets of the plane and $\sim_2$. For example, there exists a nonempty $E \subseteq \mathbb{R}^2$ such that E has a subset A such that for any κ between 2 and $2^{\aleph_0}$ inclusive, E can be split into κ sets, each of which is $\sim_2$ to A.

14.4 Cogrowth and Amenability

Though the growth of a group has connections with supramenability, work by Grigorchuk [Gri77] and Cohen [Coh82] has led to a refinement of this concept, called *cogrowth*, that has proven to be very important because it yields a characterization of amenability.

To discuss the growth of an arbitrary group, define $\gamma(G)$ to be the supremum of $\lim_{n\to\infty} \gamma_S(n)^{1/n}$ over all finite subsets S of G. By Proposition 14.23, $\gamma(G)$ is well defined and either $\gamma(G) = 1$ (G is exponentially bounded) or $\gamma(G) = \infty$ (G has exponential growth). This is a rather coarse classification, and the point of the cogrowth function is that it resolves the class of groups having exponential growth into more levels, perhaps even a continuum of levels.

Suppose a group G is *finitely presented*, which means it is given by generators and relations: $G = \langle a_1, \ldots, a_r : b_1, b_2, \ldots \rangle$ where $r < \infty$. Then $G \cong F_r/N$, where F_r is the free group with free generators $a_1, \ldots, a_r$ and N is the smallest normal subgroup of F_r containing $\{b_1, b_2, \ldots\}$. Let E_n be the set of all (reduced) words in F_r of length at most n. It is easy to see that if $r \geq 2$, E_n has $(r(2r-1)^n - 1)/(r-1)$ elements; but the important point here is that $|E_n|^{1/n} \to 2r - 1$ (even if $r = 1$). Let π be the canonical homomorphism, $\pi : F_r \to F_r/N = G$. The growth function $\gamma_S(n)$, where $S = \{a_1, \ldots, a_r\}$, is simply $|\pi(E_n)|$, the size of the image of π restricted to E_n. For finite groups the size of the kernel of a homomorphism is inversely proportional to the size of the image. But E_n is not a group, and so the size of the kernel might behave quite differently than the size of the image. This leads to the following definition.

Definition 14.34. *If G is presented as just discussed, then $\tilde{\gamma}(n)$, the* cogrowth function *of the presentation, is defined to be $|N \cap E_n|$; that is, $\tilde{\gamma}$ counts the number of words of length at most n that vanish when interpreted in G.*

As with the growth function, it will be the sequence $\tilde{\gamma}(n)^{1/n}$ and its limit that interest us. If there are no relations at all in the presentation, that is, $G = F_r$ and the presentation is the standard presentation of F_r, then $N = \{e\}$ and $\tilde{\gamma}(n) = 1$ for all n. On the other hand, suppose G is exponentially bounded. It is true for all presentations that some N-coset in F_r/N must contain at least $|E_n|/\pi(E_n)$ words of length at most n. Hence there are at least this many words in $N \cap E_n$ and so $|E_n| \geq \tilde{\gamma}(n) \geq |E_n|/\gamma(n)$. Because $\gamma(n)^{1/n} \to 1$, it follows that $\lim \tilde{\gamma}(n)^{1/n} = \lim |E_n|^{1/n} = 2r - 1$.

These two examples illustrate behavior at opposite ends of the cogrowth spectrum. It turns out that, for any presentation (with r many generators, r finite) of a group, the sequence $\tilde{\gamma}(n)^{1/n}$ tends to a limit that lies between 1 and $2r - 1$. The proof of this is elementary, but more complicated than for the ordinary growth function, and we refer the reader to [Coh82] for details. The proof yields that the limit, when not equal to 1, is at least $\sqrt{2r-1}$; that this last inequality is in fact strict is a more difficult result. The following theorem summarizes the basic facts about cogrowth.

Theorem 14.35. *Let a group G with a fixed presentation using r generators, where $2 \leq r < \infty$, be given. Then*

(a) $\lim \tilde{\gamma}(n)^{1/n}$ *exists*
(b) $1 \leq \lim \tilde{\gamma}(n)^{1/n} \leq 2r - 1$
(c) $\lim \tilde{\gamma}(n)^{1/n} = 1$ *if and only if there are no relations in the presentation (in which case $G = F_r$)*
(d) *if* $\lim \tilde{\gamma}(n)^{1/n} > 1$, *then* $\sqrt{2r-1} < \lim \tilde{\gamma}(n)^{1/n} \leq 2r - 1$

It must be emphasized that $\lim \tilde{\gamma}(n)^{1/n}$ is generally dependent on the choice of a representation. For example, let $G = \langle a_1, a_2, a_3 : a_2 a_3^{-1} \rangle$; G is isomorphic to F_2. Because $\tilde{\gamma}(n)^{1/n} \geq |E_n|/\gamma(n)$ and $\gamma(n)$ is bounded by the number of words of length at most n in the standard presentation of F_2, $\lim \tilde{\gamma}(n)^{1/n} \geq 5/3$. This shows that case (d) of Theorem 14.35 can include free groups.

Now we can define η, the *cogrowth* of a presentation, by $\eta = \log(\lim \tilde{\gamma}(n)^{1/n}) / \log(2r - 1)$. It then follows that $\eta \in \{0\} \cup (1/2, 1]$; $\eta = 0$ if and only if there are no relations in the presentation (assuming $r \geq 2$); and $\eta = 1$ if G is exponentially bounded. Thus the standard presentation of a free group has zero cogrowth (no words get killed), whereas an exponentially bounded group has full cogrowth (the number of words in E_n that get killed is, in an asymptotic sense, the same as the total number of words in E_n). Now, here is the remarkable theorem.

Theorem 14.36 (AC). *Let G be a finitely generated group. Then G is amenable if and only if $\eta = 1$ for all presentations of G with finitely many generators.*

Thus, for a group G presented as in the theorem, amenability is characterized by the asymptotic behavior of the number of words of length at most n that collapse to the identity. This number, as n approaches infinity, must be very large. Theorem 14.36 is due to Grigorchuk [Gri77] and, independently, Cohen [Coh82]. The proof of both directions appears in [Coh82] and is moderately difficult, using techniques from C^*-algebras and Kesten's characterization of amenability (discussed after the proof of Theorem 12.11). It would be nice if the method of paradoxical decompositions, which yields (Thm. 14.21's proof) such a succinct proof that nonamenable groups have exponential growth, could be used to obtain a proof that a nonamenable group has cogrowth less than 1.

One consequence of Theorem 14.36 is that for a finitely generated amenable group, the cogrowth is independent of the presentation, provided the latter uses finitely many generators. Another consequence concerns a notion of dimension introduced by Cohen [Coh82]. Given a presentation using r generators, we may metrize F_r by setting $d(u, v)$, in the case $u \neq v$, equal to $(2r-1)^{-k}$, where k is the smaller of the lengths of u and v. Then η is the entropic dimension of N as a subspace of F_r (see [Coh82] for a definition and proof). This leads to a natural definition of the dimension of a presentation as $1-\eta$. Thus a finitely generated group is amenable if and only if all of its presentations are zero-dimensional.

To apply these ideas to all groups, simply let $\eta(G)$ be defined to be the greatest lower bound of η taken over all presentations with r generators, $2 \le r < \infty$, of a subgroup of G.

Corollary 14.37 (AC). *A group G is amenable iff $\eta(G) = 1$; G has no free subgroup of rank 2 iff and only if $\eta(G) \ge 1/2$; G has a free subgroup of rank 2 if and only if $\eta(G) = 0$.*

Although it is not known whether $\eta(G)$ can take on all values in $\left[\frac{1}{2}, 1\right]$, Ol'shanskii [Ols80] has shown that for a certain periodic group G, $\eta(G) = \frac{1}{2}$. Such a group is nonamenable but has no free subgroup of rank 2. Although it had been conjectured that $AG = NF$ when restricted to finitely presented groups, that is false. In 2003 Ol'shanskii and Sapir [OS03] found an example of a finitely presented group in $NF \setminus AG$. And there has been much work since then. Lodha and Moore [LM∞], building on work of Monod [Mon13] and others, gave a specific and simply described example of a group in $NF \setminus AG$: it has a succinct finite presentation with three generators and nine relations. A torsion-free group is one having no nonidentity elements of finite order.

Theorem 14.38. *Let G_{LM} be the group generated by the following three homeomorphisms of the real line:*

$$a(t) = t+1; \quad b(t) = \begin{cases} t & \text{if } t \le 0, \\ \frac{t}{1-t} & \text{if } 0 \le t \le \frac{1}{2}, \\ 3 - \frac{1}{t} & \text{if } \frac{1}{2} \le t \le 1, \\ t+1 & \text{if } 1 \le t; \end{cases} \quad c(t) = \begin{cases} \frac{2t}{1+t} & \text{if } 0 \le t \le 1, \\ t & \text{otherwise.} \end{cases}$$

Then G_{LM} is nonamenable, has no free subgroup of rank 2, is torsion-free, and has the finite presentation given by setting the following nine words in $a^{\pm 1}, b^{\pm 1}, c^{\pm 1}$ to the identity; where $[\cdot]$ denotes the commutator:

$$\begin{array}{lll} [ba^{-1}, a^{-1}ba], & [ba^{-1}, a^{-2}ba^2], & [c, a^2b^{-1}a^{-1}], \\ [c, ab^2a^{-1}b^{-1}ab^{-1}a^{-1}], & [c, a^{-1}ba], & [c, a^{-2}ba^2], \\ [c, aca^{-1}], & [c, a^2ca^{-2}], & \\ \multicolumn{3}{l}{c^{-1}b\,[b, a^{-1}]\,cb^{-2}ab^{-1}c^{-1}b\,[a^{-1}, b]\,ab^{-1}cba^{-1}ba^{-1}.} \end{array}$$

The Lodha–Moore example is closely related to the Thompson group; $F = \langle a, b : [ab^{-1}, a^{-1}ba] = [ab^{-1}, a^{-2}ba^2] = e \rangle$, though it has other, more natural representations. Richard Thompson put forward F as a possible example of a nonamenable group in NF. That is still unresolved. It is noteworthy that the subgroup of G_{LM} generated by a and b is isomorphic to F (and so G_{LM} has a free subsemigroup of rank 2).

Conjecture 14.39 (R. Thompson). The Thompson group F is nonamenable.

Notes

The definition of supramenable groups was introduced by Rosenblatt [Ros74], but the idea goes back to Tarski and Lindenbaum. The idea of using a free subsemigroup to construct a paradox was first used in the Sierpiński–Mazurkiewicz Paradox (Thm 1.7). However, Proposition 14.3 was not explicitly stated until the paper of Rosenblatt [Ros74], who, in fact, proved the stronger form given in Corollary 14.21. This latter result was discovered independently by Sherman [She75, Thm. 1.7].

The fact that Abelian groups are supramenable (Thm. 14.4(b)) was first proved by Lindenbaum and Tarski in the 1920s. A proof appears in [Tar49, pp. 224–227]. Their proof is essentially the same as the one presented here (Thm. 14.21), and the only property of the Abelian group that is used is the fact that it is exponentially bounded. Moreover, Tarski was aware that the result is valid for more than just Abelian groups; in [Tar38a, p. 223], he points out that groups satisfying some “schwachere, aber kompliziertere Voraussetzungen” are supramenable. So even though growth conditions in groups were not introduced until much later, the result that exponentially bounded groups are supramenable must be credited to Tarski and Lindenbaum. Explicit use of the notion of exponential boundedness first appeared in a paper by Adelson-Velsky and Shreider [AS59], who proved that exponentially bounded groups are amenable. The result also appears in [Mil68b] (where Kesten’s characterization of amenability is used) and in [Cho80] and [Ros74] (where linear functionals are used).

Question 14.5 is due to Rosenblatt [Ros74]. The second part of Corollary 14.25, asserting that no nonempty subset of the real line is paradoxical, is due to Sierpiński [Sie54, Thm. 19], and his proof is essentially the same as that presented here; that is, it uses the fact that G_1 is exponentially bounded. Theorem 14.14 was first proved by Lindenbaum [Lin26, p. 218n1]; the proof presented here was discovered by Hadwiger and Debrunner [HDK64, p. 80]. Theorem 14.15 is due to G. Tomkowicz.

The concept of the growth of a group first appeared in the paper of Adelson-Velsky and Shreider [AS59], but Milnor and Wolf provided the first detailed analysis of growth conditions in groups. Milnor [Mil268b] proved Proposition 14.23 and introduced the term *exponential growth* for finitely generated groups such that $\lim \gamma_S(n)^{1/n} > 1$. Wolf [Wol68] initiated the study of groups with polynomial growth, and generalizing an example of Milnor [Mil268b], he proved the fundamental result that almost nilpotent groups have polynomial growth. In fact, it had been proved earlier by A. Hulanicki [Hul66].

The proof of Proposition 14.24(c) is essentially due to Wolf [Wol68]. Milnor [Mil68a] conjectured that the degree of growth of a group that has polynomial growth is a well-defined integer; that is, that for constants c_1 and c_2, there is an integer d such that $c_1 n^d < \gamma_S(n) < c_2 n^d$. He also raised the possibility that all finitely generated groups of polynomial growth are almost nilpotent. Bass [Bas72] improved the results of Wolf to settle Milnor's first conjecture for almost nilpotent groups, and Milnor's second conjecture was proved by Gromov [Gro81]. A useful appendix to [Gro81] by Tits contains proofs of enough of the work of Milnor and Wolf to make Gromov's paper independent of theirs. The appendix also contains a proof of Bass's result. Gromov's characterization had been known in some special cases: Its validity for solvable groups was proved by Milnor and Wolf [Mil68c, Wol68], and Chou [Cho80] proved it for elementary groups. Indeed, Chou [Cho80], extending results of Rosenblatt [Ros74], proved the even stronger Theorem 14.29. An elementary proof of a different sort of special case of Gromov's Theorem—where it is assumed that the growth function is linear—is given by Wilkie and van den Dries [WD84].

Milnor had conjectured that all groups of polynomial growth are almost nilpotent, and Wolf [Wol68] conjectured even more, that all finitely generated exponentially bounded groups are almost nilpotent. By Theorem 14.27, this is equivalent to the Milnor–Wolf Conjecture as stated here. Theorem 14.28, which provides a counterexample to this conjecture, was announced by Grigorchuk at the 1983 International Congress of Mathematicians in Warsaw; a sketch of his proof appears in [Gri83].

Theorem 14.31 and Corollary 14.32 are due to Rosenblatt [Ros74]. Theorem 14.33 is due to Mycielski [Myc56].

The notion of cogrowth, which is more important for the theory of amenable groups than is growth, was introduced independently by Grigorchuk [Gri77] and Cohen [Coh82]. Cohen conjectured that Theorem 14.36 could be used to construct a nonamenable group that has no free subgroup of rank 2, and as stated by Ol'shanskii [Ols80], this has turned out to be the case.

See [Boz80] for further work on growth conditions and amenability.

15

The Role of the Axiom of Choice

The Banach–Tarski Paradox is so contrary to our intuition that it must have some implications for the foundations of mathematics and the unrestricted use of the Axiom of Choice (AC). In this final chapter, we give a detailed account of such implications and discuss various technical points related to the use of AC. Ever since its discovery, the paradox has caused some mathematicians to look critically at AC. Indeed, as soon as the Hausdorff Paradox was discovered, it was challenged because of its use of that axiom; E. Borel [Bor14, p. 256] objected because the choice set was not explicitly defined. We address these criticisms in more detail in §15.3, but we start with several technical points that are essential to understanding the connection between AC and the paradox.

15.1 The Axiom of Choice Is Essential

The standard axioms for set theory are the Zermelo–Fraenkel Axioms; they are called ZF. When the Axiom of Choice is included, the theory is called ZF + AC, or usually just ZFC.

Results of modern set theory can be used to show that AC is indeed necessary to obtain the Banach–Tarski Paradox, in the sense that the paradox is not a theorem of ZF alone. Before we can explain why this is so, we need to introduce some notation and discuss some technical points of set theory. If T is a collection of sentences in the language of set theory, for example, $T = \mathrm{ZF}$ or $T = \mathrm{ZF} + \mathrm{AC}$, then Con($T$) is the assertion, also a statement of set theory in fact, that T is consistent, that is, that a contradiction cannot be derived from T using the usual methods of proof. We take Con(ZF) as an underlying assumption in all that follows. Gödel proved in 1938 that Con(ZF) implies (and so is equivalent to) Con(ZF + AC); thus AC does not contradict ZF (see [Jec73, Jec78]). Let LM denote the assertion that all sets of reals are Lebesgue measurable; of course, LM contradicts AC (Cor. 1.6). Because LM implies that all subsets of each $\mathbb{R}^n$ are Lebesgue measurable (see the proof of Thm. 8.17(a)), ZF + LM implies that there is no Banach–Tarski Paradox.

Ever since nonmeasurable sets were first constructed, mathematicians wondered whether it was possible to construct a nonmeasurable set without using the Axiom of Choice. All known constructions used AC in apparently unavoidable ways. The feeling that AC was necessary was confirmed in 1964 when Solovay, using Paul Cohen's just-discovered method of forcing, proved, with a complication to be discussed, Con(ZF + LM) (see [Jec78]). It follows that ZF+"There is no Banach–Tarski Paradox" is consistent; therefore the Banach–Tarski Paradox is not a theorem of ZF.

To understand the extra complication in Solovay's result, we must introduce inaccessible cardinals. A cardinal κ is *inaccessible* if it is not the sum of fewer than κ cardinals, each of which is less than κ, and it is greater than 2^η for any cardinal $\eta < \kappa$. For example, $\aleph_1$ is not greater than $2^{\aleph_0}$ and so is not inaccessible; $\aleph_\omega$ is the sum of the $\aleph_n$, $n = 0, 1, \ldots$ and so is not inaccessible. An example of an inaccessible cardinal is $\aleph_0$; any others must be very large. However, the existence of an uncountable inaccessible cardinal cannot be proved in ZF or ZFC. To see this, let IC denote the assertion that an uncountable inaccessible cardinal exists, and suppose that IC can be proved in ZFC. Then if κ_0 denotes the least uncountable inaccessible cardinal, the sets of rank less than κ_0 form a model of ZFC in which IC is false (see [Dra74, Jec78]), contradiction. Even more is true: Con(ZF + IC) cannot be proved from Con(ZF); for if ZF + Con(ZF) implies Con(ZF + IC), then Con(ZF + Con(ZF)), contradicting Gödel's second incompleteness theorem. This contrasts IC with AC: ZF can neither prove nor disprove AC, and ZF cannot prove IC, but it is not known (and cannot be proved) that ZF cannot disprove IC. Thus ZF + IC is a substantial strengthening of ZF; although it is presumed to be a consistent extension, this cannot be proved. Much study has been devoted to inaccessible cardinals, and it would be a great shock if in ZF it could be proved that uncountable inaccessible cardinals do not exist. Indeed, set theorists have studied much larger cardinals (e.g., measurable cardinals) whose existence would imply the existence of many inaccessible cardinals, and it is generally felt that these larger large cardinals yield consistent extensions of ZF. But, as pointed out, the consistency of ZF + IC cannot be proved, and one cannot completely ignore the possibility that IC contradicts ZF.

One technical point arises when discussing Lebesgue measure in the absence of AC. Without any form of Choice at all, Lebesgue measure might behave in a way that makes it an object much less worthy of study in the first place. For instance, the property that a countable union of countable sets is countable uses the Axiom of Choice in a subtle but necessary way: It is consistent with ZF that $\mathbb{R}$ is a countable union of countable sets! This assertion implies that Lebesgue measure is not countably additive. Thus it is customary to allow a modest amount of Choice, enough so that measure and category behave as expected, but not so much that the reals can be well-ordered or a nonmeasurable set constructed. One possibility is to add the Axiom of Countable Choice, which allows selections to be made from an arbitrary countable collection of sets and yields the countable additivity of both Lebesgue measure and the ideal of meager sets. But a slightly stronger axiom, the Axiom of Dependent Choice (DC), which guarantees the existence of a countable choice set $\{a_n\}$, where a_{n+1} depends on a_n, turns out to be more useful. It allows more of the classical arguments of analysis and descriptive set theory to be retained but does not yield a well-ordering of the reals. Thus the assertion LM is usually considered in the context of ZF + DC rather than just ZF. See [Jec78, Mor82] for more details on these weaker forms of AC and further references to a variety of consistency results.

With the preceding set-theoretical points in hand, we can state the fundamental result that connects IC and LM.

Theorem 15.1. Con(ZF + IC) *is equivalent to* Con(ZF + DC + LM).

Some remarks on the history of this remarkable result are in order. Soon after Cohen invented the method of forcing to prove the independence of both AC and the Continuum Hypothesis from the ZF axioms, Solovay [Sol70] saw how to obtain Con(ZF + DC + LM). The complication of Solovay's result alluded to before is that he needed to assume that the existence of an uncountable inaccessible cardinal was noncontradictory, that is, Con(ZF + IC). With this extra assumption, it follows from Solovay's result that ZF + DC is not strong enough to produce a nonmeasurable set.

For many years it was felt that the hypothesis regarding IC would eventually be eliminated from this result. But in 1980, S. Shelah (see [Rai84, She84]) proved the reverse direction of Theorem 15.1, thus establishing the necessity of Solovay's assumption. Although it is most likely that both ZF + IC and ZF + DC + LM are consistent, it is remarkable that these two questions should be so inextricably linked. Reinterpreting Shelah's theorem slightly, it states that if someone does succeed in proving from ZF that uncountable inaccessible cardinals do not exist, then it will follow that there is also a proof from ZF + DC that a nonmeasurable set exists.

Theorem 15.1 shows that, assuming Con(ZF + IC), the Banach–Tarski Paradox is not a theorem of ZF alone, nor of ZF + DC. In fact, this conclusion can be derived without any additional metamathematical assumptions, thus separating it from the status of inaccessible cardinals. This follows from a different, less well known, measure-theoretic consistency result of Solovay, one that deals with measures more general than Lebesgue measure.

Let GM denote the assertion that there is a countably additive, isometry-invariant measure on all subsets of $\mathbb{R}^n$ that normalizes the unit cube. Recall that the classic example of a nonmeasurable set actually negates GM, not just LM (Cor 1.6). Thus ZF + AC yields that GM is false. Note also that by Proposition 11.18, a general measure as in GM must agree with Lebesgue measure on the Lebesgue measurable subsets of $\mathbb{R}^n$. If, however, the Axiom of Choice is not assumed, then LM and GM differ dramatically; the latter is a much weaker statement.

Theorem 15.2. Con(ZF) *is equivalent to* Con(ZF + DC + GM).

Theorem 15.2 shows that while the theory ZF + DC + LM is equiconsistent with ZF + IC, the theory ZF + DC + GM is equiconsistent with ZF, a theory strictly weaker than ZF + IC in consistency strength. Many consequences of ZF + DC + LM can be derived from ZF + DC + GM, because the latter yields measures defined on all subsets of $\mathbb{R}^n$. In particular, GM is sufficient to guarantee the nonexistence of the Banach–Tarski Paradox.

Corollary 15.3. *The Banach–Tarski Paradox is not a theorem of ZF, nor of* ZF + DC.

Proof. Suppose ZF + DC is sufficient to derive the Banach–Tarski Paradox (of a ball in $\mathbb{R}^3$). Then ZF + DC would imply that there is no finitely additive, isometry-invariant measure defined on all subsets of $\mathbb{R}^3$ and normalizing the unit cube. This contradicts Theorem 15.2, which guarantees that even GM, which provides a countably additive invariant measure, is not inconsistent with ZF + DC. □

The proofs of Theorem 15.2 and the forward direction of Theorem 15.1 are similar in that both use forcing with a measure algebra to construct the desired models. The proof of the forward direction of Theorem 15.1 can be found in [Jec78]. The proof of Theorem 15.2 starts by forming $L[G]$, the generic extension of the constructible universe L that adds $\aleph_1$ "random reals," a variation on Cohen's original forcing technique due to Solovay. One then forms the submodel N consisting of all sets in $L[G]$ that are hereditarily ordinal definable over the subclass of $L[G]$ consisting of all countable sequences of ordinals. Then N is a model of ZF + DC [Jec78, p. 546], and Solovay proved that GM is true in N. Note that by Theorem 15.1, LM is necessarily false in N; indeed, the set of constructible reals is a nonmeasurable set [Jec78, p. 568]. For more details of the proof of Theorem 15.2, see [PS77, Sac69].

The preceding discussion shows that some form of Choice is necessary to obtain the Banach–Tarski Paradox. The same is true for many of the results in the other direction, for example, the existence of a finitely additive, translation-invariant measure on $\mathcal{P}(\mathbb{R})$ that normalizes [0, 1]. We shall give some indication of how such unprovability results are obtained by showing why ZF + DC is not strong enough to yield the Measure Extension Theorem or the amenability of Abelian groups.

At the same time as he proved Theorem 15.2, Solovay showed that Con(ZF + IC) implies Con(ZF + DC + PB), where PB denotes the assertion that all sets of reals have the Property of Baire. (Recall that AC yields a set without the Property of Baire; see the discussion following Thm. 3.12.) As with LM, it was generally expected that the seemingly extraneous hypothesis about inaccessible cardinals would be eliminated; unlike the case with LM, this expectation turned out to be correct. The following theorem of Shelah (see [Rai84]), together with Theorem 15.2, shows that there is a deep and unexpected metamathematical distinction between measure and category. This distinction also arises from the Dougherty–Foreman work (§11.2): There is a Banach–Tarski Paradox using pieces with the

Property of Baire, but no such paradox can exist using Lebesgue measurable sets. A proof of Solovay's version of the PB result, assuming Con(ZF + IC), can be found in [Jec78]; Shelah's work is in [She84].

Theorem 15.4. Con(ZF) *implies* Con(ZF + DC + PB).

This theorem asserts that ZF + DC is not strong enough to yield a set of reals that fails to have the Property of Baire. We now show that certain statements about measures—for example, the Measure Extension Theorem or the amenability of Abelian groups—do yield a set without the Property of Baire. It follows that the measure theory statements, which are theorems of ZF + AC, are not theorems of ZF + DC. The next result shows why certain measures yield topologically bad sets.

Theorem 15.5. *If there is a σ-algebra $\mathcal{A}$ that bears a finitely additive measure of total measure 1 that fails to be countably additive, then there is a set of reals that does not have the Property of Baire.*

Proof. Given μ, a measure on $\mathcal{A}$ as hypothesized, let $\{A_n : n \in \mathbb{N}\}$ witness the failure of countable additivity. Because μ is finitely additive, it follows that $a = \mu(\bigcup A_n) - \sum \mu(A_n) > 0$. Now, define ν on $\mathcal{P}(\mathbb{N})$ by $\nu(X) = (1/a)(\mu(\bigcup\{A_n : n \in X) - \sum\{\mu(A_n) : n \in X\})$; ν is a finitely additive measure on $\mathcal{P}(\mathbb{N})$ having total measure 1 and vanishing on singletons. The proof will be complete once it is shown that the collection $\mathcal{I}$ of subsets of $\mathbb{N}$ with ν-measure zero fails to have the Property of Baire when viewed as a subset of the space $2^{\mathbb{N}}$, topologized as a product. (In this proof, we identify 2 with $\{0, 1\}$ and m with $\{0, 1, \ldots, m-1\}$.) For it then follows that the subset of $[0, 1]$ consisting of reals whose binary representation correspond to the sequences of $\mathcal{I}$ fails to have the Property of Baire.

Suppose, by way of contradiction, that $\mathcal{I}$ does have the Property of Baire. Because ν vanishes on finite subsets of $\mathbb{N}$, $\mathcal{I}$ is a tail set in $2^{\mathbb{N}}$ and hence [Oxt71, pp. 84–85] either $\mathcal{I}$ is meager or $\mathcal{I}^c$, the complement of $\mathcal{I}$ in $2^{\mathbb{N}}$, is meager. The latter possibility cannot occur, for if $\mathcal{I}^c$ is meager, then so is $\{\mathbb{N} \setminus A : A \notin \mathcal{I}\}$, because this set is obtained by simply switching 0s and 1s in the sequences in $\mathcal{I}^c$, and this operation preserves nowhere dense sets. But then $\mathcal{I}^c \cup \{\mathbb{N} \setminus A : A \notin \mathcal{I}\}$, which equals all of $2^{\mathbb{N}}$, is meager, which contradicts the Baire Category Theorem.

Thus we may assume that $\mathcal{I}$ is meager, say, $\mathcal{I} = \bigcup\{X_n : n \in \mathbb{N}\}$, where each X_n is a nowhere dense subset of $2^{\mathbb{N}}$. We shall obtain a contradiction by constructing uncountably many sets in $\mathcal{P}(\mathbb{N}) \setminus \mathcal{I}$ with finite pairwise intersections. For such a family, there must be a positive integer n such that uncountably many sets in the family have ν-measure at least $1/n$, and this contradicts $\nu(\mathbb{N}) = 1$.

Consider $2^{\mathbb{N}}$ as the set of all branches (i.e., maximal totally ordered subsets) through the full binary tree. Recall that a basis for the product topology on $2^{\mathbb{N}}$ consists of all sets of the form $[s] = \{b \in 2^{\mathbb{N}} : b \text{ extends } s\}$, where $s : m \to 2$ for some finite m. Each X_n, being nowhere dense, has the property that any $s \in 2^m$ may be extended to a longer sequence t such that $[t] \cap X_n = \varnothing$. Now, build a subtree of the full binary tree as follows. Extend the sequence $\langle 0\rangle$ to some t with $[t]$ disjoint from X_0. Then extend $\langle 1\rangle$ to $\langle 100\ldots 0\rangle$, with the same length as t, and extend this new sequence further to t', where $[t']$ is disjoint from X_0. Then replace t by the sequence obtained by adding 0s to t to bring its length up to that of t'. Note that no extension of t or of t' lies in X_0. Now, consider the two sequences $t^\frown 0$ and $t^\frown 1$ and extend them in the same way that $\langle 0\rangle$ and $\langle 1\rangle$ were extended, but this time avoid X_1 rather than X_0. Then add enough 0s to t' to bring its length up to that of the two sequences just constructed, split the resultant sequence by adding a 0 and a 1, and extend these two sequences in the same way as $t^\frown 0$ and $t^\frown 1$ were extended, that is, to avoid X_1. Then bring the two extensions of $t^\frown 0$ and $t^\frown 1$ up to the same length by adding zeros. We now have four sequences of the same length, and no extension of any of them lies in $X_0 \cup X_1$. Moreover, the insertion of all the 0s guarantees that any two branches through the subtree under construction will yield subsets of $\mathbb{N}$ that are disjoint past the level at which the branches diverge.

Continue the construction, dodging each X_n in turn, and always adding enough 0s to preserve the property that each level of the tree has at most one 1. This yields a subtree of $2^{\mathbb{N}}$ with the property that every node has an extension that splits. It follows that there are $2^{\aleph_0}$ branches through the subtree, and by the construction the subsets of $\mathbb{N}$ corresponding to these branches have finite pairwise intersections and fail to lie in $\bigcup X_n = \mathcal{I}$, as desired. □

Corollary 15.6. *The following assertions, which are theorems of* ZF + AC, *are not theorems of* ZF + DC*:*

(a) *The Measure Extension Theorem (Thm. 12.7).*
(b) *Abelian groups are amenable (Thm. 12.4(b)).*
(c) *A direct union of a directed system of amenable groups is amenable (Thm. 12.4(f)).*

Proof. Because of Theorems 15.4 and 15.5, it is sufficient to show that each of (a)–(c) yields the existence of a measure as in the hypothesis of Theorem 15.5. For (a), let $\mathcal{A} = \mathcal{P}(\mathbb{N})$ and let $\mathcal{A}_0$ be the subalgebra of $\mathcal{A}$ consisting of all finite sets and their complements. If μ is the $\{0, 1\}$-valued measure on $\mathcal{A}_0$ that vanishes on finite sets, then the Measure Extension Theorem yields an extension of μ to all of $\mathcal{A}$. This finitely additive extension vanishes on singletons and so fails to be countably additive, as required. For (b), simply consider $\mathbb{Z}$. Any measure witnessing the amenability of $\mathbb{Z}$ must, by invariance and finite additivity, vanish on singletons and so is not countably additive. For (c), let G be the direct union of $\aleph_0$ copies of $\mathbb{Z}_2$. Then G is the direct union of groups isomorphic to $\mathbb{Z}_2$, $\mathbb{Z}_2 \times \mathbb{Z}_2$, $\mathbb{Z}_2 \times \mathbb{Z}_2 \times \mathbb{Z}_2, \ldots$ each of which is finite and hence amenable. But a measure witnessing G's amenability vanishes on each summand, and so countable additivity would imply it vanishes on G. □

The proof of Corollary 15.6 shows that ZF + DC does not yield the amenability of $\mathbb{Z}$. A much stronger result, due to Pincus and Solovay [PS77], is that

ZF + DC + NM is consistent, where NM states that no infinite set bears a finitely additive measure of total measure 1 that is defined on all subsets and vanishes on singletons. This yields the unprovability in ZF + DC of the existence of a finitely additive, translation-invariant measure on $\mathcal{P}(\mathbb{R}^1)$ or $\mathcal{P}(\mathbb{R}^2)$ that normalizes an interval (Cor. 12.9): The restriction of such a measure to subsets of the interval would violate NM. Their result also yields the unprovability in ZF + DC of the existence of an infinite amenable group: NM implies that *AG* consists of precisely the finite groups.

The proofs of three of the six closure properties of the class of amenable groups (Thm. 12.4) use AC. Corollary 15.6 shows that two of the three are not theorems of ZF + DC. The remaining assertion—that a subgroup of an amenable group is not amenable—must be handled differently. Solovay's original work on Theorems 15.1 and 15.4 showed that the assertions were jointly consistent; that is, he proved that Con(ZF + IC) implies Con(ZF + DC + LM + PB). Now, it is a consequence of ZF + DC + LM that $\mathbb{S}^1$ is amenable: Lebesgue measure witnesses amenability because all subsets of the circle are measurable. But, as shown in Theorem 15.5 and Corollary 15.6, ZF + DC + PB implies that $\mathbb{Z}$ is not amenable. Because $\mathbb{S}^1$ has a subgroup isomorphic to $\mathbb{Z}$, this means that it is a theorem of ZF + DC + LM + PB that some amenable group has a nonamenable subgroup. Hence, assuming the consistency of an uncountable inaccessible cardinal, Theorem 12.4(c) is not a theorem of ZF + DC. In fact, inaccessible cardinals are not needed for this, because Solovay has shown that a Cohen extension of the constructible universe followed by a random real extension can be used to get a model of GM where $\mathbb{Z}$ is not amenable.

These ideas also yield the unprovability of Tarski's Theorem (Thm. 11.1 and Cor. 11.3) in ZF. For let G be the group of all permutations of $\mathbb{Z}$ that equal the identity on all but a finite set. It is easy to see that for each $n = 1, 2, \ldots$, $(n+1)\mathbb{Z} \nleq n\mathbb{Z}$ in $\mathcal{S}(\mathbb{Z})$, the type semigroup for G-equidecomposability. But, assuming ZF + DC + PB, there is no finitely additive, G-invariant measure on $\mathcal{P}(\mathbb{Z})$ with total measure 1, because such a measure would vanish on singletons.

In analyzing the necessity of the Axiom of Choice for certain results, two approaches are possible. The first, which we have followed in the previous discussion, is to show that the result is not provable in ZF alone. The second approach is to show that the result is provably equivalent to AC (in ZF). It is not reasonable to expect that the Banach–Tarski Paradox is fully equivalent to AC because it requires only a well-ordering of $\mathbb{R}$, while AC is equivalent to the assertion that *all* sets can be well-ordered. But the question arises whether some of the more general theorems, such as the Measure Extension Theorem, are equivalent to AC. Such questions have been well studied, and the situation is summarized in Figure 15.1. The statements in each group are provably equivalent, and no statement implies a statement of a group above it; by Corollary 15.6, none of the statements are theorems of ZF + DC. For proofs of these and many other related results, see [Jec73].

Axiom of Choice Zorn's Lemma Every set can be well-ordered $\|X^2\| = \|X\|$ for all infinite sets X Tychonoff's Theorem (a product of compact topological spaces is compact)

$\Downarrow \not\Uparrow$

Boolean Prime Ideal Theorem (all Boolean algebras bear a 2-valued measure) A product of compact Hausdorff spaces is compact $[0, 1]^X$ is compact for all sets X Stone Representation Theorem (any Boolean algebra is isomorphic to a field of sets) Compactness Theorem for first-order logic

$\Downarrow \not\Uparrow$

Hahn–Banach Theorem Measure Extension Theorem All Boolean algebras admit a $[0, 1]$-valued finitely additive measure

$\Downarrow$

The Banach–Tarski Paradox

Figure 15.1. The logical relationship among variations of the Axiom of Choice.

A natural question arising from the nonimplications in Figure 15.1 is whether the classic Banach–Tarski Paradox can be derived from a theory weaker than ZFC. Inspired by work of Foreman and Wehrung [FW91], who showed that the Hahn–Banach Theorem yields a nonmeasurable set, J. Pawlikowski [Paw89] showed that this can be done: The paradox is a consequence of the Hahn–Banach Theorem. The assertion about free groups in the next result is the only place in the classic paradox where AC is used.

Theorem 15.7. *If the Hahn–Banach Theorem is true and F, a free group of rank 2, acts on X with no nontrivial fixed points, then X is F-paradoxical using six pieces.*

Proof. Assume the action is on the right: $xg \in X$; to convert a left action to a right one, just set xg to be $g^{-1}x$. Let σ and τ generate F. Partition X into F-orbits and, for an orbit Z, let $\mathcal{A}_Z = \mathcal{P}(Z)$. Let $\mathcal{B}$ be the free product of the algebras $\mathcal{A}_Z$; this is the power-set algebra $\mathcal{P}(\prod \mathcal{A}_Z)$. So if Y is contained in an orbit Z, then $\overline{Y}$ is the element of $\mathcal{B}$ consisting of all sequences $(*, *, \ldots, *, Y, *, *, \ldots)$, where $*$ is any possible subset of one of the other orbits.

A consequence of the Hahn–Banach Theorem (see [Lux69]) is that every Boolean algebra admits a finitely additive measure of total measure 1; let μ be such a measure on $\mathcal{B}$. Working as in Theorem 1.2, where $W(\alpha)$ denotes "words whose leftmost entry is α," let $A_1 = W(\sigma)$, $A_2 = W(\tau)$, $A_3 = W(\sigma^{-1})$, and $A_4 = W(\tau^{-1})$. These sets are pairwise disjoint and

$$A_3 \cup \sigma^{-1}A_1 = A_4 \cup \tau^{-1}A_2 = A_1 \cup \sigma A_3 = A_2 \cup \tau A_4 = F.$$

If $A \subseteq F$ and $x \in X$, let xA denote $\{xg : g \in A\}$, and then, for $1 \leq i \leq 4$, let $X_i = \{x \in X : \mu(\overline{xA_i}) > \frac{1}{2}\}$, where $\overline{Y}$ is the natural image of Y in $\mathcal{B}$.

Claim 1. $X = X_1\sigma \cup X_2\tau \cup X_3\sigma^{-1} \cup X_4\tau^{-1}$.

Proof of claim. The sets $\overline{xA_i}$ are pairwise disjoint because of the fixed point hypothesis and the disjointness of the A_i. So one of these sets has μ-measure less than 1/2. Suppose $\mu(\overline{xA_1}) < 1/2$ (the other cases are similar). Because $A_1 \cup \sigma A_3 = F$, we have that, in $\mathcal{B}$, $\overline{xA_1} \vee \overline{x\sigma A_3} = \mathbf{1}$. Therefore $\mu(\overline{x\sigma A_3}) > 1/2$, so $x\sigma \in X_3$, so $x \in X_3\sigma^{-1}$.

Claim 2. $X_1 \subseteq X_1\sigma \cap X_2\tau \cap X_4\tau^{-1}$; $X_2 \subseteq X_1\sigma \cap X_2\tau \cap X_3\sigma^{-1}$; $X_3 \subseteq X_2\tau \cap X_3\sigma^{-1} \cap X_4\tau^{-1}$; $X_4 \subseteq X_1\sigma \cap X_3\sigma^{-1} \cap X_4\tau^{-1}$.

Proof of claim. We prove the first only; the rest are similar. First show $X_1 \subseteq X_1\sigma$. Because $A_1 \subseteq F = A_3 \cup \sigma^{-1}A_1$, we have $A_1 \subseteq \sigma^{-1}A_1$. Now, if $x \in X_1$, then $\mu(\overline{xA_1}) > 1/2$, so $\mu(\overline{x\sigma^{-1}A_1}) > 1/2$, so $x\sigma^{-1} \in X_1$, so $x = (x\sigma^{-1})\sigma \in X_1\sigma$. The other three containments are similar; and the cases of X_2, X_3, X_4 are similar.

To conclude, let $Y_1 = X \setminus (X_1\sigma \cup X_2\tau)$ and $Y_2 = X \setminus (X_3\sigma^{-1} \cup X_4\tau^{-1})$. Then $X_1\sigma \cup X_2\tau \cup Y_1 = X_3\sigma^{-1} \cup X_4\tau^{-1} \cup Y_2 = X$. Now to get the paradox, it suffices to show that the sets $Y_1, Y_2, X_1, X_2, X_3, X_4$ are pairwise disjoint. Claim 1 immediately implies that $Y_1 \cap Y_2 = \varnothing$. The definition of X_i and the disjointness of xA_i and xA_j for $i \neq j$ (which follows from the action's lack of fixed points) imply that the X_i are pairwise disjoint. The first assertion of claim 2 immediately implies that X_1 is disjoint from Y_1 and Y_2; the cases of the other X_i are similar. □

A curiosity about the preceding proof is that a measure of one type is used to prove—by yielding a paradox—that a measure of another type cannot exist.

15.1.1 The Axiom of Determinacy

The Solovay model (Thm. 15.1) is a possibility for those who wish to have a theory in which all sets are Lebesgue measurable (and so the counterintuitive Banach–Tarski Paradox disappears). But there are other approaches, a noteworthy one involving the Axiom of Determinacy. Here we will describe the theory ZF + DC + AD and a certain model for it called $L(\mathbb{R})$. First we need some background from game theory. Let $X \subseteq \{0, 1\}^\omega$, and consider the game played by two players Alice (A) and Bob (B). The game starts with Alice choosing a_0 in $\{0, 1\}$. Then Bob chooses b_0, Alice chooses a_1, and so on. If the resulting sequence $a_0, b_0, a_1, b_1, \ldots$ is an element of X, then Alice wins; otherwise, Bob wins. A strategy for either Alice or Bob is a rule that specifies which element to choose.

The nth move described by a strategy depends on the moves up to the current stage of the game.

So a strategy for Alice is a function $S_A : \bigcup_{n\in\omega}\{0, 1\}^n \to \{0, 1\}$ so that Alice's play at the nth move is given by $a_n = S_A(b_0, b_1, \ldots, b_{n-1})$; similarly, a strategy for Bob tells him what to do given what has been done. A strategy is a *winning strategy* if it yields a win for the player regardless of the opponent's moves. A game is *determined* if one of the players has a winning strategy. The following axiom was introduced by J. Mycielski and H. Steinhaus in [MS62].

Axiom of Determinacy (AD). *For every $A \subseteq \{0, 1\}^\omega$, the game defined by A is determined.*

This axiom has many interesting consequences. Using AC, it is not hard to show (by a standard diagonalization technique over the set of possible strategies, which has size continuum) that some game is not determined (see [JW96, pp. 148–149] for a different simple proof). Therefore ZF + AD implies that the Axiom of Choice is false. Much more is true: Determinacy yields that all sets of reals are Lebesgue measurable and have the Property of Baire, and also all uncountable subsets of reals have perfect subsets. Therefore ZF + AD eliminates the Banach–Tarski Paradox.

Under the assumption of the existence of some uncountable large cardinal analogous to $\aleph_0$ (in particular, strongly compact cardinals, also known as Tychonoff cardinals), the theory ZF + DC + AD is consistent. This is because a difficult theorem of Steel and Woodin [Lar04] and [Ste96] asserts that, under the large cardinal assumption, these axioms are true in $L(\mathbb{R})$, the least class of sets that is a model of ZF and contains of all the real and ordinal numbers.

Taking these facts and theorems into account, Mycielski [Myc06, MT∞b] suggested that the theory ZF + DC + AD + $V{=}L(\mathbb{R})$ is a reasonable axiomatization of set theory (and hence mathematics) for use in science. Severely counterintuitive results such as the Banach–Tarski Paradox are false in this theory because all sets are Lebesgue measurable; there is enough choice so that Lebesgue measure works as expected; the universe does not contain any "unnecessary sets": It contains only the sets that must exist given that all reals and all ordinals are present; and all sets have the Property of Baire.

The theory ZF + DC + AD + $V{=}L(\mathbb{R})$ has an interesting connection to the Ruziewicz Problem. Recall that it is a deep theorem of Drinfeld and Margulis (Thm. 13.13) that Lebesgue measure is unique as a finitely additive measure on the measurable subsets of $\mathbb{R}^n$, $n \geq 3$. In ZF + DC + AD + $V{=}L(\mathbb{R})$, the result becomes very simple [MT∞b]. Here is the proof, which uses PB, a consequence of AD.

Theorem 15.8. ZF + DC + PB *implies that Lebesgue measure is unique as a finitely additive measure on the measurable subsets of $\mathbb{R}^n$ (resp., $\mathbb{S}^n$) that normalizes the unit cube (resp., sphere).*

Proof. Consider $\mathbb{R}^n$ first. Let μ be a finitely additive, isometry-invariant measure on $\mathcal{L}$ that normalizes the unit cube. Because of PB, Theorem 15.5 implies that μ is countably additive. Now just use Proposition 11.18. In the case of $\mathbb{S}^2$, all we need is to show that μ agrees with λ on the spherical triangles, for then we can apply the same reasoning as in Proposition 11.18. But it follows from the proof of Girard's theorem that the supposed measure agrees with Lebesgue measure on spherical triangles, which are intersections of some three hemispheres. The proof for general n is similar, using spherical simplices, which are intersections of $n+1$ hemispheres. □

Note that the preceding result fails in ZFC (there are exotic Banach measures, such as the Marczewski measures; see §13.1). But in ZF + DC + PB, the result is true in all dimensions: Lebesgue measure is unique. So while one imagines that Lebesgue, who did not believe in the Axiom of Choice or nonmeasurable sets, would be a proponent of the theory ZF + DC + LM (see §15.3 for why ZF + LM would not suffice for Lebesgue), in fact it is ZF + DC + PB that gives the uniqueness of Lebesgue measure. Of course, one can have it both ways because there are models of ZF + DC + LM + PB, either Solovay's model or $L(\mathbb{R})$ under a large cardinal assumption.

Because Proposition 11.18 is valid for Borel sets in place of measurable sets, the preceding proof works also when restricted to the Borel sets (in ZFC, that question is open; see Question 13.15).

15.2 The Axiom of Choice Can Sometimes Be Eliminated

The previous discussion concerns the noneliminability of AC from various results concerning measures. But AC can be eliminated in some special cases of a geometric nature. The most important consequence of Corollary 12.9 for paradoxical decompositions is that neither an interval in $\mathbb{R}^1$ nor a square in $\mathbb{R}^2$ is paradoxical. We have already seen (Cor. 14.25) that a proof of the linear case can be given in ZF, using the fact that G_1 is exponentially bounded. In fact, as we shall show, AC can also be eliminated from the proof that a square is not paradoxical. The main idea is the observation that one does not require a measure on all of $\mathcal{P}(\mathbb{R}^2)$, but only on a countable subalgebra, and such a measure can be constructed without having to use the Axiom of Choice. The proof requires the following result that shows that, despite the inability of ZF to yield the amenability of Abelian groups, ZF is strong enough to get a weakened form of amenability for solvable groups.

Theorem 15.9. *If G is a solvable group and $\mathcal{A}$ is a countable left-invariant subalgebra of $\mathcal{P}(G)$, then there exists $\mu: \mathcal{A} \to [0, 1]$, a left-invariant, finitely additive measure with $\mu(G) = 1$.*

Proof. Let $\{e\} = H_0 \lhd H_1 \lhd \cdots \lhd H_n = G$, where each H_i/H_{i-1} is Abelian, witness the solvability of G. We use induction on n. Because $\mathcal{A}_0 = \{A \cap H_{n-1} : A \in \mathcal{A}\}$ is a left-invariant subalgebra of $\mathcal{P}(H_{n-1})$, we may assume that there is an

H_{n-1}-invariant measure $\nu: \mathcal{A}_0 \to [0, 1]$. Define $\overline{\nu}$ on $\mathcal{A}$ by $\overline{\nu}(A) = \nu(A \cap H_{n-1})$; $\overline{\nu}$ is an H_{n-1})-invariant measure on $\mathcal{A}$.

Now, it is sufficient to show that for each finite $W \subseteq G$ and each $\epsilon > 0$, there is a measure $\mu_{W,\epsilon}$ on $\mathcal{A}$ that is invariant within ϵ for left multiplication by members of W. For then the usual compactness technique can be applied to the family of subsets $\mathcal{M}_{W,\epsilon}$ of $[0, 1]^{\mathcal{A}}$ (see the proof of Thm. 12.4(b)) to obtain the desired measure. Note that because $\mathcal{A}$ is countable, the compactness of $[0, 1]^{\mathcal{A}}$ does not require any Choice (Prop. 11.4).

The key point in the construction of $\mu_{W,\epsilon}$ is the observation that for each $A \in \mathcal{A}$ and $g_1, g_2 \in G$, $\overline{\nu}(g_1 g_2 A) = \overline{\nu}(g_2 g_1 A)$. This holds because the commutativity of G/H_{n-1} implies that $H_{n-1} g_1 g_2 = H_{n-1} g_2 g_1$, whence $g_1 g_2 g_1^{-1} g_2^{-1} \in H_{n-1}$. Now,

$$\begin{aligned}\overline{\nu}(g_1 g_2 A) &= \nu(H_{n-1} \cap g_1 g_2 A) = \nu(H_{n-1} \cap g_1 g_2 g_1^{-1} g_2^{-1} g_2 g_1 A)\\ &= \nu(g_1 g_2 g_1^{-1} g_2^{-1}(H_{n-1} \cap g_2 g_1 A) g_2 g_1 A)\\ &= \nu(H_{n-1} \cap g_2 g_1 A) = \overline{\nu}\,(g_2 g_1 A).\end{aligned}$$

We may now construct $\mu_{W,\epsilon}$ in a manner similar to the proof of Theorem 12.4(b). Let $W = \{g_1, \ldots, g_m\}$ and choose N so that $2/N \leq \epsilon$. Then define $\mu_{W,\epsilon}(A)$ to be $(1/N^m) \sum\{\overline{\nu}(g_1^{i_1} \cdots g_m^{i_m} A) : 1 \leq i_1, \ldots, i_m \leq N\}$. This yields a finitely additive measure, and because $\overline{\nu}(g_i g_j A) = \overline{\nu}(g_j g_i A)$, $\mu_{W,\epsilon}(g_k A)$ differs from $\mu_{W,\epsilon}(A)$ by no more than

$$\frac{1}{N^m} \sum_{1 \leq k \leq m} \mu(g_1^{i_1} \cdots g_{k-1}^{i_{k-1}} g_{k+1}^{i_{k+1}} \cdots g_m^{i_m} A) + \mu(g_1^{i_1} \cdots g_k^{N+1} \cdots g_m^{i_m} A),$$

which is at most $2N^{m-1}/N^m = 2/N \leq \epsilon$. This completes the proof. □

The previous result can be used to obtain a Choiceless version of the Invariant Extension Theorem for solvable groups.

Theorem 15.10. *Suppose G is a solvable group of automorphisms of a countable Boolean algebra $\mathcal{A}$, and suppose $\mu_0: \mathcal{A}_0 \to [0, \infty]$ is a G-invariant measure on $\mathcal{A}_0$, a G-invariant subalgebra of $\mathcal{A}$. Then μ_0 has a G-invariant extension to $\mathcal{A}$.*

Proof. Because $\mathcal{A}$ is countable, we may use the Theorem 12.7 (see the remarks following its proof) to extend μ_0 to a measure $\overline{\mu}: \mathcal{A} \to [0, \infty]$. Now, for $a \in \mathcal{A}$ and rationals r, s, let $S_{a,r,s} = \{g \in G : r \leq \overline{\mu}(g^{-1} a) < s\}$, and let $\mathcal{C}$ be the subalgebra of $\mathcal{P}(G)$ generated by these countably many sets $S_{a,r,s}$. Then $\mathcal{C}$ is countable, and because $h S_{a,r,s} = S_{ha,r,s}$, $\mathcal{C}$ is G-invariant. By Theorem 15.9, then, there is a left-invariant measure $\nu: \mathcal{C} \to [0, 1]$.

Now, for any $a \in \mathcal{A}$, let $f_a: G \to [0, \infty]$ be defined by $f_a(g) = \overline{\mu}(g^{-1} a)$. By the way $\mathcal{C}$ was defined, f_a is a $\mathcal{C}$-measurable function, and so we may define $\mu(a)$ to be $\int f_a \, d\nu$ if f_a is bounded and let $\mu(a) = \infty$ otherwise. It is easy to check that μ is a measure on $\mathcal{A}$, μ extends μ_0, and μ is G-invariant. □

By applying the preceding theorem to the (countable) algebra generated by the pieces of a purported paradoxical decomposition, one obtains a proof in ZF that a square is not paradoxical.

Corollary 15.11. *If G is a solvable group of isometries of $\mathbb{R}^n$ and A, B are two measurable sets that are G-equidecomposable, then $\lambda(A) = \lambda(B)$. In particular, a square in the plane is not G_2-paradoxical.*

Proof. Suppose the equidecomposability of A and B uses m pieces and is witnessed by $A = \bigcup A_i$ and $B = \bigcup g_i(A_i)$. Let H be the subgroup of G generated by $\{g_i : 1 \leq i \leq m\}$ (H is countable and solvable), and let $\mathcal{A}$ be the H-invariant subalgebra of $\mathcal{P}(G)$ generated by the $2m$ sets A_i, $g_i(A_i)$. Then $\mathcal{A}$ contains A, B, and all the pieces of the decomposition. By Theorem 15.10, Lebesgue measure on the H-invariant subalgebra $\mathcal{A} \cap \mathcal{L}$ may be extended to a measure on all of A. Because A and B are H-equidecomposable in $\mathcal{A}$, this extension must assign them the same measure, and therefore $\lambda(A) = \lambda(B)$. □

The preceding proof also yields that $\mathbb{R}^n$ is not G-paradoxical if G is any solvable group acting on $\mathbb{R}^n$. If a paradox exists, let $\mathcal{A}$ be the H-invariant algebra generated by the pieces and extend the two-valued measure on $\{\varnothing, \mathbb{R}^n\}$ to an H-invariant measure on $\mathcal{A}$.

15.3 Foundational Implications of the Banach–Tarski Paradox

What are the implications of the Banach–Tarski Paradox for the axiomatic basis of set theory? Does the paradox, so clearly false in physical reality, mean that the Axiom of Choice yields unreliable results and should be discarded? Many critics of AC have buttressed their arguments by citing the Banach–Tarski Paradox. For example, E. Borel [Bor46, p. 210], using probability theory to make his point, argued as follows: "Hence we arrive at the conclusion that the use of the Axiom of Choice and a standard application of the calculus of probabilities to the sets A, B, C [of the Hausdorff Paradox of $\mathbb{S}^2$] that this axiom allows to be defined, lead to a contradiction: therefore the Axiom of Choice must be rejected." Before turning to a defense of AC, we point out how the technical results of this chapter bear upon the issue.

To those who feel that the Banach–Tarski Paradox is absurd, it seems evident that the Axiom of Choice is the culprit. The whole situation would have to be considered in an entirely different light if somehow AC could be completely eliminated from the proof and the paradox turned out to be a theorem of ZF. That this cannot happen is precisely the content of Corollary 15.3. Thus it is fair to say that AC is indeed to blame for the classic paradox. But it is not true that the rejection of AC yields a system that is free of counterintuitive geometric paradoxes. Much of the Dougherty–Foreman work (§11.2) is valid in ZF, and their results yield a striking paradox (Thm. 11.15) for bodies in 3-space that is almost as counterintuitive as the classic Banach–Tarski Paradox: These are paradoxes in which one ignores certain nowhere dense sets.

In their original paper [BT24, p. 245], Banach and Tarski anticipated the controversy that their counterintuitive result would spawn, and they analyzed their use of the Axiom of Choice as follows: "It seems to us that the role played by the Axiom of Choice in our reasoning deserves attention. Indeed, consider the following two theorems, which are consequences of our research:

I. Any two polyhedra are equivalent by finite decomposition.
II. Two different polygons, with one contained in the other, are never equivalent by finite decomposition.

Now, it is not known how to prove either of these theorems without appealing to the Axiom of Choice: neither the first, which seems perhaps paradoxical, nor the second, which agrees fully with intuition. Moreover, upon analyzing their proofs, one could state that the Axiom of Choice occurs in the proof of the first theorem in a more limited way than in the proof of the second."

Thus Banach and Tarski pointed out that if AC is discarded, then not only would their paradox be lost, but also the result that such paradoxes do not exist in the plane. The last sentence of the excerpt refers to the fact that statement II uses choices from a larger family of sets than does statement I. But Corollary 15.11, which was proved by A. P. Morse [Mor49] in 1949, shows that statement II is indeed provable in ZF and is therefore not relevant to a discussion of AC. The exact role of AC can therefore be loosely summarized as follows. It is necessary to disprove the existence of various invariant measures on $\mathcal{P}(\mathbb{R}^n)$ and to construct such measures, but it is not necessary to disprove the existence of paradoxes.

Despite the Banach–Tarski Paradox and other objections to the Axiom of Choice, often focusing on its nonconstructive nature, the great majority of contemporary mathematicians fully accept the use of AC. It is generally understood that nonmeasurable sets (either of the Vitali type or of the sort that arise in the Banach–Tarski Paradox) lead to curious situations that contradict physical reality, but the mathematics that is brought to bear on physical problems is almost always mathematics that takes place entirely in the domain of measurable sets. And even in this restricted domain of sets, the Axiom of Choice is useful, and ZFC provides a more coherent foundation than does ZF alone. One can also wonder whether the paradox has any implications for our physical world. B. Augenstein presented in [Aug84] some speculations linking hadronic physics and paradoxes like the Banach–Tarski Paradox.

Perhaps the most important use of AC in the domain of measurable sets has to do with Lebesgue measure. As pointed out earlier in this chapter, it is consistent with ZF that Lebesgue measure fails to be countably additive; some form of Choice is needed to make Lebesgue measure the useful concept that it is. While either DC or the Axiom of Countable Choice suffices, these are somewhat unnatural weakenings of AC. If some nonconstructive choice is allowed, why not permit all nonconstructive choice? One fear of this liberal attitude, that unlimited choice will lead to a contradiction, is unfounded because of Gödel's famous theorem, using the constructible universe L, that ZFC is noncontradictory. Of course, choosing AC over DC means that duplication of the sphere is possible, but this is simply not very relevant to the mathematics of the Lebesgue measurable sets.

It is worth noting that the fact that Choice creeps into the basic results on Lebesgue measure is a subtle one, and it went unnoticed by the early practitioners in the field, some of whom (Lebesgue and Borel) went on to become strong critics of the use of nonconstructive choice. Moore [Moo83] points out that Lebesgue's "work reveals how a mathematician of the first rank may subtly fail to see that he is fundamentally violating his philosophical scruples in his own work."

The preceding remarks should not be construed as implying that the countable additivity of Lebesgue measure is more self-evident than other alternatives. The point is that Lebesgue measure, in its usual form, has become a valuable tool of modem mathematics and the use of spaces of Lebesgue integrable functions provides a simpler and clearer approach to topics of classical analysis, for example, Fourier series. An analogous problem arises with the question of the existence of infinite sets. Suppose all mathematicians and scientists become firmly convinced that all aspects of our universe are finite, that a true continuum of points simply does not exist. This does not mean that infinite sets or real numbers must be discarded as mathematical tools. It may well be that calculus is best understood and taught in the context of the real numbers, whether or not the set $\mathbb{R}$ has physical existence. The case for AC is similar: ZFC is a natural and simple foundation for mathematics that is rich enough to provide mathematicians and scientists with the tools that have proved most useful in practice.

Despite the general acceptance of AC, it is recognized that this axiom has a different character than the others. Because of the nonconstructivity it introduces, AC is avoided when possible, and proofs in ZF are considered more basic (and are often more informative) than proofs using AC. A purist might argue that once accepted as an axiom, AC should be accorded the same status as the other axioms: A statement that is proved in ZFC is just as mathematically true as one proved in ZF. But there are sometimes beneficial side effects of a discriminating attitude toward the use of AC. For instance, consider the Schröder–Bernstein Theorem for cardinality, which is much easier to prove in ZFC than in ZF (see remarks following Cor. 10.23). But the ZF-proof, essentially presented here in Theorem 3.6, is much more valuable because it generalizes in a way that the ZFC-proof does not. In particular, one obtains the Banach–Schröder–Bemstein Theorem for the equidecomposability relation in an arbitrary countably complete Boolean algebra. On the other hand, the Cancellation Law for equidecomposability (using arbitrary pieces) uses the Axiom of Choice (despite the fact that the corresponding result for cardinality avoids AC), and therefore it is not clear that it can be generalized to more general equidecomposability, such as the case where the pieces are Borel sets.

The interplay between ZF and ZFC works the other way as well. For instance, the ZF-proof that a square is not paradoxical (Cor. 15.11) was found by closely analyzing the original ZFC-proof. If a more restrictive attitude toward AC had prevailed, the original ZFC-proof might not have been discovered so soon, and the proof (in ZF) that squares are not paradoxical might have been delayed.

Despite how often it is cited in discussions of the Axiom of Choice, the Banach–Tarski Paradox is more important to pure mathematics than it is to foundational questions. The role of the Banach–Tarski Paradox in discussions about AC is, for the most part, the same as the role of nonmeasurable sets; indeed, the paradox of the sphere simply reinforced the views of those whose ideas were shaped by the question of nonmeasurable subsets of the real line. But, as we have emphasized throughout this book, the Banach–Tarski Paradox has spawned some valuable mathematical ideas (principally amenability in groups, but also additional ideas both geometric and algebraic). Thus the Banach–Tarski Paradox has a historical importance for mathematics that is completely independent of foundational questions and the axioms of set theory.

Notes

See [Jec78] for an exposition of several fundamental consistency results such as Con(ZF + AC), Con(ZF + ¬AC), Con(ZFC + IC), Con(ZF + $\mathbb{R}$ is a countable union of countable sets), and a proof of Solovay's theorem: Con(ZF + IC) implies Con(ZF + DC + LM + PB). More technical consistency results concerning the interplay between AC, its variants, and certain mathematical statements (such as those in Fig. 15.1) may be found in [Jec73]. For a discussion of the role of DC and the Axiom of Countable Choice in the theory of measure and category, see [Moo83]. This topic is also discussed in Moore's book [Moo82], which provides a valuable historical account of many of the controversies concerning AC.

Solovay's theorem (the forward direction of Thm. 15.1) and the result of Theorem 15.4 using Con(ZF + IC) was proved in 1964; complete details were published in [Sol70]. The problem of using forcing to prove the consistency of LM was suggested by Paul Cohen, the discoverer of forcing. It is noteworthy that Solovay first proved, assuming only Con(ZF), Con(ZF + LM) (see [Moo82, p. 304]), but DC failed in his model of ZF + LM. Thus he persisted to get Con(ZF + DC + LM), which, however, required the consistency of Con(ZF + IC). Shelah's refinements of Solovay's results (the reverse direction of Thm. 15.1 and Thm. 15.4 without IC) were proved in 1980 [She84]. His proof that Con(ZF + DC + LM) implies Con(ZF + IC) was improved by Raisonnier [Rai84] (see also [Rai82]). It is worth noting that the full strength of LM is not required. Shelah proved that ZF + DC+ "Every $\mathbf{\Sigma}^1_3$ set of reals is Lebesgue measurable" yields that $\aleph_1$ is inaccessible in L. Moreover, Raisonnier showed that

ZF + DC + Every $\mathbf{\Sigma}^1_2$ set is Lebesgue measurable and every $\mathbf{\Sigma}^1_3$ set has the Property of Baire also yields the inaccessibility of $\aleph_1$ in L. The latter result is noteworthy because Con(ZF + DC + PB) can be proved assuming only Con(ZF); this is Theorem 15.4.

Theorem 15.2, which is important for us because it yields Corollary 15.3, was proved by Solovay in 1964. A proof appeared in [Sac69], while a proof of a similar result, which contains arguments very similar to those in the sketch of Theorem 15.2's proof given here, appears in [PS77]. Originally Solovay proved only Con(ZF + DC + GM_1), where GM_n denotes the assertion of GM for $\mathbb{R}^n$ only. But, as he pointed out to the authors, only a slight modification is needed to get that each GM_n holds in the model N discussed after Corollary 15.3.

Corollary 15.6 was known to Solovay [Sol70, p. 3]. A proof of Theorem 15.5 along a somewhat different line than the proof presented here (which is due to A. Taylor) is presented by Pincus [Pin74].

The Axiom of Determinacy was first presented by Mycielski and Steinhaus in 1962 [MS62], where it is shown that AD implies LM. Mycielski [Myc64b] showed that AC implies PB.

Corollary 15.11 is due to A. P. Morse [Mor49]. The proof presented here was inspired by an argument of Mycielski [Myc79, Thm. 3.6], who showed that the conclusion of Theorem 15.7 is valid for countable groups satisfying Følner's Condition.

Appendices

A

Euclidean Transformation Groups

The types of transformations that are used to produce paradoxes in Euclidean spaces and on spheres are usually the Euclidean isometries, but occasionally more general affine maps arise. Because the affine group is useful in studying and classifying isometries, we summarize the relevant facts about affine transformations. The book by Hausner [Hau65] is a good reference for a more detailed presentation.

Definition A.1. *A bijection* $f:\mathbb{R}^n \to \mathbb{R}^n$ *is called* affine *if for all* $P, Q \in \mathbb{R}^n$ *and reals* α, β with $\alpha + \beta = 1$, $f(\alpha P + \beta Q) = \alpha f(P) + \beta f(Q)$. *The affine transformations of* $\mathbb{R}^n$ *form a group, which is denoted by* $A_n(\mathbb{R})$.

Geometrically, a bijection is affine if and only if it carries lines to lines and preserves the ratio of distances along a line. Any nonsingular linear transformation is affine, because a linear transformation satisfies Definition A.1 for all α, β, not just pairs summing to 1. The group of nonsingular linear transformations of $\mathbb{R}^n$ is denoted by $GL_n(\mathbb{R})$ (general linear group). Linear maps leave the origin fixed, but affine maps need not do so; all translations of $\mathbb{R}^n$ are affine. Let T_n denote the group of translations of $\mathbb{R}^n$. Then T_n is isomorphic to the additive group of $\mathbb{R}^n$ because composition of translations corresponds to addition of the translation vectors. It is an extremely useful fact that every affine map has a canonical representation in terms of linear maps and translations.

Theorem A.2. *If* $f \in A_n(\mathbb{R})$, *then there are uniquely determined maps* $\tau \in T_n$, *and* $\ell \in GL_n(\mathbb{R})$ *such that* $f = \tau\ell$. *Moreover, the map* $\pi : A_n \to GL_n(\mathbb{R})$ *defined by* $\pi(f) = \ell$ *is a group homomorphism with kernel* T_n. *Hence* T_n *is a normal subgroup of* $A_n(\mathbb{R})$ *and* $A_n(\mathbb{R})/T_n \cong GL_n(\mathbb{R})$.

To prove the first part of this theorem, one shows τf is linear, where τ is the translation by $-f(\mathbf{0})$; the rest follows in a straightforward manner. (For details, see [Hau65, Chap. 8].) A useful consequence of this theorem and the fact that a member of $GL_n(\mathbb{R})$ is determined by its values at n linearly independent vectors, is that an affine map is completely determined by its values at $n+1$ points, no three of which are collinear.

Theorem A.2 allows us to define the determinant of an affine map to be the determinant of the corresponding linear map, that is, $\det f = \det \pi(f)$. Let $SL_n(\mathbb{R})$ (special linear group) denote the group of linear transformations of determinant 1; $SA_n(\mathbb{R})$ the group of affine maps of determinant 1. Note that $T_n \subseteq SA_n(\mathbb{R})$.

Definition A.3. *An* isometry *of* $\mathbb{R}^n$ *(or of any metric space) is a distance-preserving bijection of* $\mathbb{R}^n$ *to itself. Let* G_n *denote the n-dimensional isometry group.*

We are using the word *isometry* in a global sense. Occasionally we refer to *partial isometries*: bijections from a subset A to a subset B that preserve distance. In Euclidean space, and on spheres, any partial isometry may be extended to an isometry.

Theorem A.4. (*See [Hau65, §9.4]*) *Every isometry of* $\mathbb{R}^n$ *is affine; that is,* $G_n \subseteq A_n(\mathbb{R})$.

The group of isometries of $\mathbb{R}^n$ that are also linear transformations is quite important, because it is the group of isometries of the unit sphere, $\mathbb{S}^{n-1}$. Such linear transformations are characterized by their representation by *orthogonal matrices*, matrices A for which $A^T = A^{-1}$ (A^T denotes the transpose of A). Hence $O_n(\mathbb{R})$ is used to denote this group; that is, $O_n(\mathbb{R}) = G_n \cap GL_n(\mathbb{R})$, and $O_n(\mathbb{R})$ is called the *orthogonal group*. The transformations in $O_n(\mathbb{R})$ are exactly those linear transformations of $\mathbb{R}^n$ that preserve the dot product, that is, $\ell(P) \cdot \ell(Q) = P \cdot Q$. Because $\det A = \det A^T$, and if A is orthogonal, $AA^T = I$, it follows that if $\ell \in O_n(\mathbb{R})$, then $(\det \ell)^2 = 1$. This yields the following theorem.

Theorem A.5. *If* $\ell \in O_n(\mathbb{R})$, *then* $\det \ell = \pm 1$.

The homomorphism π takes the isometry group G_n to $O_n(\mathbb{R})$, and because $T_n \subseteq G_n$, it follows that T_n is a normal subgroup of G_n and $G_n/T_n \cong O_n(\mathbb{R})$. Thus every isometry is an orthogonal transformation followed by a translation. We use $SO_n(\mathbb{R})$ (special orthogonal) to denote the orthogonal transformations having determinant $+1$; these are the orientation-preserving orthogonal transformations. By analogy with $\mathbb{R}^2$, transformations in $SO_n(\mathbb{R})$ are sometimes called rotations, and $SO_n(\mathbb{R})$ may be called the rotation group of the sphere $\mathbb{S}^{n-1}$. It is sometimes useful to have a notation for the orientation-preserving isometries of $\mathbb{R}^n$; thus let $SG_n = \{\sigma \in G_n : \det \sigma = +1\}$.

Because for any $f \in A_n(\mathbb{R})$, $|\det f|$ is the factor by which f changes area (or Lebesgue measure; see [Hau65, §9.2] or [Wei73, §6.3]), Theorem A.5 implies that distance-preserving maps preserve area as well. Thus G_n and $SA_n(\mathbb{R})$ contain only measure-preserving transformations. Because det is a homomorphism, $O_n(\mathbb{R})/SO_n(\mathbb{R})$ and G_n/SG_n are each isomorphic to $\mathbb{Z}_2$. This allows us to extend results proved for certain maps of determinant $+1$ to maps of the same sort, but with determinant ± 1.

A useful consequence of the preceding representation of G_n is that G_n is isomorphic to a subgroup of $SL_{n+1}(\mathbb{R})$. If $\sigma \in G_n$ has the form $\tau\ell$ where τ is the translation by $(v_1, \ldots, v_n)$ and ℓ is represented by the orthogonal matrix (a_{ij}), then let $M(\sigma)$ be the $(n+1) \times (n+1)$ matrix:

$$\begin{bmatrix} & & & v_1 \\ & & & v_2 \\ & a_{ij} & & \vdots \\ & & & v_n \\ 0 & 0 \cdots 0 & & 1 \end{bmatrix}.$$

The mapping M is an isomorphism of G_n with a subgroup of $SL_{n+1}(\mathbb{R})$.

In low dimensions we can be more explicit about these groups; proofs may be found in [Hau65, Chap. 9].

In $\mathbb{R}^1$ the only orthogonal maps are the identity, I, and the flip taking x to $-x$. Hence $SO_1(\mathbb{R}) = \{I\}$ and $O_1(\mathbb{R}) = \{I, x \mapsto -x\}$. By Theorem A.2 and Theorem A.4, it follows that $G_1 = \{x \mapsto \pm x + a : a \in \mathbb{R}\}$. Because $G_1/T_1 \cong O_1(\mathbb{R}) \cong \mathbb{Z}_2$ and $T_1 \cong \mathbb{R}$, the normal series $\{I\} \lhd T_1 \lhd G_1$ shows that G_1 is solvable. Because $A_1(\mathbb{R})$ consists of transformations of the form $x \mapsto ax + b$, it follows that the length-preserving affine maps (those with $|a| = 1$) are just the isometries.

Let ρ_θ denote the counterclockwise rotation about the origin in $\mathbb{R}^2$ through θ radians; ρ_θ is represented by $\left[\begin{smallmatrix} \cos\theta & -\sin\theta \\ \sin\theta & \cos\theta \end{smallmatrix}\right]$. Let ϕ_θ denote the reflection in the line through the origin making an angle θ with the x-axis. Then $SO_2(\mathbb{R}) = \{\rho_\theta : 0 \leqslant \theta < 2\pi\}$, which is Abelian and isomorphic to the circle group (the set of complex numbers of unit modulus under multiplication), and $O_2(\mathbb{R}) = SO_2(\mathbb{R}) \cup \{\phi_\theta : 0 \leqslant \theta < \pi\}$. To see that G_2 is solvable, consider the sequence $\{I\} \lhd T_2 \lhd SG_2 \lhd G_2$. Because T_2 is the kernel of $\pi : SG_2 \to SO_2(\mathbb{R})$, SG_2/T_2 is isomorphic to the Abelian group $SO_2(\mathbb{R})$. Because $G_2/SG_2 \cong \mathbb{Z}_2$, this sequence witnesses the solvability of G_2.

A planar isometry, ψ, may be written uniquely as $\tau\ell$ with $\tau \in T_2$ and $\ell \in O_2(\mathbb{R})$, and ψ can be characterized according to the choice of τ and ℓ:

- If $\ell = 1$, then ψ is a translation.
- If $\ell = \rho_\theta$, then ψ is a counterclockwise rotation through θ radians about some point in $\mathbb{R}^2$.
- If $\ell = \phi_\theta$, then ψ is a *glide reflection*, that is, a reflection in some line followed by a translation in the direction parallel to the line of reflection. If the component of τ in that direction is the identity (i.e., τ is a translation perpendicular to the line of reflection), then ψ is simply a reflection.

If we let $SL_2(\mathbb{Z})$ denote the subgroup of $SL_2(\mathbb{R})$ that carries $\mathbb{Z} \times \mathbb{Z}$ to itself (matrices with coefficients in $\mathbb{Z}$), then there are elements of $SL_2(\mathbb{Z})$ that are not orthogonal. For example, $\left[\begin{smallmatrix}1 & 1\\ 0 & 1\end{smallmatrix}\right] \in SL_2(\mathbb{Z}) \setminus O_2(\mathbb{R})$. Thus, unlike $\mathbb{R}^1$, the area-preserving affine transformations consist of more than just the isometries. In fact, G_2 is solvable, while $SL_2(\mathbb{Z})$ has a free non-Abelian subgroup (see §8.1).

A solvable group cannot contain a free group of rank 2, because it follows from the characterization of solvability in terms of the finiteness of the series of commutator subgroups that a solvable group universally satisfies a nontrivial relation $w = 1$, where w is a nontrivial reduced word in the variables x, x^{-1}, y, y^{-1}. But we can be more explicit for G_1 and G_2. Because $\sigma^2 \in T_1$ for any $\sigma \in G_1$, the equation $x^2y^2x^{-2}y^{-2} = 1$ is satisfied by all pairs x, y in G_1. In G_2, the normal series $\{I\} \lhd T_2 \lhd SG_2 \lhd G_2$ yields that $\sigma^2\rho^2\sigma^{-2}\rho^{-2}$ lies in T_2 for any $\sigma, \rho \in G_2$. Hence $\sigma^2\rho^2\sigma^{-2}\rho^{-2}$ commutes with $\sigma^{-2}\rho^{-2}\sigma^2\rho^2$, yielding a nontrivial relation in G_2. This shows directly why neither G_1 nor G_2 contains a pair of independent elements.

A complete classification of isometries gets more complicated as the dimension increases. In $\mathbb{R}^3$, we do not need such a classification, but we do require more information about $SO_3(\mathbb{R})$. Using the fact that any element of $SO_3(\mathbb{R})$ has $+1$ as an eigenvalue, it can be shown (see [Hau65]) that $SO_3(\mathbb{R})$ consists of rotations about a line (axis) through the origin. Hence the composition of two rotations of $\mathbb{R}^3$ that fix the origin is another such rotation. Moreover, using the fact that orthogonal transformations preserve angles, it follows that if $\rho, \sigma \in SO_3(\mathbb{R})$, then $\rho\sigma\rho^{-1}$ is a rotation whose rotation angle is the same as that of σ. Unlike $SO_1(\mathbb{R})$ and $SO_2(\mathbb{R})$, $SO_3(\mathbb{R})$ is not solvable, because it contains a free subgroup of rank 2 (see Thm. 2.1).

The matrix representation of an element of $SO_3(\mathbb{R})$ with respect to an orthonormal basis whose first element is the x-axis is simply

$$\begin{bmatrix} 1 & 0 & 0 \\ 0 & \cos\theta & -\sin\theta \\ 0 & \sin\theta & \cos\theta \end{bmatrix},$$

where θ is the angle of rotation. In fact, one can compute the axis and the angle of a rotation from its matrix representation with respect to any orthonormal basis, as the following theorem shows.

Theorem A.6. *Let (a_{ij}) be the matrix of a rotation ρ of $\mathbb{R}^3$ through an axis containing* **0**. *Let θ be ρ's angle of rotation ($0 \leqslant \theta < 2\pi$), and let $\vec{A}$ be a unit vector along ρ's axis so oriented that the rotation obeys the right-hand rule. Then $2\vec{A}\sin\theta = (a_{32} - a_{23}, a_{13} - a_{31}, a_{21} - a_{12})$.*

Proof. Let $\vec{A} = (b_1, b_2, b_3)$ and ρ_1 be the rotation represented by

$$\begin{bmatrix} b_1 & 0 & b \\ b_2 & -b_3/b & -b_1b_2/b \\ b_3 & b_2/b & -b_1b_3/b \end{bmatrix},$$

where $b = \sqrt{b_2^2 + b_3^2}$; ρ_1 takes $(1, 0, 0)$ to (b_1, b_2, b_3). Then

$$\rho = \rho_1 \begin{bmatrix} 1 & 0 & 0 \\ 0 & \cos\theta & -\sin\theta \\ 0 & \sin\theta & \cos\theta \end{bmatrix} \rho_1^{-1},$$

and a computation, using the fact that ρ_1, is orthogonal and hence $\rho_1^{-1} = \rho_1^T$, shows that the right-hand side of the desired equation is

$$(2(b_1 b_2^2 + b_1 b_3^2)\sin\theta/b^2, 2b_2 \sin\theta, 2b_3 \sin\theta),$$

which equals $2\vec{A}\ \sin\theta$, as required. □

One important fact about orientation-preserving isometries of $\mathbb{R}^3$ is that it is easy to recognize when there is a fixed point. For if $\sigma \in SG_3$ has the representation $\tau\ell$ where τ is the translation by a vector $\vec{V}$ and ℓ is a rotation about the axis $\vec{A}$, then a has a fixed point in $\mathbb{R}^3$ if and only if $\vec{V}$ is perpendicular to $\vec{A}$.

Finally, we note that the groups considered here contain the corresponding groups in lower dimensions. If $m < n$, then simply decompose $\mathbb{R}^n$ into $\mathbb{R}^m \times \mathbb{R}^{n-m}$ and extend the transformations of $\mathbb{R}^m$ by letting them be the identity in the $n - m$ new coordinates.

Occasionally we will use the fact that $GL_n(\mathbb{R})$ is a topological group, where the topology is obtained by embedding the group in $\mathbb{R}^{n^2}$. This implies that any subgroup of G_n is a topological group with topology inherited from G_n.

B

Jordan Measure

This appendix contains some of the basic facts about Jordan measure, which is a more elementary notion than Lebesgue measure. Lebesgue measure uses countable collections of intervals (or cubes) to cover a set, but Jordan measure uses only finite collections of intervals. We use υ, for volume, to denote Jordan measure

Definition B.1. *Let A be a bounded subset of $\mathbb{R}^n$. Define $\upsilon^*(A)$ to be $inf\{\Sigma_{i=1}^n$ volume(K_i) : $\{K_i\}$ is a pairwise interior-disjoint collection of finitely many cubes such that $A \subseteq \bigcup K_i\}$ and $\upsilon_*(A)$ to be* sup$\{\Sigma_{i=1}^n$ *volume(K_i) : $\{K_i\}$ is a pairwise interior-disjoint collection of finitely many cubes such that $\bigcup K_i \subseteq A\}$. If $\upsilon^*(A) = \upsilon_*(A)$, then A is called* Jordan measurable *and this common value, the* Jordan measure *of A, is denoted by $\upsilon(A)$. The collection of bounded Jordan measurable sets is denoted by $\mathcal{J}$.*

Some authors use rectangles rather than cubes in the definition of υ^* and υ_*, but because any rectangle may be approximated arbitrarily closely by finitely many cubes, the two definitions are equivalent. Because

$$\upsilon_*(A) \leqslant \lambda_*(A) \leqslant \lambda^*(A) \leqslant \upsilon^*(A),$$

where λ^*, λ_* denote Lebesgue outer and inner measure, respectively, $\mathcal{J} \subseteq \mathcal{L}$, the collection of Lebesgue measurable sets, and λ agrees with υ on the sets in $\mathcal{J}$. But $\mathcal{J}$ is much smaller than $\mathcal{L}$: If A is the set of rationals in [0, 1], then $\upsilon_*(A) = 0$ and $\upsilon^*(A) = 1$. Or, choose an open set E containing A with $\lambda(E) < 1$; then $\upsilon_*(E) < 1 = \upsilon^*(E)$, so $\mathcal{J}$ does not even contain all open sets. The following characterization is central to the study of Jordan measurable sets; for a proof, see [Olm59, §1305]. The boundary of a set A (i.e., $\overline{A} \setminus \mathrm{I}nt(A)$) is denoted by ∂A.

Theorem B.2. *A bounded subset A of $\mathbb{R}^n$ is in $\mathcal{J}$ if and only if $\upsilon(\partial A) = 0$.*

One consequence of this characterization is that if $A, B \in \mathcal{J}$, then $A \cup B$, $A \setminus B \in \mathcal{J}$; that is, $\mathcal{J}$ is a subring of $\mathcal{P}(\mathbb{R}^n)$. Moreover, if $A \in \mathcal{J}$, then $\overline{A} \setminus A$, which is contained in ∂A, is nowhere dense; therefore A differs from $\overline{A}$ by a meager set, which implies that $A \in \mathcal{B}$, the collection of sets with the Property of Baire.

So $\mathcal{J} \subseteq \mathcal{L} \cap \mathcal{B}$. It was pointed out before that $\mathcal{J}$ does not contain the Borel sets; but neither is $\mathcal{J}$ contained in the Borel sets. If C is the Cantor subset of [0,1], then the usual proof that $\lambda(C) = 0$ shows that, in fact, $\upsilon(C) = 0$. It follows that all subsets of C lie in $\mathcal{J}$; but $|C| = 2^{\aleph_0}$, so the number of subsets of C is greater than $2^{\aleph_0}$, the number of Borel sets.

From the definition of υ_*, it follows that a set $A \in \mathcal{J}$ has Jordan measure zero if and only if A has no interior. Because for $A \in \mathcal{J}$, $\upsilon(A) = \upsilon(\overline{A})$, this yields that if $A \in \mathcal{J}$, then $\upsilon(A) = 0$ if and only if $\lambda(A) = 0$ if and only if A is nowhere dense if and only if A is meager. Thus υ is unbiased as regards Lebesgue measure and category. A set in $\mathcal{J}$ has Jordan measure zero if either the set has Lebesgue measure zero or the set is meager.

See Proposition 11.8 for a result about Jordan measure's uniqueness. The historical importance of Jordan measure stems from its connection with the Riemann integral. A function is Jordan measurable if and only if it is Riemann integrable (if and only if the set of points at which it is discontinuous has Lebesgue measure zero).

C

Graph Theory

A *graph* consists of a vertex set V and an edge set E, where each edge, written $u \leftrightarrow v$, consists of two vertices; a vertex in an edge is said to be a *neighbor* of the other vertex. A *simple graph* is one in which there are no multiple edges (any edge occurs at most once in E) and no loops ($u \leftrightarrow u$). In this book, *graph* means *simple graph*. If multiple edges are allowed, the term *multigraph* is commonly used, though in this book we will say only that multiple edges are allowed. If the edges have a direction, then the graph is called a *directed graph* and edges are written $u \rightarrow v$. This book makes much use of infinite graphs, where the vertex set is infinite. The *degree* of a vertex is the number of edges containing it.

A *path* in a graph is a (possibly infinite) sequence of vertices $(v_1, v_2, \ldots)$ such that each $v_i \leftrightarrow v_{i+1}$ is an edge. A *cycle* is a path that is finite and whose last vertex is the same as the first. The vertices in a path or cycle are generally assumed to be distinct (except for the first and last vertex of a cycle). A *tree* is a graph with no cycles. A graph is *connected* if for every two vertices there is a path from one to the other. Any graph can be partitioned into maximal connected subgraphs, called the *connected components*.

A graph is *k-regular* if all vertices have degree k. A finite graph that is 2-regular can be partitioned into disjoint cycles. A graph in which every vertex has degree 1 or 2 splits into cycles and paths. A degree-1 vertex is often called a *leaf*. A graph is *locally finite* if every vertex has finite degree; for such a graph, any connected component is countable.

A *bipartite graph* is one whose vertices split into A and B so that every edge has one end in A and the other in B. It is easy to prove that a graph is bipartite iff every cycle has even length.

A *matching* in a graph is a set of disjoint edges. A *perfect matching* is one whose ends cover all vertices; a graph with an odd number of vertices cannot have a perfect matching. We care most about infinite graphs, because matchings can yield equidecomposabilty of sets. This next theorem, due to P. Hall and R. Rado, is also known as the Marriage Theorem; the condition about neighbors is the *marriage condition*. For the rest of this section, we allow multiple edges; this has no impact on the next results because the removal of duplicate edges does not affect the marriage condition.

Theorem C.1. *A finite bipartite graph with parts A, B has a matching that covers all vertices in A iff, for any k, every set of k vertices in A has at least k neighbors in B.*

A proof can be found in most graph theory texts (see, e.g., [BM76]). The extension to the infinite case was first done by Marshall Hall.

Theorem C.2 (Hall–Rado–Hall Theorem) (AC). *A bipartite graph with parts A, B so that every vertex in A has finite degree admits a matching of every vertex in A iff for any finite k, every set of k vertices in A has at least k neighbors.*

Proof. Give B the discrete topology (all sets are open; compact sets are the finite sets) and for $a \in A$, let $N_a \subseteq B$ be the neighbors of a. Then, by Tychonoff's Theorem and the compactness of each N_a, the product $Y = \Pi_{a \in A} N_a$ is a compact subspace of B^A, the space of all functions from A to B. Now, for any finite nonempty $H \subseteq A$, let $M(H)$ be all $f \in B^A$ such that each $f(a) \in N(a)$ and f is a matching when restricted to H. By Theorem C.1, each $M(H) \neq \varnothing$. Also each $M(H)$ is a closed subset of Y (because not being in $M(H)$ depends only on the behavior at two points, $Y \backslash M(H)$ can be written as a union of open sets in the product topology). And the family $\{M(H)\}$ has the finite intersection property because, again by Theorem C.1, a matching exists for any finite set. Therefore there is some $f \in \cap\{M(H) : H$ finite and nonempty$\}$; such an f is a matching of all of A. □

Theorem C.3 (Two-Sided Marriage Theorem) (AC). *A locally finite bipartite graph that satisfies the marriage condition for finite subsets of either part admits a perfect matching.*

Proof. Let A and B be the two parts of the graph. By Theorem C.2, there is a matching M_A covering A, and another one M_B covering B. Start constructing the perfect matching by using the edges of $M_A \cap M_B$, removing them from consideration. Now the graph determined by the remaining edges in $M_A \cup M_B$ has maximum degree 2. It therefore decomposes into even cycles and singly or doubly infinite paths. Each such admits a perfect matching, and so we have the desired matching covering the whole graph. □

Historically, these marriage theorems were preceded by König's Theorem, which is closely related: A k-regular bipartite graph with k finite has a perfect matching. The preceding results have what might be called bigamist's variations; the proof is easy and the extension is very useful for equidecomposability (see §12.4.1).

Theorem C.4 (AC). (a) *Suppose G is bipartite on A and B, every vertex in A has finite degree, and, for a fixed positive integer d, the marriage condition holds in the stronger form: Every set of k vertices in A has at least dk neighbors. Then* *there are d matchings of the vertices in A so that the B-ends of the edges in all the matchings are distinct.*

(b) *Suppose G is locally finite and bipartite, the marriage condition of (a) holds, and, in addition, the classic marriage condition holds for vertices in B. Then there are d matchings, as in (a), and every vertex of B is matched.*

Proof. For (a), replace each $a_i \in A$ by d copies, connecting each to all the neighbors of a_i; the condition then becomes the normal marriage condition, and Theorem C.2 can be applied. For (b), the same proof idea works, using Theorem C.3. □

Bibliography

[Ada54] Adams, J. F., On decompositions of the sphere, *J. London Math. Soc.* **29** (1954), 96–99.

[AS59] Adelson-Velsky, G. M., and Yu. A. Shreider, The Banach mean on groups (Russian), *Uspehi Mat. Nauk* **12**:78 (1957), 131–136. (*Math. Rev.* **20** (1959) #1238.)

[Adi79] Adian, S. I., *The Burnside Problem and Identities in Groups*, Springer, 1978.

[Adi83] Adian, S. I., Random walks on free periodic groups, *Math. U.S.S.R. Izvestiya* **21** (1983), 425–434.

[AM38] Agnew, R. P., and A. P. Morse, Extensions of linear functionals, with applications to limits, integrals, measures, and densities, *Ann. Math.* **39** (1938), 20–30.

[Ake81] Akemann, C. A., Operator algebras associated with Fuchsian groups, *Houston J. Math.* **7** (1981), 295–301.

[Aug84] Augenstein, B. W., Hadron physics and transfinite set theory, *Int. J. Theoret. Phys.* **23** (1984), 1197–1205.

[BM57] Balcerzyk, S., and J. Mycielski, On the existence of free subgroups in topological groups, *Fund. Math.* **44** (1957), 303–308.

[BM61] Balcerzyk, S., and J. Mycielski, On faithful representations of free products of groups, *Fund. Math.* **50** (1961), 63–71.

[Ban23] Banach, S., Sur le problème de la mesure, *Fund. Math.* **4** (1923), 7–33. Reprinted in S. Banach, *Oeuvres*, vol. I, Éditions Scientifiques de Pologne, 1967.

[Ban24] ———, Un théorème sur les transformations biunivoques, *Fund. Math.* **6** (1924), 236–239. Reprinted, loc. cit.

[BT24] Banach, S., and A. Tarski, Sur la decomposition des ensembles de points en parties respectivement congruents, *Fund. Math.* **6** (1924), 244–277. Reprinted, loc. cit.

[BU48] Banach, S., and S. Ulam, Problème 34, *Coll. Math.* **1** (1948), 152–153.

[Ban83] Bandt, C., Metric invariance of Haar measure, *Proc. Am. Math. Soc.* **87** (1983), 65–69.

[BB86] Bandt, C., and G. Baraki, Metrically invariant measures on locally homogeneous spaces and hyperspaces, *Pac. J. Math.* **121** (1986), 13–28.

[Bas72] Bass, H., The degree of polynomial growth of finitely generated nilpotent groups, *Proc. London Math. Soc.* **25** (1972), 603–614.

[Bea83] Beardon, A. F., *The Geometry of Discrete Groups*, Springer, 1983.

[BK96] Becker, H., and A. S. Kechris, *The Descriptive Set Theory of Polish Group Actions*, London Math. Soc. Lecture Notes Series (1996), vol. 232.

[Bek98] Bekka, M. B., On uniqueness of invariant means, *Proc. Am. Math. Soc.* **126** (1998), 507–514.

[BHV08] Bekka, B., P. de la Harpe, and A. Valette, *Kazhdan's Property (T)*, Cambridge Univ. Press, New Mathematical Monographs (2008), vol. 11.

[BP82] Belley, J.-M., and V. S. Prasad, A measure invariant under group endomorphisms, *Mathematika* **29** (1982), 116–118.

[Ben00] Bennett, C., A paradoxical decomposition of Escher's Angels and Devils (Circle Limit IV), *Math. Intelligencer* **22**:3 (2000), 39–46.

[BH15] Berger, A., and T. P. Hill, *An Introduction to Benford's Law*, Princeton Univ. Press, 2015.

[Ber05] Bernstein, F., Untersuchungen aus der Mengenlehre, *Math. Ann.* **61** (1905), 117–155.

[Bol78] Boltianskii, V. G., *Hilbert's Third Problem*, trans. R. Silverman, Winston, 1978.

[Bol50] Bolzano, B., *Paradoxes of the Infinite*, trans. D. Steele, Routledge and Kegan Paul, 1950. Originally, *Paradoxien des Unendlichen*, Reclam, 1851.

[BM76] Bondy, J. A., and U. S. R. Murty, *Graph Theory with Applications*, American Elsevier, 1976.

[Bor14] Borel, É., *Leçons sur la Théorie des Fonctions*, 2nd ed., Gauthier-Villars, 1914.

[Bor46] ______, *Les Paradoxes de l'Infini*, 3rd ed., Gallimard, 1946.

[Bor83] Borel, A., On free subgroups of semi-simple groups, *Ens. Maths.* **29** (1983), 151–164.

[Bot57] Bottema, O., Orthogonal isomorphic representations of free groups, *Nieuw Arch. v. Wiskunde* (3) **5** (1957), 71–74.

[Boz80] Bożejko, M., Uniformly amenable discrete groups, *Math. Ann.* **251** (1980), 1–6.

[Bre55] Brenner, J. L., Quelques groupes libre de matrices, *C. R. Acad. Sci. Paris* **241** (1955), 1689–1691.

[Bur04] Burke, M. R., Paradoxical decompositions of planar sets of positive outer measure, *J. Geom.* **79** (2004), 56–58.

[Cas57] Cassels, J. W. S., *An Introduction to Diophantine Approximation*, Cambridge Univ. Press, 1957.

[Cat81] Cater, F. S., Problem 284, *Can. Math. Bull.* **24** (1981), 382–383.

[CGH99] Ceccherini-Silberstein, T., R. I. Grigorchuk, and P. de la Harpe, Amenability and paradoxical decompositions for pseudogroups and discrete metric spaces, *Trudy Mat. Inst. Steklova* **224** (1999), 68–111.

[CJR58] Chang, B., S. A. Jennings, and R. Ree, On certain pairs of matrices which generate free groups, *Can. J. Math.* **10** (1958), 279–284.

[Che78] Chen, S., On nonamenable groups, *Internat. J. Math. and Math. Sci.* **1** (1978), 529–532.

[Cho76] Chou, C., The exact cardinality of the set of means on a group, *Proc. Am. Math. Soc.* **55** (1976), 103–106.

[Cho80] ______, Elementary amenable groups, *Ill. J. Math.* **24** (1980), 396–407.

[Chu69] Chuaqui, R. B., Cardinal algebras and measures invariant under equivalence relations, *Trans. Am. Math. Soc.* **142** (1969), 61–79.

[Chu73] ______, The existence of an invariant countably additive measure and paradoxical decompositions, *Not. Am. Math. Soc.* **20** (1973), A-636–637.

[Chu76] ______, Simple cardinal algebras and their applications to invariant measures, *Notas Math. Univ. Católica Chile*, **6** (1976), 106–131.

[Chu77] ______, Measures invariant under a group of transformations, *Pac. J. Math.* **68** (1977), 313–329.

[Coh82] Cohen, J., Cogrowth and amenability of discrete groups, *J. Func. Anal.* **48** (1982), 301–309.

[Coh80] Cohn, L., *Measure Theory*, Birkhäuser, 1980.

[CN74] Comfort, W., and S. Negrepontis, *The Theory of Ultrafilters*, Springer, 1974.

[CFW81] Connes, A., J. Feldman, and B. Weiss, An amenable equivalence relation is generated by a single transformation, *Ergod. Theoret. Dyn. Syst.* **1** (1981), 431–450.

[CW80] Connes, A., and B. Weiss, Property *T* and asymptotically invariant sequences, *I sr. J. Math.* **37** (1980), 209–210.

[Dan85] Dani, S. G., On invariant finitely additive measures for automorphism groups acting on tori, *Trans. Am. Math. Soc.* **287** (1985), 189–199.

[Dan85b] Dani, S. G., A note on invariant finitely additive measures, *Proc. Am. Math. Soc.* **93** (1985), 67–72.

[DO79] Davies, R. O., and A. Ostaszewski, Denumerable compact metric spaces admit isometry-invariant finitely additive measures, *Mathematika* **26** (1979), 184–186.

[Day49] Day, M. M., Means on semigroups and groups, *Bull. Am. Math. Soc.* **55** (1949), 1054–1055.

[Day57] ———, Amenable semigroups, *Ill. J. Math.* **1** (1957), 509–544.

[Day61] ———, Fixed-point theorems for compact convex sets, *Ill. J. Math.* **5** (1961), 585–590. Correction, *Ill. J. Math.* **8** (1964), 713.

[Day64] ———, Convolutions, means and spectra, *Ill. J. Math.* **8** (1964), 100–111.

[Dek56a] Dekker, T. J., Decompositions of sets and spaces, I, *Indag. Math.* **18** (1956), 581–589.

[Dek56b] ———, Decompositions of sets and spaces, II, *Indag. Math.* **18** (1956), 590–595.

[Dek57] ———, Decompositions of sets and spaces, III, *Indag. Math.* **19** (1957), 104–107.

[Dek58a] ———, *Paradoxical Decompositions of Sets and Spaces*, Academisch Proefschrift, Amsterdam: van Soest, 1958.

[Dek58b] ———, On free groups of motions without fixed points, *Indag. Math.* **20** (1958), 348–353.

[Dek59a] ———, On reflections in Euclidean spaces generating free products, *Nieuw Arch. v. Wiskunde* **7**:3 (1959), 57–60.

[Dek59b] ———, On free products of cyclic rotation groups, *Can. J. Math.* **11** (1959), 67–69.

[DG54] Dekker, T. J., and J. de Groot, Decompositions of a sphere, *Proc. Int. Math. Cong.* **2** (1954), 209.

[DG56] ———, Decompositions of a sphere, *Fund. Math.* **43** (1956), 185–194.

[DS83] Deligne, P., and D. Sullivan, Divison algebras and the Hausdorff–Banach–Tarski Paradox, *End. Math.* **29** (1983), 145–150.

[JR79] del Junco, A., and J. Rosenblatt, Counterexamples in ergodic theory and number theory, *Math. Ann.* **245** (1979), 185–197.

[DSS95] Deuber, W., M. Simonovits, and V. T. Sós, A note on paradoxical metric spaces, *Studia Sci. Math. Hungar.* **30** (1995), 17–23.

[DR11] Devadoss, S. L., and J. O'Rourke, *Discrete and Computational Geometry*, Princeton Univ. Press, 2011.

[Dix50] Dixmier, J., Les moyennes invariantes dans les semigroupes et leur applications, *Acta Sci. Math. (Szeged)* **12** (1950), 213–227.

[Dix73] Dixon, J., Free subgroups of linear groups, *Lecture Notes in Math.* **319** (1973), 45–55.

[DF92] Dougherty, R., and M. Foreman, Banach–Tarski Paradox using pieces with the Property of Baire, *Proc. Nat. Acad. Sci.* **89** (1992), 10726–10728.

[DF94] ———, Banach–Tarski decompositions using sets with the Property of Baire, *J. Am. Math. Soc.* **7**:1 (1994), 75–124.

[Dou90] van Douwen, E. K., Measures invariant under actions of F_2, *Topol. Appl.* **34** (1990), 53–68.

[Dra74] Drake, F. R., *Set Theory: An Introduction to Large Cardinals*, North-Holland, 1974.

[Dri85] Drinfield, V. G., Finitely additive measures in S^2 and S^3 invariant with respect to rotations, *Func. Anal. Appl.* **18** (1985), 245–246.

[DHK63] Dubins, L., M. Hirsch, and J. Karush, Scissor congruence, *Isr. J. Math.* **1** (1963), 239–247.

[DS67] Dunford, N., and J. Schwartz, *Linear Operators, Part I*, Wiley Interscience, 1967.

[Ede88] Edelstein, M., On isometric complementary subsets of the unit ball, *J. London Math. Soc.* **37**:2 (1988), 158–163.

[Egg63] Eggleston, H. G., *Convexity*, Cambridge Univ. Press, 1963.

[EW15] Elgersma, M., and S. Wagon, Closing a Platonic gap, *Math. Intelligencer* **37**:1 (2015), 54–61.

[Emc46] Emch, A., Endlichgleiche Zerscheidung von Parallelotopen in gewöhnlichen und höhern Euklidischen Räumen, *Comm. Math. Helv.* **18** (1946), 224–231.

[Eme78] Emerson, W., Characterizations of amenable groups, *Trans. Am. Math. Soc.* **241** (1978), 183–194.

[Eme79] ———, The Hausdorff Paradox for general group actions, *J. Func. Anal.* **32** (1979), 213–227.

[End77] Enderton, H. B., *Elements of Set Theory*, Academic Press, 1977.

[EM76] Erdös, P., and R. D. Mauldin, The nonexistence of certain invariant measures, *Proc. Am. Math. Soc.* **59** (1976), 321–322.

[Eve63] Eves, H., *A Survey of Geometry*, vol. 1, Allyn and Bacon, 1963.

[Fol55] Følner, E., On groups with full Banach mean value, *Math. Scand.* **3** (1955), 243–254.

[For89] Foreman, M., Amenable group actions on the integers; an independence result, *Bull. Am. Math. Soc. (N.S.)* **21** (1989), 237–240.

[FW91] Foreman, M., and F. Wehrung, The Hahn–Banach Theorem implies the existence of a non-Lebesgue measurable set, *Fund. Math.* **138** (1991), 13–19.

[Fre78] Freyhoffer, H., *Medidas en Grupos Topólogicos, Notas Math. Univ. Catolica Chile* **8** (1978).

[Gal14] Galileo, *Dialogues Concerning Two New Sciences*, Macmillan, 1914. Trans. of original text published in 1638.

[Gar85a] Gardner, R. J., Convex bodies equidecomposable by locally discrete groups of isometries, *Mathematika* **32** (1985), 1–9.

[Gar85b] ———, A problem of Sallee on equidecomposable convex bodies, *Proc. Am. Math. Soc.* **94** (1985), 329–332.

[Gar89] ———, Measure theory and some problems in geometry, *Atti Sem. Mat. Fis. Univ. Modena* **39** (1985), 39–60.

[GL90] Gardner, R. J., and M. Laczkovich, The Banach–Tarski Theorem on polygons, and the cancellation law, *Proc. Am. Math. Soc.* **109** (1990), 1097–1102.

[GW89] Gardner, R. J., and S. Wagon, At long last, the circle has been squared, *Am. Math. Soc. Notices* **36** (1989), 1338–1343.

[Ger33] Gerwien, P., Zerschneidung jeder beliebigen Anzahl von gleichen geradlinigen Figuren in dieselben Stücke, *J. für die Reine und Angew. Math. (Crelle's J.)* **10** (1833), 228–234.

[Ger83] ———, Zerschneidung jeder beliebigen Menge verschieden gestalteter Figuren von gleichem Inhalt auf der Kugelfläche in dieselben Stücke, *J. für die Reine und Angew. Math. (Crelle's J.)* **10** (1883), 235–240.

[GN57] Goldberg, K., and M. Newman, Pairs of matrices of order two which generate free groups, *Ill. J. Math.* **1** (1957), 446–448.

[GMP∞a] Grabowski, Ł. A. Máthé, and O. Pikhurko, Measurable circle-squaring, preprint.

[GMP∞b] Grabowski, Ł. A. Máthé, and O. Pikhurko, Measurable equidecompositions for group actions with an expansion property, preprint.

[Gre80] Greenberg, M. J., *Euclidean and Non-Euclidean Geometries*, 2nd ed., Freeman, 1980.

[Gra63] Granirer, E., On amenable semigroups with a finite-dimensional set of invariant means, I, *Ill. J. Math.* **7** (1963), 32–48.

[Gre69] Greenleaf, F. P., *Invariant Means on Topological Groups*, Van Nostrand, 1969.

[Gri77] Grigorchuk, R. I., Symmetric random walks on discrete groups (in Russian), *Uspekhi Mat. Nauk* **32**:6 (1977), 217–218.

[Gri83] ———, On Milnor's problem of group growth, *Soviet Math. Dokl.* **28** (1983), 23–26.

[Gri87] ———, On a problem of Day on nonelementary amenable groups in the class of finitely defined groups, *Mathematical Notes (trans. of Matematicheskie Zametki)* **60** (1996), 580–583.

[Gri96] ———, Supramenability and the problem of occurrence of free semigroups, *Func. Anal. Appl.* **21** (1987), 64–66.

[Gri15] ———, personal communication (2015).

[Gro56] de Groot, J., Orthogonal isomorphic representations of free groups, *Can. J. Math.* **8** (1956), 256–262.

[GD54] de Groot, J., and T. Dekker, Free subgroups of the orthogonal group, *Comp. Math.* **12** (1954), 134–136.

[Gro81] Gromov, M., Groups of polynomial growth and expanding maps, with an appendix by J. Tits, *Inst. Hautes Etudes Sci. Publ. Math.* **53** (1981), 53–78.

[GS86] Grünbaum, B., and G. C. Shephard, *Tilings and Patterns*, Freeman, 1986.

[Gus51] Gustin, W., Partitioning an interval into finitely many congruent parts, *Ann. Math.* **54** (1951), 250–261.

[Gys49] Gysin, W., Aufgabe 51: Lösung, *Elem. Math.* **4** (1949), 140.

[Had57] Hadwiger, H., *Vorlesungen über Inhalt, Oberfläche und Isoperimetrie*, Springer, 1957.

[HDK64] Hadwiger, H., H. Debrunner, and V. Klee, *Combinatorial Geometry in the Plane*, Holt, Rinehart, and Winston, 1964.

[Han57] Hanf, W., On some fundamental problems concerning isomorphism of Boolean algebras, *Math. Scand.* **5** (1957), 205–217.

[HWHSW08] Hardy, G. W., E. M. Wright, R. Heath-Brown, J. Silverman, and A. Wiles, *An Introduction to the Theory of Numbers*, Oxford Univ. Press, 2008

[Har83] de la Harpe, P., Free groups in linear groups, *Ens. Math.* **29** (1983), 129–144.

[HS86] de la Harpe, P., and G. Skandalis, Un résultat de Tarski sur le actions moyennables de groupes et le partitions paradoxales, *Ens. Math.* **32** (1986), 121–138.

[Hau14a] Hausdorff, F., *Grundzüge der Mengenlehre*, Veit, 1914. Repr., New York: Chelsea.

[Hau14b] Hausdorff, F., Bemerkung über den Inhalt von Punktmengen, *Math. Ann.* **75** (1914), 428–433.

[Hau57] ———, *Set Theory*, trans. J. Aumann et al., Chelsea, 1957.

[Hau65] Hausner, M., *A Vector Space Approach to Geometry*, Prentice Hall, 1965.

[Hea56] Heath, T. L., *The Thirteen Books of Euclid's Elements*, vol. 3, Dover, 1956.

[HW83] Henle, J., and S. Wagon, Problem 6353, *Am. Math. Monthly* **90** (1983), 62–63.

[HR63] Hewitt, E., and K. Ross, *Abstract Harmonic Analysis*, vol. 1, Springer, 1963.

[HT48] Horn, A., and A. Tarski, Measures in Boolean algebras, *Trans. Am. Math. Soc.* **64** (1948), 467–497.

[Hul66] Hulanicki, A., On the spectral radius of Hermitian elements in group algebras, *Pac. J. Math.* **18** (1966), 277–287.

[Jac12] Jackson, W. H., Wallace's theorem concerning plane polygons of the same area, *Am. J. Math.* **34** (1912), 383–390.

[Jec73] Jech, T. J., *The Axiom of Choice*, North-Holland, 1973.

[Jec78] ———, *Set Theory*, Academic Press, 1978.

[Jor77] Jørgensen, T., A note on subgroups of *SL*(2,C), *Q. J. Math. Oxford* (2) **28** (1977), 209–211.

[Jus87] Just, W., A bounded paradoxical subset of the plane, *Bull. Polish Acad. Sci.* **36** (1988), 1–3.

[JW96] Just, W., and M. Weese, *Discovering Modern Set Theory. I: The Basics*, American Mathematical Society, 1996.

[Juz82] Jůza, M., Remarque sur la mesure intérieure de Lebesgue, *Časopis pro Pěstováni Matematiky* **107** (1982), 422–424.

[KV15] Kandola, S., and S. Vandervelde, A paradoxical decomposition of the real line, preprint, 2015; http://arxiv.org/abs/1511.01019/.

[Kel00] Kellerhals, R., Old and new about Hilbert's Third Problem, in *European Women in Mathematics (Loccum, 1999)*, 179–187, Hindawi, 2000.

[Kes59a] Kesten, H., Symmetric random walks on groups, *Trans. Am. Math. Soc.* **92** (1959), 336–354.

[Kes59b] ———, Full Banach mean values on countable groups, *Math. Scand.* **7** (1959), 146–156.

[Kaz67] Kazhdan, D. A., Connection of the dual space of a group with the structure of its closed subgroups, *Func. Anal. Appl.* **1** (1967), 63–65. English trans. of *Funktsional'nyi Analizi Ego Prilozheniya* **1** (1967), 71–74.

[Kin53] Kinoshita, S., A solution of a problem of R. Sikorski, *Fund. Math.* **40** (1953), 39–41.

[KL11] Kiss, G., and M. Laczkovich, Decomposition of balls into congruent pieces, *Mathematika* **57** (2011), 89–107.

[KF90] Klein, F., and R. Fricke, *Vorlesungen über die Theorie der Elliptischen Modulfunctionen*, vol. 1, Teubner, 1890.

[Kle54] Klee, V., Invariant extensions of linear functionals, *Pac. J. Math.* **4** (1954), 37–46.

[Kle79] Klee, V., Some unsolved problems in geometry, *Math. Mag.* **52** (1979), 131–145.

[Kon16] König, D., Über graphen und ihre Anwendung auf Determinantentheorie und Mengenlehre, *Math. Ann.* **77** (1916), 454–465.

[Kon26] ———, Sur les correspondances multivoque des ensembles, *Fund. Math.* **8** (1926), 114–134.

[Kon27] ———, Über eine Schlussweise aus dem Endlichen uns Unendliche, *Acta Sci. Math. (Szeged)* **3** (1927), 121–130.

[KV25] König, D., and S. Valkó, Über mehrdeutige Abbildungen von Mengen, *Math. Ann.* **95** (1925), 135–138.

[Kra88] Krasa, S., The action of a solvable group on an infinite set never has a unique invariant mean, *Trans. Am. Math. Soc.* **305** (1988), 369–376.

[Kri71] Krivine, J.-L., *Introduction to Axiomatic Set Theory*, trans. D. Miller, Reidel, 1971.

[KN74] Kuipers, L., and H. Niederreiter, *Uniform Distribution of Sequences*, John Wiley, 1974.

[Kum56] Kummer, H., Translative Zerlegungsgleichheit k-dimensionaler Parallelotopen, *Arch. Math.* **7** (1956), 219–220.

[Kun80] Kunen, K., *Set Theory: An Introduction to Independence Proofs*, North-Holland, 1980.

[Kur51] Kuranishi, M., On everywhere dense imbedding of free groups in Lie groups, *Nag. Math. J.* **2** (1951), 63–71.

[Kur24] Kuratowski, K., Une propriété des correspondances biunivoques, *Fund. Math.* **6** (1924), 240–243.

[Kur66] Kuratowski, K., *Topology*, vol. 1, trans. J. Jaworowski, Academic Press, 1966.

[KM68] Kuratowski, K., and A. Mostowski, *Set Theory*, trans. M. Mączyński, North-Holland, 1968.

[Kur56] Kurosh, A. G., *The Theory of Groups*, vol. 2, trans. K. Hirsch, Chelsea, 1956.

[Lac88a] Laczkovich, M., Von Neumann's Paradox with translations, *Fund. Math.* **131** (1988), 1–12.

[Lac88b] ———, Closed sets without measurable matching, *Proc. Am. Math. Soc.* **103** (1988), 894–896.

[Lac90] ———, Equidecomposability and discrepancy; a solution of Tarski's circle-squaring problem, *J. Reine Angew. Math.(Crelle's J.)* **404**, 77–117.

[Lac91a] ———, Invariant signed measures and the cancellation law, *Proc. Am. Math. Soc.* **111** (1991), 421–431.

[Lac91b] ———, Equidecomposability of sets, invariant measures, and paradoxes, *Rendiconti dell'Istituto di Matematica dell'Univ. Trieste* **23** (1991), 145–176.

[Lac92a] ———, Uniformly spread discrete sets in $\mathbb{R}^d$, *J. London Math. Soc.* **46** (1992), 39–57.

[Lac92b] ———, Decomposition of sets with small boundary, *J. London Math. Soc.* **46** (1992), 58–64.

[Lac92c] ———, Paradoxical decompositions using Lipschitz functions, *Mathematika* **39** (1992), 216–222.

[Lac93] ———, Decomposition of sets with small or large boundary, *Mathematika* **40** (1993), 290–304.

[Lac99] ———, Paradoxical sets under $SL_2(\mathbb{R})$, *Ann. Univ. Sci. Budapest* **42** (1999), 141–145.

[Lac01] ———, On paradoxical spaces, *Stud. Sci. Math. Hungar.* **38** (2001), 267–271.

[Lac03] ———, Equidecomposability of Jordan domains under groups of isometries, *Fund. Math.* **177** (2003), 151–173.

[Lan70] Lang, R., *Kardinalalgebren in der Masstheorie*, Diplomarbeit, Universität Heidelberg, 1970.

[Lar04] Larson, B., *The Stationary Tower: Notes on a Course of W. Hugh Woodin*, University Lecture Series **32**, American Mathematical Society, 1996.

[Leb04] Lebesgue, H., *Leçons sur l'Intégration et la Recherche des Fonctions Primitives*, Gauthier-Villars, 1904.

[Leh64] Lehner, J., *Discontinuous Groups and Automorphic Functions*, American Mathematical Society, 1964.

[Leh66] ———, *A Short Course in Automorphic Functions*, Holt, Rinehart, and Winston, 1966.

[Lin26] Lindenbaum, A., Contributions a l'étude de l'espace métrique I, *Fund. Math.* **8** (1926), 209–222.

[LT26] Lindenbaum, A., and A. Tarski, Communication sur les recherches de la théorie des ensembles, *C. R. Séances Soc. Sci. Lettres Varsovie*, **19** (1926), 299–330.

[LM∞] Lodha, Y., and J. T. Moore, A finitely presented group of piecewise projective homeomorphisms, preprint.

[Lux69] Luxemburg, W. A. J., Reduced direct products of the real number system and equivalents of the Hahn-Banach extension theorem, in *Applications of Model Theory to Algebra, Analysis, and Probability*, ed. W. A. J. Luxemburg, Holt, Rinehart, and Winston, 1969.

[LU68] Lyndon, R. C., and J. L. Ullman, Pairs of real 2-by-2 matrices that generate free products, *Mich. Math. J.* **15** (1968), 161–166.

[LU69] ———, Groups generated by two parabolic linear fractional transformations, *Can. J. Math.* **21** (1969), 1388–1403.

[LR51] Łoś, J., and C. Ryll-Nardzewski, On the application of Tychonoff's Theorem in mathematical proofs, *Fund. Math.* **38** (1951), 233–237.

[LR81] Losert, V., and H. Rindler, Almost invariant sets, *Bull. London Math. Soc.* **13** (1981), 145–148.

[Mab10] Mabry, R., Stretched shadings and a Banach measure that is not scale-invariant, *Fund. Math.* **209** (2010), 95–113.

[Mag73] Magnus, W., Rational representations of Fuchsian groups and non-parabolic subgroups of the modular group, *Nachr. Akad. Wiss. Göttingen Math.-Phys. Kl. II* (1973), 179–189.

[Mag74] ———, *Noneuclidean Tessellations and Their Groups*, Academic Press, 1974.

[Mag75] ———, Two generator subgroups of *PSL*(2,C), *Nachr. Akad. Wiss. Göttingen, Math.-Phys. Kl. II* (1975), 81–94.

[MKS66] Magnus W., A. Karass, and D. Solitar, *Combinatorial Group Theory*, Interscience, 1966.

[Mar80] Margulis, G. A., Some remarks on invariant means, *Mh. Math.* **90** (1980), 233–235.

[Mar82] ______, Finitely additive invariant measures on Euclidean spaces, *Ergod. Theoret. Dyn. Syst.* **2** (1982), 383–396.

[MU∞] Marks, A., and S. Unger, Baire measurable paradoxical decompositions via matchings, preprint.

[Mas72] Mason, J. H., Can regular tetrahedra be glued together face to face to form a ring?, *Math. Gaz.* **56** (1972), 194–197.

[Mat95] Mattila, P., *Geometry of Sets and Measures in Euclidean Spaces: Fractals and Rectifiability*, Cambridge Univ. Press, 1995.

[Mau81] Mauldin, R. D., *The Scottish Book*, Birkhäuser, 1981.

[Maz20] Mazurkiewicz, S., Problème 8, *Fund. Math.* **1** (1920), 224.

[Maz21] Mazurkiewicz, S., Sur la décomposition d'un segment en une infinité d'ensembles non mesurables superposables deux a deux, *Fund. Math.* **2** (1921), 8–14.

[MS14] Mazurkiewicz, S., and W. Sierpiński, Sur un ensemble superposables avec chacune de ses deux parties, *C. R. Acad. Sci. Paris* **158** (1914), 618–619.

[Mes66] Meschkowski, H., *Unsolved and Unsolvable Problems in Geometry*, trans. J. Burlak, Ungar, 1966.

[MP81] Millman, R. S., and G. D. Parker, *Geometry: A Metric Approach with Models*, Springer, 1981.

[Mil68a] Milnor, J., Problem 5603, *Am. Math. Monthly* **75** (1968), 685–686.

[Mil68b] ______, A note on curvature and fundamental group, *J. Diff. Geom.* **2** (1968), 1–7.

[Mil68c] ______, Growth of finitely generated solvable groups, *J. Diff. Geom.* **2** (1968), 447–449.

[Moi74] Moise, E. E., *Elementary Geometry from an Advanced Standpoint*, 2nd ed., Reading, MA: Addison-Wesley, 1974.

[Mon13] Monod, N., Groups of piecewise projective homeomorphisms, *Proc. Am. Math. Soc.* **110** (2013), 4524–4527.

[Moo13] Moore, J. T., Amenability and Ramsey theory, *Fund. Math.* **220** (2013), 263–280.

[Moo82] Moore, G. H., *Zermelo's Axiom of Choice*, Springer, 1982.

[Moo83] ______, Lebesgue's measure problem and Zermelo's axiom of choice: The mathematical effects of a philosphical dispute, *Ann. N. Y. Acad. Sci.* **412** (1983), 129–154.

[Mor49] Morse, A. P., Squares are normal, *Fund. Math.* **36** (1949), 35–39.

[Myc54] Mycielski, J., On a problem of Sierpiński concerning congruent sets of points, *Bull. Acad. Pol. Sc. Cl. III* **2** (1954), 125–126.

[Myc55a] ______, About sets with strange isometrical properties (I), *Fund. Math.* **42** (1955), 1–10.

[Myc55b] ______, On the paradox of the sphere, *Fund. Math.* **42** (1955), 348–355.

[Myc56] ______, On the decompositions of Euclidean spaces, *Bull. Acad. Pol. Sc. Cl. III* **4** (1956), 417–418.

[Myc57a] ______, On the decomposition of a segment into congruent sets and related problems, *Coll. Math.* **5** (1957), 24–27.

[Myc57b] ______, Problème 166, *Coll. Math.* **4** (1957), 240.

[Myc58a] ______, About sets with strange isometrical properties (II), *Fund. Math.* **45** (1958), 292–295.

[Myc58b] ______, About sets invariant with respect to denumerable changes, *Fund. Math.* **45** (1958), 296–305.

[Myc64] ______, Independent sets in topological algebras, *Fund. Math.* **55** (1964), 139–147.

[Myc64b] ______, On the axiom of determinateness, *Fund. Math.* **53** (1964), 205–224.

[Myc73] ______, Almost every function is independent, *Fund. Math.* **81** (1973), 43–48.

[Myc74] ______, Remarks on invariant measures in metric spaces, *Coll. Math.* **32** (1974), 105–112.

[Myc77a] ———, Two problems on geometric bodies, *Am. Math. Monthly* **84** (1977), 116–118.

[Myc77b] ———, Can one solve equations in groups?, *Am. Math. Monthly* **84** (1977), 723–726.

[Myc79] ———, Finitely additive measures, I, *Coll. Math.* **42** (1979), 309–318.

[Myc80] ———, Problems on finitely additive invariant measures, in *General Topology and Modern Analysis*, ed. McAuley and Rao, 431–436, Academic Press, 1980.

[Myc89] ———, The Banach–Tarski Paradox for the hyperbolic plane, *Fund. Math.* **132** (1989), 143–149.

[Myc98] ———, Non-amenable groups with amenable action and some paradoxical decompositions in the plane, *Coll. Math.* **75** (1998), 149–157.

[Myc06] ———, A system of axioms of set theory for the rationalists, *Notices Am. Math. Soc.* **53** (2006), 206–213.

[MS62] Mycielski, J., and H. Steinhaus A mathematical axiom contradicting the axiom of choice, *Bull. Acad. Pol. Sc. Cl. III* **10** (1962), 1–3.

[MS58] Mycielski, J., and S. Świerczkowski, On free groups of motions and decompositions of the Euclidean space, *Fund. Math.* **45** (1958), 283–291.

[MT13] Mycielski, J., and G. Tomkowicz, The Banach–Tarski Paradox for the hyperbolic plane (II), *Fund. Math.* **222** (2013), 289–290.

[MT∞a] ———, On small subsets in Euclidean spaces (with an Appendix by K. Nowak and G. Tomkowicz), preprint.

[MT∞b] ———, Remarks on the universe $L(\mathbb{R})$ and the Axiom of Choice, preprint.

[MW84] Mycielski, J., and S. Wagon, Large free groups of isometries and their geometrical uses, *Ens. Math.* **30** (1984), 247–267.

[Nam64] Namioka, I., Følner's condition for amenable semi-groups, *Math. Scand.* **15** (1964), 18–28.

[Nas67] Nash-Williams, C. St. J. A., Infinite graphs—a survey, *J. Comb. Th.* **3** (1967), 286–301.

[Neu33] Neumann, B., Über ein gruppentheoretisch-arithmetisch Problem, *Sitzungber. Preuss. Akad. Wiss. Phys.-Math. Klasse* No. X (1933), 18 pp.

[Neu28] von Neumann, J., Die Zerlegung eines Intervalles in abzählbar viele kongruente Teilmengen, *Fund. Math.* **11** (1928), 230–238. Reprinted, loc. cit.

[Neu29] ———, Zur allgemeinen Theorie des Masses, *Fund. Math.* **13** (1929), 73–116. Reprinted, loc. cit.

[Neu61] von Neumann, J., Ein System algebraisch unabhängiger Zahlen, *Math. Ann.* **99** (1928), 134–141. Reprinted in J. von Neumann, *Collected Works*, vol. I, Pergamon, 1961.

[Nis40] Nisnewitsch, V. L., Über gruppen, die durch matrizen über einem kommutativen Feld isomorph darstellbar sind, *Mat. Sbornik* **50** (N.S.8) (1940), 395–403. Russian, with German summary.

[NT63] Neumann, B. H., and T. Taylor, Subsemigroups of nilpotent groups, *Proc. Roy. Soc. London Ser. A* **274** (1963), 1–4.

[Niv67] Niven, I., *Irrational Numbers*, Carus Mathematical Monographs, No. 11, Mathematical Association of America. Distributed by John Wiley, 1967.

[Oh05] Oh, H., The Ruziewicz problem and distributing points on homogeneous spaces of a compact Lie group, *Isr. J. Math.*, **149** (2005), 301–316.

[Olm59] Olmsted, J. M. H., *Real Variables*, Appleton-Century-Crofts, 1959.

[Ols80] Ol'shanskii, A. Y., On the problem of the existence of an invariant mean on a group, *Russian Math. Surveys* (*Uspekhi*) **35**:4 (1980), 180–181.

[OS03] Ol'shanskii, A., and M. Sapir, Non-amenable finitely presented torsion-by-cyclic groups, *Inst. Hautes Etudes Sci. Publ. Math.* No. 96 (2003), 43–169.

[Osg03] Osgood, W. F., A Jordan curve of positive area, *Trans. Am. Math. Soc.* **4** (1903), 107–112.

[Oso76] Osofsky, B., Problem 6102, *Am. Math. Monthly* **83** (1976), 572.

[Oso78] Osofsky, B., and S. Adams, Problem 6102 and solution, *Am. Math. Monthly* **85** (1978), 504.

[Oxt71] Oxtoby, J. C., *Measure and Category*, Springer, 1971.

[Pat86] Paterson, A. L. T., Nonamenability and Borel paradoxical decompositions for locally compact groups, *Proc. Am. Math. Soc.* **96** (1986), 89–90.

[Pat88] Paterson, A. L. T., *Amenability*, American Mathematical Society, 1988.

[Paw91] Pawlikowski, J., The Hahn–Banach Theorem implies the Banach–Tarski Paradox, *Fund. Math.* **138** (1991), 20–21.

[Pen91] Penconek, M., On nonparadoxical sets, *Fund. Math.* **139** (1991), 177–191.

[Pie84] Pier, J.-P., *Amenable Locally Compact Groups*, John Wiley, 1984.

[Pin72] Pincus, D., Independence of the prime ideal theorem from the Hahn-Banach Theorem, *Bull. Am. Math. Soc.* **78** (1972), 766–770.

[Pin74] ______, The strength of the Hahn–Banach Theorem, *Lecture Notes Math.* **369** (1974), 203–248.

[PS77] Pincus, D., and R. Solovay, Definability of measures and ultrafilters, *J. Sym. Logic* **42** (1977), 179–190.

[Pla31] Playfair, J., *Elements of Geometry*, 8th ed., with additions by William Wallace, Edinburgh, 1831.

[Pro83] Promislow, D., Nonexistence of invariant measures, *Proc. Am. Math. Soc.* **88** (1983), 89–92.

[Rai82] Raisonnier, J., Ensembles non mesurables et filtres rapides, *C. R. Acad. Sci. Paris* **294** (1982), 285–287.

[Rai84] ______, A mathematical proof of S. Shelah's theorem on the measure problem and related results, *Isr. J. Math.* **48** (1984), 48–56.

[RW68] Rajagopalan, M., and K. G. Witz, On invariant means which are not inverse invariant, *Can. J. Math.* **20** (1968), 222–224.

[Ree61] Ree, R., On certain pairs of matrices which do not generate a free group, *Can. Math. Bull.* **4** (1961), 49–52.

[Ric67] Rickert, N. W., Amenable groups and groups with the fixed point property, *Trans. Am. Math. Soc.* **127** (1967), 221–232.

[Ros74] Rosenblatt, J., Invariant measures and growth conditions, *Trans. Am. Math. Soc.* **193** (1974), 33–53.

[Ros79] ______, Finitely additive invariant measures, II, *Coll Math.* **42** (1979), 361–363.

[Ros81] ______, Uniqueness of invariant means for measure-preserving transformations, *Trans. Am. Math. Soc.* **265** (1981), 623–636.

[RT81] Rosenblatt, J., and M. Talagrand, Different types of invariant means, *J. London Math. Soc.* **24** (1981), 525–532.

[Roy68] Royden, H. L., *Real Analysis*, 2nd ed., Macmillan, 1968.

[Rus10] Russell, B., *The Philosophy of Logical Atomism*, Routledge, 2010.

[Ruz21] Ruziewicz, S., Sur un ensemble non dénombrable de points, superposables avec les moitiés de sa parties aliquote, *Fund. Math.* **2** (1921), 4–7.

[Ruz24] ______, Une application de l'équation functionelle $f(x+y) = f(x) + f(y)$ à la décomposition de la droite en ensembles superposables non mesurables, *Fund. Math.* **5** (1924), 92–95.

[Rob47] Robinson, R. M., On the decomposition of spheres, *Fund. Math.* **34** (1947), 246–260.

[Sac69] Sacks, G., Measure theoretic uniformity in recursion theory and axiomatic set theory, *Trans. Am. Math. Soc.* **142** (1969), 381–420.

[Sag75] Sageev, G., An independence result concerning the Axiom of Choice, *Ann. Math. Logic* **8** (1975), 1–184.

[Sah79] Sah, C. H., *Hilbert's Third Problem: Scissors Congruence*, Research Notes in Mathematics 33, Pitman, 1979.

[San47] Sanov, I. N., A property of a representation of a free group (in Russian), *Doklady Akad. Nauk SSSR (N.S.)* **57** (1947), 657–659.

[Sat95] Satô, K., A free group acting without fixed points on the rational unit sphere, *Fund. Math.* **148** (1995), 63–69.

[Sat97] Satô, K., A free group of rotations with rational entries on the 3-dimensional unit sphere, *Nihonkai Math. J.* **8** (1997), 91–94.

[Sat99] Satô, K., A free group acting on $\mathbb{Z}^2$ without fixed points, *Ens. Math.* **45** (1999), 189–194.

[Sat03] Satô, K., A locally commutative free group acting on the plane, *Fund. Math.* **180** (2003), 25–34.

[Sch80] Schmidt, K., Asymptotically invariant sequences and an action of SL(2, **Z**) on the 2-sphere, *Isr. J. Math.* **37** (1980), 193–208.

[Sch81] ———, Amenability, Kazhdan's property *T*, strong ergodicity and invariant means for ergodic group-actions, *Ergod. Theory Dyn. Syst.* **1** (1981), 223–236.

[Sch64] Schmidt, W., Metrical theorems on fractional parts of sequences, *Trans. Am. Math. Soc.* **110** (1964), 493–518.

[Sha06] Shalom, Y., The algebraization of Kazhdan's property (T), *Int. Cong. Math., Eur. Math. Soc. Zurich* **2** (2006), 1283–1310.

[She84] Shelah, S., Can you take Solovay's inaccessible away? *Isr. J. Math.* **48** (1984), 1–47.

[She75] Sherman, J., *Paradoxical sets and amenability in groups*, PhD diss., UCLA, 1975.

[She79] ———, A new characterization of amenable groups, *Trans. Am. Math. Soc.* **254** (1979), 365–389.

[She90] Sherman, G. A., On bounded paradoxical subsets of the plane, *Fund. Math.* **136** (1990), 193–196.

[She91] Sherman, G. A., Properties of paradoxical sets in the plane, *J. Geom.* **40** (1991), 170–174.

[Sie50] Siegel, C. L., Bemerkung zu einem Satze von Jakob Nielsen, *Mat. Tidsskrift, Ser. B* (1950), 66–70.

[Sie22] Sierpiński, W., Sur l'égalité $2m = 2n$ pour les nombres cardinaux, *Fund. Math.* **3** (1922), 1–6.

[Sie45a] ———, Sur le paradoxe de MM. Banach et Tarski, *Fund. Math.* **33** (1945), 229–234.

[Sie45b] ———, Sur le paradoxe de la sphère, *Fund. Math.* **33** (1945), 235–244.

[Sie46] ———, Sur la congruence des ensembles de points et ses généralisations, *Comm. Math. Helv.* **19** (1946–1947), 215–226.

[Sie47] ———, Sur un ensemble plan qui se décompose en $2^{\aleph_0}$ ensembles disjoints superposables avec lui, *Fund. Math.* **34** (1947), 9–13.

[Sie48a] ———, Sur l'équivalence des ensembles par décomposition en deux parties, *Fund. Math.* **35** (1948), 151–158.

[Sie48b] ———, Sur un paradoxe de M. J. von Neumann, *Fund. Math.* **35** (1948), 203–207.

[Sie50a] ———, Sur quelques problèmes concernant la congruence des ensembles de points, *Elem. Math.* **5** (1950), 1–4.

[Sie50b] ———, Sur un ensemble plan singulier, *Fund. Math.* **37** (1950), 1–4.

[Sie54] ———, *On the Congruence of Sets and Their Equivalence by Finite Decomposition*, Lucknow, 1954. Reprinted, Chelsea, 1967.

[Sie55] ———, Sur une relation entre deux substitutions linéaires, *Fund. Math.* **41** (1955), 1–5.

[Sik69] Sikorski, R., *Boolean Algebras*, 3rd ed., Springer, 1969.

[Sil56a] Silverman, R. J., Invariant linear functions, *Trans. Am. Math. Soc.* **81** (1956), 411–424.

[Sil56b] ______, Means on semigroups and the Hahn–Banach Extension Property, *Trans. Am. Math. Soc.* **83** (1956), 222–237.

[Sol70] Solovay, R. M., A model of set-theory in which every set of reals is Lebesgue measurable, *Ann. Math.* **92** (1970), 1–56.

[Ste96] Steel, J., *Core Model Iterability Problem*, Lecture Notes in Logic; Association of Symbolic Logic, Springer, 1996.

[Str57] Straus, E. G., On a problem of W. Sierpiński on the congruence of sets, *Fund. Math.* **44** (1957), 75–81.

[Str59] ______, On Sierpiński sets in groups, *Fund. Math.* **46** (1959), 332–333.

[Str79] Stromberg, K., The Banach–Tarski Paradox, *Am. Math. Monthly* **86** (1979), 151–161.

[Sul81] Sullivan, D., For $n > 3$, there is only one finitely-additive measure on the n-sphere defined on all Lebesgue measurable sets, *Bull. Am. Math. Soc.* (*N.S.*) **1** (1981), 121–123.

[Swi58] Świerczkowski, S., On a free group of rotations of the Euclidean space, *Indag. Math.* **20** (1958), 376–378.

[Swi07] Świerczkowski, S., *Looking Astern: A Life of Adventure on Five Continents*, 2007.

[Tao04] Tao, T., The Banach–Tarski Paradox, notes, https://www.math.ucla.edu/~tao/preprints/Expository/banach-tarski.pdf, 2004.

[Tar24a] Tarski, A., O równoważności wielokatów (in Polish, with French summary), *Przeglad Matematyczno-Fizyczny* No. 1–2 (1924), 54.

[Tar24b] Tarski, A., Sur quelques théorèmes qui équivalent à l'axiome du choix, *Fund. Math.* **5** (1924), 147–154.

[Tar25] ______, Problème 38, *Fund. Math.* **7** (1925), 381.

[Tar29] ______, Sur les fonctions additives dans les classes abstraites et leur applications au problème de la mesure, *C. R. Séances Soc. Sci. Lettres Varsovie, Cl. III* **22** (1929), 114–117.

[Tar38a] ______, Über das absolute Mass linearen Punktmengen, *Fund. Math.* **30** (1938), 218–234.

[Tar38b] ______, Algebraische Fassung des Massproblems, *Fund. Math.* **31** (1938), 47–66.

[Tar49] ______, *Cardinal Algebras*, Oxford Univ. Press, 1949.

[Tit72] Tits, J., Free subgroups in linear groups, *J. Alg.* **20** (1972), 250–270.

[Tru90] Truss, J. K., The failure of cancellation laws for equidecompsability types, *Can. J. Math.* **42** (1990), 590–606.

[Tom11] Tomkowicz, G., A free group of piecewise linear transformations, *Coll. Math.* **125** (2011), 141–146.

[Tom∞] ______, On decompositions of the hyperbolic plane satisfying many congruences, preprint.

[TW14] Tomkowicz, G., and S. Wagon, Visualizing paradoxical sets, *Math. Intelligencer* **36**:3 (2014), 36–43.

[Ula30] Ulam, S., Zur Masstheorie in der allgemeinen Mengenlehre, *Fund. Math.* **16** (1930), 140–150.

[Ula60] ______, *A Collection of Mathematical Problems*, Interscience, 1960.

[Ver82] Vershik, A., Amenability and approximation of infinite groups, *Sel. Math. Sov.* **2** (1982), 311–330. Revision of an appendix to the Russian translation of [Gre69].

[Vio56] Viola, T., Su un problema riguardante le congruenze degli insiemi di punta, I, II, *Rendic. Accad. Naz. Lincei, Ser. VIII*, **20** (1956), 290–293, 431–438.

[Vit05] Vitali, G., *Sul problema della mesura dei gruppi di punti di una retta*, Bologna, 1905.

[Wae49] van der Waerden, B. L., Aufgabe 51, *Elem. Math.* **4** (1949), 18.

[Wag81a] Wagon, S., Invariance properties of finitely additive measures in $\mathbf{R}^n$, *Ill. J. Math.* **25** (1981), 74–86.

[Wag81b] ———, Circle-squaring in the twentieth century, *Math. Intelligencer* **3**:4 (1981), 176–181.

[Wag82] ———, The use of shears to construct paradoxes in $\mathbf{R}^2$, *Proc. Am. Math. Soc.* **85** (1982), 353–359.

[Wag83] ———, Partitioning intervals, spheres and balls into congruent pieces, *Can. Math. Bull.* **26** (1983), 337–340.

[Wag07] ———, The Banach–Tarski Paradox, Wolfram Demonstrations Project, http://demonstrations.wolfram.com/TheBanachTarskiParadox/, 2007.

[Wan69] Wang, S. P., The dual space of semi-simple Lie groups, *Am. J. Math.* **91** (1969), 921–937.

[Wan74] ———, On the first cohomology group of discrete groups with Property (T), *Proc. Am. Math. Soc.* **42** (1974), 621–624.

[Wan75] ———, On isolated points in the dual spaces of locally compact groups, *Math. Ann* **218** (1975), 19–34.

[Wan81] ———, A note on free subgroups in linear groups, *J. Alg.* **71** (1981), 232–234.

[Wap05] Wapner, L. M., *The Pea and the Sun: A Mathematical Paradox*, A K Peters, 2005.

[Weh73] Wehrfritz, B. A. F., *Infinite Linear Groups*, Ergeb. der Math. vol. 76, Springer, 1973.

[Weh94] Wehrung, F., Baire paradoxical decompositions need at least six pieces, *Proc. Am. Math. Soc.* **121** (1994), 643–644.

[Wei73] Weir, A. J., *Lebesgue Integration and Measure*, Cambridge Univ. Press, 1973.

[WD84] Wilkie, A. J., and L. van den Dries, An effective bound for groups of linear growth, *Arch. Math.* **42** (1984), 391–396.

[Wil88] Willis, G. A., Continuity of translation invariant linear functionals on $C_0(G)$ for certain locally compact groups G, *Mh. Math.* **105** (1988), 161–164.

[Wil05] Wilson, T. M., A continuous movement version of the Banach–Tarski Paradox: A solution to De Groot's problem, *J. Symb. Log.* **70**:3 (2005), 946–952.

[Wol68] Wolf, J., Growth of finitely-generated solvable groups and curvature of Riemannian manifolds, *J. Diff. Geom.* **2** (1968), 421–446.

[Yan91] Yang, Z., Action of amenable groups and uniqueness of invariant means, *J. Func. Anal.* **97** (1991), 50–63.

[Zak93a] Zakrzewski, P., The existence of invariant probability measures for a group of transformations, *Isr. J. Math.* **83** (1993), 343–352.

[Zak93b] ———, The existence of invariant σ-finite measures for a group of transformations, *Isr. J. Math.* **83** (1993), 275–287.

List of Symbols

λ	Lebesgue measure
λ_*	Lebesgue inner measure
λ^*	Lebesgue outer measure
ν	Jordan measure
$\mathbb{R}^n$	Euclidean n-space
$\mathbb{S}^n$	The unit sphere in $\mathbb{R}^{n+1}$
$\mathbb{H}^n$	Hyperbolic n-space
$\mathbb{L}^n$	Elliptic n-space
J	The unit cube in $\mathbb{R}^n$
$\mathbb{C}$	The field of complex numbers
$\mathbb{N}$	The natural numbers (the nonnegative integers)
$\mathbb{Z}$	The set of all integers
$\mathcal{L}$	The set of Lebesgue measurable subsets of $\mathbb{R}^n$ or $\mathbb{S}^n$
$\mathcal{M}$	The family of meager subsets of $\mathbb{R}^n$
$\mathcal{N}$	The family of null sets (measure zero sets) of $\mathbb{R}^n$
$\mathcal{J}$	The set of Jordan measurable subsets of $\mathbb{R}^n$
$\mathcal{B}$	The set of subsets of $\mathbb{R}^n$ or $\mathbb{S}^n$ having the Property of Baire
$\mathcal{LB}$	The set of subsets of $\mathbb{R}^n$ or $\mathbb{S}^n$ that are Lebesgue measurable and have the Property of Baire
$\mathcal{R}$	The Boolean ring of bounded, regular-open subsets of $\mathbb{R}^n$ or $\mathbb{S}^n$
$\mathcal{R}_1$	The Boolean ring of bounded, regular-open, Jordan measurable subsets of $\mathbb{R}^n$ or $\mathbb{S}^n$
$\mathcal{A}_a$	The relativization of a Boolean algebra $\mathcal{A}$ to an element a
$\aleph_0$	The cardinality of $\mathbb{N}$
$2^{\aleph_0}$	The cardinality of the continuum
$\sim$	The equidecomposability relation
$\sim_n$	Equidecomposability using n pieces
$X \sim_T Y$	X and Y are equidecomposable using translations
$A \preccurlyeq B$	A is equidecomposable to a subset of B

$\mathcal{S}$	The semigroup of equidecomposability types
e	The identity in a group
I	The identity matrix
T_n	The group of all translations of $\mathbb{R}^n$
$O_n(\mathbb{R})$	The group of $n \times n$ real orthogonal matrices
$SO_n(\mathbb{R})$	The group of matrices in $O_n(\mathbb{R})$ having determinant $+1$
$SO_n(\mathbb{Q})$	The group of matrices in $SO_n(\mathbb{R})$ having rational entries
G_n	The group of all isometries of $\mathbb{R}^n$
SG_n	The group of all orientation-preserving isometries of $\mathbb{R}^n$
$A_n(\mathbb{R})$	The group of all affine transformations of $\mathbb{R}^n$
$SA_n(\mathbb{R})$	The group of all affine transformations of $\mathbb{R}^n$ having determinant $+1$
$GL_n(\mathbb{R})$	The group of all real matrices with nonzero determinant
$SL_n(\mathbb{R})$	The group of all real matrices with determinant $+1$
π	The canonical homomorphism from $A_n(\mathbb{R})$ to $GL_n(\mathbb{R})$
$SL_n(\mathbb{Z})$	The group of matrices in $SL_n(\mathbb{R})$ with integer entries
$SL_n(\mathbb{C})$	The group of complex matrices with determinant $+1$
$PSL_n(\mathbb{Z})$	$SL_n(\mathbb{Z})/\{\pm I\}$
$PSL_n(\mathbb{C})$	$SL_n(\mathbb{C})/\{\pm I\}$
AG	The class of amenable groups
EG	The class of elementary groups
NF	The class of groups without a free subgroup of rank 2
SG	The class of supramenable groups
NS	The class of groups without a free subsemigroup of rank 2
EB	The class of exponentially bounded groups
$B(m, n)$	The Burnside group with m generators and having exponent n
$W(\sigma)$	A word beginning on the left with σ
ZF	The Zermelo–Fraenkel axioms of set theory
AC	The Axiom of Choice
ZFC	ZF + AC
LM	The assertion "all sets of reals are Lebesgue measurable"
PB	The assertion "all sets of reals have the Property of Baire"
DC	The Axiom of Dependent Choice
IC	The assertion "that an uncountable inaccessible cardinal exists"
$\text{Int}(A)$	The interior of the set A
$\overline{A}$	The closure of the set A
∂A	The boundary of the set A
$\text{dom}(f)$	The domain of the function f
$\mathbf{0}$	The origin of $\mathbb{R}^n$
$\text{frac}(x)$	The fractional part of the real number x
$\lVert \cdot \rVert$	The Euclidean norm in $\mathbb{R}^n$ or a matrix norm
$G * H$	The free product of two groups
F_n	The free group of rank n
$\vec{a}$	A vector in $\mathbb{R}^n$

$\langle x \rangle$	The distance from the real number x to the nearest integer
$\lfloor x \rfloor$	The integer part of the real number x
$D(S, H)$	The discrepancy of a finite set $S \subset [0, 1)^n$ of cardinality N with respect to a Lebesgue measurable subset H of $[0, 1)^n$ and expressed by the formula $D(S, H) = \left\lvert \frac{1}{N} \lvert S \cap H \rvert - \lambda(H) \right\rvert$
$\Delta(S, H)$	The expression $\Delta(S, H) = \lvert \lvert S \cap H \rvert - \lambda(H) \rvert$, where S is a discrete subset of $\mathbb{R}^2$ and H is a bounded Lebesgue measurable subset of $\mathbb{R}^2$
$s(Q)$	The side-length of the square Q
$\ln x$	The logarithm of x using base e
$\log_{10} x$	The logarithm of x using base 10
$\log_2 x$	The logarithm of x using base 2
$\mathrm{diam}(A)$	The diameter of set A
χ_A	The characteristic function of the set A
$[g, h] = ghg^{-1}h^{-1}$	The commutator of the elements g and h in a group
${}_g f$	The function $x \mapsto f(g^{-1}x)$ for a real-valued function f and a group element g

Index

Printed in the United States
By Bookmasters